OPTICAL IMAGING AND SPECTROSCOPY

OPTICAL IMAGING AND SPECTROSCOPY

DAVID J. BRADY
Duke University, Durham, North Carolina

OPTICAL SOCIETY OF AMERICA

A JOHN WILEY & SONS, INC., PUBLICATION

Co-published by John Wiley & Sons, Inc., and The Optical Society of America

Library of Congress Cataloging-in-Publication Data:

Brady, David Jones, 1961-
Optical imaging and spectroscopy / David Jones Brady.
p. cm.
Includes bibliographical references and index.
ISBN 978-0-470-04823-8 (cloth)
1. Optical spectroscopy. 2. Imaging systems. 3. Image processing. 4. Optical instruments
I. Title.
QC454.O66B73 2008
621.36—dc22

2008036200

Printed in the United States of America

10 9 8 7 6 5 4 3 2 1

To the Taxpayers of the
United States of America

CONTENTS

PREFACE

In an age of ubiquitous information; where any question can instantly be answered at the click of a mouse, it is important to remember that some questions require book-length answers. This book answers the question, "What are the design principles of computational optical sensors?" This book is not a homage to old ideas, in fact it celebrates the death of the photochemically recorded image. But it does honor the ancient concept of the book.

A book-length idea requires a narrative, with protagonists (such as the intrepid photon, speeding information from object to data), antagonists (such as the fickle photon, arriving when it pleases with no consideration to resulting signal fluctuations), and a satisfying denouement. Careful contemplative research is necessary to develop such a narrative. For nearly a century, the Optical Society of America has fostered the research that provides the basis for this book's story. Books and professional societies are as alive and essential to advanced science and engineering in this century as in the last.

With this in mind, it is particularly satisfying that this book is produced under the joint Wiley-OSA imprint. I knew from the moment the series was announced that this would be the perfect venue for "Optical Imaging and Spectroscopy." While there have been many twists and turns in the text's plot over the intervening years, including numerous delays as I struggled to resolve the narrative, these have been the natural struggles of an author. OSA's reviewers provided essential early feedback to the structure and thrust of the text and Wiley has been a consistent and solid supporter of its editorial development.

I know that there are excellent books coming in this series and I look forward to reading those stories. For my part, given a year or two to recover I may have yet another story to tell. Try googling "What are the design methods for optical components?"

David J. Brady

Durham, North Carolina

ACKNOWLEDGMENTS

I began serious work on this text on the Fourth of July, 2004. I thank Rachael, Katherine, and Alexander for letting me write on the 4th (and occasionally on other holidays) and for their patience, love, and support over the 4 years absorbed by this project.

Draft forms of this text were used to teach courses at Duke University, the University of Illinois, and the University of Arizona. I thank the students enrolled in those courses for their patience with the text and problems and for helping to improve the material. I am also grateful to colleagues and students at the University of Minnesota, where I developed this text as a visitor to the Institute for Mathematics and Its Applications in Fall 2005.

The students and staff of the Duke Imaging and Spectroscopy Program (DISP) and its previous incarnation as the Illinois Interferometric Imaging Initiative were essential to the development of this text. Thanks to Alan Chen, Kent Hill, Scott Basinger, Ken Purchase, Junpeng Guo, Daniel Marks, Ron Stack, Remy Tumbar, Prasant Potluri, Yunhui Zheng, Evan Cull, John Burchette, Mohan Shankar, Bob Hao, Scott McCain, Scott Witt, Rick Tarka, Rick Morrison, George Barbastathis, Michal Balberg, Jose Jimenez, Matt Fetterman, Unnikrishnan Gopinathan, Santosh Narayankhedkar, Zhaochun Xu, Shawn Kraut, Ben Hamza, Michael Gehm, Yanquia Wang, Arnab Sinha, Michael Sullivan, Renu John, Nikos Pitsianis, Paul Vosburgh, Leah Goldsmith, and Bob Guenther. Current DISPers Andrew Portnoy, Christina Fernandez, Nan Zheng, Se Hoon Lim, Ashwin Wagadarikar, Yongan Tang and Nathan Hagen were particularly helpful in reviewing sections and exercises and tolerating their advisor's book obsession. I also remember that on January 1, 2001, DISP was just Steve Feller and I.

Demetri Psaltis introduced me to this subject, and Professor Psaltis and his students have been my closest friends and colleagues for many years. Mark Neifeld is central to this story and has starred in every role, including lab mate, colleague, competitor, collaborator, and friend. Ali Adibi, Ken Hsu, Mike Haney, and Kelvin Wagner have also been influential in the development of this text. Faculty colleagues at Illinois and Duke—particularly George Papen, Margery Osborne, Eric Michielssen, David Munson, Jim Coleman, Richard Blahut, Yoram Bresler, Pankaj Agrawal, Xiaobai

Sun, Jungsang Kim, Rebecca Willett, Larry Carin, Daniel Gauthier, Joseph Izatt, Adam Wax, Curtis Taylor, Robert Clark, and Jeff Glass—have been an essential means of leading me from error and pointing me to opportunity. I also thank Kristina Johnson for luring me from my comfortable home in the Midwest to my more comfortable home in the Southeast.

Professor Dennis Healy of the University of Maryland and DARPA provided enormous inspiration and guidance for the concepts described in this text. I am also deeply grateful to Ravi Athale for philosophical guidance and for creating the DARPA MONTAGE program. Historically, imaging and spectroscopy are pursued by almost distinct communities. The link between these fields developed herein is due to Karen Petersen and her extraordinary initiative in starting the Advanced Biosensors Program at the National Institute on Alcoholism and Alcohol Abuse. The vison of Alan Craig at AFOSR and Jiri Jonas of the Beckman Institute in initially seeding my work on computational imaging and of Kent Miller of AFOSR in sustaining the vision over the years is also deeply appreciated.

Timothy Schulz has been generous and incredibly insightful in the development and execution of the Compressive Optical MONTAGE Photography Initiative. The fantastic COMPI team is too large to list comprehensively, but Alan Silver, Bob Gibbons, Mike Feldman, Bob Te Kolste, Joseph Mait, Dennis Prather, Michael Fiddy, Caihua Chen, Nicholas George, and Tom Suleski have been essential collaborators. Thanks also to the members of the computational imaging community who have given me hints, including W. Thomas Cathey, Rafael Piestun, Mark Christensen, Jim Leger, Alexander Sawchuck, Joseph Goodman, Jim Fienup, Richard Baraniuk, Kevin Kelly, Emmanuel Candes, Ronald Coifman, Jun Tanida, Joseph van der Gracht, Chuck Matson, Bob Plemmons, Sudhakar Prasad, and Shaya Fainman.

I am also grateful to the editorial staff of John Wiley & Sons for their efficient and professional assistance in the preparation of this text.

Of course, none of these individuals or institutions are reponsible for error, silliness, or other weaknessness in the text. I have written this book in the "editorial we," partly so that I need not personally discover each equation and partly to remind myself that you and I are in this together, dear reader. I hope that in these pages we discover a common passion for optical sensing.

D. J. Brady

Course materials, code used to generate figures, supplementary exercises, and other resources related to this text are available online at `www.opticalimaging.org`.

ACRONYMS

ACS	astigmatic coherence sensor
AOTF	acoustooptic tunable filter
APS	active pixel sensor
CASSI	coded aperture snapshot spectral imager
c.c.	complex conjugate
CGH	computer-generated hologram
CS	compressed sensing
CTE	charge transfer efficiency
DFT	discrete Fourier transform
DMD	digital mirror device
DOF	depth of field
EDOF	extended depth of field
EM	expectation–maximization (algorithm)
FFT	fast Fourier transform
FOV	field of view (IFOV = instantaneous FOV)
FPA	focal plane array
FSR	free spectral range
FWHM	full width at half maximum
ICA	independent component analysis
LCTF	liquid crystal tunable filter
LG	least gradient
LSI	linear shift-invariant
LSQI	least squares with quadratic inequality
LWIR	longwave infrared
MP	magnifying power
MRTD	minimum resolvable temperature difference
MSE	mean-square error
MTF	modulation transfer function
MURA	modified URA (q.v. URA, below)
NA	numerical aperture
NEP	noise equivalent power

NETD	noise equivalent temperature difference
NIR	near infrared
OCT	optical coherence tomography
OLS	ordinary least squares
OTF	optical transfer function
PCA	principal component(s) analysis
PSF	point spread function
PTF	pixel transfer function
RGB	red-green-blue
RIP	restricted isometry property
ROIC	readout integrated circuit
RSI	rotational shear interferometer/interferometry
RST	reference structure tomography
SLM	spatial light modulator
STED	stimulated emission depletion
STF	system transfer function
SVD	singular value decomposition
TCR	thermal coefficient of resistance
TOMBO	thin observation module by bound optics
TTI	total transmitted information
TV	total variation
TWIST	two-step iterative shrinkage/thresholding (algorithm)
URA	uniformly redundant array

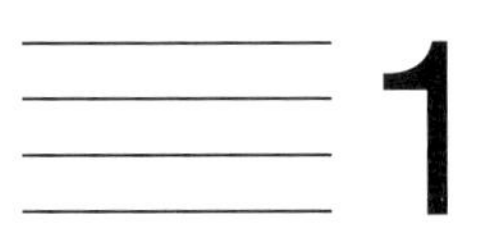

1

PAST, PRESENT, AND FUTURE

> I believe that if more effort is directed into the No-Man's land between raw sensory data and the distinguishable signals which are the starting point of statistical theory, the second decade of information theory will be as rich in practical improvements in communications techniques as the first was in intellectual clarifications.
>
> —D. Gabor [84]

1.1 THREE REVOLUTIONS

Sensing is the interface between the physical and digital worlds. This text focuses on *computational optical sensing*, by which we mean the creation of digital information from electromagnetic radiation with wavelengths ranging from 200 to 20,000 nanometers (nm). Optical sensors are incorporated in imagers, spectrometers, communication transceivers, and optical information processing devices. This text focuses on imaging and spectroscopy. Imagers include microscopes, telescopes, video- and still cameras, and machine vision systems. *Spectrometers* are sensor engines for molecular detection and imaging, chemical analysis, environmental monitoring, and manufacturing process control.

Computational sensing is revolutionizing the design and utility of optical imagers and spectrometers. In emerging applications, optical sensors are the backbone of robotics; transit control systems; security systems; medical diagnostics and genomics; and physical, chemical, and biological research. This text does not specifically consider these applications, but it does provide the reader with a solid foundation to design systems for any of them. The text focuses on

- The relationship between continuous object and optical field parameters and digital image data
- The use of coherence functions, most commonly the cross-spectral density and the power spectral density, to analyze optical systems

- Coding strategies in the design of computational sensors
- The limits of specific spectrometer and imager design strategies

Readers active in physical, chemical, or biological research or nonoptical sensor design should find these topics helpful in understanding the limits of modern sensors. Readers seeking to become expert in optical imaging system design and development will need to supplement this text with courses in digital image processing, lens system design, and optoelectronics. Optical systems is a field of stunning complexity and beauty, and we hope that the basics of system analysis presented here will draw the reader into continuing research and study.

The optical sensing problem is illustrated in Fig. 1.1. The goal is to sense a remote object using signals communicated through the optical field. The sensor consists of optical elements, optoelectronic detectors, and digital processing. In some cases, we consider the remote object to be ambiently illuminated or to be self-luminous. In other cases we may consider temporally or spatially structured illumination as part of the sensor system. The system forms an image of the object consisting of a spatial map of the object radiance or density or of spatially resolved object features such as spectral density, polarization, or even chemical composition.

Figure 1.1 illustrates the culmination of several milliennia of optical sensor system development. The history of optical sensors is punctuated by three revolutions:

1. *Optical Elements*. Optical instruments capable of extending natural vision emerged approximately 700 years ago. Early instruments included spectacles to correct natural vision and the camera obscura for convenient image tracing. Over several hundred years these instruments evolved into microscopes and telescopes. These systems used human vision to transduce light into images. Image storage and communication occurred through handmade copies or traces or through written descriptions.
2. *Automatic Image Recording*. Photochemical recording began to replace handmade images approximately 200 years ago. The first true photographic processes emerged in 1839 from Daguerre's work in France and Talbot's work in England. Each inventor worked over a decade to perfect his process. At first, long exposure times limited photographs to static scenes. Early portraits

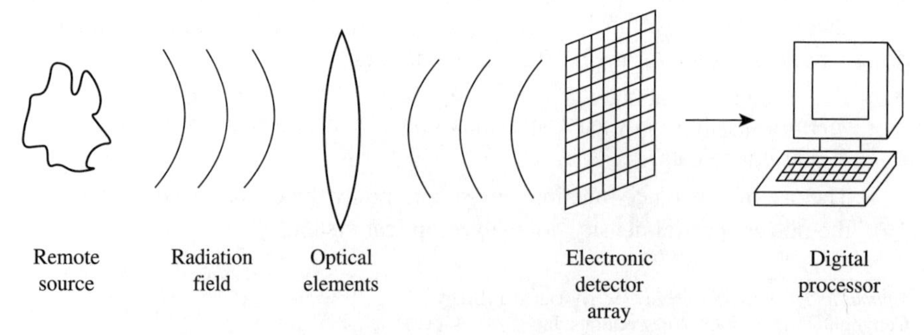

Figure 1.1 Computational optical sensor system.

required the subject to remain stationary for several minutes. Daguerre's famous image shot in 1838 from his laboratory overlooking the Boulevard du Temple in Paris is generally considered the first photograph of a human subject, a man standing still to have his shoes shined. Photographs of dynamic scenes emerged over succeeding decades with the development of flash photography, faster optical elements, and faster photochemistry. Consider, however, the revolutionary impact of the introduction of photography. Images recorded prior to 1839 have been "retouched" by the human hand. Kings are taller and cleaner-looking than they really were. Commoners are not recorded at all. Only since 1839 can one observe true snapshots of history.

3. *Computational Imaging.* Electronic imaging began about 80 years ago with the development of video capture systems for television. As with early optics, the first systems enabled people to see the previously unseen, in this case images of remote places, but did not record images for prosperity. True computational imaging requires three twentieth-century inventions: (a) optoelectronic signal transduction; (b) signal recording, communication, and digitization; and (c) digital signal processing. Signal transduction began with television, but the first electronic recording system, the Ampex VR-1000 magnetic tape deck, was not introduced until 1956. Digital signal processing emerged during World War II. Initial computational imaging applications emerged from *ra*dio *d*etecting *a*nd *r*anging (radar) applications. Electronic systems continued to emerge through the 1970s with the development of deep-space imaging and facsimile transmission. The period from 1950 through 1980 was also rich in the development of medical imaging based on x-ray and magnetic resonance tomography. The most important inventions for optical imaging during this period included semiconductor focal planes, microprocessors, and memories. These developments resulted in the first digital optical imaging systems by the mid-1980s. These systems have continued to evolve as computational hardware has gone from 1970s-style building-scale data centers, to 1980s-style desktop personal computers, to 1990s-style microprocessors in embedded microcameras.

At the moment of this writing the displacement of photochemical recording by optoelectronics is nearly complete, but the true implications of the third revolution are only just emerging. Just as the transition from an image that one could see through a telescope to an image that one could hold in one's hand was profound, the transition from analog photography to digital imaging is not about making old technology better, but about creating new technology. One hopes that this text will advance the continuing process of invention and discovery.

1.2 COMPUTATIONAL IMAGING

The transition from imaging by photochemistry to imaging by computer is comparable to the transition from accounting by abacus to accounting by computer. Just as computational accounting enables finance on a scale unimaginable in the paper era, computational imaging has drastically expanded the number of imaging systems, the number of images captured, and the utility of images—and yet, what has really changed? Isn't a picture recorded on film or on an electronic focal plane basically the same thing? The electronic version can be stored and recalled automatically, but the film version generally has comparable or better resolution, dynamic range, and sensitivity. How is being digital different or better?

In contrast to a physical object consisting of patterns on paper or film, a digital object is a mathematical entity. The digital object is independent of its physical instantiation in silicon, magnetic dipoles, or dimples on a disk. With proper care in coding and transmission, the digital object may be copied infinitely many times without loss of fidelity. A physical image, in contrast, looses resolution when copied and degrades with time. The primary difference between an analog image and a computational image is that the former is a tangible thing while the latter is an algebraic object.

Early applications exploited the mathematical nature of electronic images by enabling nearly instantaneous image transmission and storage, by creating images of multidimensional objects or invisible fields and by creating automated image analysis and enhancement systems. New disciplines of *computer vision* and *digital image processing* emerged to computationally analyze and enhance image data.

Excellent texts and a strong literature exist in support of computer vision and digital image processing. This text focuses on the tools and methods of an emerging community at the interface between digital and physical imaging and sensing system design. Computational sensing does not replace computer vision or digital image processing. Rather, by providing a more powerful and efficient physical layer, computational sensing provides new tools and options to the digital image processing and interpretation communities.

The basic issue addressed by this text is that the revolutionary opportunity represented by electronic detection and digital signal processing has yet to be fully exploited in sensor system design. The only difference between analog and digital cameras in many cases is that an electronic focal plane as replaced film. The differences between conventional design and computational sensor design are delineated as follows:

- The goal of conventional optical systems, even current electronic cameras and spectrometers, is to create an isomorphism. These systems rely on analog processing by lenses or gratings to form the image. The image is digitized after analog processing. Only modest improvements are made to the digitized image.
- The goal of computational sensor design, in contrast, is to jointly design analog preprocessing, analog-to-digital conversion, and digital postprocessing to optimize image quality or utility metrics.

Computational imaging systems may not have a "focal plane" or may deliberately distort focal plane data to enhance postprocessing capacity.

The central question, of course, is: *How might computational optical sensing improve the performance and utility of optical systems?* The short answer to this question is *in every way!* Computational design improves conventional image metrics, the utility of images for machine vision and the amenity of images for digital processing. Specific opportunities include the following:

1. *Image Metrics.* Computational sensing can improve depth of field, field of view, spatial resolution, spectral resolution, signal fidelity, sensitivity, and dynamic range. Digital systems to the time of this writing often compromised image quality to obtain the utility of digital signals, but over time digital images will increasingly exceed analog performance on all metrics.
2. *Multidimensional Imaging.* The goal of a multidimensional imaging system is to reconstruct a digital model of objects in their native embedding spaces. Conventional two-dimensional (2D) images of three-dimensional (3D) objects originate in the capacity of lens and mirror systems to form physical isomorphisms between the fields on two planes. With the development of digital processing, tomographic algorithms have been developed to transform arrays of 2D images into digital 3D object models. Integrated physical and digital design can improve on these methods by eliminating dimensional tradeoffs (such as the need to scan in time for tomographic data acquisition) and by enabling reconstruction of increasingly abstract object dimensions (space–time, space–spectrum, space–polarization, etc.).
3. *Object Analysis and Feature Detection.* The goal of object analysis is to abstract nonimage data from a scene. In emerging applications, sensors enable completely automated tasks, such as robotic positioning and control, biometric recognition, and human–computer interface management. Current systems emphasize heuristic analysis of images. Integrated design allows direct measurement of low-level physical primitives, such as basic object size, shape, position, polarization, and spectral radiance. Direct measurement of significant primitives can dramatically reduce the computational cost of object analysis. On a deeper level, one can consider object abstraction as measurement on generalized object basis states.
4. *Image Compression and Analysis.* The goal of image compression is to represent the digital model of an object as compactly as possible. One can regard the possibility of digital compression as a failure of sensor design. If it is possible to compress measured data, one might argue that too many measurements were taken. As with multidimensional imaging and object analysis, current compression algorithms assume a 2D focal model for objects. Current technology seeks a compressed linear basis or a nonlinear feature map capable of efficiently representing a picture. Integrated physical and digital design implements generalized bases and adaptive maps directly in the optical layer. One has less freedom to implement algorithms in the physical layer than in the digital system, but early data reduction enables both simpler and lower-power acquisition platforms and more efficient data processing.
5. *Sensor Array Data Fusion and Analysis.* Multiaperture imaging is common in biological systems but was alien to artificial imaging prior to the computational age. Modern computational systems will dramatically surpass the multiaperture capabilities of biology by fusing data from many subapertures spanning broad spectral ranges.

1.3 OVERVIEW

An optical sensor estimates the state of a physical object by measuring the optical field. The state of the object may be encoded in a variety of optical parameters, including both spatial and spectral features or functions of these features.

Referring again to Fig. 1.1, note that an optical sensing system includes

1. An *embedding space* populated by target objects
2. *A radiation model* mapping object properties onto the optical signal
3. *A propagation model* describing the transmission of optical signals across the embedding space
4. *A modulation model* describing the coding of optical signals by optical elements
5. *A detection model* describing transduction of optical signals at electronic interfaces
6. *An image model* describing the relationship of transduced and processed digital data to object parameters

Considerable analytical and physical complexity is possible in each of these system components. The radiation model may range from simple scattering or fluorescence up to sophisticated quantum mechanical field–matter interactions. As this is an optics text, we generally ignore the potential complexity of the object–field relationship and simply assume that we wish to image the field itself.

This text considers three propagation models:

- *Geometric fields* propagate along rays. A *ray* is a line between a point on a radiating object and a measurement sensor. In geometric analysis, light propagates in straight lines until it is reflected, refracted, or detected. Geometric fields are discussed in Chapter 2.
- *Wave fields* propagate according to physical wave equations. Wave fields add diffractive effects to the geometric description and enable physical description of the state of the field at any point in space. After review of basic mathematical tools in Chapter 3, we analyze wave fields in Chapter 4.
- *Correlation fields* propagate according to models derived from wave fields, but focus on transformations of optical observables rather than the generally unobservable electric fields. Correlation field analysis combines wave analysis with a simple model of the quantum process of optical detection. After reviewing detection processes in Chapter 5, we develop correlation field analysis in Chapter 6.

The progression from geometric to wave to correlation descriptions involves increasing attention to the physical details of the object-measurement mapping system. The geometric description shows how one might form isomorphic and encoded image capture devices, but cannot account for diffractive, spectral, or interferometric artifacts in these systems. The wave model describes diffraction, but cannot explain interferometry, noise, or spectroscopy. The correlation model accounts for these effects, but would need augmentation in analysis of quantum coherence and nonlinear optical effects. We develop optical modulation and detection models for optical sensors consistent with each propagation model in the corresponding chapters.

After establishing basic physical models for field propagation, modulation, and detection, we turn to the object model in Chapter 7, which focuses on the transformation from continuous fields to digital data, and Chapter 8, which focuses on object data coding and estimation. Discrete representation is the hallmark of digital optical sensors. In discrete analysis, the object state is represented by a vector of coefficients **f** and the measurement state is represented by a vector of coefficients **g**. We consider three different relationships between **f** and **g**.

- *Isomorphic mappings* form a one-to-one correspondence between components of **g** and components of **f**. Examples include focal imaging systems and dispersive spectrometers. As discussed in Chapter 7, computational design and analysis is helpful even for isomorphic systems.
- *Dimension preserving mappings* capture measurements **g** embedded in a space of similar dimension with the object embedding space. One normally considers objects distributed over a 2D or 3D embedding space. Sensors based on convolutions, radon transformations, or Fourier transformations do not capture isomorphic data, but simple inversions are available to restore isomorphism.
- *Discrete mappings* assume no underlying embedding space for the measurements **g**. Measurements under discrete mappings consist of linear or nonlinear projections of the object state.

The inversion algorithm applied in any specific context is determined by both the nature of the object parameters of interest and the physical mapping implemented by the sensor system.

Having completed a survey of the tools needed to analyze and design computational optical sensors in Chapters 2–8, we put the tools to use in Chapters 9 and 10 in describing specific design strategies and opportunities.

In offering the text as a one-semester course, a quick survey of Chapter 2 introduces the basic concepts of optical imaging (using ray tracing) and of computational imaging (using coded aperture imaging). Coded aperture imaging is not of great practical importance, but it provides an instructive and accessible introduction to issues that recur throughout the text. Chapters 3 and 4 present a straightforward course in Fourier optics augmented by wavelet analysis and linear spaces. While we make relatively modest direct use of wavelets in the rest of the book, the student will find wavelets of high utility for system modeling in Chapters 7–10 and will find the general concepts of vector spaces and multiscale analysis essential. Students with prior experience in signal processing may find Chapter 3 unnecessary, I hope that optics students will find the presentation of wavelets more accessible here than in the signal processing literature. Similarly, optics students with previous Fourier optics experience may find Chapter 4 unnecessary. Chapter 5 is a brief overview of optical detectors sufficient for a discussion of system design. This chapter is left for self-study in the one-semester course. Overall, the author hopes that upper-level engineering, physics, mathematics, and computer science undergraduates will find Chapters 2–5 an accessible introduction to basic optical systems. While familiarity with the material in Chapters 2–5 is essential to understanding what comes later, the reader leaving the course after Chapter 5 would be missing the most critical concepts in optical sensing.

The core of the course begins in Chapter 6, where the text considers statistical fields created by natural sources. A course that hurries through the early chapters should arrive with time to spend on this chapter and the remainder of the text. Optical coherence theory is wonderfully developed by Wolf [252], Mandel and Wolf [165], and Goodman [99], but I hope that the reader will find the focus on imaging system analysis and coherence measurement presented in Chapter 6 unique and useful. Similarly, the discussion on sampling in Chapter 7 covers issues that are also covered elsewhere, but I hope that the simple and direct treatment of isomorphic sampling is clearer than other treatments. The discussion of generalized sampling in Section 7.5 covers emerging concepts.

Chapter 8 covers algorithms and coding issues covered elsewhere, although coding strategies are uniquely colored by the understanding of optical fields and generalized sampling developed to this point. If nothing else, the reader should leave Chapter 8 with reduced faith in least-square estimators and mean-square error metrics.

Many texts conclude with an optional chapter or two on advanced topics. That is not the case here. I cannot imagine that a reader would learn the tools in Chapters 2–8 without experiencing the joy of applying them in Chapters 9 and 10.

1.4 THE FOURTH REVOLUTION

The first revolution in optical sensing, the development of optical elements, was based on glass, skilled artisans, and markets for consumer goods. This required a civilized society with advanced materials and manufacturing capabilities. The transition from spectacles to telescopes and microscopes required the existence of a sophisticated scientific community. These developments took many generations of human activity. Could early optical scientists foresee the next revolution? I expect that they could, and how often they must have wished for an automated mechanism for recording images observed by the unaided eye.

The next revolution, photochemistry, emerged nearly simultaneously with the birth of electronic communications. The inventor of electronic communications, Samuel Morse, visited Daguerre's laboratory shortly after the Boulevard du Temple was recorded and described the image in an April 20, 1839 article in the *New York Observer*. Both inventors knew well the tortured process of invention and the faith of the inventor in the previously impossible. The idea of automated image transmission was not far behind the idea of automated recording. In the grand scheme of history, the 75 years between the first photochemical images to television were brief.

The revolutionary transition from photochemistry to computational imaging is nearing completion. The necessary devices first emerged about 25 years ago (i.e., in the early 1980s); one expects that another quarter century will complete this revolution. Optical scientists and engineers now wonder, is there a fourth revolution? As an author one may hope for stasis, such that the words and analysis herein may live forever. Being more scientist and engineer than author, however, I am happy to report that a fourth revolution has already begun.

The fourth revolution will be the age of optical circuits and antennas. As discussed in Chapter 5, the bedrock assumption of modern optics is that electronic detectors measure the time-averaged irradiance of the optical field. The fourth revolution will discard this assumption. Optical design is currently profoundly influenced by the incoherent interface between optical signals and digital data. Within the next decade (i.e., by 2018), coherent coupling between optical and electronic states in nanostructured and plasmonic devices will be combined with quantum interference in electronic states to produce optical coherence sensors. These systems will be combined with complex 3D optics to produce integrated transducers. 3D optics is represented in nascent form by photonic crystal materials, but advanced modeling, 3D fabrication techniques and materials will produce imaging systems and spectrometers with very different noise characteristics and form factors.

A new revolution sometimes kills the old, as digital imaging has killed photochemical imaging, and sometimes feeds the old, as digital imaging has increased demand for optical elements. Happily, I believe that the fourth revolution will only increase the need to understand the content of this book. The basic approaches to sampling, field analysis, and signal analysis outlined herein are necessary to both the present and the future. Most significantly, limits on the bandwidth of the optical system, the significance of these limits for image metrics, and strategies to surpass the naive limits will remain the same even as the physical nature of the optical analog-to-digital interface evolves. With an eye on both the present and the future, therefore, read on, dear reader.

PROBLEMS

1.1 *Imaging and Processing*. Estimate the number of calculations performed per person worldwide in 50-year increments from 1800 to the present. Estimate the number of images photochemically and electronically recorded per person over the same time period.

1.2 *Digital Data*. Estimate the worldwide fraction of stored digital data that is image data.

1.3 *Digital Images*. Estimate the ratio of the number of images stored photochemically to the number of images stored electronically in 1960, 1980,

2000, and 2020. Explain your reasoning. What if only still or only moving pictures are considered?

1.4 *Persistence*. Estimate the lifetime of a film image and of a digital image. Discuss factors that might, over time, lead to the degradation of such images.

1.5 *Weighing Design*. Suppose that you are given 12 gold coins. Exactly one of the coins is counterfeit and weighs more or less than the rest. You have a sensitive two-pan balance, which reports only which pan is heavier. How many measurements do you need on the balance to find the counterfeit coin and determine whether it is lighter or heavier? Describe your measurement strategy. How might this problem be relevant to optical sensor design?

1.6 *Boulevard du Temple*. Consider Nicholas Jenkins' analysis of the number of people in Daguerre's Boulevard du Temple presented online at `http://www.stanford.edu/~njenkins/archives/2007/08/traces.html`. How many people do you observe in the image? What is your estimate of the exposure time?

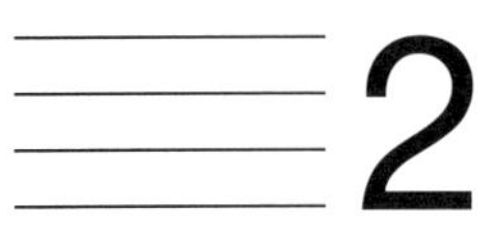

2

GEOMETRIC IMAGING

> The study of imaging now embraces many major areas of modern technology, especially the several disciplines within electrical engineering, and will be both the stimulus for and the recipient of, new advances in information science, computer science, environmental science, device and materials science, and just plain high-speed computing. It can be confidently recommended as a fertile subject area for students entering upon a career in engineering.
>
> —R. N. Bracewell [26]

2.1 VISIBILITY

This chapter introduces the radiation field as a relationship between luminous objects and optical detectors. Visibility and the modulation of visibility using optical elements are the most important concepts of the chapter. *Visibility* is a relationship between points in a space. Two points are visible to each other if light radiated from one illuminates the other. In three dimensions, visibility is a six-dimensional relationship between each pair of points. The concept of visibility in the present chapter anticipates the impulse response function in Chapter 3, the point spread function in Chapter 4, and coherence response kernels in Chapter 6.

We consider systems consisting of objects, object spaces, radiation fields, optical elements, and detectors, as illustrated in Fig. 2.1. The goal of the system is to estimate object parameters from the detector state. Most commonly, the object parameters of interest consist of an *image* of the object. An image is an object density function $f(x)$, defined at points on the object embedding space; $f(x)$ may represent the luminence, spectral density, or scattering density of an object. Quite often we are interested only in the object distribution over a subspace of a higher-dimensional embedding space. For example, in typical imaging systems the image is a distribution over a 2D plane projected from 3D Euclidean space.

Optical Imaging and Spectroscopy. By David J. Brady

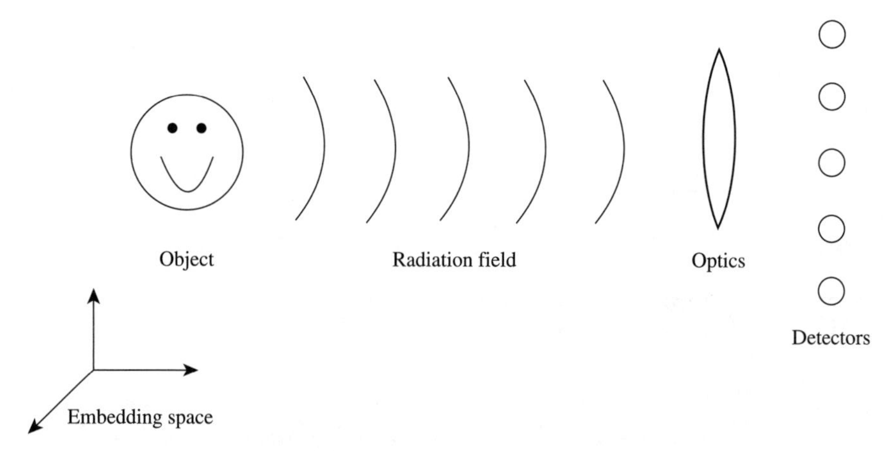

Figure 2.1 Sensor system environment.

Spaces, spatial distributions, spatial transformations, and mappings between continuous distributions and discrete states are fundamental to the analysis of modern optical sensor systems. In optical sensing we are most often interested in objects embedded in the n-dimensional Euclidean space $\mathbb{R}^n$. Images are typically 2D or 3D; spectra are often one-dimensional. We are sometimes interested in 3D or 4D spectral images or even 5D spatial temporal spectral images. We will also find opportunities to consider much higher-dimensional function spaces. For example, an image with $N \times N$ pixels may be considered a point in $(N \times N)$-dimensional space.

An object density function is mapped onto an optical field and propagates through the embedding space to a detector array. The field is also a function over the embedding space, but the field at diverse points is related by propagation rules. The radiation field associates points in the embedding space. Two points in the object space are said to be visible to each other if light from an omnidirectional point source at one of the points illuminates the other point. The visibility $v(A, B)$ is a commutative function, meaning that if point A is visible to B, then point B is visible to A, for example, $v(A, B) = v(B, A)$. Visibility in unmodulated free space is illustrated in Fig. 2.2, which shows the field radiating from A as a set of rays in all directions. The ray AB is drawn between the points.

In uniform free space, all points are mutually visible, meaning that $v(A, B) = 1$ for all A and B. Universal visibility may sound useful, but from a sensor perspective it is hopeless. As illustrated in Fig. 2.1, sensing is achieved by placing discrete detectors in the object space. A detector at point B integrates the field incident on B from all visible points. The detector measurement at B can be modeled as

$$g(B) = \int v(A, B) f(A) dA \tag{2.1}$$

where $f(A)$ is the density of the object at A and the integral is over all points in the object space. With no optical elements in the embedding space, $g(B) = \int f(A) dA$ is

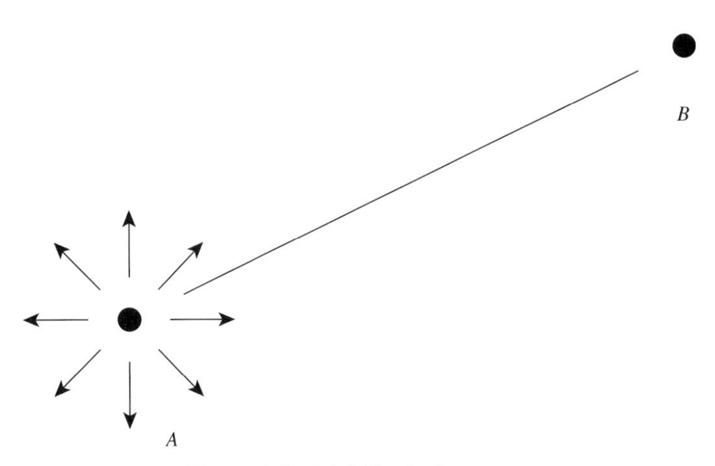

Figure 2.2 Visibility in free space.

equal to the integrated object density independent of the position of the detector. Since all detectors measure the same quantity, only the presense or absense of the object can be determined; no object features or parameters may be estimated.

The simplest modulation is an absorbing reference object, as illustrated in Fig. 2.3. The figure shows two detectors and a single obscurant. We assume that the obscurant perfectly absorbs all the radiation that strikes it. The obscurant divides the object space into four regions:

1. BC, the region visible to the detectors at both points B and C
2. $B\overline{C}$, the region visible to the detector at B but not to the detector at C
3. $\overline{B}C$, the region visible to the detector at C but not to the detector at B
4. $\overline{B}\,\overline{C}$, the region invisible to both detectors

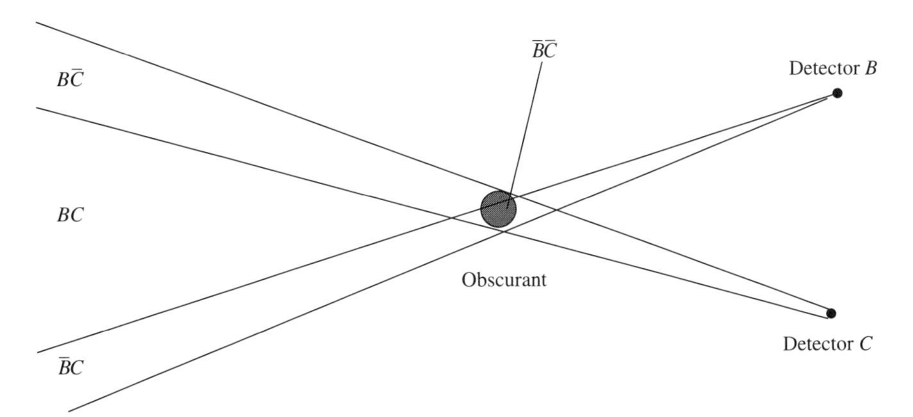

Figure 2.3 Visibility modulated by an obscurant.

The detector measurements are

$$g(B) = f_{BC} + f_{B\overline{C}} \tag{2.2}$$

$$g(C) = f_{BC} + f_{\overline{B}C} \tag{2.3}$$

where, for example, $f_{BC} = \int_{BC} f(A)dA$. Equations (2.2) and (2.3) reflect a transformation of the continuous object distribution $f(A)$ onto discrete descriptors f_{BC}. The tranformation between the discrete descriptors and the discrete measurements g are often written as a single equation of the form

$$\mathbf{g} = \mathbf{Hf} \tag{2.4}$$

where the vector **g** consists of the set of measurements made on the field and the vector **f** is a set of discrete descriptors of the object. **H** is a matrix mapping the object state onto the measurement state.

The vector **f** cannot completely specify the continuous distribution $f(A)$. Although the goal is often to describe the continuous field, all computational sensor systems ultimately are based on discrete measurements. Primary challenges of system design include inverting Eqn. (2.4) to estimate the vector **f** and relating the discrete object features to the continuous object distribution $f(x)$.

One difficulty associated with estimation of **f** is that Eqn. (2.4) is generally ill-conditioned for inversion. Such is the case, for example, in Eqn. (2.2), which obtains only two measurements for three unknown object parameters. One may attempt to overcome this barrier by increasing the number of measurements, but for the particular sensor geometry sketched in Fig. 2.3 the number of distinct source regions may grow faster than the number of measurements as additional measurement points are introduced. Solution to this problem involves the introduction of discrete object parameterizations independent of the measurement system.

We repeatedly return to linear mappings from the object state represented by **f** to the measurement state represented by **g** over the course of this text. Critical questions in the design of this mapping focus on the range of mappings **H** that one can physically implement, how to choose from among the realizable mappings the one yielding the "best" measured data, and how to invert the mapping to estimate **f** or features of **f**. In the current example, **H** consists of binary visibility values over regions identified by physical obscurants. The nature of the object parameterization and of the mappings **H** that one can implement with such obscurants are considered in Section 2.7.

More generally, the visibility is modulated by diverse optical elements (lenses, prisms, mirrors, etc.). Visibility operations implemented by these elements are considered in the next section.

2.2 OPTICAL ELEMENTS

Optical elements that redirect and combine ray bundles are substantially more useful than the simple obscurant of Fig. 2.3. Useful devices include *refractive elements*, such as lenses and prisms; *reflective elements*, such as mirrors; *diffractive elements*, such as holograms and gratings; and *interferometric devices*, such as thin film filters and micro cavities. Much of this text focuses on the capacity of these elements for visibility modulation. Our ability to analyze and design these elements will grow with the sophistication of our field model. In anticipation of more advanced models, we delay discussion of interferometry and diffraction until later chapters.

Reflection and refraction may be considered using geometric redirection of rays at interfaces, as shown in Fig. 2.4. The interface is a boundary between a dielectric medium with index of refraction n_1 and a dielectric with index n_2. A line normal to the plane of incidence is shown in the figure. An incident ray makes an angle θ_i with respect to the surface normal. As shown in the figure, the reflected ray also makes an angle $\theta_r = \theta_i$ with respect to the surface normal. The refracted ray satisfies Snell's law:

$$n_1 \sin \theta_1 = n_2 \sin \theta_2 \tag{2.5}$$

A planar dielectric interface bends the rays that join each pair of points and introduces a second, reflected, path between points on the same side of the interface. With a single interface or a stack of parallel interface planes, however, all points in a space remain mutually visible (see Problem 2.1).

A *prism* is a wedge of dielectric material. A prism redirects the propagation direction of incident rays. As illustrated in Fig. 2.5, a prism bifurcates the object space such

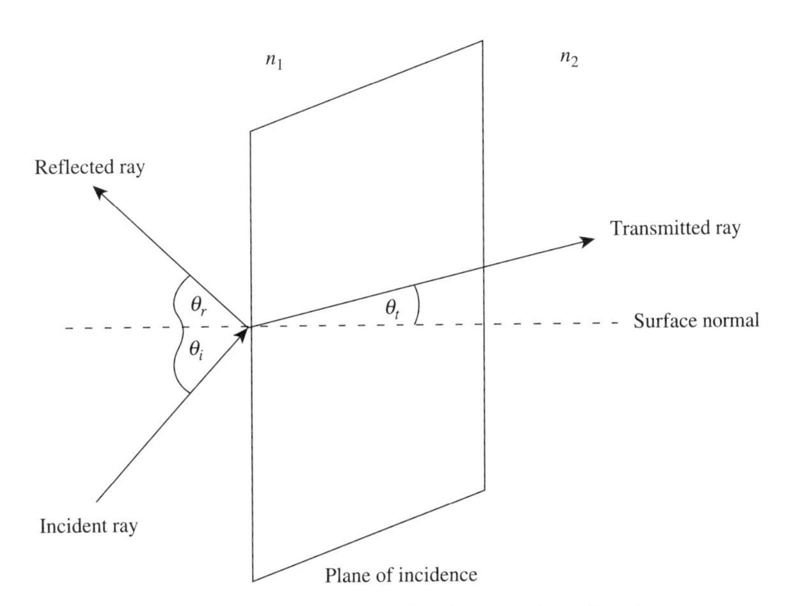

Figure 2.4 Reflection and refraction at a planar interface.

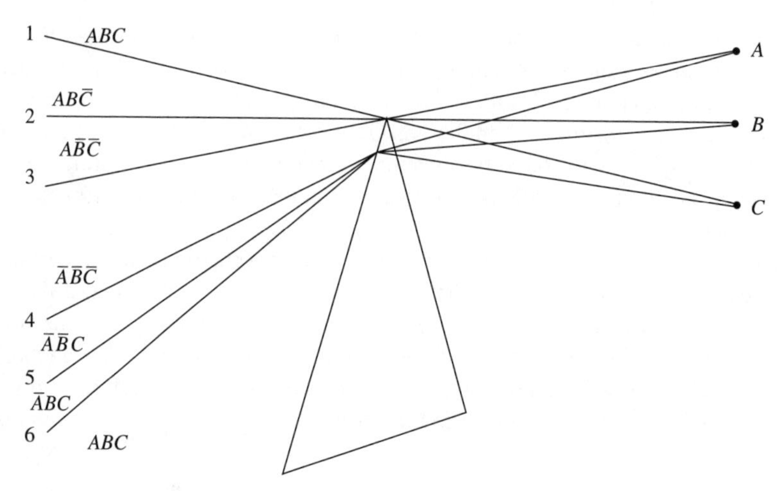

Figure 2.5 Segmentation of the object space by three points refracted through a prism.

that a region to the left of the prism is invisible to an observation point. All points above line 1, connecting point C and the vertex, are visible at C. Rays incident on point C from points in the half-space below line 1 are refracted by the prism. Rays from points on line 4 strike the prism so as to be refracted toward point C. All rays striking the prism from points above line 4 are refracted to pass below point C. This region is invisible at point C and is labeled $\overline{C}$. Similar analysis applies to lines 2 and 5 for point B and lines 3 and 6 for point A. With respect to the observation points A, B, and C, region ABC is visible to all three points. $\overline{A}BC$ is visible to points B and C but not point A. The prism produces a linear mapping between the object density integrated over the six distinct regions of Fig. 2.5 similar to the mapping of Eqn. (2.4).

Prisms are often used in optical systems to redirect, fold, or rotate ray bundles. For example, prism assemblies enable compact binocular designs. Prism assemblies are also used in interferometers, such as the rotational shear interferometer described in Section 6.3.3. Prisms are also commonly used as spectral dispersion elements. Spectral dispersion consists of refraction of rays corresponding to different colors in different directions. Dispersion occurs in prisms because of the dependence of the index of refraction on wavelength. Prisms are also sometimes used as evanescent wave couplers for waveguides or biosensors. Evanescent waves arise because some incident rays are trapped within the prism as a result of *total internal reflection* at the second interface.

To understand the action of a prism more quantitatively, consider the wedge illustrated in Fig. 2.6. Our goal is to find the relationship between the direction of incident rays and refracted rays. For simplicity, we limit our consideration to rays confined to a plane perpendicular to the prism surfaces. We also assume that surface normals of the prism lie in a single plane. θ_a is the angle the incident ray makes with the horizontal in plane of incidence, θ_b is the angle the transmitted ray makes with the horizontal; $\theta_{\perp 1}$

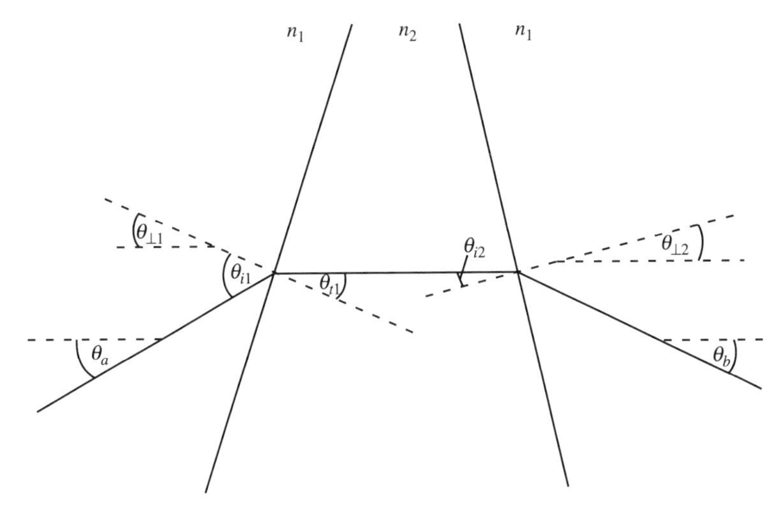

Figure 2.6 Geometry for prism refraction as described by Eqn. (2.8).

is the angle that the normal to the first interface makes with respect to the horizontal, and $\theta_{\perp 2}$ is the corresponding angle for the second interface. The angle of incidence referred to in Snell's law is then

$$\theta_{i1} = \theta_a - \theta_{\perp 1} \tag{2.6}$$

Relative to the horizontal, the angle of transmittance through the first interface is

$$\theta_{t1} = \arcsin\left(\frac{n_1}{n_2}\sin(\theta_a - \theta_{\perp 1})\right) + \theta_{\perp 1} \tag{2.7}$$

where θ_{t1} is the incidence angle, relative to the horizontal for the second interface. The final transmittance angle is thus

$$\theta_b = \theta_{\perp 2} + \arcsin\left(\frac{n_2}{n_1}\sin\left(\arcsin\left(\frac{n_1}{n_2}\sin(\theta_a - \theta_{\perp 1})\right) + \theta_{\perp 1} - \theta_{\perp 2}\right)\right) \tag{2.8}$$

Figure 2.7 shows θ_b as a function of θ_a for various values of n_2/n_1. Notice that while θ_b is a monotonic function of θ_a, the function is somewhat nonlinear.

As an example of spectral dispersion using a prism, Fig. 2.8 plots θ_b as a function of wavelength for fixed θ_a. Notice that the angular dispersion is quite small. Prisms are useful for highly efficient but small-angle dispersion. Less efficient but much faster dispersion is obtained using diffractive elements.

Parallel rays (i.e., rays with identical values of θ_a) are refracted into parallel rays by a prism. A *lens* refracts parallel rays such that they cross at a focal point. As illustrated

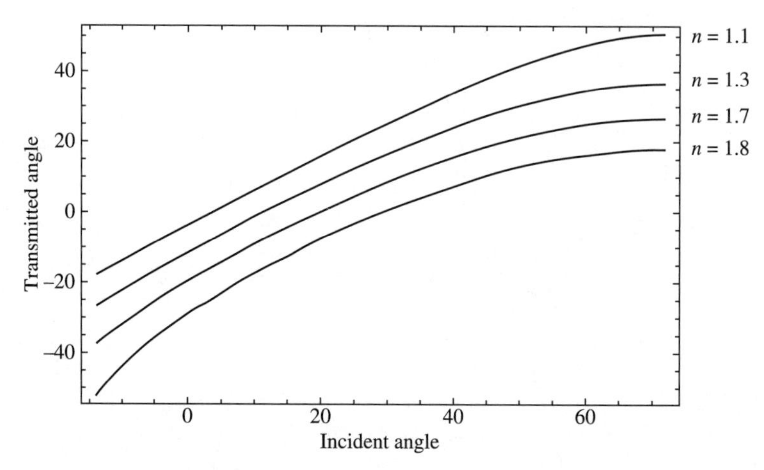

Figure 2.7 Transmission angle θ_b as a function of the incidence angle θ_a for $\theta_{\perp 1} = \pi/10$, $\theta_{\perp 2} = -\pi/10$ for $n_1 = 1$, and various values of n_2.

in Fig. 2.9, a lens may be regarded as a prism with curved surfaces. At the center of the lens, opposite faces are parallel and a normally incident ray is undeflected. As one moves above the center axis of the lens in the figure, opposite faces tilt up such that a ray incident along the horizontal axis is refracted down (under the assumption that the index of refraction of the lens is greater than the index of the surrounding medium). If the tilt of the lens surface increases linearly away from the axis, the refraction angle increases linearly such that all the refracted rays cross at a point. Similarly, the lens surface tilts down below the axis such that rays are refracted up through the focal point.

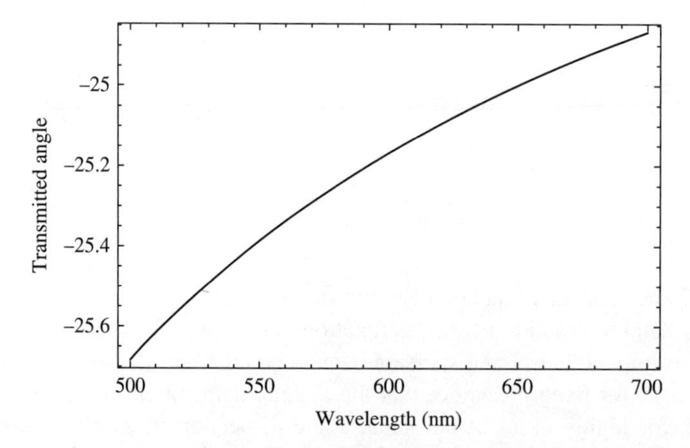

Figure 2.8 Transmission angle θ_b as a function of the wavelength of the incident light for $\theta_a = 0$ and $\theta_{\perp 1} = \pi/10$, $\theta_{\perp 2} = -\pi/10$. The index of refraction and dispersion parameters are those of flint glass F2.

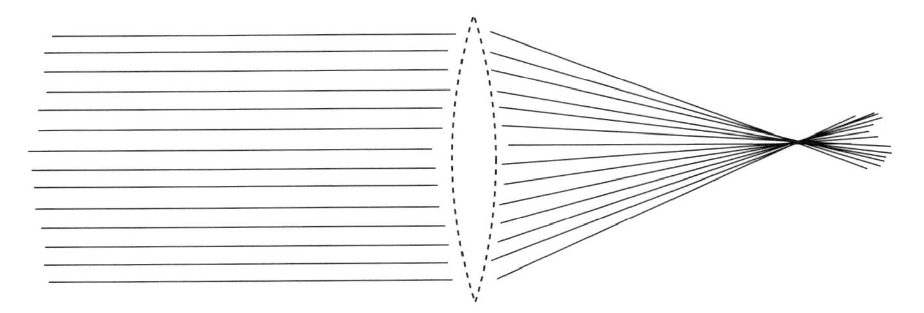

Figure 2.9 A lens as a prism with spatially varying interface orientation.

We analyze refraction through a lens using the geometry of Fig. 2.10. The z axis is a line through the center of the lens (the optical axis). Lens surfaces are generally sections of a sphere because it is easier to polish a surface into spherical shape. For radius of curvature R_1, the first surface is described by

$$x^2 + y^2 + (z - d_1)^2 = R_1^2 \tag{2.9}$$

Under the assumption that $x, y \ll z, R_1$, this surface reduces to the paraboloid

$$z_1 = \frac{x^2 + y^2}{2R_1} - d_1 \tag{2.10}$$

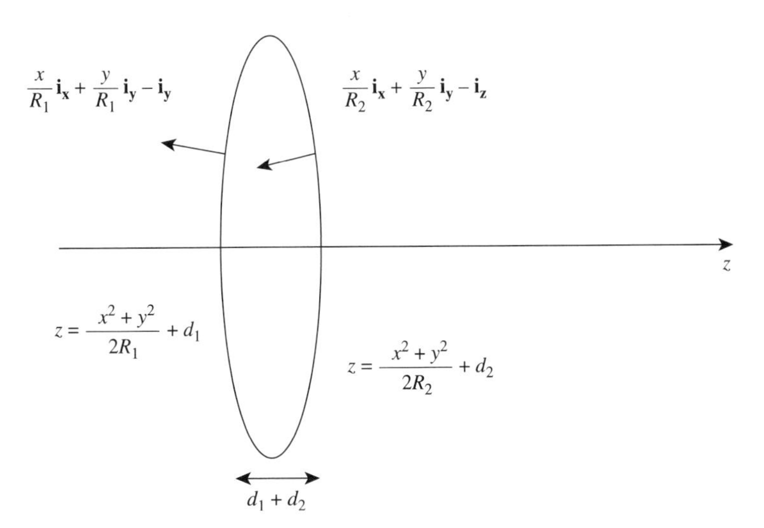

Figure 2.10 Lens surfaces and surface normals.

Similarly, we suppose that the second interface is $z_2 = (x^2 + y^2)/2R_2 + d_2$. The surface normals are described by the vectors

$$\begin{aligned} \mathbf{i}_{n1} &= \frac{x}{R_1}\mathbf{i}_x + \frac{y}{R_1}\mathbf{i}_y - \mathbf{i}_z \\ &= \frac{\rho}{R_1}\mathbf{i}_\rho - \mathbf{i}_z \\ \mathbf{i}_{n2} &= \frac{\rho}{R_2}\mathbf{i}_\rho - \mathbf{i}_z \end{aligned} \tag{2.11}$$

where $\mathbf{i}_\rho = (x\mathbf{i}_x + y\mathbf{i}_y)/\sqrt{x^2 + y^2}$. Suppose that the lens is illuminated by parallel rays incident along $\mathbf{i}_z$. The refracted wave at each point on the lens lies in the plane of the surface normal and the z axis. Noting that the surface normal of both surfaces lies in the plane spanned by $\mathbf{i}_\rho$ and $\mathbf{i}_z$, a vector along the direction of propagation for the refracted wave after the first interface is $\mathbf{i}_{t1}(\rho) = \alpha\,\mathbf{i}_z + \beta(\rho/R_1)\mathbf{i}_\rho$ for some α and β and for $\rho = \sqrt{x^2 + y^2}$. Application of Snell's law tells us that $n_1\mathbf{i}_z \times \mathbf{i}_{n1} = n_2\mathbf{i}_{t1} \times \mathbf{i}_{n1}$ or

$$\frac{n_1\rho}{R_1}\mathbf{i}_z \times \mathbf{i}_\rho = \frac{n_2\rho}{R_1}(\alpha + \beta)\mathbf{i}_z \times \mathbf{i}_\rho \tag{2.12}$$

which reduces to $\alpha + \beta = n_1/n_2$. We solve for α and β using the normalization condition $|\mathbf{i}_{t1}| = 1$, which yields

$$\beta = \frac{(n_1/n_2) - \sqrt{1 + (\rho/R_1)^2 - (\rho/R_1)^2 (n_1/n_2)^2}}{1 + (\rho/R_1)^2} \tag{2.13}$$

Under the approximation that $\rho/R_1 \ll 1$, the solutions to α and β reduce to lowest order as $\beta \approx (n_1/n_2) - 1$ and $\alpha \approx 1$.

Under this approximation, the rate of refraction varies linearly as a function of ρ and, as illustrated in Fig. 2.11, the refracted rays cross at a focal point in the higher-index material at a distance F_1 from the surface of the lens. Under the approximation that the lens surface is at $z = 0$, a vector from a point on the surface of the lens to the focal point is $F_1\mathbf{i}_z - \rho\mathbf{i}_\rho$. This vector is parallel to $\mathbf{i}_{t1}$ if $F_1/\alpha = -R_1/\beta$, which yields

$$F_1 = \frac{n_2 R_1}{n_2 - n_1} \tag{2.14}$$

Similar analysis allows us to discover the focus and focal length for the two surface lenses. Assuming that each lens is thin enough that there is no displacement between the position in x and y of the ray through the first interface and the ray through the second interface, a vector along the transmitted ray is $\mathbf{i}_{t2} = \gamma\mathbf{i}_z + \delta(\rho/R_2)\mathbf{i}_\rho$ for constants γ and δ. Application of Snell's law at this interface yields $n_2\mathbf{i}_{t1} \times \mathbf{i}_{n2} = n_1\mathbf{i}_{t2} \times \mathbf{i}_{n2}$ or $n_2(\alpha - \beta R_2/R_1) = n_1(\gamma - \delta)$. Projecting the transmitted ray $\mathbf{i}_{t2}$ to the focus yields the overall focal length $F = -R_2\gamma/\delta$. Assuming, as with

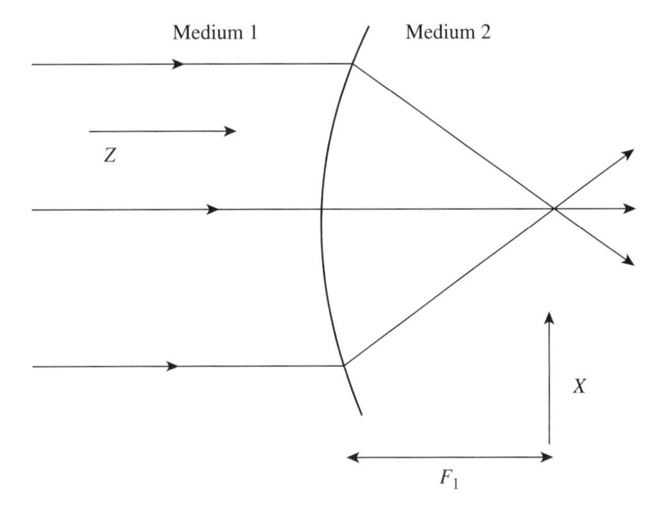

Figure 2.11 Refraction by a parabolic surface.

α, that $\gamma \approx 1$, we obtain

$$n_2 - (n_1 - n_2)\frac{R_2}{R_1} = n_1 - n_1\frac{R_2}{F} \tag{2.15}$$

or

$$\frac{1}{F} = \left(\frac{n_2}{n_1} - 1\right)\left(\frac{1}{R_1} - \frac{1}{R_2}\right) \tag{2.16}$$

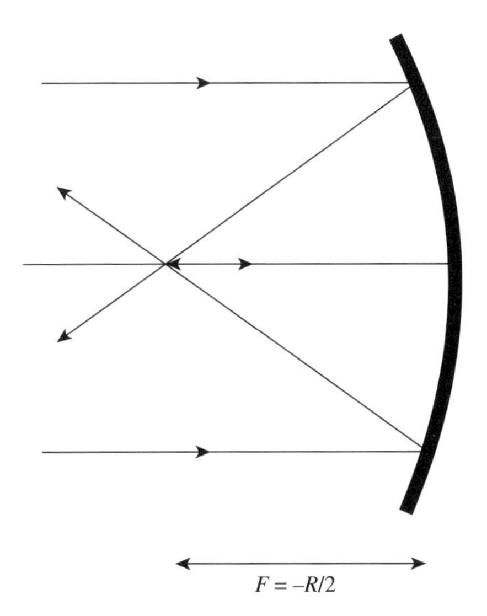

Figure 2.12 Focusing by a parabolic mirror.

Equation (2.16) is the *lensmaker's formula*, describing the relationship between material composition, surface curvature, and focal length. The radii of curvature R_1 and R_2 may be positive or negative. For the convex lens of Fig. 2.10, R_1 is positive and R_2 is negative. For $n_2 > n_1$, this produces a positive focal length lens. It is possible, of course, for the focal length F to be negative, which corresponds to a virtual focus to the left of the lens for an incident parallel ray bundle. Ray analysis can also be used to find the focus of a curved mirror, as illustrated in Fig. 2.12. A parabolic mirror with radius of curvature R focuses parallel rays at a distance $R/2$ in front of the mirror. A negative curvature produces a virtual focus behind the mirror.

2.3 FOCAL IMAGING

The focal properties of a thin lens for rays incident along the optical axis may be used for ray tracing in arbitrary imaging systems based on three simple rules:

1. A ray parallel to the optical axis is refracted through the back (front) focal point of a converging (diverging) lens. This rule was derived in Section 2.2 and is illustrated in Fig. 2.13.
2. A ray through the front (back) focal point of a converging (diverging) lens is refracted parallel to the optical axis. This rule is based on reprocity. A light ray propagates along the same path both backward and forward. This rule is illustrated in Fig. 2.14.

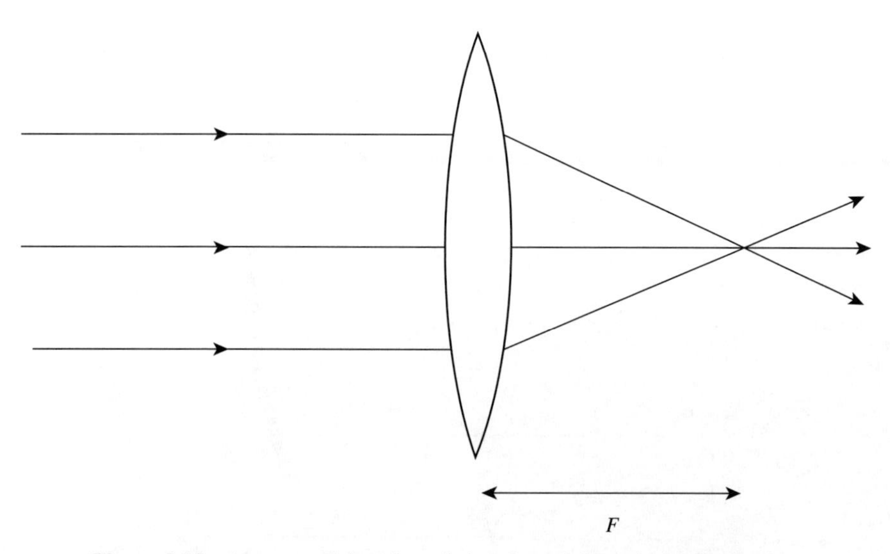

Figure 2.13 A ray parallel to the axis is refracted through the focal point.

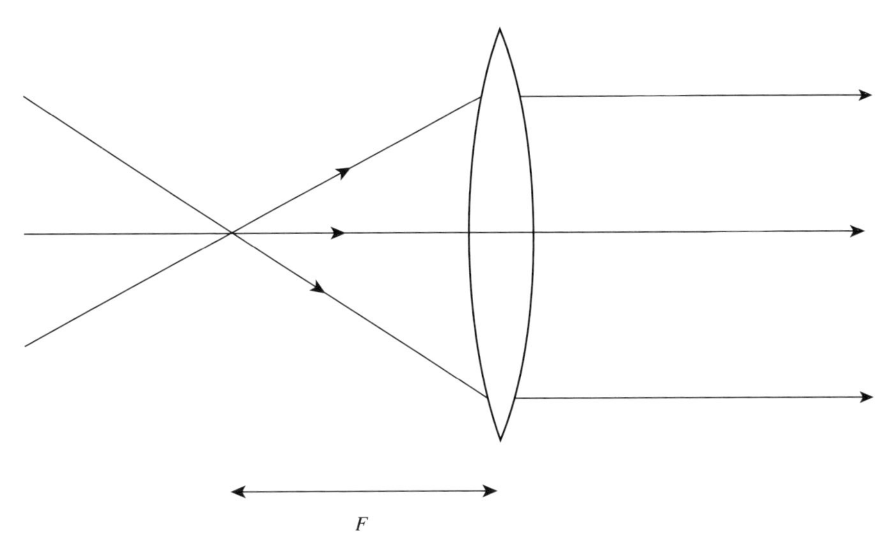

Figure 2.14 A ray through the focal point is refracted parallel to the axis.

3. A ray through the center of the lens is undeflected. A ray going through the center hits both interfaces at the same angle and is refracted out parallel to itself. This rule is illustrated in Fig. 2.15.

Let's use these rules to analyze some example systems. Consider an object at distance d_o from a lens at point $(x_o, 0)$ in the transverse plane, as illustrated in Fig. 2.16.

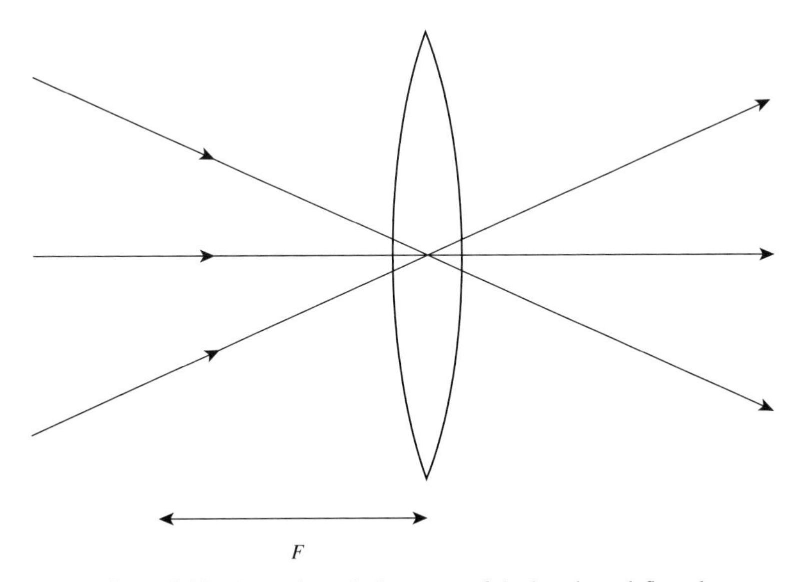

Figure 2.15 A ray through the center of the lens is undeflected.

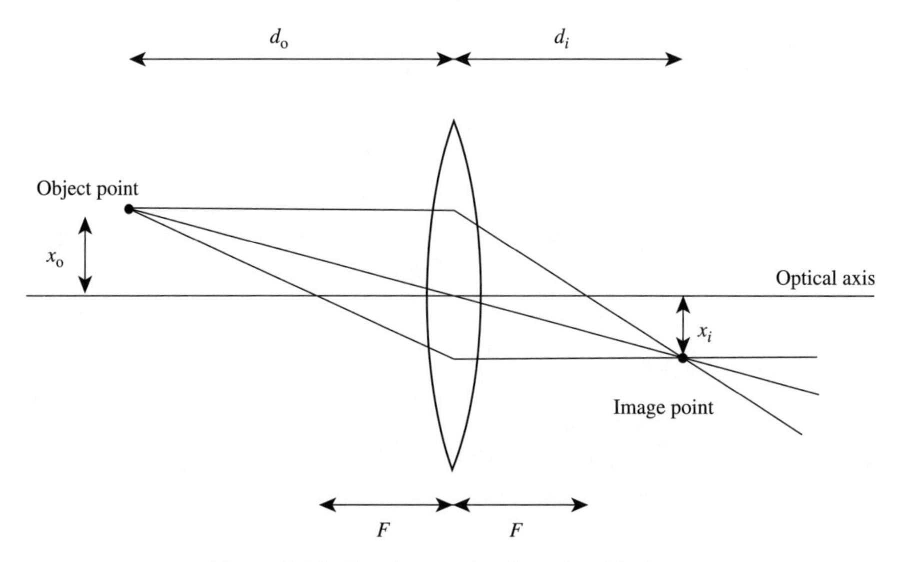

Figure 2.16 Imaging a point through a thin lens.

The lines associated with the three rules are drawn in the figure. The three rays from the source point cross in an image point. Our goal here is to show that there is in fact such an image point and to discover where it is. Let the distance from the lens to the image point be d_i and assume that the transverse position of the image point is $(x_i, 0)$. By comparing congruent triangles, we can see from rule 1 that $x_i/(d_i - F) = x_o/F$, from rule 2 that $x_i/F = x_o/(d_o - F)$, and from rule 3 that $x_o/d_o = -(x_i/d_i)$. Any two of these conditions can be used to produce the thin-lens imaging law

$$\frac{1}{d_o} + \frac{1}{d_i} = \frac{1}{F} \tag{2.17}$$

The magnification from the object to the image is $M = x_i/x_o = -d_i/d_o$, where the minus sign indicates that the image is inverted.

Virtual images and objects are important to system analysis. A virtual image has a negative image distance relative to a lens. The lens produces a ray pattern from a virtual image as though the image were present to the left of the lens. Similarly, an object with negative range illuminates the lens with rays converging toward an object to the right. The virtual image concept is illustrated in Fig. 2.17, which shows a real object with $d_o < F$ illuminating a positive focal length lens. Following our ray tracing rules, a ray parallel to the axis is refracted through the back focal point and a ray through the center is undiverted. These rays do not cross to the right of the lens, but if we extend them to the left of the object, they do cross at a virtual image point. A ray emanating from the object point as though it came from the front focal point is refracted parallel to the optical axis. This ray

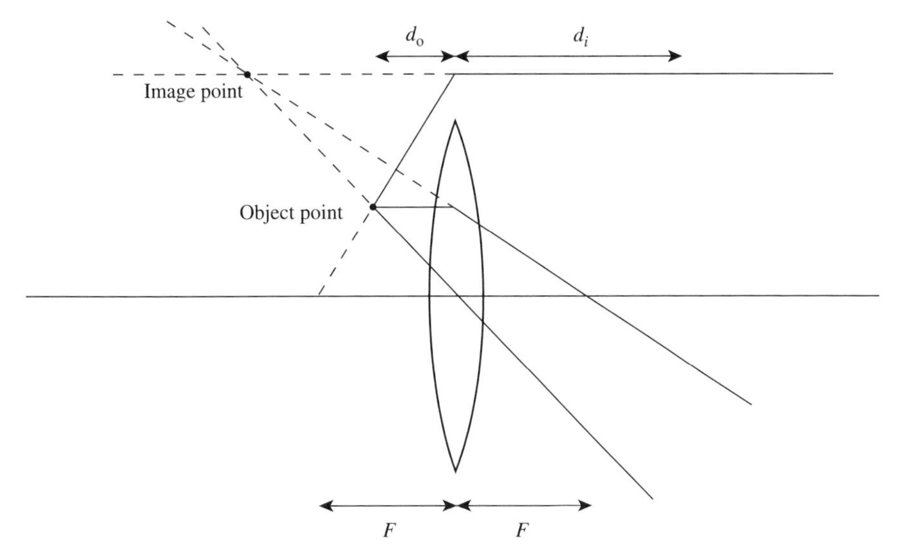

Figure 2.17 A positive focal length lens forms an erect virtual image of an object inside its focal length.

intersects the first two rays when extended to the left. Our ray trace $d_i < 0$ for $0 < d_o < F$ is consistent with the thin-lens image law result:

$$d_i = \frac{d_o F}{d_o - F} \tag{2.18}$$

Since d_i is opposite in sign to d_o, the magnification is positive for this system and the image is erect.

Figure 2.18 illustrates ray tracing with a negative focal length lens. The concave element forms an erect virtual image of an object to the left of the focal point. A horizontal ray for this system refracts as though coming from the negative focal point. A ray through the center passes through the center, and a ray incident on the negative front focal point refracts horizontal to the axis. These rays do not meet to the right of the lens, but do meet in the virtual image when extended to the left. An observer looking through the lens would see the virtual image.

Focal reflective optics are analyzed using very similar ray propagation rules. A mirror of radius R has a focal length of $F = 2/R$, where R is considered positive for a concave mirror and negative for a convex mirror. A ray striking the mirror parallel to the optical axis is reflected through the focal point. A ray striking the center of the mirror is reflected at an angle equal to the angle of incidence. A ray passing through the focal point is reflected back parallel to the optical axis. As illustrated in Fig. 2.19, image formation using a parabolic mirror may be analyzed using these simple ray tracing rules or the thin-lens imaging law.

Ray tracing may be extended to analyze multiple element optical systems. For example, Fig. 2.20 illustrates image formation using a two-lens system. To ray

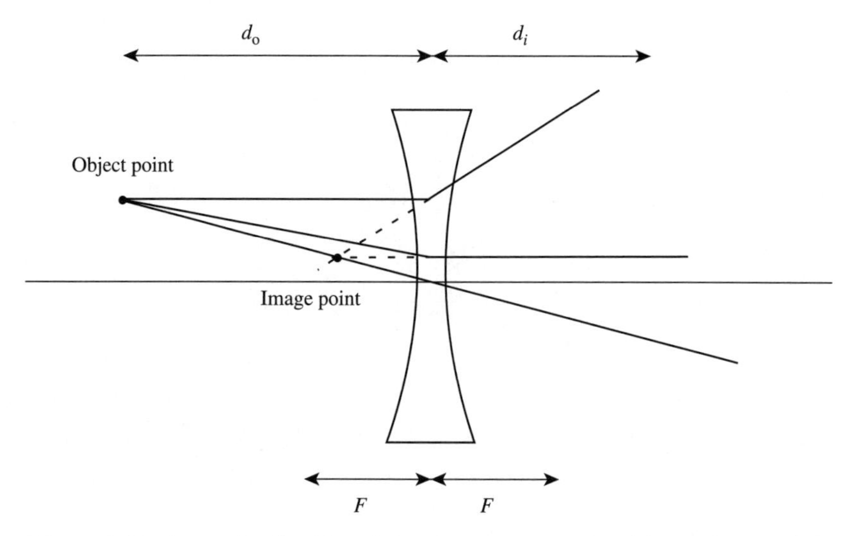

Figure 2.18 A negative focal length lens forms an erect virtual image for $d_o > |F|$.

trace this system, one first traces the rays for the first lens as though the second element were not present. The image for the first lens forms to the right of the second lens. Treating the intermediate image as a virtual object with $d_o < 0$, ray tracing bends the ray incident on the virtual object along the optical axis through the back focal point, the incident ray through the center of the second lens is undeflected and the ray incident from the front focal point is refracted parallel to the

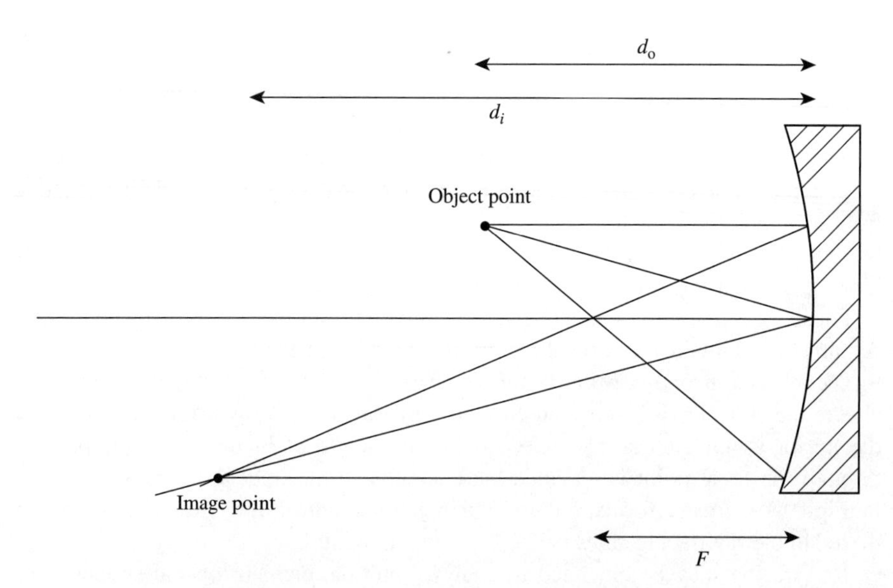

Figure 2.19 Image formation by a parabolic mirror.

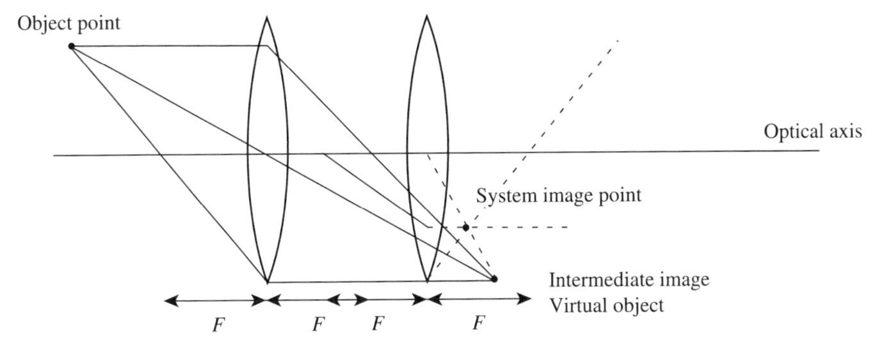

Figure 2.20 Image formation by a lens system.

optical axis. These rays meet in a real inverted image. The magnification of the compound system is the product of the magnifications of the individual systems, $M = d_{i1} d_{i2}/d_{o1} d_{o2}$.

We made a number of approximations to derive the lensmakers formula for parabolic surfaces. With modern numerical methods, these approximations are unnecessary. Ray tracing software to map focal patterns (spot diagrams) for thick and compound optical elements is commonly used in optical design. As an introduction to ray tracing, Problem 2.7 considers exact analysis of the lens of Fig. 2.10. Problem 2.8 requires the student to write a graphical program to plot rays through an arbitrary sequence of surfaces.

Computational ray tracing using digital computers is the foundation of modern lens design. Ray tracing programs project discrete rays from surface to surface. Simple ray tracing for thin elements, on the other hand, may be analyzed using "*ABCD*" matrices, which implement simple plane-to-plane optical transformations, as illustrated in Fig. 2.21. Plane-to-plane, or "paraxial," system analysis is a consistent theme in this text. We describe the ray version of this approach here, followed by wave field versions in Section 4.4 and the statistical field version in Section 6.2.

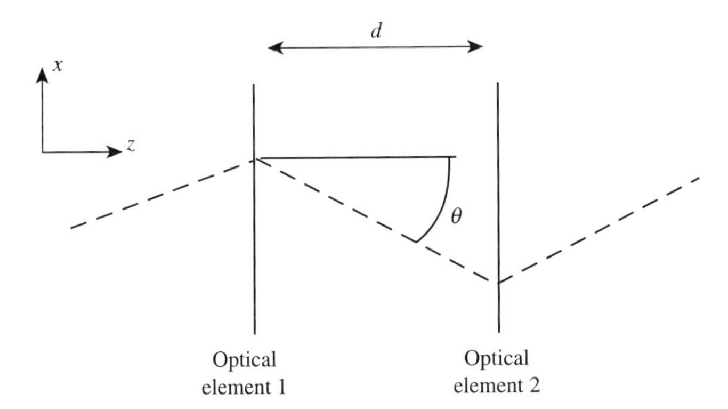

Figure 2.21 Paraxial ray tracing using *ABCD* matrices. The dashed line represents a ray path.

The state of a ray for planar analysis is described by the position x at which the ray strikes a given plane and the slope $\theta = dx/dz$ of the incident ray. The state of the ray is represented by a vector

$$\begin{bmatrix} x \\ \theta \end{bmatrix} \tag{2.19}$$

The slope of the ray is invariant as it propagates through homogeneous space; the transformation of the ray on propagating a distance d through free space is

$$\begin{bmatrix} x' \\ \theta' \end{bmatrix} = \begin{bmatrix} 1 & d \\ 0 & 1 \end{bmatrix} \begin{bmatrix} x \\ \theta \end{bmatrix} \tag{2.20}$$

On striking a thin lens, the slope of a ray is transformed to $\theta' = \theta - x/F$, but the position of the ray is left unchanged. The *ABCD* matrix for a thin lens is accordingly

$$\begin{bmatrix} A & B \\ C & D \end{bmatrix} = \begin{bmatrix} 1 & 0 \\ -\frac{1}{F} & 1 \end{bmatrix} \tag{2.21}$$

One may use these matrices to construct simple models of the action of lens systems. For example, the ray transfer matrix from an object plane a distance d_o in front of a lens to the image plane a distance d_i behind the lens is

$$\begin{bmatrix} 1 & d_i \\ 0 & 1 \end{bmatrix} \begin{bmatrix} 1 & 0 \\ -1/F & 1 \end{bmatrix} \begin{bmatrix} 1 & d_o \\ 0 & 1 \end{bmatrix} = \begin{bmatrix} 1 - \frac{d_i}{F} & d_o + d_i - \frac{d_i d_o}{F} \\ -\frac{1}{F} & 1 - \frac{d_o}{F} \end{bmatrix} \tag{2.22}$$

If $B = d_o + d_i - (d_i d_o/F) = 0$ then the thin-lens imaging law is satisfied and, as expected, the output ray position x' is independent of the slope of the input ray. In this case, $A = 1 - d_i/F = -d_i/d_o$ and $x' = Ax$ is magnified as expected.

Paraxial ray tracing is often used to roughly analyze optical systems. *ABCD* ray tracing is also used to model the transformation of Gaussian beams (Section 3.5) and to propagate radiance (Section 6.7).

2.4 IMAGING SYSTEMS

Most imaging systems consist of sequences of optical elements. While details of such systems are most conveniently analyzed using modern ray tracing and design software, it is helpful to have a basic concept of the design philosophy of common instruments, such as cameras, microscopes, and telescopes. A camera records images on film or electronic focal planes. We have much to say about camera and spectrometer design in Chapters 9 and 10. A microscopes makes small close objects appear larger to the eye and may be combined with a camera to record magnified images. A

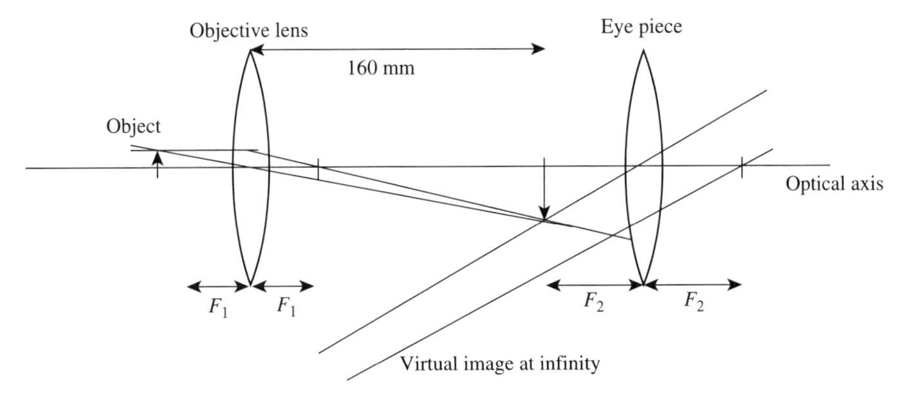

Figure 2.22 Ray diagram for a compound microscope.

telescope makes large distant objects appear larger and may be combined with a camera to record enlarged, if still demagnified, images.

As sketched in Fig. 2.22, a basic microscope consists of an objective lens and an eyepiece. The object is placed just outside the objective focal point, yielding a highly magnified real image just in front of the eyepiece. The distance from the objective to the eyepiece is typically enclosed in the microscope body and is called the *tube length*. In the most common convention, the tube length is 160 mm. This enables one to swap objective lenses and eyepieces within standard microscope bodies. A 10× objective has a focal length of 16 mm, producing a real image magnified by 10 at the eyepiece focal point. A 40× objective has a focal length of 4 mm. From the eyepiece focal plane one may choose to relay the magnified image onto a recording focal plane and/or through the eyepiece for visual inspection. The eye focuses most naturally on objects essentially at infinity. Thus the eyepiece is situated to form a virtual image at infinity. The virtual image is greatly magnified.

As a basic measure of the performance of a microscope, one may compare the angular size of the object as observed through the eyepiece to the angular size of the object viewed without the microscope. One observing the object without the microscope would hold it at one's near point (the closet point to the eye on which one can focus). While the near point varies substantially with age, 254 mm is often given as a standard value. An object of extent x_o thus subsumes an angle of $x_o/254$ when observed without the microscope. This object is magnified to a real image of size $x_i = -160x_o/f_o$ by the objective lens. The angular extent of the object at infinity viewed through the eyepiece is x_i/f_e. Thus the magnifying power (MP) of the microscope is

$$\mathrm{MP} = -\left(\frac{160}{f_o}\right)\left(\frac{254}{f_e}\right) \tag{2.23}$$

for f_o and f_e in mm.

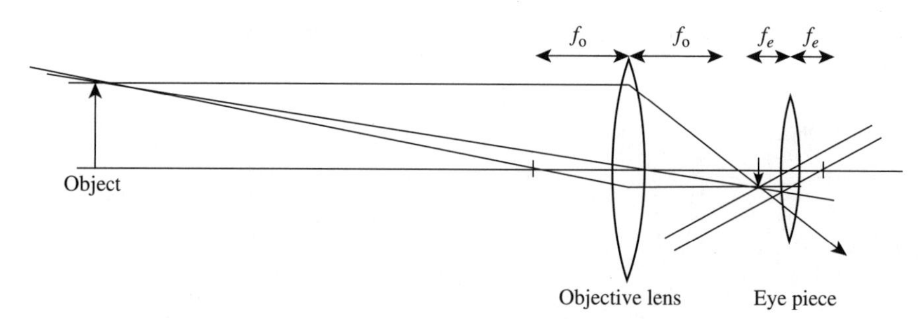

Figure 2.23 Ray diagram for a refractive telescope.

Modern microscopes incorporate many optical components and multiple beam-paths within the system body, rendering a standard tube length and fixed objective and eyepiece positions archaic. These systems use "infinity corrected" objectives to project parallel rays from the objective lens for processing by optical components within the tube. In many cases, of course, the goal is to form an enlarged real image on an electronic focal plane rather than a virtual image for human consumption.

A *telescope* demagnifies an object but increases the angular range that the object subsumes at the eye. (In this sense the telescope is the reverse of a microscope. The microscope reduces the angular range of rays from the object but increases its scale.) A refractive telescope design is sketched in Fig. 2.23.

The angular extent subsumed by a distant object without a telescope is x_o/R, where x_o is the transverse extent of the object and R is the object range. The real image formed by the objective lens is of extent $f_o x_o/R$. The angular extent of the image after passing through the telescope is thus $(f_o x_o/Rf_e)$. The angular extent has thus been magnified by the factor (f_o/f_e).

Reflective elements are commonly used for telescopes because one can build much larger apertures at much less cost with mirrors. (A mirror only needs a high-quality surface; a lens needs high-quality volume.) As an example, a Cassegrain reflecting telescope is illustrated in Fig. 2.24. A large concave primary mirror

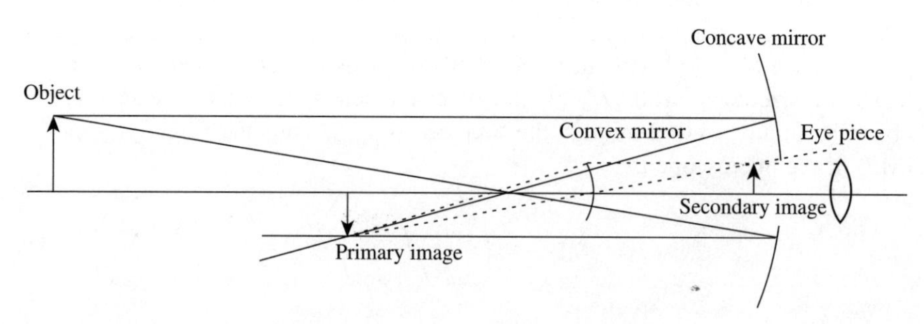

Figure 2.24 A Cassegrain telescope.

focuses light on a convex secondary mirror. The combination of the primary and the secondary form a real image through a hole in the center of the primary. The telescope is designed to produce a virtual image at infinity through the eyepiece and to magnify the angular size of the object. The fact that reflective optics are dispersion-free is an additional benefit in telescope design.

2.5 PINHOLE AND CODED APERTURE IMAGING

In a focal imaging system, all rays passing through point x_o in the object plane pass through point x_i in the image plane. The visibility between points in the object plane and points in the image plane is $v(x_o, y_o, x_i, y_i) = \delta(x_i - Mx_o, y_i - MY_o)$, where M is the magnification. We refer to systems implementing such point-to-point visibilities as *isomorphic* systems. In general, imaging based on isomorphisms is extremely attractive. However, we find throughout this text that substantial design power is released by relaxing the isomorphic mapping. Motivations for considering nonisomorphic, or *multiplex*, imaging include

- Isomorphisms are often physically impossible. The isomorphism of focal imaging applies only between planes in 3D space. Real objects are distributed over 3D space and may be described by spectral and polarization features over higher-dimensional spaces. One must not let the elegance of the focal mapping distract one from the fact that objects span 3D.
- Details of physical optics and optoelectronic sampling interfaces may make multiplex systems attractive or essential. For example, lenses are not available in all frequency ranges.
- Multiplex systems may enable higher system performance relative to data efficiency, resolution, depth of field, and other parameters.

The next three sections present an introduction to multiplex imaging systems using examples drawn from geometric analysis. The first example, coded aperture imaging, is illustrative of the significance of the continuous-to-discrete mapping in digital sensor systems and of the role of code design in multiplex measurement. The second example, computed tomography, introduces multidimensional imaging. The third example, reference structure tomography, introduces representation spaces and projection sampling.

Coded aperture imaging was developed as a means of imaging fields of high-energy photons. Refractive and reflective imaging elements are unavailable at high energies, so focal imaging is not an option for these systems. Pinhole imaging, which dates back over 500 years in the form of *camera obscura* and related instruments, is the precursor to coded aperture imaging.

A pinhole imaging system is illustrated in Fig. 2.25. The pinhole is a small hole of diameter d in an otherwise opaque screen. The light observed on a measurement screen a distance l behind the pinhole consists of projections of the incident light through the pinhole. The pinhole is described by the function $\mathrm{circ}(x/d, y/d)$,

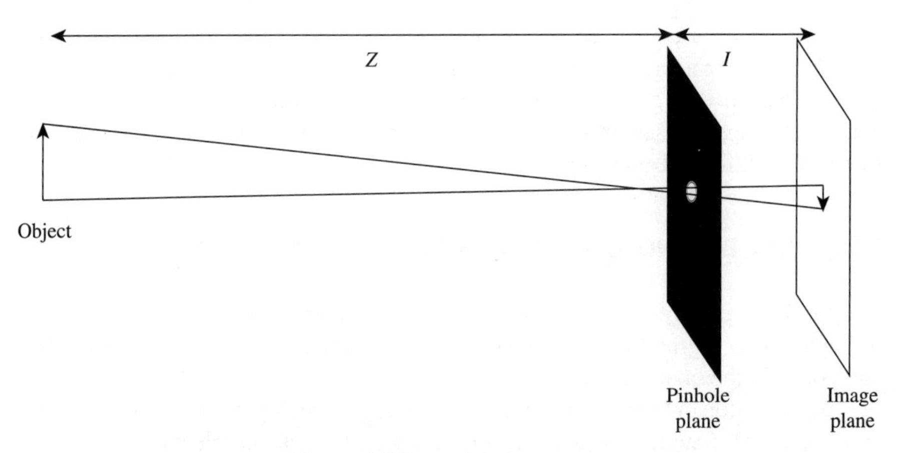

Figure 2.25 A pinhole camera.

where

$$\mathrm{circ}(x, y) = \begin{cases} 1 & \sqrt{x^2 + y^2} \leq 0.5 \\ 0 & \text{otherwise} \end{cases} \tag{2.24}$$

A point source at position (x_o, y_o, z_o) in front of the pinhole is visible at position (x_i, y_i) on the measurement plane if the ray from (x_o, y_o, z_o) to (x_i, y_i) passes through the pinhole without obscuration. A bit of geometry convinces one that the visibility for the pinhole camera is

$$h(x_i, y_i) = \mathrm{circ}\left(\frac{x_i + (l/z_o)x_o}{d + (ld/z_o)}, \frac{y_i + (l/z_o)y_o}{d + (ld/z_o)}\right) \tag{2.25}$$

Refering to Eqn. (2.1), the mapping between the measurement field and the object field for this visibility is

$$g(x_i, y_i) = \iiint f(x_o, y_o, z_o)\mathrm{circ}\left(\frac{x_i + (l/z_o)x_o}{d + (ld/z_o)}, \frac{y_i + (l/z_o)y_o}{d + (ld/z_o)}\right) dx_o\, dy_o\, dz_o \tag{2.26}$$

Equation (2.26) is a convolution and one might attempt decovolution to recover the original source distribution. In practice, however, one is more apt to consider $g(x_i, y_i)$ as the image of the source projected in two dimensions or to use the tomographic methods discussed in the next section to recover an estimate of the 3D distribution. $g(x_i, y_i)$ is inverted with respect to the object and the magnification for an object at range z_o is $M = -l/z_o$. The *resolution* of the reconstructed source in isomorphic imaging systems is defined by the spatial extent of the point visibility, which in this case is $d + (ld/z_o)$.

A smaller pinhole improves resolution, but decreases the optical energy reaching the focal plane from the object. The primary motivation in developing coded aperture imaging is to achieve the imaging functionality and resolution of a pinhole system without sacrificing photon throughput. Diffraction also plays a role in determining optimal pinhole size. We discuss diffraction in detail in Chapter 4, but it is helpful to note here that the size of the projected pinhole will increase because of diffraction by approximately $l\lambda/d$, where λ is the wavelength of the optical field. On the basis of this estimate, the actual resolution of the pinhole camera is

$$\Delta x = d + \frac{ld}{z_o} + \frac{l\lambda}{d} \tag{2.27}$$

Δx is minimized by the selection $d_{\text{opt}} = \sqrt{l z_o \lambda/(z_o + l)}$. Assuming that $z_o \gg l$, $d_{\text{opt}} \approx \sqrt{l\lambda}$ and

$$\Delta x_{\min} \approx 2\sqrt{l\lambda} \tag{2.28}$$

As an example, the optimal pinhole size is approximately 100 μm for l equal to 1 cm and λ equal to one micrometer. The angular resolution of this system is approximately 20 milliradians (mrad).

Coded aperture imaging consists of replacing the pinhole mask with a more complex pattern, $t(x, y)$. Increasing the optical throughput without reducing the system resolution is the primary motivation for coded aperture imaging. As illustrated in Fig. 2.26, each object point projects the shadow of the coded aperture. The overlapping projections from all of the object points are integrated pixel by pixel on the sensor plane. The visibility for the coded aperture system is

$$h(x_i, y_i) = t\left(\frac{x_i + (l/z_o)x_o}{1 + (l/z_o)}, \frac{y_i + (l/z_o)y_o}{1 + (l/z_o)}\right) \tag{2.29}$$

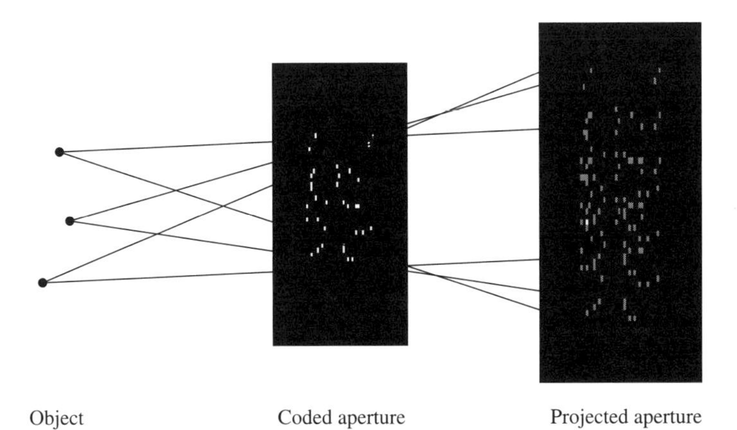

Figure 2.26 Coded aperture imaging geometry.

and the input-output transformation is

$$g(x_i, y_i) = \iiint f(x_o, y_o, z_o) t\left(\frac{x_i + (l/z_o)x_o}{1 + (l/z_o)}, \frac{y_i + (l/z_o)y_o}{1 + (l/z_o)}\right) dx_o\, dy_o\, dz_o \qquad (2.30)$$

In contrast with Eqn. (2.26), $g(x_i, y_i)$ in Eqn. (2.30) is not isomorphic to $f(x_o, y_o, z_o)$.

The coded aperture system is illustrative of several important themes in multiplex imaging, including the following:

- *Sampling.* Computational processing is required to produce an image from $g(x_i, y_i)$. Computation operates on discrete measurement values rather than continuous fields. The process of turning continuous distributions into discrete samples is a central focus of imaging system design and analysis.
- *Coding.* While some constraints exist on the nature of the coded aperture visibility and resolution, the system designer has great freedom in the selection of $t(x, y)$. Design of the aperture pattern is a coding problem. One seeks a code to maximize information transfer between object features and the detection system.
- *Inversion.* Even after $t(x, y)$ is specified, many different algorithms may be considered for estimation of $f(x_o, y_o, z_o)$ from $g(x_i, y_i)$. Algorithm design for this situation is an *inverse problem.*

For simplicity, we initially limit our analysis to 2D imaging. Equation (2.30) is reduced to a 2D imaging transformation under the assumption $l/z_o \ll 1$ using the definitions $\theta_x = x_o/z_o$, $\theta_y = y_o/z_o$, and

$$\hat{f}(\theta_x, \theta_y) = \int f(z_o\theta_x, z_o\theta_y, z_o) dz_o \qquad (2.31)$$

In this case

$$g(x_i, y_i) = \iint \hat{f}(\theta_x, \theta_y) t\big(x_i + l\theta_x, y_i + l\theta_y\big) d\theta_x\, d\theta_y \qquad (2.32)$$

The continuous distribution $g(x_i, y_i)$ is reduced to discrete samples under the assumption that one measures the output plane with an array of optoelectronic detectors. The spatial response of the (ij)th detector is modeled by the function $p_{ij}(x, y)$ and the discrete output data array is

$$g_{ij} = \int g(x_i, y_i) p_{ij}(x_i, y_i) dx_i\, dy_i \qquad (2.33)$$

Sampling is typically implemented on a rectangular grid using the same pixel function for all samples, such that

$$p_{ij}(x, y) = p(x - i\Delta, y - j\Delta) \qquad (2.34)$$

where Δ is the pixel pitch.

Code design consists of the selection of a set of discrete features describing $t(x, y)$. We represent t discretely as

$$t(x, y) = \sum_{ij} t_{ij} \tau(x - i\Delta, y - j\Delta) \tag{2.35}$$

As with the sampling system, we will limit our consideration to rectangularly spaced functions as the expansion basis. But, again as with the sampling system, it is important to note for future consideration that this is not the only possible choice.

Substituting Eqns. (2.33) and (2.35) into Eqn. (2.32), we find

$$\begin{aligned} g_{i'j'} = \sum_{ij} t_{ij} \int\int \hat{f}(\theta_x, \theta_y) \tau\left(x' + l\theta_x - i\Delta_t, y' + l\theta_y - j\Delta_t\right) \\ \times p(x' - i'\Delta, y' - j'\Delta) d\theta_x\, d\theta_y\, dx' dy' \end{aligned} \tag{2.36}$$

Assuming that the sampling rate on the coded aperture is the same as the sampling rate on the sensor plane, we define

$$\hat{p}_{i-i',j-j'}(\theta_x, \theta_y) = \int \tau\left(x' + l\theta_x - i\Delta, y' + l\theta_y - j\Delta_t\right) h(x' - i'\Delta, y' - j'\Delta) dx'\, dy' \tag{2.37}$$

and

$$\hat{f}_{i,j} = \int \hat{f}(\theta_x, \theta_y) \hat{p}_{i,j}(\theta_x, \theta_y) d\theta_x\, d\theta_y \tag{2.38}$$

such that

$$g_{i'j'} = \sum_{ij} t_{ij} \hat{f}_{i-i',j-j'} \tag{2.39}$$

where $\hat{f}_{i,j}$ is interpreted as a discrete estimate for $\hat{f}(\theta_x, \theta_y)$. $\hat{p}_{i-i',j-j'}(\theta_x, \theta_y)$ is the sampling function that determines the accuracy of the assumed correspondence. As an example, we might assume that

$$\tau(x, y) = p(x, y) = \text{rect}\left(\frac{x}{\Delta}\right) \text{rect}\left(\frac{y}{\Delta}\right) \tag{2.40}$$

where

$$\text{rect}(x) = \begin{cases} 1 & |x| \leq 0.5 \\ 0 & \text{otherwise} \end{cases} \tag{2.41}$$

Evaluating Eqn. (2.37) for this aperture and sensor plane sampling function, we find that $\hat{p}_{i-i',j-j'}(\theta_x, \theta_y)$ is the product of triangle functions in θ_x and θ_y, as illustrated in Fig. 2.27. The extent of the sampling function is Δ/l along the θ_x and θ_y directions,

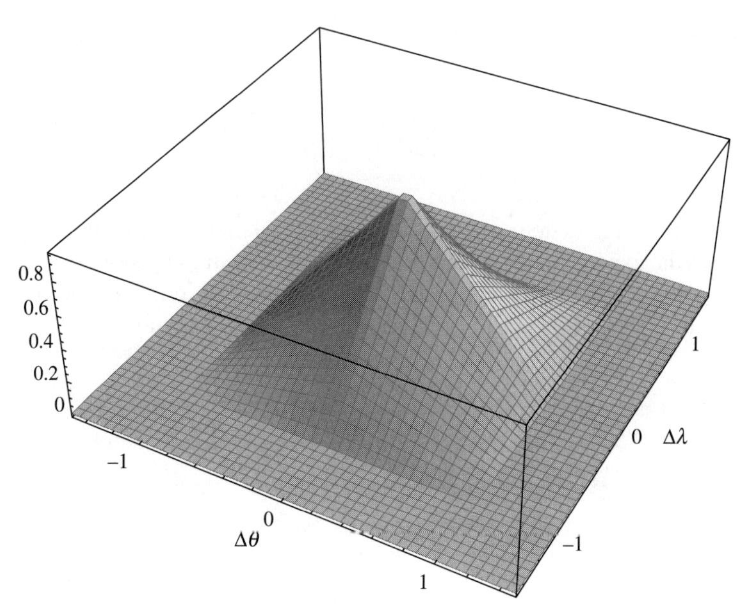

Figure 2.27 Sampling function $\hat{p}_{i+i'=0,\, j+j'=0}(\theta_x, \theta_y)$ for rect pattern coded aperture and sensor plane functions. $\hat{p}_{i,j}(\theta_x, \theta_y)$ is a weighting function for producing the discrete measurement $\hat{f}_{i,j}$ from the continuous object distribution.

meaning that the angular resolution of the imaging system is approximately Δ/l. The center of the sampling function is at $\theta_x = (i - i')\Delta/l$, $\theta_y = (j - j')\Delta/l$.

The challenge of coding and inversion consists of selecting coefficients t_{ij} and an algorithm for estimating $\hat{f}_{i,j}$ from Eqn. (2.39). Equation (2.39) is a correlation between the object and the coded aperture and may be compactly expressed as

$$\mathbf{g} = \mathbf{t} \star \mathbf{f} \tag{2.42}$$

This linear transformation of **f** may be inverted by linear or nonlinear methods. In general, one seeks an inversion method optimizing some system measure, such as the mean-square error $\|\mathbf{f}_e - \mathbf{f}\|$ between the estimated object $\mathbf{f}_e$ and the actual object **f**. In the case of coded aperture imaging, one may attempt to optimize estimation over both code design (i.e., selection of **t**) and image recovery algorithms.

Circulant linear transformations such as correlation and convolution are inverted by circulant transformations, meaning that the linear inverse of Eqn. (2.42) is also a correlation. Representing the inverting matrix as $\hat{\mathbf{t}}$, the linear estimation algorithm takes the form

$$\begin{aligned}\mathbf{f}_e &= \breve{\mathbf{t}} \star \mathbf{G} \\ &= \breve{\mathbf{t}} \star (\mathbf{t} \star \mathbf{f}) + \breve{\mathbf{t}} \star \mathbf{b}\end{aligned} \tag{2.43}$$

where we have accounted for noise in the measurement by adding a noise component $\mathbf{b}$ to $\mathbf{g}$. The goals of system design for linear inversion are to select $\breve{\mathbf{t}}$ and $\mathbf{t}$ such that

- $\mathbf{t}$ is physically allowed.
- $\breve{\mathbf{t}} \star \mathbf{t}\star$ is an identity operator.
- The effect of $\breve{\mathbf{t}} \star \mathbf{b}$ in signal ranges of interest is minimized.

The physical implementation of the coded mask is usually taken to be a pinhole array, in which case the components t_{ij} of t are either 0, for opaque regions of the mask; or 1, for the pinholes. Somewhat better noise rejection would be achieved if t_{ij} could be selected from 1 and -1. In some systems this is achieved by subtracting images gathered from complementary coded apertures. Using such bipolar codes it is possible to design delta-correlated t_{ij}. Codes satisfying $\breve{\mathbf{t}} \star \mathbf{T} = \delta$ are termed *perfect sequences* [130] or *nonredundant arrays*.

A particularly effective approach to coded aperture design based on *uniformly redundant arrays* was proposed by Fenimore and Cannon [73]. The URAs of Fenimore and Cannon and Gottesman and Fenimore [73,102] are based on quadratic residue codes. An integer q is a quadratic residue modulo an integer p if there exists an integer x such that

$$x^2 = q(\text{mod } p) \tag{2.44}$$

If p is a prime number such that $p(\text{mod } 4) = 1$, then a uniformly redundant array is generated by letting

$$t_{ij} = \begin{cases} 0 & \text{if } i = 0 \\ 1 & \text{if } j = 0,\ i \neq 0 \\ 1 & \text{if } i \text{ AND } j \text{ are quadratic residues modulo } p \\ 1 & \text{if neither } i \text{ nor } j \text{ are quadratic residues modulo } p \\ 0 & \text{otherwise} \end{cases} \tag{2.45}$$

The decoding matrix $\breve{\mathbf{t}}$ is defined according to

$$\hat{t}_{ij} = \begin{cases} +1 & \text{if } i = j = 0 \\ +1 & \text{if } t_{ij} = 1 \\ -1 & \text{if } t_{ij} = 0,\ (i, j \neq 0) \end{cases} \tag{2.46}$$

This choice of $\mathbf{t}$ and $\breve{\mathbf{t}}$ is referred to as the *modified uniformly redundant array* (MURA) by Gottesman and Fenimore [102]. $\breve{\mathbf{t}}$ and $\mathbf{t}$ are delta-correlated, with a peak correlation value for zero shift equal to the number of holes in the aperture and with zero correlation for other shifts. To preserve the shift-invariant assumption that a shadow of the mask pattern is cast on the detector array from all angles of incidence, it is necessary to periodically tile the input aperture with the code $\mathbf{t}$. Figures 2.28, 2.29, and 2.30 show the transmission codes for $p = 5$, 11, and 59, respectively. The cross-correlation $\breve{\mathbf{t}} \star \mathbf{t}$ is also shown. The cross-correlation is

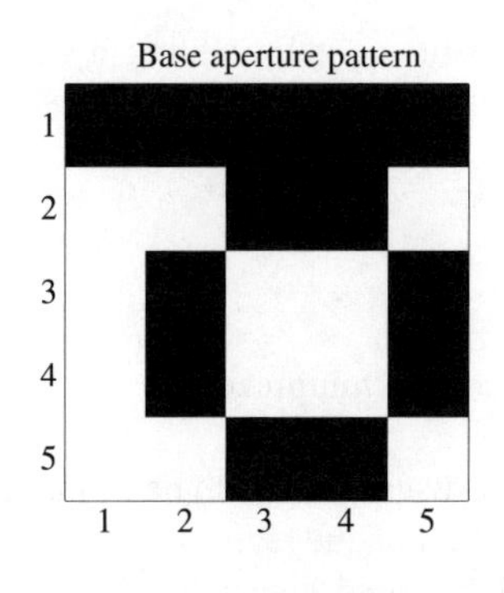

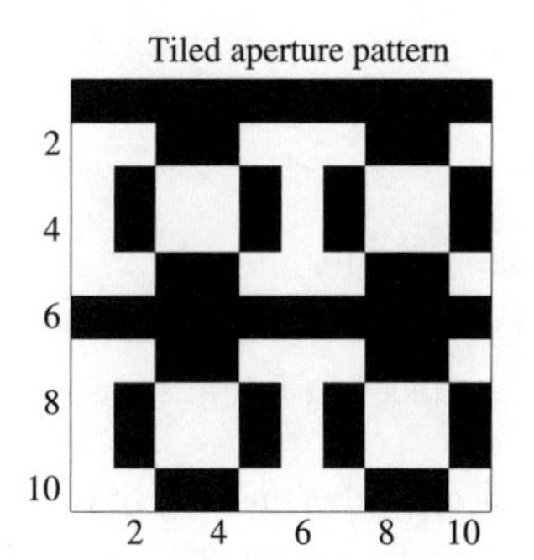

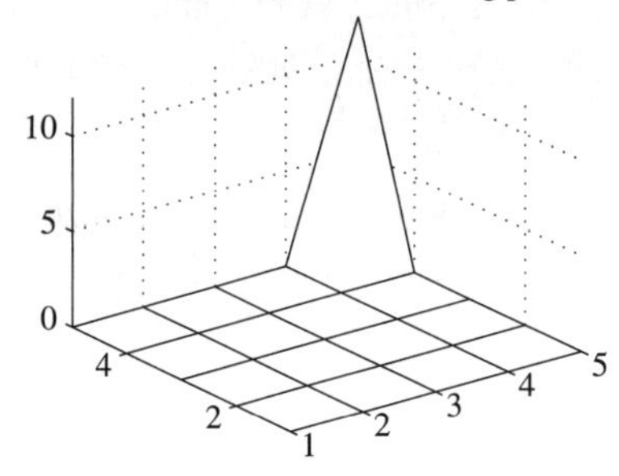

Figure 2.28 Base transmission pattern, tiled mask, and inversion deconvolution for $p = 5$.

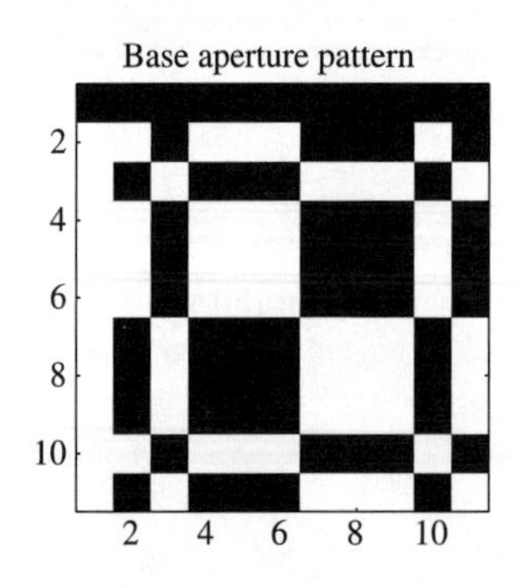

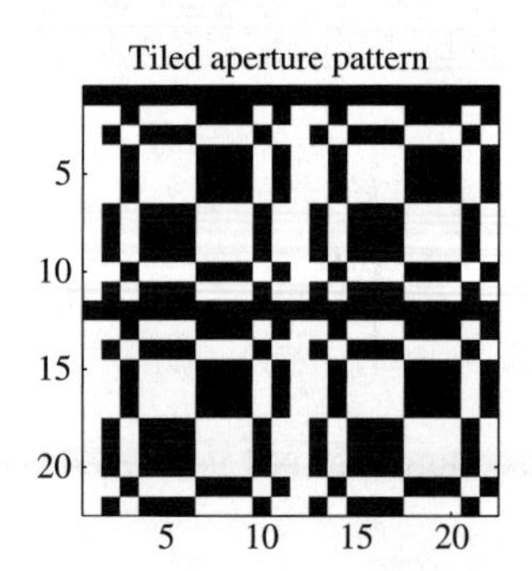

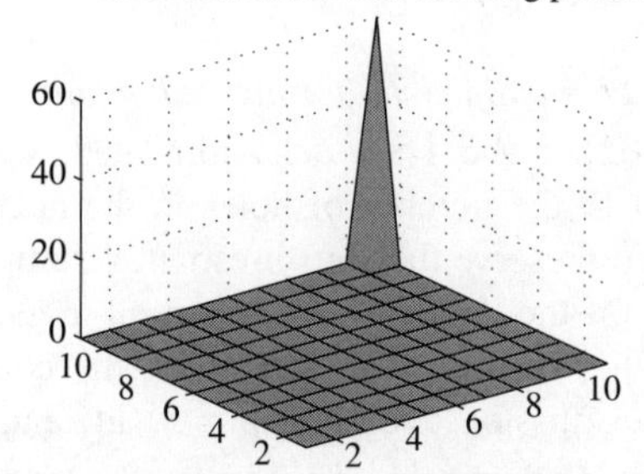

Figure 2.29 Base transmission pattern, tiled mask, and inversion deconvolution for $p = 11$.

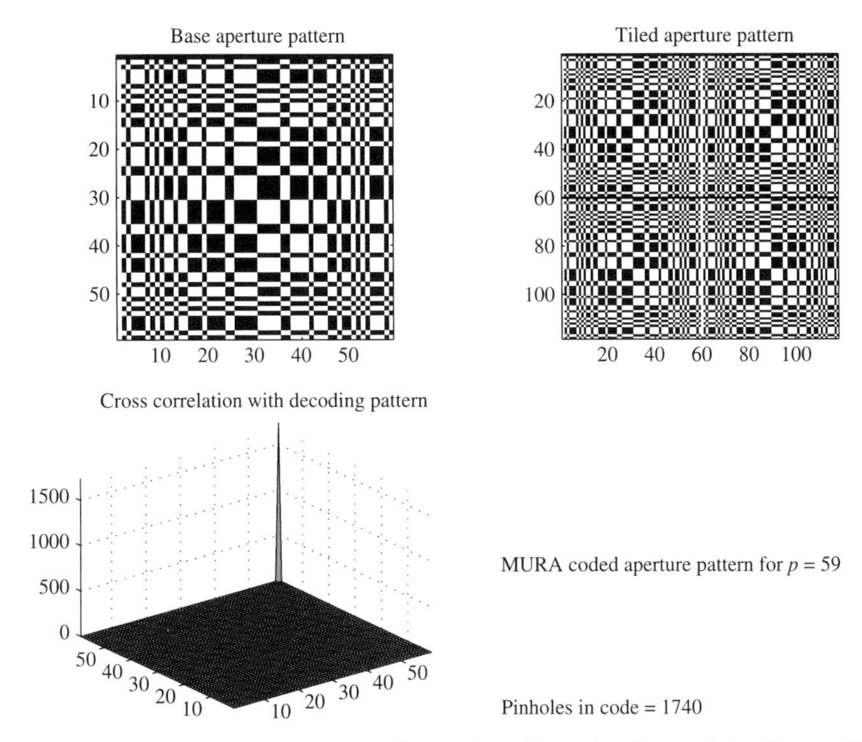

Figure 2.30 Base transmission pattern, tiled mask, and inversion deconvolution for $p = 59$.

implemented under cyclic boundary conditions rather than using zero padding. In contrast with a pinhole system, the number of pixels in the reconstructed coded aperture image is equal to the number of pixels in the base transmission pattern. Figures 2.31–2.33 are simulations of coded aperture imaging with the 59×59-element MURA code. As illustrated in the figure, the measured 59×59-element data are strongly positive. For this image the maximum noise-free measurement value is 100, and the minimum value is 58, for a measurement dynamic range of <2. We will discuss noise and entropic measures of sensor system performance at various points in this text, in our first encounter with a multiplex measurement system we simply note that isomorphic measurement of the image would produce a much higher measurement dynamic range for this image.

In practice, noise sensitivity is a primary concern in coded aperture and other multiplex sensor systems. For the MURA-based coded aperture system, Gottesman and Fenimore [102] argue that the pixel signal-to-noise ratio is

$$\mathrm{SNR}_{ij} = \frac{Nf_{ij}}{\sqrt{Nf_{ij} + N\sum_{kl} f_{kl} + \sum_{kl} B_{kl}}} \tag{2.47}$$

where N is the number of holes in the coded aperture and B_{kl} is the noise in the (kl)th pixel. The form of the SNR in this case is determined by signal-dependent, or "shot," noise. We discuss the noise sources in electronic optical detectors in Chapter 5 and

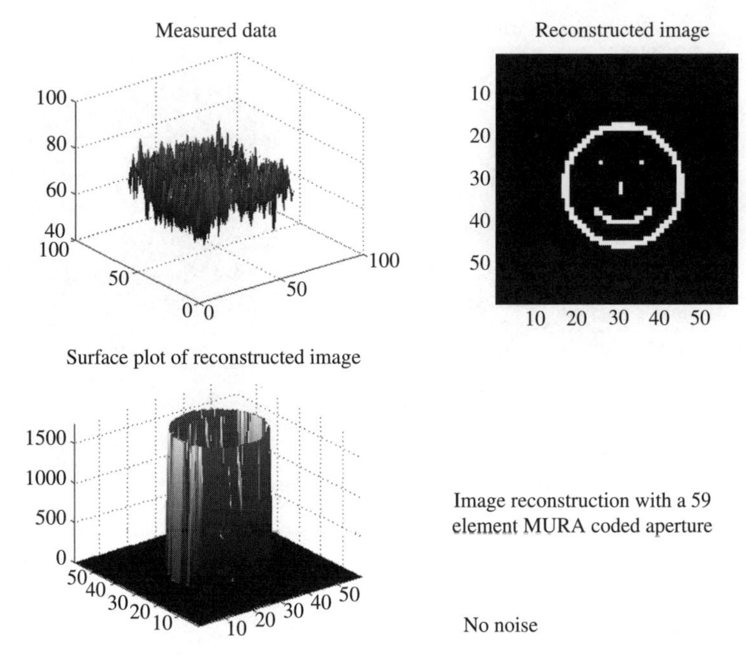

Figure 2.31 Coded aperture imaging simulation with no noise for the 59 × 59-element code of Fig. 2.30.

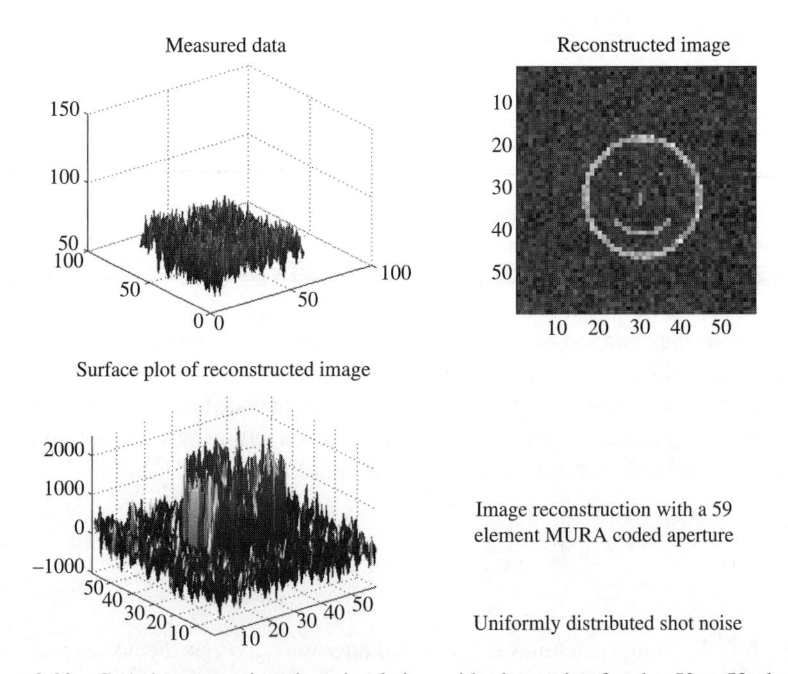

Figure 2.32 Coded aperture imaging simulation with shot noise for the 59 × 59-element code of Fig. 2.30.

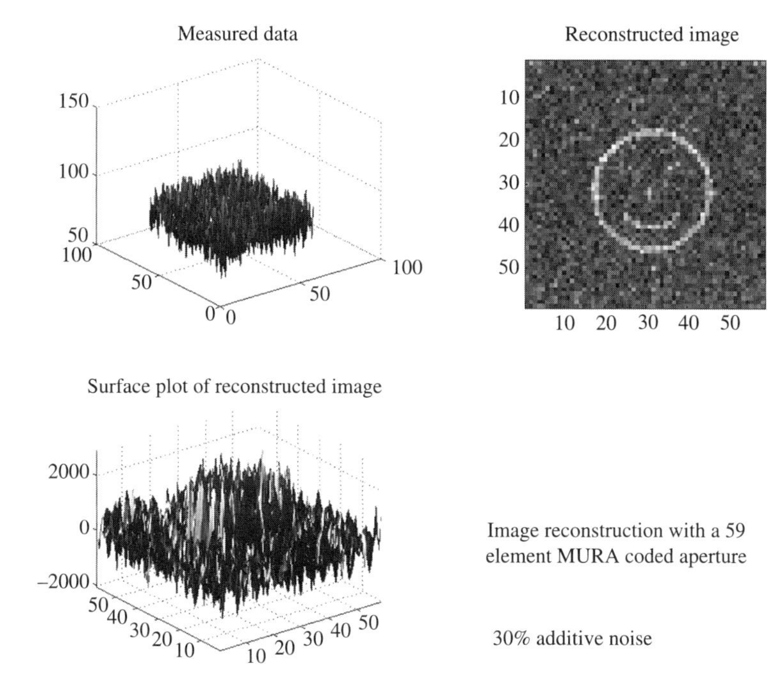

Figure 2.33 Coded aperture imaging simulation with additive noise for the 59 × 59-element code of Fig. 2.30.

derive the square-root characteristic form of shot noise in particular. For the 59 × 59 MURA aperture, $N = 1749$. If we assume that the object consists of binary values 1 and 0, the maximum pixel SNR falls from 41 for a point object to 3 for an object with 200 points active. The smiley face object of Fig. 2.31 consists of 155 points.

Dependence of the SNR on object complexity is a unique feature of multiplex sensor systems. The equivalent of Eqn. (2.47) for a focal imaging system is

$$\mathrm{SNR}_{ij} = \frac{Nf_{ij}}{\sqrt{Nf_{ij} + B_{ij}}} \tag{2.48}$$

This system produces an SNR of approximately $\sqrt{N}$ independent of the number of points in the object.

As with the canonical wave and correlation field multiplex systems presented in Sections 10.2 and 6.4.2, coded aperture imaging provides a very high depth field image but also suffers from the same SNR deterioration in proportion to source complexity.

2.6 PROJECTION TOMOGRAPHY

To this point we have considered images as two-dimensional distributions, despite the fact that target objects and the space in which they are embedded are typically three-dimensional. Historically, images were two-dimensional because focal imaging is a plane-to-plane transformation and because photochemical and electronic detector arrays are typically 2D films or focal planes. Using computational image synthesis, however, it is now common to form 3D images from multiplex measurements. Of course, visualization and display of 3D images then presents new and different challenges.

A variety of methods have been applied to 3D imaging, including techniques derived from analogy with biological stereo vision systems and actively illuminated acoustic and optical ranging systems. Each approach has advantages specific to targeted object classes and applications. Ranging and stereo vision are best adapted to opaque objects where the goal is to estimate a surface topology embedded in three dimensions.

The present section and the next briefly overview tomographic methods for multidimensional imaging. These sections rely on analytical techniques and concepts, such as linear transform theory, the Fourier transform and vector spaces, which are not formally introduced until Chapter 3. The reader unfamiliar with these concepts may find it useful to read the first few sections of that chapter before proceeding. Our survey of computed tomography is necessarily brief; detailed surveys are presented by Kak and Slaney [131] and Buzug [37].

Tomography relies on a simple 3D extension of the density-based object model that we have applied in this chapter. The word *tomography* is derived from the Greek *tomos*, meaning slice or section, and *graphia*, meaning *describing*. The word predates computational methods and originally referred to an analog technique for imaging a cross section of a moving object. While *tomography* is sometimes used to refer to any method for measuring 3D distributions (i.e., optical coherence tomography; Section 6.5), *computed tomography* (CT) generally refers to the projection methods described in this section.

Despite our focus on 3D imaging, we begin by considering tomography of 2D objects using a one-dimensional detector array. 2D analysis is mathematically simpler and is relevant to common X-ray illumination and measurement hardware. 2D slice tomography systems are illustrated in Fig. 2.34. In *parallel beam* systems, a collimated beam of X rays illuminates the object. The object is rotated in front of the X-ray source and one-dimensional detector opposite the source measures the integrated absoption along a line through the object for each ray component.

As always, the object is described by a density function $f(x, y)$. Defining, as illustrated in Fig. 2.35, l to be the distance of a particular ray from the origin, θ to be the angle between a normal to the ray and the x axis, and α to be the distance along the ray, measurements collected by a parallel beam tomography system take the form

$$g(l, \theta) = \int f(l\cos\theta - \alpha\sin\theta, l\sin\theta + \alpha\cos\theta)d\alpha \qquad (2.49)$$

where $g(l, \theta)$ is the *Radon transform* of $f(x, y)$. The Radon transform is defined for $f \in L^2(\mathbb{R}^n)$ as the integral of f over all hyperplanes of dimension $n-1$. Each

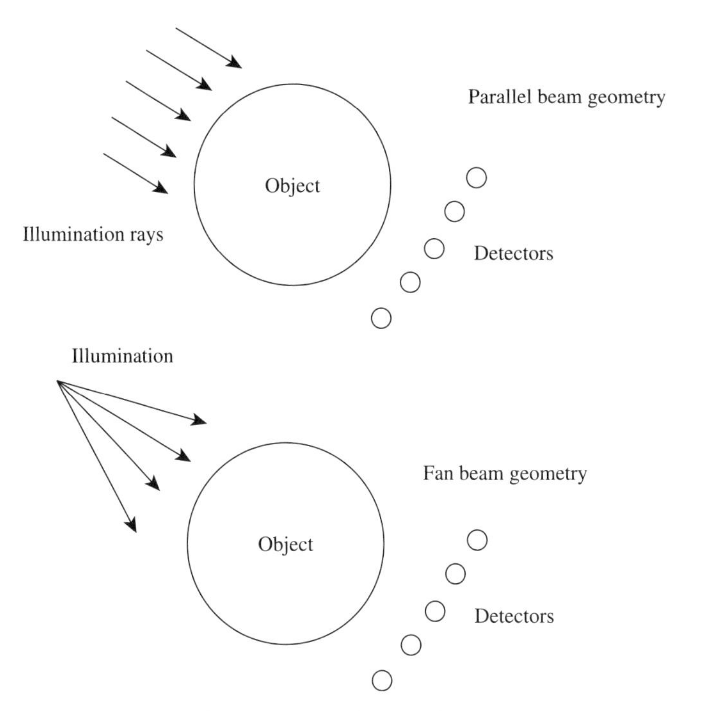

Figure 2.34 Tomographic sampling geometries.

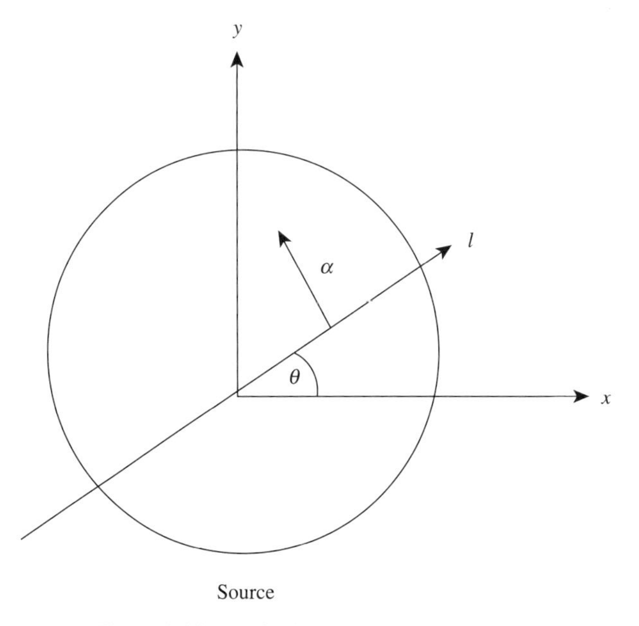

Figure 2.35 Projection tomography geometry.

hyperplane is defined by a surface normal vector $\mathbf{i_n}$ [in Eqn. (2.49) $\mathbf{i_n} = \cos\theta \mathbf{i_x} + \sin\theta \mathbf{i_y}$]. The equation of the hyperplane is $\mathbf{x} \cdot \mathbf{i_n} = l$. The Radon transform in $\mathbb{R}^n$ may be expressed

$$\mathcal{R}\{f\}(l, \mathbf{i_n}) = \int_{\mathbb{R}^{n-1}} f(l\mathbf{i_n} + \boldsymbol{\alpha})(d\alpha)^{n-1} \tag{2.50}$$

where α is a vector in the hyperplane orthogonal to $\mathbf{i_n}$. With reference to the definition given in Eqn. (3.10), the Fourier transform with respect to l of $\mathcal{R}\{f\}(l, \mathbf{i_n})$ is

$$\begin{aligned} \mathcal{F}_l\{\mathcal{R}\{f\}(l, \mathbf{i_n})\} &= \int\int \cdots \int \hat{f}(\mathbf{u}) e^{2\pi \mathbf{u} \cdot (l\mathbf{i_n} + \alpha)} e^{-2\pi i u_l l} (d\alpha)^{n-1} (du)^n dl \\ &= \hat{f}(\mathbf{u} = u_l \mathbf{i_n}) \end{aligned} \tag{2.51}$$

Equation (2.51), relating the 1D Fourier transform of the Radon transform to the Fourier transform sampled on a line parallel to $\mathbf{i_n}$, is called the *projection slice theorem.*

In the case of the Radon transform on $\mathbb{R}^2$, the Fourier transform with respect to l of Eqn. (2.49) yields

$$\hat{g}(u_l, \theta) = \int\int\int\int g(l, \theta) e^{-j2\pi u_l l} du\, dv\, dl\, d\alpha \tag{2.52}$$

$$= \hat{f}(u = u_l \cos\theta, v = u_l \sin\theta) \tag{2.53}$$

where $\hat{f}$ is the Fourier transform of f. If we sample uniformly in l space along an aperture of length R_s, then $\Delta u_l = 2\pi/R_s$. The sample period along l determines the spatial extent of the sample.

In principle, one could use Eqn. (2.52) to sample the Fourier space of the object and then inverse-transform to estimate the object density. In practice, difficulties in interpolation and sampling in the Fourier space make an alternative algorithm more attractive. The alternative approach is termed *convolution–backprojection.* The algorithm is as follows:

1. Measure the projections $g(l, \theta)$.
2. Fourier-transform to obtain $\hat{g}(u_l, \theta)$.
3. Multiply $\hat{g}(u_l, \theta)$ by the filter $|u_l|$ and inverse-transform. This step consists of convolving $g(l, \theta)$ with the inverse transformation of $|u_l|$ (the range of u_l is limited to the maximum frequency sampled). This step produces the filtered function $Q(l, \theta) = \int |u_l| \hat{g}(u_l, \theta) \exp(i2\pi u_l l) du_l$.
4. Sum the filtered functions $Q(l, \theta)$ interpolated at points $l = x\cos\theta + y\sin\theta$ to produce the reconstructed estimate of f. This constitutes the *backprojection* step.

To understand the filtered backprojection approach, we express the inverse Fourier transform relationship

$$f(x, y) = \int\int \hat{f}(u, v) e^{i2\pi(ux+vy)} du\, dv \tag{2.54}$$

in cylindrical coordinates as

$$f(x, y) = \int_0^{2\pi} \int_0^{\infty} \hat{f}(w, \theta) e^{i2\pi w(x \cos \theta + y \sin \theta)} w \, dw \, d\theta \tag{2.55}$$

where $w = \sqrt{u^2 + v^2}$. Equation (2.55) can be rearranged to yield

$$f(x, y) = \int_0^{\pi} \int_0^{\infty} \left(\hat{f}(w, \theta) e^{i2\pi w(x \cos \theta + y \sin \theta)} + \hat{f}(w, \theta + \pi) e^{-i2\pi w(x \cos \theta + y \sin \theta)}\right) w \, dw \, d\theta \tag{2.56}$$

$$= \int_0^{\pi} \int_{-\infty}^{\infty} \hat{f}(w, \theta) e^{i2\pi w(x \cos \theta + y \sin \theta)} |w| dw \, d\theta \tag{2.57}$$

where we use the fact that for real-valued $f(x, y)$, $\hat{f}(w, \theta) = \hat{f}(-w, \theta + \pi)$. This means that

$$f(x, y) = \int_0^{\pi} Q(l = x \cos \theta + y \sin \theta, \theta) d\theta \tag{2.58}$$

The convolution–backprojection algorithm is illustrated in Fig. 2.36, which shows the Radon transform, the Fourier transform $\hat{g}(u_l, \theta)$, the Fourier transform $\hat{f}(u, v)$, $Q(l, \theta)$, and the reconstructed object estimate. Note, as expected from the projection slice theorem, that $\hat{g}(u_l, \theta)$ corresponds to slices of $\hat{f}(u, v)$ "unrolled" around the origin. Edges of the Radon transform are enhanced in $Q(l, \theta)$, which is a "high-pass" version of $g(l, \theta)$.

We turn finally to 3D tomography, where we choose to focus on projections measured by a camera. A camera measures a bundle of rays passing through a principal point, (x_o, y_o, z_o). For example, we saw in Section 2.5 that a pinhole or coded aperture imaging captures $\hat{f}(\theta_x, \theta_y)$, where, repeating Eqn. (2.31)

$$\hat{f}(\theta_x, \theta_y) = \int f(z_o \theta_x, z_o \theta_y, z_o) dz_o \tag{2.59}$$

$\hat{f}(\theta_x, \theta_y)$ is the integral of $f(x, y, z)$ along a line through the origin of the (x_o, y_o, z_o) coordinate system. (θ_x, θ_y) are angles describing the direction of the line on the unit sphere. In two dimensions, a system that collects rays through series of principal points implements *fan beam tomography*. Fan beam systems often combine a point x-ray source with a distributed array of detectors, as illustrated in Fig. 2.34.

In 3D, tomographic imaging using projections through a sequence of principal points is *cone beam tomography*. Note that Eqn. (2.59) over a range of vertex points is not the 3D Radon transform. We refer to the transformation based on projections along ray bundles as the *X-ray transform*. The X-ray transform is closely related

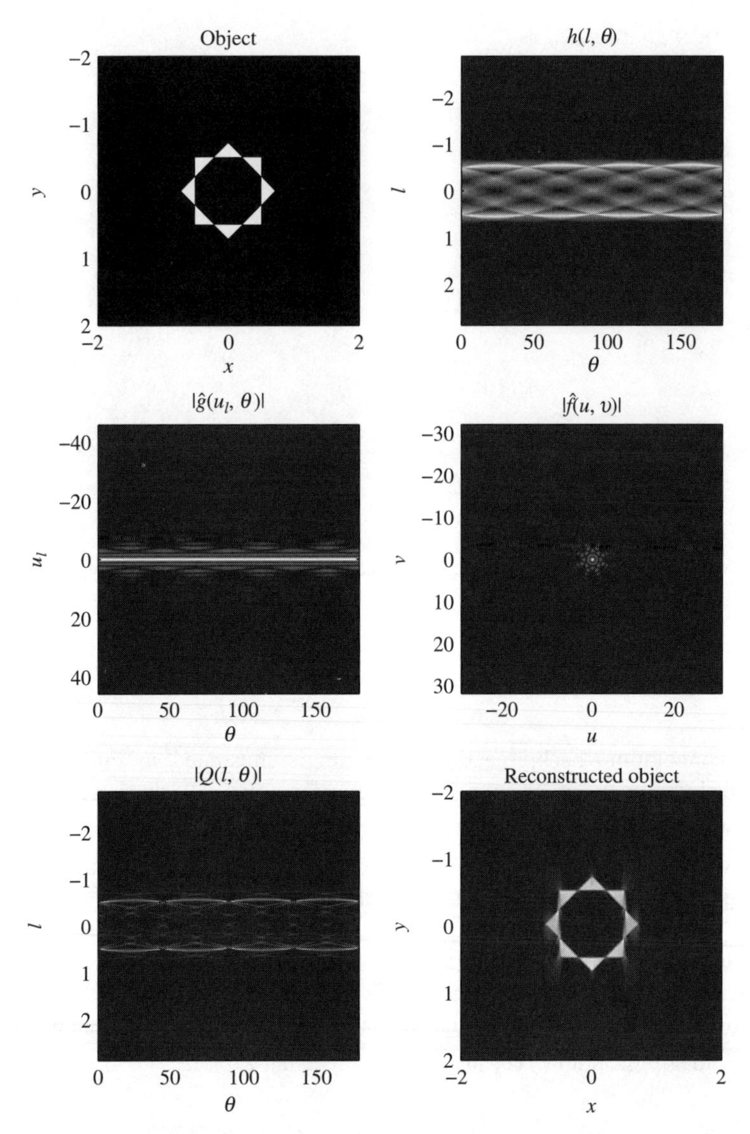

Figure 2.36 Tomographic imaging with the convolution–backprojection algorithm.

to the Radon transform, however, and can be inverted by similar means. The 4D X-ray transform of 3D objects, consisting of projections through all principal points on a sphere surrounding an object, overconstrains the object. Tuy [233] describes reduced *vertex paths* that produce well-conditioned 3D X-ray transforms of 3D objects.

A discussion of cone beam tomography using optical imagers is presented by Marks et al. [168], who apply the circular vertex path inversion algorithm

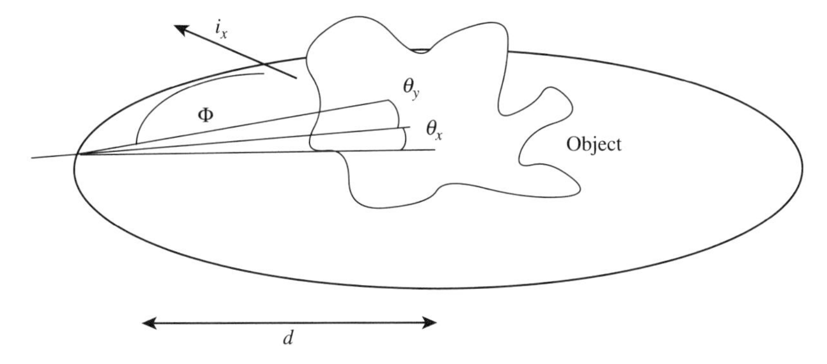

Figure 2.37 Cone beam geometry.

developed by Feldkamp et al. [68]. The algorithm uses 3D convolution–backprojection based on the vertex path geometry and parameters illustrated in Fig. 2.37. Projection data $f_\Phi(\theta_x, \theta_y)$ are weighted and convolved with the separable filters

$$h_y(\theta_y) = \int_{-\omega_{yo}}^{\omega_{yo}} d\omega|\omega| \exp\frac{i\omega\theta_y - 2|\omega|}{\omega_{yo}}$$

$$h_x(\theta_x) = \frac{\sin(\theta_x \omega_{zo})}{\pi\theta_z} \tag{2.60}$$

where ω_{yo} and ω_{zo} are the angular sampling frequencies of the camera. These filters produce the intermediate function

$$Q_\Phi(\theta_x, \theta_y) = \int\int d\theta'_x \, d\theta'_y \, h_x(\theta_x - \theta'_x) h_y(\theta_y - \theta'_y) \frac{f_\Phi(\theta'_x, \theta'_y)}{\sqrt{1 + \theta'^2_x \, \theta'^2_y}} \tag{2.61}$$

and the reconstructed 3D object density is

$$f_E(x, y, z) = \frac{1}{4\pi^2} \int \frac{d^2}{(d + x\cos\phi)^2} Q_\phi\left[\frac{y}{d + x\cos\phi}, \frac{z\sin\phi}{d + x\cos\phi}\right] d\phi \tag{2.62}$$

2.7 REFERENCE STRUCTURE TOMOGRAPHY

Optical sensor design boils down to compromises between mathematically attractive and physically attainable visibilities. In most cases, physical mappings are the starting point of design. So far in this chapter we have encountered two approaches driven primarily by ease of physical implementation (focal and tomographic imaging) and one attempt to introduce artificial coding (coded aperture imaging). The tension between physical and mathematical/coding constraints continues to develop in the remainder of the text.

The ultimate challenge of optical sensor design is to build a system such that the visibility $v(A, B)$ is optimal for the sensor task. 3D optical design, specifying the structure and composition of optical elements in a volume between the object and detector elements, is the ultimate toolkit for visibility coding. Most current optical design, however, is based on sequences of quasiplanar surfaces. In view of the elegance of focal imaging and the computational challenge of 3D design, the planar/ray-based approach is generally appropriate and productive. 3D design will, however, ultimately yield superior system performance.

As a first introduction to the challenges and opportunities of 3D optics, we consider reference structure tomography (RST) as an extension of coded aperture imaging to multiple dimensions [30]. Ironically, to ease explanation and analysis, our discussion of RST is limited to the 2D plane. The basic concept of a reference structure is illustrated in Fig. 2.38. A set of detectors at fixed points observes a radiant object through a set of reference obscurants. A ray through the object is visible at a detector if it does not intersect an obscurant; rays that intersect obscurants are not visible. The reference structure/detector geometry segments the object space into discrete visibility cells. The visibility cells in Fig. 2.38 are indexed by *signatures* $\chi_i = \ldots 1101001 \ldots$, where χ_{ij} is one if the ith cell is visible to the jth detector and zero otherwise.

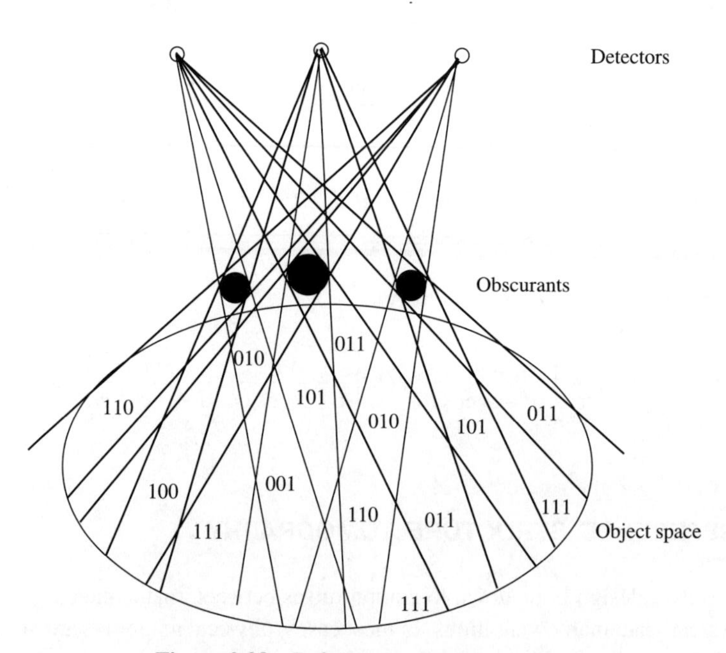

Figure 2.38 Reference structure geometry.

As discussed in Section 2.1, the object signal may be decomposed into components f_i such that

$$f_j = \int_{\text{cell}_j} f(A)dA \tag{2.63}$$

Measurements on the system are then of the form

$$g_i = \sum_j \chi_{ij} f_j \tag{2.64}$$

As is Eqn. (2.4), Eqn. (2.64) may be expressed in discrete form as $\mathbf{g} = \mathbf{Hf}$. In contrast with coded aperture imaging, the RST mapping is shift variant and the measurements are not embedded in a natural continuous space.

Interesting scaling laws arise from the consideration of RST systems consisting of m detectors and n obscurants in a d-dimensional embedding space. The number of possible signatures (2^m) is much larger than the number of distinct visibility cells ($O(m^d n^d)$). Agarwal et al. prove a lower bound of order $(mn/\log(n))^d$ exists on the minimum number of distinct signatures [2]. This means that despite the fact that different cells have the same signature in Fig. 2.38, as the scale of the system grows, each cell tends to have a unique signature.

The RST system might thus be useful as a point source location system, since a point source hidden in $(mn)^d$ cells would be located in m measurements. Since an efficient system for point source location on N cells requires $\log N$ measurements, RST is efficient for this purpose only if $m \ll n$. Point source localization may be regarded as an extreme example of *compressive sampling* or *combinatorial group testing*. The nature of the object is constrained, in this case by extreme sparsity, such that reconstruction is possible with many fewer measurements than the number of resolution cells. More generally, reference structures may be combined with compressive estimation [38,58] to image multipoint sparse distributions on the object space.

Since the number of signature cells is always greater than m, the linear mapping $\mathbf{g} = \mathbf{Hf}$ is always ill-conditioned. $\mathbf{H}$ is a $m \times p$ matrix, where p is the number of signatures realized by a given reference structure. $\mathbf{H}$ may be factored into the product $\mathbf{H} = \mathbf{U}\Sigma\mathbf{V}^\dagger$ by *singular value decomposition* (SVD). $\mathbf{U}$ and $\mathbf{V}$ are $m \times m$ and $p \times p$ unitary matrices and Σ is a $m \times p$ matrix with nonnegative singular values along the diagonal and zeros off the diagonal.

Depending on the design of the reference structure and the statistics of the object, one may find that the pseudoinverse

$$\mathbf{f_e} = \mathbf{V}\Sigma^{-1}\mathbf{U}^\dagger\mathbf{g} \tag{2.65}$$

is a good estimate of $\mathbf{f}$. We refer to this approach in Section 7.5 as projection of $\mathbf{f}$ onto the range of $\mathbf{H}$. The range of $\mathbf{H}$ is an m-dimensional subspace of the p-dimensional space spanned by the state of the signature cells. The complement of this subspace in $\mathbb{R}^p$ is the null space of **H**. The use of prior information, such as the fact that **f** consists of a point source, may allow one to infer the null space structure of **f** from measurements in the range of **H**. Brady et al. [30] take this approach a step further by assuming that **f** is described by a continuous basis rather than the visibility cell mosaic.

Analysis of the measurement, representation, and analysis spaces requires more mathematical tools than we have yet utilized. We develop such tools in the next chapter and revisit abstract measurement strategies in Section 7.5.

PROBLEMS

2.1 *Visibility*

(a) Suppose that the half space $x > 0$ is filled with water while the half-space $x < 0$ is filled with air. Show that the visibility $v(A, B) = 1$ for all points A and $B \in \mathbb{R}^3$.

(b) Suppose that the subspace $l > x > -1$ is filled with glass (e.g., a window) while the remainder of space is filled with air. Show that the visibility $v(A, B) = 1$ for all points A and $B \in \mathbb{R}^3$.

(c) According to Eqn. (2.5), the transmission angle θ_2 becomes complex for $(n_1/n_2)\sin\theta_1 > 1$. In this case, termed total internal reflection, there is no transmitted ray from region 1 into region 2. How do you reconcile the fact that if $n_1 > n_2$ some rays from region 1 never penetrate region 2 with the claim that all points in region 1 are visible to all points in region 2?

2.2 *Materials Dispersion.* The dispersive properties of a refractive medium are often modeled using the Sellmeier equation

$$n^2(\lambda) = 1 + \frac{B_1\lambda^2}{\lambda^2 - C1} + \frac{B_2\lambda^2}{\lambda^2 - C2} + \frac{B_3\lambda^2}{\lambda^2 - C3} \tag{2.66}$$

The Sellmeier coefficients for Schott glass type SF14 are given in Table 2.1.

(a) Plot $n(\lambda)$ for SF14 for $\lambda = 400\text{–}800$ nm.

(b) For $\theta_{\perp 1} = \pi/8$, $\theta_{\perp 2} = -\pi/8$, and $\theta_a = -\pi/16$, plot θ_b as a function of λ over the range 400–800 nm for an SF14 prism.

TABLE 2.1 Dispersion Coefficients of SF14 Glass

B_1	1.69182538
B_2	0.285919934
B_3	1.12595145
C_1	0.0133151542
C_2	0.0612647445
C_3	118.405242

2.3 *Microscopy.* A modern microscope using an infinity-corrected objective forms an intermediate image at a distance termed the *reference focal length.* What is the magnification of the objective in terms of the reference focal length and the objective focal length?

2.4 *Telescopy.* Estimate the mass of a 1-m aperture 1-m focal length glass refractive lens. Compare your answer with the mass of a 1-m aperture mirror.

2.5 *Ray Tracing.* Consider an object placed 75 cm in front a 50-cm focal length lens. A second 50-cm focal length lens is placed 10 cm behind the first. Sketch a ray diagram for this system, showing the intermediate and final image planes. Is the image erect or inverted, real or virtual? Estimate the magnification.

2.6 *Paraxial Ray Tracing.* What is the *ABCD* matrix for a parabolic mirror? What is the ray transfer matrix for a Cassegrain telescope? How does the magnification of the telescope appear in the ray transfer matrix?

2.7 *Spot Diagrams.* Consider a convex lens in the geometry of Fig. 2.10. Assuming that $R_1 = R_2 = 20$ cm and $n = 1.5$, write a computer program to generate a spot diagram in the nominal thin lens focal plane for parallel rays incident along the z axis. A *spot diagram* is a distribution of ray crossing points. For example, use numerical analysis to find the point that a ray incident on the lens at point (x, y) crosses the focal plane. Such a ray hits the first surface of the lens at point $z = (x^2 + y^2)/2R_1 - d_1$ and is refracted onto a new ray direction. The refracted ray hits the second surface at the point such that $(x, y, z) + C\mathbf{i}_{t1}$ satisfies $z = d_2 - (x^2 + y^2)/2R_1$. The ray is refracted from this point toward the focal plane. Plot spot diagrams for the refracted rays for diverse input points at various planes near the focal plane to attempt to find the plane of best focus. Is the best focal distance the same as predicted by Eqn. (2.16)? What is the area covered by your tightest spot diagram? (*Hint:* Since the lens is rotationally symmetric and the rays are incident along the axis, it is sufficient to solve the ray tracing problem in the xz plane and rotate the 1D spot diagram to generate a 2D diagram.)

2.8 *Computational Ray Tracing.* Automated ray tracing software is the workhorse of modern optical design. The goal of this problem is to write a ray tracing program using a mathematical programming environment (e.g., Matlab or Mathematica). For simplicity, work with rays confined to the xz plane. In a 2D ray tracing program, each ray has a current position in the plane and a current unit vector. To move to the next position, one must solve for the intercept at which the ray hits the next surface in the optical system. One draws a line from the current position to the next position along the current direction and solves for the new unit vector using Snell's law. Snell's law at any surface in the plane is

$$n_1 \mathbf{i}_\mathbf{i} \times \mathbf{i}_\mathbf{n} = n_2 \mathbf{i}_\mathbf{t} \times \mathbf{i}_\mathbf{n} \tag{2.67}$$

where $\mathbf{i_i}$ is a unit vector along the incident ray, $\mathbf{i_t}$ is a unit vector along the refracted ray and $\mathbf{i_n}$ is a unit vector parallel to the surface normal. Given $\mathbf{i_i}$ and $\mathbf{i_n}$, one solves for $\mathbf{i_t}$ using Snell's law and the normalization condition. The nth surface is defined by the curve $F_n(x, z) = 0$ and the normal at points on the surface is

$$\frac{\partial F_n}{\partial x}\mathbf{i_x} + \frac{\partial F_n}{\partial z}\mathbf{i_z} \tag{2.68}$$

which one must normalize to obtain $\mathbf{i_n}$. As an example, Fig. 2.39 is a ray tracing plot for 20 rays normally incident on a sequence of lenses. The index of refraction of each lens is 1.3. The nth surface is described by

$$z = n\Delta + \kappa(x - \alpha)(x + \alpha)x^2 \tag{2.69}$$

for n even and

$$z = (n - 1)\Delta + \delta - \kappa(x - \alpha)(x + \alpha)x^2 \tag{2.70}$$

for n odd. In Fig. 2.39, $\kappa = 1/2500$, $\alpha = 10$, $\Delta = 4$, and $\delta = 1$. (Notice that ray tracing does not require specification of spatial scale. See Section 10.4.1 for a discussion of scale and ray tracing.) Write a program to trace rays through an arbitrary sequence of surfaces in the 2D plane. Demonstrate the utility of your program by showing ray traces through three to five

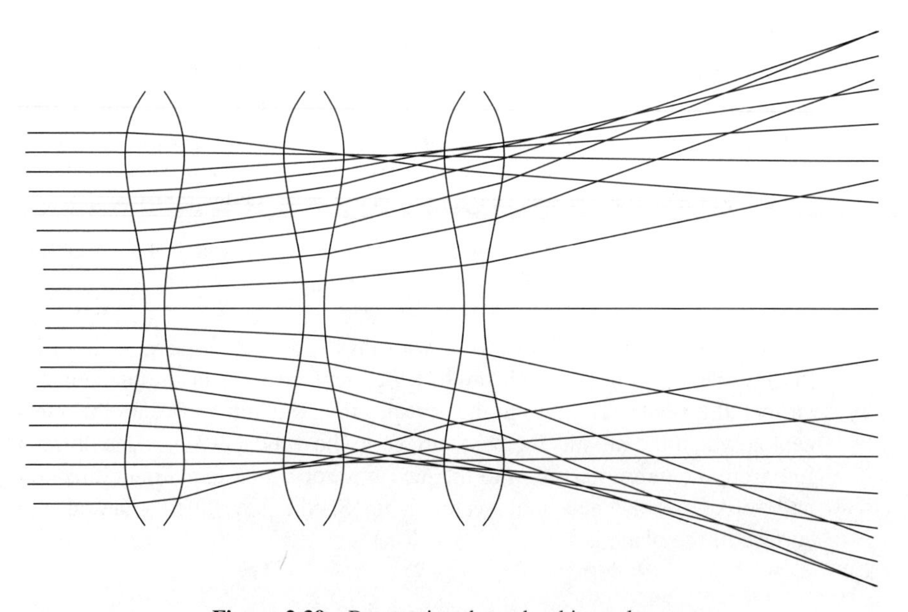

Figure 2.39 Ray tracing through arbitrary lenses.

lenses satisfying the general form of Eqn. (2.69) for various values of α, κ, δ, and Δ.

2.9 *Pinhole Visibility*. Derive the geometric visibility for a pinhole camera [Eqn. (2.25)].

2.10 *Coded Aperture Imaging*. Design and simulate a coded aperture imaging system using a MURA mask. Use a mask with 50 or more resolution cells along each axis.

(a) Plot the transmission mask, the decoding matrix, and the cross-correlation of the two.

(b) Simulate reconstruction of an image of your choice, showing both the measured data and its reconstruction using the MURA decoding matrix.

(c) Add noise to your simulated data in the form of normal and Poisson distributed values at various levels and show reconstruction using the decoding matrix.

(d) Use inverse Fourier filtering to reconstruct the image (using the FFT of the mask as an inverse filter). Plot the FFT of the mask.

(e) Submit your code, images and a written description of your results.

2.11 *Computed Tomography*

(a) Plot the functions $f_1(x, y) = \text{rect}(x)\text{rect}(y)$ and $f_2(x, y) = \text{rect}(x)\text{rect}(y)(1 - \text{rect}(x + y)\text{rect}(x - y))$.

(b) Compute the Radon transforms of $f_1(x, y)$ and $f_2(x, y)$ numerically.

(c) Use a numerical implementation of the convolution–backprojection technique described in Section 2.6 to estimate $f_1(x, y)$ and $f_2(x, y)$ from their Radon transforms.

2.12 *Reference Structure Tomography*. A 2D reference structure tomography system as sketched in Fig. 2.38 modulates the visibility of an object embedding space using n obscuring disks. The visibility is observed by m detectors.

(a) Present an argument justifying the claim that the number of signature cells is of order m^2n^2.

(b) Assuming prior knowledge that the object is a point source, estimate the number of measurements necessary to find the signature cell occupied by the object.

(c) For what values of m and n would one expect a set of measurements to uniquely identify the signature cell occupied by a point object?

ANALYSIS

...the electric forces can disentangle themselves from material bodies and can continue to subsist as conditions or changes in the state of space.

—H. Hertz [118]

3.1 ANALYTICAL TOOLS

Chapter 2 developed a geometric model for optical propagation and detection and described algorithms for object estimation on the basis of this model. While geometric analysis is of enormous utility in analysis of patterns formed on image or projection planes, it is less useful in analysis of optical signals at arbitrary points in the space between objects and sensors. Analysis of the optical system at all points in space requires the concept of "the optical field." The field describes the state of optical phenomena, such as spectra, coherence, and polarization parameters, independent of sources and detectors.

Representation, analysis, transformation, and measurement of optical fields are the focus of the next four chapters. The present chapter develops a mathematical framework for analysis of the field from the perspective of sensor systems. A distinction may be drawn between optical systems, such as laser resonators and fiber waveguides, involving relatively few spatial channels and systems, such as imagers and spectrometers, involving a large number of degrees of freedom. The degrees of freedom are typically typically expressed as pixel values, modal amplitudes or Fourier components. It is not uncommon for a sensor system to involve $10^6 - 10^{12}$ parameters. To work with such complex fields, this chapter explores mathematical tools drawn from harmonic analysis. We begin to apply these tools to physical field analysis in Chapter 4.

We develop two distinct sets of mathematical tools:

1. *Transformation tools*, which enable us to analyze propagation of fields from one space to another
2. *Sampling tools*, which enable us to represent continuous field distributions using discrete sets of numbers

A third set of mathematical tools, signal encoding and estimation from sample data, is considered in Chapters 7 and 8.

This chapter initially considers transformation and sampling in the context of conventional Fourier analysis, meaning that transformations are based on Fourier transfer functions and impulse response kernels and sampling is based on the Whittaker–Shannon sampling theorem. Since the mid-1980s, however, dramatic discoveries have revolutionized harmonic signal analysis. New tools drawn from wavelet theory allow us to compare various mathematical models for fields, including different classes of functions (plane waves, modes, and wavelets) that can be used to represent fields. Different mathematical bases enable us to flexibly distribute discrete components to optimize computational processing efficiency in field analysis.

Section 3.2 broadly describes the nature of fields and field transformations. Sections 3.3–3.5 focus on basic properties of Fourier and Fresnel transformations. We consider conventional Shannon sampling in Section 3.6. Section 3.7 discusses discrete numerical analysis of linear transformations. Sections 3.8–3.10 briefly extend sampling and discrete transformation analysis to include wavelet analysis.

3.2 FIELDS AND TRANSFORMATIONS

A *field* is a distribution function defined over a space, meaning that a field value is associated with every point in the space. The optical field is a *radiation field*, meaning that it propagates through space as a wave. Propagation induces relationships between the field at different points in space and constrains the range of distribution functions that describe physically realizable fields.

Optical phenomena produce a variety of observables at each point in the radiation space. The “data cube” of all possible observables at all points is a naive representation of the information encoded on the optical signal. Well-informed characterization of the optical signal in terms of signal values at discrete points, modal applitudes, or boundary conditions greatly reduces the complexity of representing and analyzing the field. Our goal in this chapter is to develop the mathematical tools that enable efficient analysis. While our discussion is often abstract, it is important to remember that the mathematical distributions are tied to physical observables. On the other hand, this association need not be particularly direct. We often find it useful to use field distributions that describe functions that are not directly observable in order to develop system models for ultimately observable phenomena.

The state of the optical field is described using distributions defined over various spaces. The most obvious space is Euclidean space–time, $\mathbb{R}^3 \times T$, and the most obvious field distribution is the electric field $\mathbf{E}(\mathbf{r}, t)$. The optical field cannot be an arbitrary distribution because physical propagation rules induce correlations between points in space–time. For the field independent of sources and detectors, these relationships are embodied in the Maxwell equations discussed in Chapter 4. The quantum statistical nature of optical field generation and detection is most easily incorporated in these relationships using the coherence theory developed in Chapter 6.

System analysis using field theory consists of the use of physical relationships to determine the field in a region of space on the basis of the specification of the field in a different region. The basic problem is illustrated in Fig. 3.1. The field is specified in one region, such as the (x, y) plane illustrated in the figure. This plane may correspond to the surface of an object or to an aperture through which object fields propagate. Given the field on an input boundary, we combine mathematical tools from this chapter with physical models from Chapter 4 to estimate the field elsewhere. In Fig. 3.1, the particular goal is to find the field on the output (x', y') plane.

Mathematically, propagation of the field from the input boundary to an output boundary may be regarded as a transformation from a distribution $f(x, y, t)$ on the input plane to a distribution $g(x', y', t)$ on the output plane. For optical fields this transformation is linear.

We represent propagation of the field from one boundary to another by the transformation $g(x', y') = T\{f(x, y)\}$. The transformation $\mathcal{T}\{\cdot\}$ is linear over its functional domain if for all functions $f_1(x, y)$ and $f_2(x, y)$ in the domain and for all scalars α and β

$$\mathcal{T}\{\alpha f_1(x) + \beta f_2(x)\} = \alpha\mathcal{T}\{f_1(x)\} + \beta T\{f_2(x)\} \tag{3.1}$$

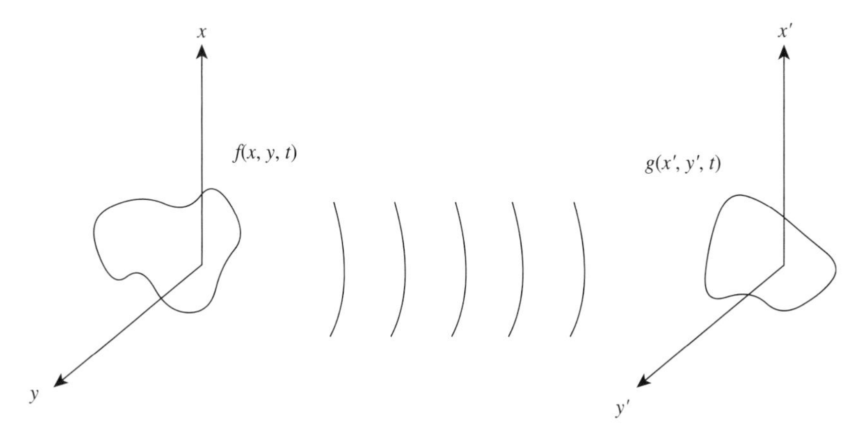

Figure 3.1 Plane-to-plane transformation of optical fields.

Linearity allows transformations to be analyzed using superpositions of basis functions. A function $f(x, y)$ in the range of a discrete basis $\{\phi_n(x, y)\}$ may be expressed

$$f(x, y) = \sum_n f_n \phi_n(x, y) \tag{3.2}$$

This chapter explores several potential basis functions ϕ_n. The advantage of this approach is the effect of a linear transformation on the basis functions is easily mapped onto transformations of arbitrary input distributions by algebraic methods. Linearity implies that

$$T\{f(x, y)\} = \sum_n f_n T\{\phi_n(x, y)\} \tag{3.3}$$

For the discrete representation of Eqn. (3.2), the range of the transformation is also spanned by a discrete basis $\psi_m(x, y)$ such that

$$T\{\phi_n(x, y)\} = \sum_m h_{nm} \psi_m(x, y) \tag{3.4}$$

Representing the input function $f(x, y)$ by a vector of discrete coefficients $\mathbf{f} = \{f_n\}$ and the output function $g(x', y')$ by a vector of discrete coefficients $\mathbf{g} = \{g_m\}$ such that

$$g(x', y') = \sum_m g_m \psi_m(x, y) \tag{3.5}$$

linearity allows us to express the transformation algebrically as $\mathbf{g} = \mathbf{Hf}$, where the coefficients of $\mathbf{H}$ are h_{nm}.

Similar representations of a linear transformation may be developed for functions represented on continuously indexed bases. The simplest example builds on the Dirac δ function. The input distribution is represented on the δ function basis as

$$f(x, y) = \iint f(x', y')\delta(x - x', y - y')dx'\,dy' \tag{3.6}$$

Using the δ function basis an arbitrary linear transformation may be expressed in integral form by noting

$$\begin{aligned}
T\{f(x, y)\} &= T\left\{\iint f(x', y')\delta(x - x', y - y')\,dx'dy'\right\} \\
&= \iint f(x', y')T\{\delta(x - x', y - y')\}dx'dy' \\
&= \iint f(x', y')h(x'', x', y'', y')\,dx'dy'
\end{aligned} \tag{3.7}$$

where $h(x'', x', y'', y') = T\{\delta(x - x', y - y')\}$ is the *impulse response* associated with the transformation $T\{\cdot\}$.

Harmonic bases are particularly useful in the analysis of *shift-invariant* linear transformations. Letting $g(x', y') = T\{f(x, y)\}$, $T\{\cdot\}$ is shift-invariant if

$$T\{f(x - x_o, y - y_o)\} = g(x' - x_o, y' - y_o) \tag{3.8}$$

The impulse response of a shift-invariant transformation takes the form $h(x', x', y', y') = h(x' - x, y' - y)$ and the integral representation of the transformation is the convolution

$$g(x', y') = T\{f(x, y)\} = \iint f(x, y)h(x' - x, y' - y)dx\,dy \tag{3.9}$$

Shift invariance means that shifts in the input distribution produce identical shifts in the output distribution. Imaging systems are often approximately shift-invariant after accounting for magnification, distortion, and discrete sampling.

Optical field propagation can usually be modeled as a linear transformation and can sometimes be modeled as a shift-invariant linear transformation. Shift invariance applies in imaging and diffraction in homogeneous media, where a shift in the input object produces a corresponding shift in the output image or diffraction pattern. Shift invariance does not apply to diffraction or refraction through structured optical scatterers or devices. Recalling, for example, field propagation through reference structures from Section 2.7, one easily sees that shift in the object field produces a complex shift-variant recoding of the object shadow.

3.3 FOURIER ANALYSIS

Fourier analysis is a powerful tool for analysis of shift-invariant systems. Since shift-invariant imaging and field propagation are central to optical systems, Fourier analysis is ubiquitous in this text. The attraction is that Fourier analysis allows the response of a linear shift-invariant system to be modeled by a multiplicative *transfer function* (*filter*). Fourier analysis also arises naturally as an explanation of resonance and color in optical cavities and quantum mechanical interactions. One might say that Fourier analysis is attractive because optical propagation is linear and because field–matter interactions are nonlinear.

Fourier transforms take slightly different forms in different texts; here we represent the Fourier transform $\hat{f}(u)$ of a function $f(x) \in L^2(\mathbb{R})$ as

$$\hat{f}(u) = \int_{-\infty}^{\infty} f(x)e^{-2\pi ixu}dx \tag{3.10}$$

with inverse transformation

$$f(x) = \int_{-\infty}^{\infty} \hat{f}(u)e^{2\pi ixu}du \tag{3.11}$$

In optical system analysis one often has recourse to multidimensional Fourier transforms, which one defines for $f \in L^2(\mathbb{R}^n)$ as

$$\hat{f}(u_1, u_2, u_3 \cdots) = \int_{-\infty}^{\infty}\int_{-\infty}^{\infty}\int_{-\infty}^{\infty} \ldots f(x_1, x_2, x_3 \ldots)\Pi_j e^{-2\pi i x_j u} dx_j \qquad (3.12)$$

Useful identities associated with the one-dimensional Fourier transformation include

1. *Differentiation*

$$\mathcal{F}\left\{\frac{df(x)}{dx}\right\} = 2\pi i u \hat{f}(u) \qquad (3.13)$$

and

$$\mathcal{F}^{-1}\left\{\frac{d\hat{f}(u)}{du}\right\} = -2\pi i x f(x) \qquad (3.14)$$

2. *Translation*

$$\mathcal{F}\{f(x - x_o)\} = e^{-2\pi i x_o u}\hat{f}(u) \qquad (3.15)$$

3. *Dilation*. For real $a \neq 0$

$$\mathcal{F}\{f(ax)\} = \frac{1}{|a|}\hat{f}\left(\frac{u}{a}\right) \qquad (3.16)$$

4. *Convolution*

$$\mathcal{F}\{f(x)g(x)\} = \int_{-\infty}^{\infty} \hat{f}(u')\hat{g}(u - u')du' \qquad (3.17)$$

and

$$\mathcal{F}\left\{\int_{-\infty}^{\infty} f(x')g(x - x')dx'\right\} = \hat{f}(u)\hat{g}(u) \qquad (3.18)$$

5. *Plancherel's theorem*. $\|f(x)\| = \|\hat{f}(u)\|$, for example

$$\int_{-\infty}^{\infty} |\hat{f}(u)|^2 du = \int_{-\infty}^{\infty} |f(x)|^2 dx \qquad (3.19)$$

Plancherel's theorem is easily extended to show that for functions $f(x) \in L^2(\mathbb{R})$ and $g(x) \in L^2(\mathbb{R})$

$$\int_{-\infty}^{\infty} \hat{f}^*(u)\hat{g}(u)\, du = \int_{-\infty}^{\infty} f^*(x)g(x)dx \tag{3.20}$$

6. *Localization*. Defining

$$\mu_f = \frac{1}{||f||^2} \int_{-\infty}^{\infty} x|f(x)|^2 dx \tag{3.21}$$

$$\sigma_f^2 = \frac{1}{||f||^2} \int_{-\infty}^{\infty} (x - \mu_f)^2 |f(x)|^2 dx \tag{3.22}$$

$$\mu_{\hat{f}} = \frac{1}{||\hat{f}||^2} \int_{-\infty}^{\infty} u|\hat{f}(u)|^2\, du \tag{3.23}$$

and

$$\sigma_{\hat{f}}^2 = \frac{1}{||\hat{f}||^2} \int_{-\infty}^{\infty} (u - \mu_{\hat{f}})^2 |\hat{f}(u)|^2 du \tag{3.24}$$

we find

$$\sigma_f^2 \sigma_{\hat{f}}^2 \geq \frac{1}{16\pi^2} \tag{3.25}$$

Equation (3.25) is often referred to as the *uncertainty relationship*, is the basis for the Heisenberg uncertainty relationship of quantum mechanics. It describes the impossibility of simultaneous signal localization in both the spatial and spectral regimes.

Proof of Eqn. (3.25) begins by noting that we can set $\mu_f = 0$ without loss of generality because by Eqn. (3.15) we know $|\hat{f}(u)|$ to be invariant under translation of the x axis. We then find that

$$\sigma_f^2 \sigma_{\hat{f}}^2 = \frac{1}{||f||^2 ||\hat{f}||^2} \int_{-\infty}^{\infty} \int_{-\infty}^{\infty} x^2 |f(x)|^2 (u - \mu_{\hat{f}})^2 |\hat{f}(u)|^2 dx\, du \tag{3.26}$$

Combining Eqns. (3.14) and (3.19), we note that

$$\int_{-\infty}^{\infty} |2\pi i x f(x)|^2 dx = \int_{-\infty}^{\infty} \left| \frac{d\hat{f}(u)}{du} \right|^2 du \tag{3.27}$$

and

$$\sigma_f^2 \sigma_{\hat{f}}^2 = \frac{1}{4\pi^2} \frac{1}{\|f\|^2 \, \|\hat{f}\|^2} \int_{-\infty}^{\infty} \int_{-\infty}^{\infty} u^2 \left| \frac{d\hat{f}(u')}{du} \right|^2 |\hat{f}(u + \mu_{\hat{f}})|^2 du' du \tag{3.28}$$

To simplify Eqn. (3.28), we note that an inner product $\langle a \mid b \rangle$, where a and b are functions in $L^2\{\mathbb{R}\}$ may be defined

$$\langle a|b \rangle = \int_{-\infty}^{\infty} a^*(x) b(x) dx \tag{3.29}$$

In terms of this inner product, Eqn. (3.28) takes the form

$$\sigma_f^2 \sigma_{\hat{f}}^2 = \frac{1}{4\pi^2} \frac{\langle (d\hat{f}/du) | (d\hat{f}/du) \rangle \langle u\hat{f} | u\hat{f} \rangle}{\langle f|f \rangle \langle \hat{f}|\hat{f} \rangle} \tag{3.30}$$

Proof of the inner product identity $[\langle a \mid a \rangle \ \langle b \mid b \rangle] \geq |\langle a \mid b \rangle|^2$ is straightforward. Applying this identity to Eqn. (3.30) yields

$$\sigma_f^2 \sigma_{\hat{f}}^2 \geq \frac{1}{4\pi^2} \frac{\left| \left\langle \frac{d\hat{f}}{du} | u\hat{f} \right\rangle \right|^2}{\langle f|f \rangle \langle \hat{f}|\hat{f} \rangle} \tag{3.31}$$

For a function $\hat{f}(u)$ in $L^2\{\mathbb{R}\}$ integration of

$$\frac{d\left(u\hat{f}^2\right)}{du} = \hat{f}^2 + u\hat{f}^* \frac{d\hat{f}}{du} + u\hat{f} \frac{d\hat{f}^*}{du} \tag{3.32}$$

yields

$$2\Re\left\{ \left\langle \frac{d\hat{f}}{du} | u\hat{f} \right\rangle \right\} = -\langle \hat{f}|\hat{f} \rangle \tag{3.33}$$

Noting that

$$\left|\left\langle \frac{d\hat{f}}{du}|u\hat{f}\right\rangle\right|^2 \geq \left|\Re\left\{\left\langle \frac{d\hat{f}}{du}|u\hat{f}\right\rangle\right\}\right|^2 = \frac{\langle \hat{f}|\hat{f}\rangle}{4} \tag{3.34}$$

we find

$$\sigma_f^2 \sigma_{\hat{f}}^2 \geq \frac{1}{16\pi^2} \tag{3.35}$$

Proof that Eqn. (3.35) is an equality for

$$f(x) = ae^{-b(x-x_o)^2} \tag{3.36}$$

is left as an exercise.

Properties 1–6 extend trivially to multidimensional Fourier transforms. Properties related to coordinate transformations are unique to multidimensional systems, however. In two dimensions, we may consider the effect of rotation as an example.

Rotation in Two Dimensions The two-dimensional Fourier transform is

$$\hat{f}(u, v) = \int_{-\infty}^{\infty}\int_{-\infty}^{\infty} f(x, y)e^{-2\pi i(xu+yv)}dx\,dy \tag{3.37}$$

Under the rotation

$$\begin{pmatrix} x' \\ y' \end{pmatrix} = \begin{pmatrix} \cos\theta & \sin\theta \\ -\sin\theta & \cos\theta \end{pmatrix}\begin{pmatrix} x \\ y \end{pmatrix} \tag{3.38}$$

we find that

$$\mathcal{F}\{f(x', y')\} = \hat{f}(u\cos\theta + v\sin\theta,\, v\cos\theta - u\sin\theta) \tag{3.39}$$

Cylindrical Coordinates Analysis of the Fourier transform in other coordinate systems is also of interest. In optical systems with a well-defined axis of propagation, cylindrical coordinates are of particular interest. In cylindrical coordinates the two-dimensional Fourier transformation is

$$\hat{f}(u, \hat{\theta}) = \int_{0}^{\infty}\int_{-\pi}^{\pi} f(\rho, \theta)e^{-2\pi i\rho u\cos(\theta-\hat{\theta})}\rho\,d\rho\,d\theta \tag{3.40}$$

If the input distribution is circularly symmetric $f(\rho, \theta) = f(\rho)$ and the Fourier transformation reduces to the Hankel transformation

$$\hat{f}(u) = 2\pi \int_0^\infty f(\rho) J_0(2\pi\rho u)\rho \, d\rho \tag{3.41}$$

where $J_0(x)$ is the zeroth-order Bessel function of the first kind and we use the identity

$$2\pi J_0(2\pi x) = \int_{-\pi}^{\pi} e^{-2\pi i x \cos(\theta)} d\theta \tag{3.42}$$

The inverse transform is the same as the forward transform:

$$f(\rho) = 2\pi \int_0^\infty \hat{f}(u) J_0(2\pi\rho u) u \, du \tag{3.43}$$

3.4 TRANSFER FUNCTIONS AND FILTERS

As mentioned previously, Fourier analysis is particularly attractive when considering linear shift-invariant (LSI) transformations arising in imaging and diffraction. The basis of this attraction arises from the convolution theorem [Eqn. (3.18)]. Recall the integral form of an LSI transformation from Eqn. (3.9):

$$g(x', y') = T\{f(x, y)\} = \iint f(x, y) h(x' - x, y' - y) dx \, dy \tag{3.44}$$

Applying the 2D form of the convolution theorem to Eqn. (3.44) yields

$$\begin{aligned} \mathcal{F}\{T\{f(x, y)\}\} &= \mathcal{F}\left\{T\left\{\iint f(x, y) h(x' - x, y' - y) dx \, dy\right\}\right\} \\ &= \hat{f}(u, v)\hat{h}(u, v) \end{aligned} \tag{3.45}$$

where $\hat{f}(u, v)$ is the 2D Fourier transform of $f(x, y)$ and $\hat{h}(u, v)$ is the Fourier transform of the impulse response. $\hat{h}(u, v)$ is called the *transfer function* and is a complete descriptor of the linear transformation. The advantage of using the transfer function rather than the impulse response in analysis is that the transformation in Fourier space is a simple product of the input function and the transfer function, rather than an integral transformation. Transfer functions are used extensively in the analysis of optical imaging systems, as discussed in Sections 4.7 and 6.4.

The transfer function plays the role of a *filter*, meaning that it modulates the spectrum of the input signal, attenuating some spatial frequencies while allowing others to pass. In a system without gain, the filter cannot increase the strength of any spectral component. In this case we require $|\hat{h}(u, v)| < 1$. As we do not consider systems with optical gain in this text, this requirement will be enforced for all the systems that we consider. The transfer function may shift the phase of the signal spectrum, however, meaning that $\hat{h}(u, v)$ is generally complex.

We consider several simple examples of transfer functions in two dimensions. A square aperture lowpass filter, for example, may be defined using rect(u), as defined in Eqn. (2.41)

$$\hat{h}(u, v) = \text{rect}(uD)\,\text{rect}(vD) \tag{3.46}$$

where D is a constant. The Fourier transform of rect(x) is

$$\text{sinc}(x) = \frac{\sin(\pi x)}{\pi x} \tag{3.47}$$

so the impulse response corresponding to the square filter is

$$h(x, y) = \frac{1}{D^2}\,\text{sinc}\,\frac{x}{D}\,\text{sinc}\,\frac{y}{D} \tag{3.48}$$

The circular filter defined by

$$\hat{h}(u, v) = \text{circ}\left(D\sqrt{u^2 + v^2}\right) \tag{3.49}$$

where

$$\text{circ}(\rho) = \begin{cases} 1 & |\rho| \leq 0.5 \\ 0 & \text{otherwise} \end{cases} \tag{3.50}$$

is also common in imaging and optical system analysis. Since the circ(ρ) is radially symmetric, the inverse Fourier transform of $\hat{h}(u, v)$ reduces to the Fourier–Bessel transform described with a Hankel transformation in Eqn. (3.41). We use the identity

$$\int_0^x uJ_0(u)du = xJ_1(x) \tag{3.51}$$

to define jinc(ρ) as

$$\begin{aligned} \text{jinc}(\rho) &= 2\pi \int_0^{0.5} J_0(2\pi\rho u)u\,du \\ &= \frac{J_1(\pi\rho)}{2\rho} \end{aligned} \tag{3.52}$$

and the impulse response corresponding to the circular filter

$$h(x, y) = \frac{1}{D^2} \text{jinc}\left(\frac{\sqrt{x^2 + y^2}}{D} \right) \tag{3.53}$$

As a third example, consider the Gaussian filter

$$\hat{h}(u, v) = e^{-\pi D^2 (u^2 + v^2)} \tag{3.54}$$

The Gaussian is the fundamental order of the family of *Hermite–Gaussian functions*

$$\phi_n(x) = e^{-\pi x^2} H_n\left(\sqrt{2\pi} x\right) \tag{3.55}$$

where $H_n(x)$ is the nth-order Hermite polynomial. These functions have the useful property of being eigenfunctions of the Fourier transformation such that

$$\mathcal{F}\{\phi_n(x)\} = i^n \phi_n(u) \tag{3.56}$$

Because of the close relationships between the Fourier and Fresnel transformations, this property is central to the utility of Hermite–Gaussian functions in the description of optical beams in laser resonators under Fresnel propagation assumptions. The Hermite–Gaussian functions are the basis functions of choice in analyzing shift-variant paraxial modes in laser systems. They are of less utility in imaging systems, where shift-invariant modes are needed.

For the present purpose it is sufficient to note that in the $n = 0$ case the Gaussian function is itself an eigenfunction of the Fourier transformation, specifically

$$\mathcal{F}\{\phi_o(u)\} = e^{-\pi x^2} \tag{3.57}$$

so the impulse response corresponding to the Gaussian filter is

$$h(x, y) = \frac{1}{D^2} e^{-(\pi/D^2)(x^2 + y^2)} \tag{3.58}$$

Square, circular, and Gaussian filters are illustrated in Fig. 3.2. Characteristic features include the strong sidelobes of the rectangular and circular impulse responses in comparison to the monotonic rise and fall of the Gaussian impulse. The sidelobes are characteristic of filters with discontinuities, which introduce high-frequency features in the transform space.

The anisotropic structure of the *separable* rectangular filter and impulse as compared to the radially symmetric structure of the other patterns is also noticeable. The impulse corresponding to the circular filter is called the *Airy pattern* and is somewhat broader in distance to the first zero than the sinc impulse. (The first zero occurs at $\rho = 1.22D$, as compared to $x = 1$ for the sinc pattern.) One can overemphasize the significance of the first zero, however; the Gaussian pattern has no zeros at all.

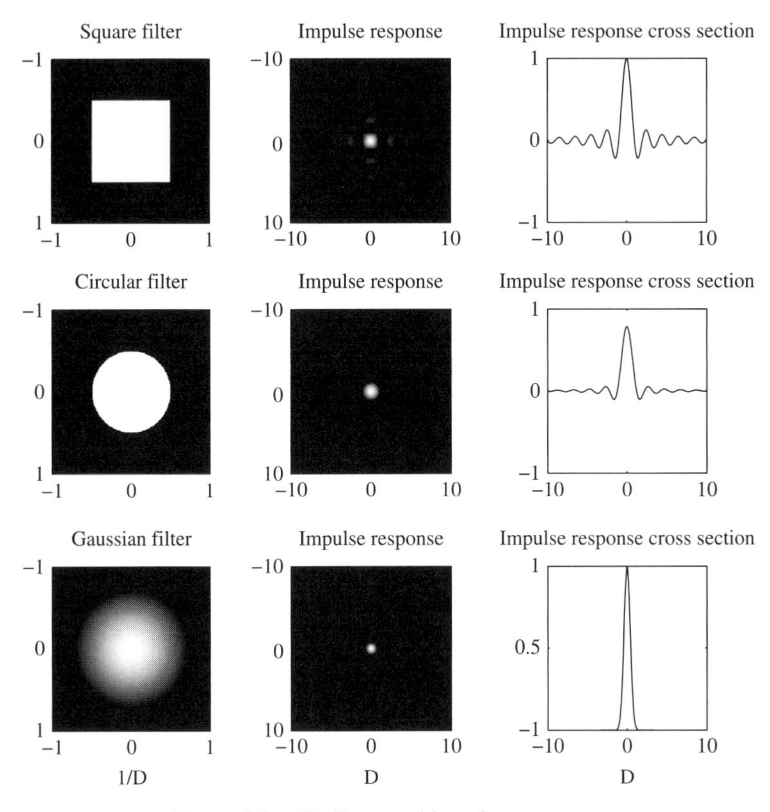

Figure 3.2 2D filters and impulse responses.

3.5 THE FRESNEL TRANSFORMATION

The Fresnel transformation is the most common linear shift-invariant transformation in optical system analysis. As discussed in Section 4.7, the Fresnel transformation is descriptive of optical propagation in free space or homogeneous materials. The one-dimensional Fresnel transformation with real parameter $\tau > 0$ is defined as

$$\begin{aligned} \tilde{f}_\tau(x) &= \frac{1}{\tau}\int_{-\infty}^{\infty} f(x')e^{i\pi((x-x')^2/\tau^2)}\,dx' \\ &= (f * h_\tau)(x) \end{aligned} \tag{3.59}$$

where $*$ is the convolution operator and $h_\tau(x) = e^{i\pi(x/\tau)^2}/\tau$ is the Fresnel kernel. The Fresnel transformation has the inverse kernel

$$h_\tau^{-1}(x) = \frac{e^{-i\pi(x/\tau)^2}}{\tau} \tag{3.60}$$

such that

$$(\tilde{f}_\tau * h_\tau^{-1})(x) = ((f * h_\tau) * h_\tau^{-1})(x) = f(x) \tag{3.61}$$

Since the Fresnel transformation is linear and shift-invariant, it is described by a transfer function in the Fourier domain:

$$\hat{h}_\tau(u) = e^{i(\pi/4)} e^{-i\pi\tau^2 u^2} \tag{3.62}$$

Note that if we rescale the output variable (e.g., by selecting $u' = u/\tau^2$), the Fresnel kernel is an eigenfunction of the Fourier transformation. The close association between the Fresnel and Fourier transformations leads to several interesting and useful identities, such as

$$\hat{f}\left(u = \frac{x}{\tau^2}\right) = e^{i(\pi/4)} h_\tau(x) * \left[h_\tau^{-1}(x)((f * h_\tau)(x))\right] \tag{3.63}$$

and

$$\tilde{f}_\tau(x = u\tau^2) = \tau h_\tau(x = u\tau^2)\mathcal{F}\{\tau h_\tau(x) f(x)\} \tag{3.64}$$

where $\mathcal{F}\{\cdot\}$ represents the Fourier transform.

A localization theorem may be derived for the Fresnel transform in analogy with the localization theorem presented as Eqn. (3.25) for the Fourier transform. Defining

$$\mu_{\tilde{f}} = \frac{1}{\|\tilde{f}\|^2} \int_{-\infty}^{\infty} x|\tilde{f}(x)|^2 dx \tag{3.65}$$

and

$$\sigma_{\tilde{f}}^2 = \frac{1}{\|\tilde{f}\|^2} \int_{-\infty}^{\infty} (x - \mu_{\tilde{f}})^2 |\tilde{f}(x)|^2 dx \tag{3.66}$$

we find with σ_f defined as in Eqn. (3.22)

$$\sigma_f^2 \sigma_{\tilde{f}}^2 \geq \frac{\tau^4}{16\pi^2} \tag{3.67}$$

with equality for

$$f(x) = a e^{-b(x-x_o)^2} e^{-i\pi(x/\tau)^2} \tag{3.68}$$

Proof of Eqn. (3.67) combines the Fourier uncertainty relationship [Eqn. (3.35)] with the Fourier–Fresnel relationship expressed in Eqn. (3.64). According to

Eqn. (3.64), $\hat{g}(u) = \tilde{f}(x = u\tau^2)\exp(i\pi u^2\tau^2)/\tau$ and $g(x) = \tau\exp(i\pi x^2/\tau^2)f(x)$ form a Fourier transform pair, meaning that $\sigma_g^2\sigma_{\hat{g}}^2 \geq 1/16\pi^2$. But $\sigma_g = \tau\sigma_f$ and, assuming zero means

$$\begin{aligned}\sigma_{\hat{g}}^2 &= \int_{-\infty}^{\infty} u^2|\hat{g}(u)|^2 du \\ &= \frac{1}{\tau^2}\int_{-\infty}^{\infty} u^2|\tilde{f}(x = u\tau^2)\exp(i\pi u^2\tau^2)|^2 du \\ &= \frac{1}{\tau^6}\int_{-\infty}^{\infty} x^2|\tilde{f}(x)|^2 dx \\ &= \frac{\sigma_{\tilde{f}}^2}{\tau^6}\end{aligned} \tag{3.69}$$

Multiplying Eqn. (3.69) by $\sigma_g^2 = \tau^2\sigma_f^2$ yields Eqn. (3.67).

The Fresnel transformation is particularly interesting in Fourier analysis of optical systems because (1) the Fresnel transform is the basis of a reasonable model for optical diffraction and (2) the Fourier transform kernel is an eigenfunction of the Fresnel transform. We demonstrate point 1 in Section 4.4. We can prove point 2 now. In one dimension, the Fourier transform kernel is

$$h_u(x) = e^{-2\pi iux} \tag{3.70}$$

The Fresnel transform of $h_u(x)$ is

$$\begin{aligned}\tilde{h}_u(x) &= \frac{1}{\tau}\int_{-\infty}^{\infty} e^{-2\pi ix u}e^{i\pi((x-x')^2/\tau^2)}dx' \\ &= e^{i(\pi/4)}e^{-i\pi u^2\tau^2}e^{-i2\pi ux} \\ &= e^{i(\pi/4)}e^{-i\pi u^2\tau^2}h_u(x)\end{aligned} \tag{3.71}$$

Thus, we see that $h_u(x)$ is an eigenfunction of the Fresnel transform with eigenvalue $e^{i(\pi/4)}e^{-i\pi u^2\tau^2}$. We apply this property in analyzing electromagnetic diffraction in Section 4.4.

The Fresnel transform of the Hermite–Gaussian functions is also useful in optical system analysis. For example, the Fresnel transform of the fundamental Gaussian

$\phi_o(x) = \exp(-\pi x^2)$ is

$$
\begin{aligned}
\tilde{\phi}_{0\tau}(x) &= \frac{1}{\tau} \int_{-\infty}^{\infty} e^{-\pi x'^2} e^{i\pi((x-x')^2/\tau^2)} dx' \\
&= \mathcal{F}^{-1}\left\{ e^{i(\pi/4)} e^{-\pi u^2} e^{-i\pi\tau^2 u^2} \right\} \\
&= \frac{e^{i(\pi/4)}}{\sqrt{1+i\tau^2}} e^{-\pi(x^2/(1+i\tau^2))} \\
&= \frac{e^{i(\pi/4)}}{\sqrt{1+i\tau^2}} e^{i\pi(x^2\tau^2/(1+\tau^4))} \phi_o\left(\frac{x}{\sqrt{1+\tau^4}} \right)
\end{aligned}
\tag{3.72}
$$

where we use the identity for $a > 0$

$$
\mathcal{F}\left\{ e^{-\pi(a+ib)x^2} \right\} = \frac{1}{\sqrt{a+ib}} e^{-\pi(u^2/(a+ib))} \tag{3.73}
$$

The Gaussian distribution is thus also an *eigenfunction* of the Fresnel transformation with a complex change in scale. This result can be extended to the full range of Hermite–Gaussian functions using the identity

$$
\sqrt{2\pi}\phi_n(x) = 2\pi x \phi_{n-1}(x) - \frac{d}{dx}\phi_{n-1}(x) \tag{3.74}
$$

which is proved in Problem 3.4. Using Eqn. (3.74), we find the Fresnel transform of the nth-order Hermite–Gaussian function is

$$
\begin{aligned}
\tilde{\phi}_{n\tau}(x) &= \frac{1}{\tau} \int_{-\infty}^{\infty} \phi_n(x) e^{i\pi((x-x')^2/\tau^2)} dx' \\
&= \mathcal{F}^{-1}\left\{ i^n e^{i(\pi/4)} e^{-i\pi\tau^2 u^2} \phi_n(u) \right\} \\
&= \frac{1}{\sqrt{2\pi}} \mathcal{F}^{-1}\left\{ i^n e^{i(\pi/4)} e^{-i\pi\tau^2 u^2} \left[2\pi u \phi_{n-1}(u) - \frac{d}{du}\phi_{n-1}(u) \right] \right\} \\
&= \frac{1}{\sqrt{2\pi}} \mathcal{F}^{-1}\left\{ 2\pi i u (1 - i\tau^2) \mathcal{F}\{\tilde{\phi}_{(n-1)\tau}\} - i\frac{d}{du}\mathcal{F}\{\tilde{\phi}_{(n-1)\tau}\} \right\} \\
&= \frac{1}{\sqrt{2\pi}} \left[2\pi x \tilde{\phi}_{(n-1)\tau}(x) - (1 - i\tau^2)\frac{d}{dx}\tilde{\phi}_{(n-1)\tau}(x) \right]
\end{aligned}
\tag{3.75}
$$

where we apply the Fourier differentiation identities Eqns. (3.13) and (3.14).

The hypothesis that

$$
\tilde{\phi}_{n\tau}(x) = e^{i(\pi/4)} \frac{(1 - i\tau^2)^{n/2}}{(1 + i\tau^2)^{(n+1)/2}} e^{i\pi(x^2\tau^2/(1+\tau^4))} \phi_n\left(\frac{x}{\sqrt{1+\tau^4}} \right) \tag{3.76}
$$

is immediately proved from Eqn. (3.75) by induction on Eqn. (3.72). The family of functions $\phi_n(x)$ form a *complete orthogonal system* on L^2 ($\mathbb{R}$), meaning that an arbitrary function $f(x) \in L^2(\mathbb{R})$ may be expressed as

$$f(x) = \sum_{n=-\infty}^{\infty} f_n \phi_n(x) \tag{3.77}$$

This expansion is most useful in laser cavity analysis, where the number of modes in the expansion is often quite small. Of course, once $f(x)$ is expanded in the Hermite–Gaussian basis, the expansion coefficients are invariant under propagation, meaning that

$$\tilde{f}_\tau(x) = \sum_{n=-\infty}^{\infty} f_n \tilde{\phi}_{n\tau}(x) \tag{3.78}$$

Since $\tilde{\phi}_{n\tau}(x)$ is known from Eqn. (3.76), Eqn. (3.78) converts the integral Fresnel transform into an algebraic transform if the expansion coefficients f_n are known. Since the Hermite–Gaussian functions are orthogonal, the expansion coefficients are

$$f_n = \int f(x)\phi_n(x)dx \tag{3.79}$$

Before leaving our initial discussion of the Fresnel transform, note that the transform extends trivially to two dimensions. The 2D Fresnel transformation is

$$\tilde{f}_\tau(x, y) = \frac{1}{\tau^2} \int\int_{-\infty}^{\infty} f(x', y') e^{i\pi((x-x')^2/\tau^2)} e^{i\pi((y-y')^2/\tau^2)} dx' dy' \tag{3.80}$$

Since many cavities have a well-defined axis of symmetry, it is sometimes useful to consider the Fresnel transform in cylindrical coordinates:

$$\tilde{f}_\tau(\rho, \phi) = \frac{e^{i\pi(\rho^2/\tau^2)}}{\tau^2} \int_{-\pi}^{\pi}\int_{0}^{\infty} f(\rho', \phi') e^{i2\pi(\rho\rho' \cos(\phi-\phi')/\tau^2)} e^{i\pi(\rho'^2/\tau^2)} \rho' \, d\rho' \, d\phi \tag{3.81}$$

The Laguerre–Gaussian functions

$$\psi_{mn}(\rho, \phi) = (\sqrt{\pi}\rho)^n e^{-\pi\rho^2} e^{-in\phi} L_m^n(2\pi\rho^2) \tag{3.82}$$

where $L_m^n(2\rho^2)$ is a generalized Laguerre polynomial, play a similar role in cylindrical coordinates to the Hermite–Gaussian functions in Cartesian coordinates. Specifically, the Fresnel transform of $\psi_{mn}(\rho, \phi)$ is [65]

$$\tilde{\psi}_{mn\tau}(\rho, \phi) = \frac{e^{i(\pi/4)} e^{-in\phi} e^{-i(n+m+1)\arctan(\tau^2)} (\sqrt{\pi}\rho)^n}{(1+\tau^4)^{\frac{n+1}{2}}} e^{-\pi(\rho^2/(1+i\tau^2))} L_m^n\left(\frac{2\pi\rho^2}{1+\tau^4}\right) \tag{3.83}$$

3.6 THE WHITTAKER–SHANNON SAMPLING THEOREM

The idea of representing continuous functions using discrete values, as in Eqn. (3.77), dates back to Fourier. The basic concept is to describe the continuous function $f(x)$ as a superposition of *basis* functions $\{\phi_n(x)\}$ such that

$$f(x) = \sum_{n=-\infty}^{\infty} \alpha_n \phi_n(x). \tag{3.84}$$

The coefficients α_n form a discrete vector α that represents $f(x)$. The harmonic functions $\sin(2\pi nx/X)$ and $\cos(2\pi nx/X)$ or $\exp(2\pi inx/X)$ are canonical examples of basis functions. A *Fourier series* represents any periodic continuous function $f(x)$ of period X on the basis of harmonic functions as

$$f(x) = \sum_{n=-\infty}^{\infty} a_n e^{i2\pi(nx/X)} \tag{3.85}$$

where

$$a_n = \frac{1}{X} \int_{-X/2}^{X/2} f(x) e^{-2\pi i(nx/X)} dx \tag{3.86}$$

The Fourier series illustrates two concepts common to the process of representing continuous functions. The first point is that the represented functions lie in a specific vector space. For the Fourier series, the vector space is the space of square integrable periodic functions. A function $f(x)$ is periodic with period X if $\forall x \in \mathbb{R}\; f(x+X) = f(x)$. A periodic function $f(x)$ is square integrable if the integral of $|f(x)|^2$ over one period is finite. The vector space associated with a given basis is represented by V. A *vector space* is a set that is closed under vector addition and scalar multiplication. For a set of functions, these properties may be stated

$$\begin{aligned} &\text{If } f_1(x) \in V \quad \text{and} \quad f_2(x) \in V \\ &\text{then } f_1(x) + f_2(x) \in V \end{aligned} \tag{3.87}$$

and

$$\begin{aligned} &\text{If } f(x) \in V \\ &\text{then } \alpha f(x) \in V \quad \forall \alpha \in \mathbb{C} \end{aligned} \tag{3.88}$$

Both of these properties trivially hold for V equal to the set of periodic functions of period X.

The second point illustrated in Fourier series decomposition is that an inner product, $\langle f|g \rangle$, between vectors in V is needed to determine the representation coefficients. For periodic functions a suitable inner product is

$$\langle f|g\rangle = \int_{-X/2}^{X/2} f^*(x)g(x)dx \tag{3.89}$$

Referring to Eqn. (3.86), we see that $a_n = \langle \exp(2\pi inx/X)|f(x)\rangle$. The requirement that $f(x)$ be square integrable ensures that a_n is well defined and finite. The requirement means that the vector norm $\|f\| = \langle f|f\rangle$ is well defined. A vector space with a norm suitably measuring "distance" between vectors is called a *Hilbert space*. The space of square integrable periodic functions using the inner product of Eqn. (3.89) is a Hilbert space.

Many families of basis functions have been developed to represent continuous functions. In addition to the harmonic functions, the Hermite–Gaussian functions mentioned in previous sections have been particularly popular in optics. Families of *special functions*, such as Legendre polynomials and spherical harmonics, are popular in representing distributions in electromagnetic and quantum theory. Since the mid-1980s, the generation of functional bases has become an advanced science in its own right in the context of wavelet theory. We consider wavelet representations in Section 3.8.

This section describes a startling approach to discrete representation developed by Shannon [219] in 1949. Under the *Whittaker–Shannon sampling theorem*, a continuous function is represented on a basis such that the representation coefficients are equal to periodically spaced values of the function itself. The core idea of the sampling theorem is that a signal may be fully represented by capturing a discrete sequence of signal values.

Detailed consideration of the relationship between sampling in measurement, analysis, and display is critical to understanding optical sensors. It is not possible in practice to actually measure the field at specific points in space or time. (Real detectors integrate field values over finite space–time windows.) In Chapter 7 we examine the tension between measurement sampling and analytic sampling in more detail and consider the process of measurement itself as an inner product with the field. For the present purpose of exploring the mathematical tools needed to analyze the field, however, it is important to understand the classical sampling theorem.

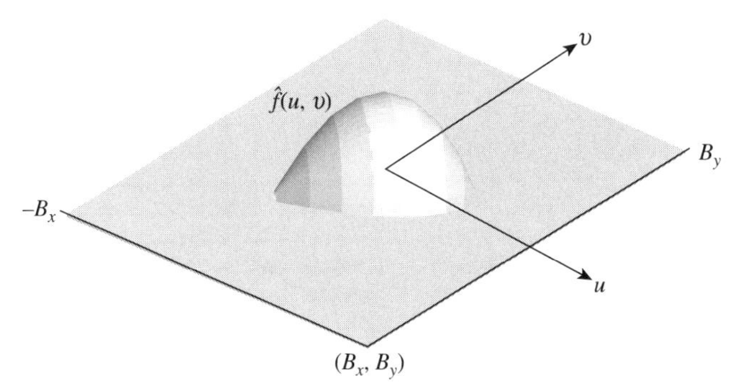

Figure 3.3 $\hat{f}(u, v)$ is limited to a *band area* bounded by $(\pm B_x, \pm B_y)$.

In deriving the sampling theorem, Shannon limited his attention to bandlimited functions. As we are interested in images, our discussion of Shannon's result will focus on the 2D function $f(x, y)$. By *bandlimited*, we mean that the Fourier transform of f, $\hat{f}(u, v)$, vanishes outside a certain range. Specifically, we assume that $|\hat{f}(u, v)| = 0$ for $|u| > B_x$ or $|v| > B_y$. Since $\hat{f}(u, v)$ exists, we may also say that $f \in L^2(\mathbb{R}^2)$, the space of 2D functions that are square integrable in Lebesgue's sense. The Hilbert space V_B associated with the Shannon basis is thus the space of bandlimited functions in $L^2(\mathbb{R}^2)$.

The function $\hat{f}(u, v)$ within the bandlimit is sketched in Fig. 3.3. Consider a function $\hat{f}_t(u, v)$, which is a periodic tiling of the Fourier plane with copies of $\hat{f}(u, v)$, as sketched in Fig. 3.4. Since $\hat{f}_t(u, v)$ is a periodic function, it can be expressed as the Fourier series

$$\hat{f}_t(u, v) = \sum_{n=-\infty}^{\infty} \sum_{m=-\infty}^{\infty} a_{nm} \exp\left[-i\pi\left(\frac{nu}{B_x} + \frac{nv}{B_y}\right)\right] \tag{3.90}$$

Over the range of $\hat{f}(u, v)$, $\hat{f}(u, v) = \hat{f}_t(u, v)$. This means that

$$\begin{aligned} f(x, y) &= \int_{-B_x}^{B_x} \int_{-B_y}^{B_y} \hat{f}_t(u, v) e^{2\pi i(ux+vy)} du\, dv \\ &= \sum_{n=-\infty}^{\infty} \sum_{m=-\infty}^{\infty} a_{nm} \int_{-B_x}^{B_x} \int_{-B_y}^{B_y} \exp\left[-i\pi\left(\frac{nu}{B_x} + \frac{nv}{B_y}\right)\right] du\, dv \\ &= \sum_{n=-\infty}^{\infty} \sum_{m=-\infty}^{\infty} 4B_x B_y a_{nm} \operatorname{sinc}(2B_x x - n) \operatorname{sinc}(2B_y y - m) \end{aligned} \tag{3.91}$$

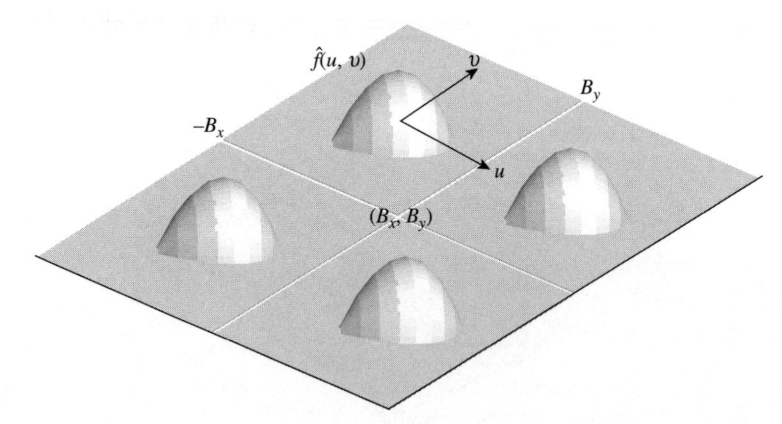

Figure 3.4 Tiling of the Fourier plane with copies of $\hat{f}(u, v)$.

Evaluating Eqn. (3.91), at $x = n/2B_x$ and $y = m/2B_y$, we find $a_{nm} = f((n/2B_x), (m/2B_y))/4B_xB_y$, which yields the Whittaker–Shannon sampling theorem:

$$f(x, y) = \sum_{n=-\infty}^{\infty} \sum_{m=-\infty}^{\infty} f\left(\frac{n}{2B_x}, \frac{m}{2B_y}\right) \operatorname{sinc}(2B_x x - n) \operatorname{sinc}(2B_y y - m) \tag{3.92}$$

The sampling theorem implies that the continuous function $f(x, y)$ can be fully represented by the discrete samples $f(n/2B_x, m/2B_y)$. The sampling rates $2B_x$ and $2B_y$ are necessary in x and y to accurately reconstruct $f(x, y)$ from discrete measurements. Sampling at these rates is called "Nyquist" sampling, the corresponding periods between samples are $\Delta_x = \frac{1}{2}B_x$ and $\Delta_y = \frac{1}{2}B_y$.

In cases of particular interest, we focus on the value of $f(x, y)$ only over some finite aperture. If we leave $f(x, y)$ unconstrained outside this aperture, only a finite number of samples are needed to characterize the function. If the extent of the aperture of interest is $(-X, X)$ and $(-Y, Y)$, the number of samples needed is $16B_xB_yXY$.

3.7 DISCRETE ANALYSIS OF LINEAR TRANSFORMATIONS

The discrete Fourier transform (DFT) of a two-dimensional array of samples is defined as

$$\hat{f}_{n'm'} = \frac{1}{\sqrt{NM}} \sum_{n=-(N/2)}^{(N/2)-1} \sum_{m=-(M/2)}^{(M/2)-1} f_{nm} e^{i2\pi(nn'/N)} e^{i2\pi(mm'/M)} \tag{3.93}$$

This transformation can be inverted to yield

$$f_{nm} = \frac{1}{\sqrt{NM}} \sum_{n'=-(N/2)}^{(N/2)-1} \sum_{m'=-(M/2)}^{(M/2)-1} \hat{f}_{n'm'} e^{-i2\pi(nn'/N)} e^{-i2\pi(mm'/M)} \tag{3.94}$$

The discrete Fourier transform is often used to approximate the Fourier transform in considering continuous functions. From Eqn. (3.90), we know that

$$\hat{f}(u, v) = \frac{1}{4B_xB_y} \operatorname{rect}\left(\frac{u}{2B_x}\right) \operatorname{rect}\left(\frac{v}{2B_y}\right) \sum_{n=-\infty}^{\infty} \times \sum_{m=-\infty}^{\infty} f\left(\frac{n}{2B_x}, \frac{m}{2B_y}\right) exp\left[-i\pi\left(\frac{nu}{B_x} + \frac{mv}{B_y}\right)\right] \tag{3.95}$$

While Eqn. (3.95) represents an exact model of $\hat{f}(u, v)$, it also involves a sum over an infinite number of samples. An approximate numerical analysis uses a finite number of samples $f_{nm} = f(n/2B_x, m/2B_y)$ based on truncating the series at some approximate maximum spatial extent $N/2B_x = 2X$. The approximate Fourier transform is

$$\hat{f}_{\text{approx}}(u, v) = \frac{1}{4B_xB_y}\text{rect}\left(\frac{u}{2B_x}\right)\text{rect}\left(\frac{v}{2B_y}\right)\sum_{n=-(N/2)}^{(N/2)-1}$$
$$\times \sum_{m=-(N/2)}^{(N/2)-1} f_{nm} \exp\left[-i\pi\left(\frac{nu}{B_x} + \frac{mv}{B_y}\right)\right] \tag{3.96}$$

The DFT of the truncated set of samples $\{f_{nm}\}$ is

$$\hat{f}_{n'm'} = \frac{1}{N}\sum_{n=-(N/2)}^{(N/2)-1}\sum_{m=-(N/2)}^{(N/2)-1} f_{nm} \exp\left[-i2\pi\left(\frac{nn'}{N} + \frac{mm'}{N}\right)\right] \tag{3.97}$$

where $N = 4BX$ and, for simplicity, we set $B = B_x = B_y$, $X = Y$. The DFT samples approximate the Fourier transform of $f(x,y)$ at sample points such that $\hat{f}_{n'm'} = 4B^2N\hat{f}_{\text{approx}}(u = n'/2X, v = m'/2X)$. While truncation of nonzero terms means that $\hat{f}_{n'm'}$ is not exactly equal to $\hat{f}(u = n'/2X, v = m'/2X)$, the inverse DFT of $\hat{f}_{n'm'}$ does produce exact values of the sampled function at sample points within the truncation window.

Consider, as an example, the bandlimited one-dimensional function $f(x) = \text{sinc}(x - \delta)$, for $\delta \ll 1$. In this case $\hat{f}(u) = e^{-2\pi i\delta u}\text{rect}(u)$. For this pair of functions, $B = \frac{1}{2}$, $f_0 = \text{sinc}(\delta)$ and, for $n \neq 0$, $f_n = f(n) \approx (-1)^n\delta/n$. Sampling this function over the window from $x = -N/2$ to $x = N/2 - 1$ produces N samples spaced by 1 in the spatial domain. The DFT produces N samples in Fourier space between $u = -\frac{1}{2}$ and $u = \frac{1}{2} - 1/N$ spaced by $1/N$. The error between the samples $\hat{f}_n$ and $\hat{f}(u)$ for various sampling ranges X for this function is shown in Fig. 3.5. Note that the error does not decrease at the edges of the sampling band, a phenomenum common in Fourier analysis of discontinuous functions. However, the error between the numerical spectrum and the actual spectrum does decrease as the sampling window increases over much of the Fourier passband.

Despite our assumption that $f(x)$ is bandlimited in deriving the sampling theorem, it is not uncommon to attempt numerical Fourier analysis of functions with infinite support in both x and u. In numerically analyzing the Fourier transform of such a function, one selects a window size X and a sampling period Δ to obtain $N = 2X/\Delta$ samples $f_n = f(n\Delta)$. The DFT of the samples f_n produces N discrete samples $\hat{f}_{n'}$ nominally corresponding to values of $\hat{f}(u = n'/2X)$ covering the frequency range $(1/2X - 1/\Delta) \leq u \leq 1/\Delta$. The sampling period over the range of u is $1/2X$. As an example of these scaling laws, the DFT of the Gaussian function $f(x) = \exp(-\pi x^2)$, for which $\hat{f}(u) = \exp(-\pi u^2)$, is illustrated in Fig. 3.6.

Discrete Fourier analysis is attractive in considering linear transformations because shift-invariant transformations are modeled by simple multiplicative filter functions in Fourier transform space and because fast numerical algorithms for modeling

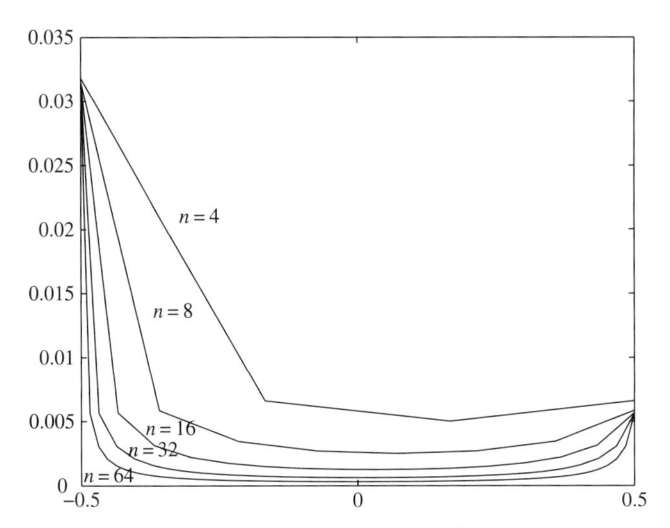

Figure 3.5 Magnitude of the difference between $\hat{f}(u)$ and $\hat{f}_{n=uX}$ over the bandpass of $f(x) = \mathrm{sinc}(x - \delta)$ for various values of $N = 4XB$ and $\delta = 0.01$.

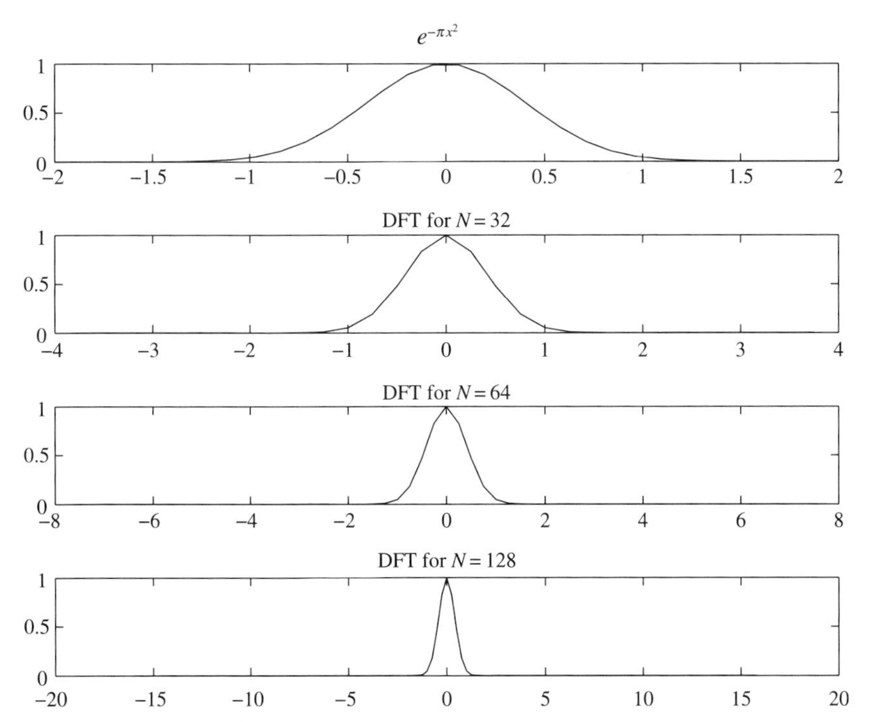

Figure 3.6 DFT of $e^{-\pi x^2}$ sampled uniformly over the window $-2 \leq x \leq 2$ for various values of N. As we increase the number of samples while keeping X constant, the sample period in Fourier space stays constant and the number of samples within the significant region of the signal remains fixed. Increasing the sampling window would increase the Fourier resolution. Both the sampling window and the sampling rate must be increased to maintain resolution in both spaces.

Fourier transforms are readily available. As we briefly consider through wavelet theory in the second half of this chapter, Fourier analysis is not a unique or universally attractive technique for modeling linear transformations.

As an example of the first attraction of Fourier analysis, consider again the Fresnel transform. As described in Eqn. (3.62), the transfer function for the Fresnel transform is $\hat{h}_\tau(u) = \exp(i(\pi/4))\exp(-i\pi\tau^2u^2)$. We may model the Fresnel transform on a function $f(x)$ by modulating the DFT of a sampled version of $f(x)$ by this transfer function and then applying an inverse DFT. Returning to the example of a Gaussian signal, the analytic form of the Fresnel transform is given in Eqn. (3.72). As illustrated in Fig. 3.7, a numerical estimate of the Fresnel transform is obtained by multiplying the DFT of the sampled function by the transfer function and inverse transforming. The spatial window spans $|x| < 20$. In total, 4096 samples were used with a sample spacing of 0.0098. Each plot shows the absolute value of the transformed signal as well as the real and imaginary components. Values of τ are 0 at top running through 0.5, 1, 2, 4, and 8 at the bottom. As τ increases, the transformed signal becomes increasingly diffuse. For $\tau = 8$, the numerically estimated transform encounters large errors as the transformed signal extends beyond the range of the window.

Numerical estimation of the transform is reasonably correct up to values of τ such that the transformed signal extends beyond the input window. Note that both f_{nm} and $\hat{f}_{n'm'}$ act as Fourier series components when reconstructing the approximate

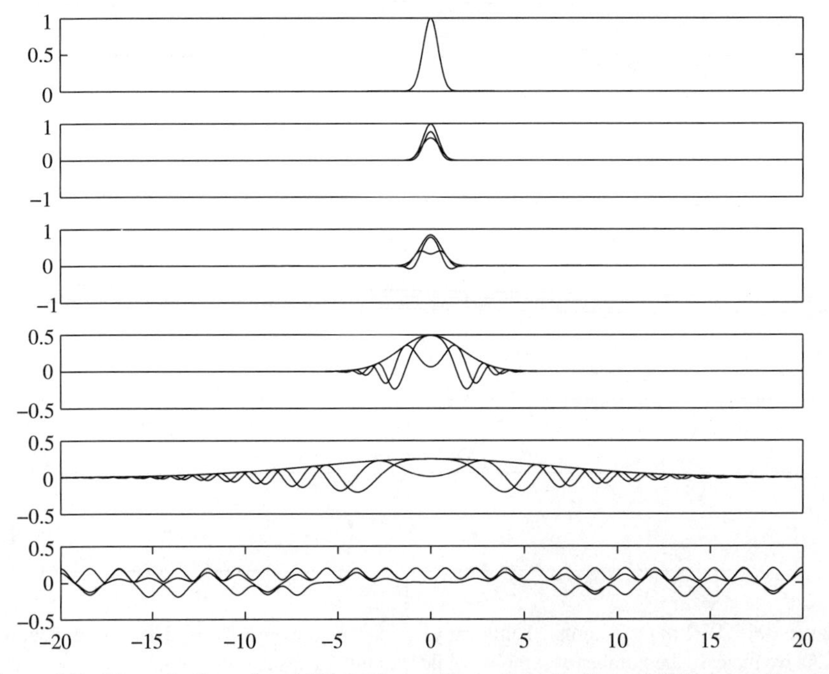

Figure 3.7 Numerically estimated Fresnel transform of the fundamental Hermite–Gaussian mode.

continuous functions. The series reconstructions are periodic, when the reconstruction extends beyond the transform window periodicity produces interference, or aliasing, between bandpass or spatial windows.

The second motivation for Fourier methods in linear systems analysis focuses on the computational efficiency of computing the Fourier transform. Nominally, the DFT of a one-dimensional N element dataset is represented by a $N \times N$ transformation matrix multiplying a length N dataset. This transformation would require $O(N^2)$ operations. In practical systems, the fast Fourier transform (FFT) is used to greatly reduce the number of computational operations required. Hierarchical decimation is the heart of the FFT algorithm. The one-dimensional DFT of length N, defined as

$$\hat{f}(n') = \frac{1}{\sqrt{N}} \sum_{n=-(N/2)}^{(N/2)-1} f_n e^{i2\pi(nn'/N)} \tag{3.98}$$

is decimated into two DFTs of length $N/2$ by the arrangement

$$\begin{aligned}\hat{f}_{n'} &= \frac{1}{\sqrt{2}\sqrt{N/2}} \sum_{n=-(N/4)}^{(N/4)-1} f_{2n} e^{i2\pi(nn'/(N/2))} + \frac{e^{(i2\pi n'/N)}}{\sqrt{2}\sqrt{N/2}} \sum_{n=-(N/4)}^{(N/4)-1} f_{(2n+1)} e^{i2\pi(nn'/(N/2))} \\ &= \frac{\hat{f}_{en'}}{\sqrt{2}} + e^{(i2\pi n')/N} \frac{\hat{f}_{on'}}{\sqrt{2}} \end{aligned} \tag{3.99}$$

where $\hat{f}_{en'}$ and $\hat{f}_{on'}$ are the length $N/2$ DFTs of the even and odd coefficients of f_n. Since the two shorter transformations each require $O(N^2/4)$ operations, decimation reduces the number of operations required by a factor of 2. If N is a power of 2, recursive decimation reduces the number of operations required from $O(N^2)$ to $O(N \log_2 N)$.

For two-dimensional Fourier transforms, FFT algorithms are separably applied along rows or columns of data. For an $N \times N$ dataset, the FFT reduces the computational order from $O(N^4)$ to $O(N^2 \log_2^2 N)$. For images with $N = 1024$, the FFT algorithm reduces computational complexity by four orders of magnitude.

3.8 MULTISCALE SAMPLING

Concepts of discrete representation and sampling have evolved considerably in the half-century since Shannon presented the sampling theorem [234]. The evolution of sampling theory has accelerated in the past quarter-century with the development of a generalized methodology for developing bases and representation spaces. Wavelet theory is the most elegant means of understanding emerging strategies.

The primary goal of this text is to develop a framework for analysis and design of physical/digital interfaces in optical sensor systems. Since more sophisticated models of signal sampling and analysis are enabling in pursuit of this goal, we present a brief introduction to wavelets and generalized sampling in this chapter. Our hope is to be accessible to the optical engineer without unnecessarily insulting the mathematician. We refer the mathematically inclined reader to the mathematics and signal processing literature [53,164].

As discussed in Chapter 7, modern sampling theory must distinguish between sampling in the sense of measurement and in the sense of signal analysis and representation. Sampling theory is applied to

1. *Signal Analysis*. In the sampling theorem we have shown that the signal space V_B is spanned by the function sinc(B_x). We can easily analyze linear transformations of signals on this space by analyzing transformations of the basis function.
2. *Signal Estimation*. Equation (3.100) is a recipe for estimating the signal value of $f(x)$ at any point x from discrete samples.
3. *Sensor System Design*. If we measure a signal at discrete points in space or time, the sampling theorem informs the rate at which the signal must be measured for accurate representation.

In practice, challenges arise to the application of the Shannon sampling approach to each of these uses. In cases 1 and 2, the fact that the function sinc(x) does not have compact support makes computation and analysis expensive. In the third case, one must account for the fact that it is not generally possible to measure functions to infinite spatial or temporal resolution, meaning that true measurements of $f(x)$ are not generally available. We discuss details of actual measurements in optical systems in Chapters 5 and 7. For present purposes, it is helpful to simply consider the possibility that while expansion coefficients are somehow related to local features of the signal, they need not correspond to actual signal values. As a first step to resolving all three of these challenges, it is helpful to consider sampling strategies that are not based on sinc(x).

For simplicity, we consider wavelet representations of one-dimensional functions. The 1D version of Eqn. (3.92) is

$$f(x) = \sum_{n=-\infty}^{\infty} f\left(\frac{n}{2B}\right) \operatorname{sinc}(2Bx - n) \tag{3.100}$$

where we assume $f(x) \in V_B$ and V_B is the subspace of bandlimited functions in $L^2(\mathbb{R})$. To address the challenges described above, we maintain the concept of representation of $f(x)$ by a discretely shift-invariant localized function, but we replace sinc(x) with a *scaling function* $\phi(x)$. We imagine representing $f(x)$ in terms of the scaling function as

$$f_\phi(x) = \sum_{n=-\infty}^{\infty} c_n \phi(x - n) \tag{3.101}$$

$\phi(x)$ is the *generating function* for the vector space $V(\phi)$ is spanned by $\{\phi(x-n)\} \in \mathbb{Z}$.

As an example, we consider a scaling function and decomposition discovered in 1910 by Haar [108]. The Haar scaling function is $\beta^0(x) = \text{rect}(x - \frac{1}{2})$. Decomposition on this function takes the form

$$f_0(x) = \sum_{n=-\infty}^{\infty} c_n \beta^0(x - n) \tag{3.102}$$

With the restriction that $\sum_{n\in\mathbb{Z}} |c_n|^2$ is finite, $f_0(x) \in L^2(\mathbb{R})$ and the family of functions $\{\phi_n(x) = \beta^0(x - n)\}$ forms an orthonormal basis of a subspace $V_0 \subset L^2(\mathbb{R})$. Evaluating the inner product $\langle \beta^0(x - m)|f_0 \rangle$ using the orthogonality relationship $\langle \beta^0(x - m)|\beta^0(x - n) \rangle = \delta_{nm}$ we see from Eqn. (3.102) that

$$c_n = \int_{-\infty}^{\infty} \beta^0(x - n) f_0(x) dx \tag{3.103}$$

The function $f(x)$ may be decomposed into two components, $f_0 \in V_0$ and $f_\perp \notin V_0$, such that $f(x) = f_0(x) + f_\perp(x)$. For all functions $g(x) \in V_0$, $\langle g|f_\perp \rangle = 0$. Thus for the orthonormal basis $\{\beta^0(x - n)\}$, we obtain

$$\langle \beta^0(x - n)|f \rangle = \langle \beta^0(x - n)|(f_0 + f_\perp) \rangle = \langle \beta^0(x - n)|f_0 \rangle \tag{3.104}$$

and $c_n = \langle \beta^0(x - n)|f \rangle$. $f_0(x)$ is the *projection* of $f(x)$ onto V_0, $P_{V_0}f$. An example of $P_{V_0}f$ for $f(x) = x^2/10$ is shown in Fig. 3.8.

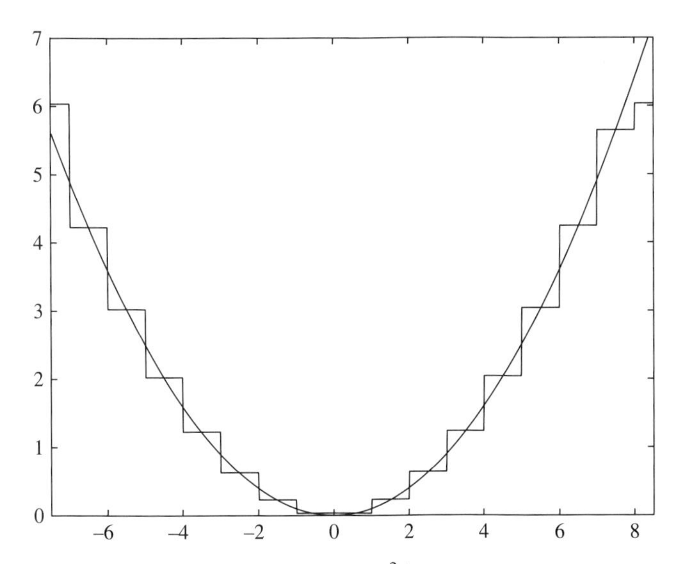

Figure 3.8 Projection of $f(x) = x^2/10$ onto the Haar basis.

Less fidelity is observed in the projection of a more complex function onto V_0 in Fig. 3.9(a). To improve the fidelity of the representation, we are tempted to use a narrower generating function, for example, $\phi_{-1,n}(x) = \sqrt{2}\beta^0(2x - n)$. As illustrated in Fig. 3.9(b), this rescaled generating function does, in fact, improve the representation fidelity. Continuing on this train of thought, we choose to define families of sampling functions on scales j such that

$$\phi_{j,n}(x) = \frac{1}{\sqrt{2^j}}\beta^0\left(\frac{x}{2^j} - n\right) \tag{3.105}$$

Each rescaled generating function corresponds to a new Hilbert space of functions $V_j \in L^2(\mathbb{R})$. Note that the basis functions for the space V_j can be expressed in the space V_{j-1} as

$$\phi_{j,n}(x) = \frac{1}{\sqrt{2}}\left[\phi_{j-1,2n}(x) + \phi_{j-1,2n+1}(x)\right] \tag{3.106}$$

This means that $V_{j+1} \subset V_j$. We observe, of course, that estimation of $f(x)$ is more accurate on V_j than on V_{j+1}. For the Haar scaling function this refinement process continues indefinitely until in the limit

$$\lim_{j \mapsto -\infty} V_j = L^2(\mathbb{R}) \tag{3.107}$$

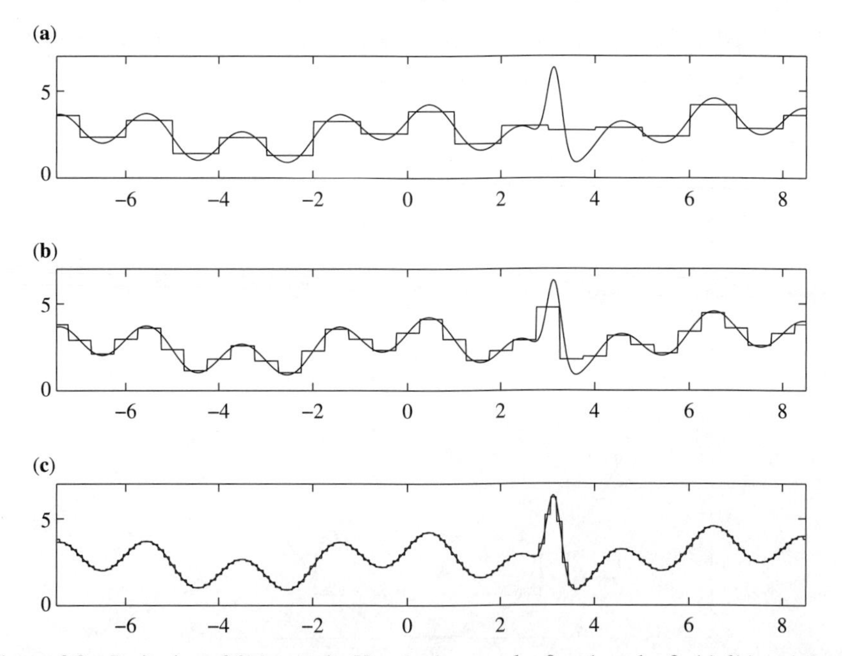

Figure 3.9 Projection of $f(x)$ onto the Haar basis on scales 0, −1, and −3: (a) $f(x)$ and $P_{V_0}f$; (b) $f(x)$ and $P_{V_{-1}}f$; and (c) $f(x)$ and $P_{V_{-3}}f$.

Nonredundant representation of $f(x)$ on multiple scales is a goal of wavelet theory. We achieve this goal in considering the subspace W_j, the orthogonal complement of V_j in V_{j-1}. By "orthogonal" we mean $V_j \cap W_j = \{0\}$. By design, V_j and W_j span V_{j-1}, for example

$$V_{j-1} = V_j \oplus W_j \tag{3.108}$$

The *wavelet* corresponding to the scaling function $\phi(x)$ is the generating function for the basis of W_0. For example, the wavelet corresponding to the Haar scaling function is

$$\begin{aligned} \psi(x) &= \beta^0(2x) - \beta^0(2x-1) \\ &= \phi_{-1,0}(x) - \phi_{-1,1}(x) \end{aligned} \tag{3.109}$$

The scaling function can also be expressed in the basis V_{-1} as

$$\begin{aligned} \phi(x) &= \beta^0(x) \\ &= \phi_{-1,0}(x) + \phi_{-1,1}(x) \end{aligned} \tag{3.110}$$

The wavelet and scaling functions for this case are shown in Fig. 3.10. Since the basis functions for the V_{-1} can be expressed in terms of the scaling and wavelet functions as

$$\phi_{-1,n}(x) = \begin{cases} \dfrac{1}{\sqrt{2}}[\phi(x-n) + \psi(x-n)] & \text{for } n \text{ even} \\ \dfrac{1}{\sqrt{2}}[\phi(x-n) - \psi(x-n)] & \text{for } n \text{ odd} \end{cases} \tag{3.111}$$

we see that the linear combination of the bases $\phi_{0,n}(x)$ and $\psi(x-n)$ span the space V_{-1}. Since $\psi(x)$ is orthogonal to all of the basis vectors $\phi_{0,n}(x)$, $\psi(x) \in V_0$. $\psi(x-n)$ is also orthogonal to $\psi(x-m)$ for $m \neq n$. Thus, $\psi(x-m)$ is an orthonormal basis for W_0.

By scaling the wavelet function, one arrives at a basis for W_j. In exact correspondence to Eqn. (3.105) the basis for W_j is

$$\psi_{j,n}(x) = \frac{1}{\sqrt{2^j}}\psi\left(\frac{x}{2^j} - n\right) \tag{3.112}$$

There is a substantial difference between the subspaces generated by the scaling function and the subspaces generated by the wavelet, however. While

$$\{0\} \subset \cdots \subset V_2 \subset V_1 \subset V_o \subset V_{-1} \subset V_{-2} \subset \cdots \subset L^2(\mathbb{R}) \tag{3.113}$$

the wavelet subspaces are not similarly nested. Specifically, $W_j \subset V_{j-1}$ but $W_j \not\subset W_{j-1}$. In fact, all of the wavelet subspaces are orthogonal. This means that $\cap_{j=-\infty}^{\infty} W_j = \{0\}$ and $\cup_{j=-\infty}^{\infty} W_j = L^2(\mathbb{R})$. This distinction is illustrated in Fig. 3.11,

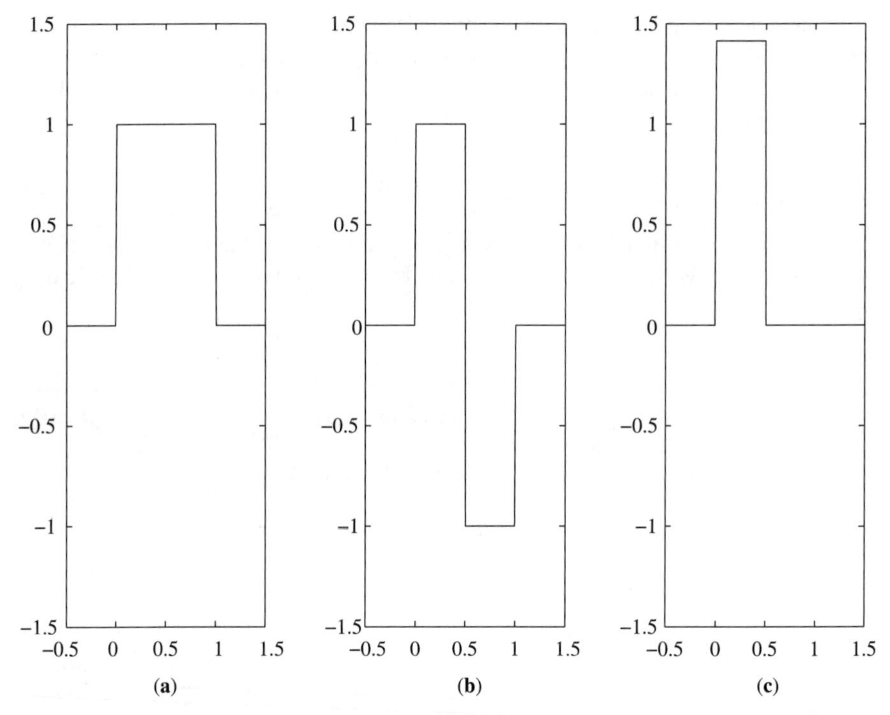

Figure 3.10 Scaling, wavelet, and $\phi_{-1,0}(x)$ functions for the Haar basis: (a) $\phi(x) = \beta^0(x)$; (b) $\psi(x) = \beta^0(2x) - \beta^0(2x-1)$; (c) $\phi_{-1,0}(x) = 2^{0.5}\beta^0(2x)$.

which shows the scaling function and the wavelet on the $j = 0, -1, -2, -3$ scales. Note that while the scaling functions on various scales are not orthogonal, the wavelets on each scale are orthogonal with regard to both shift and change in scale. This orthogonality among the wavelets may be expressed

$$\langle \psi_{j,n} | \psi_{j',n'} \rangle = \delta_{jj'} \delta_{nn'} \tag{3.114}$$

Since the wavelet subspaces are mutually orthogonal and since their union covers $L^2(\mathbb{R})$, for $f(x) \in L^2(\mathbb{R})$ we find that the wavelet analog to Eqn. (3.92) is

$$f(x) = \sum_{j \in \mathbb{Z}} \sum_{n \in \mathbb{Z}} f_{j,n} \psi_{j,n}(x) \tag{3.115}$$

where $f_{j,n} = \langle f | \psi_{j,n} \rangle$. Differences with the sampling theorem include the facts that $f_{j,n}$ is no longer a sample value of $f(x)$ and that there is no longer any constraint on the bandwidth of $f(x)$. Since the wavelet function has compact support, however, $f_{j,n}$ is similar to a sample because it is related to signal values at a specific region in space. Of course, the wavelet decomposition is a "multiscale" decomposition, and the localization of $f_{j,n}$ varies from one scale to the next.

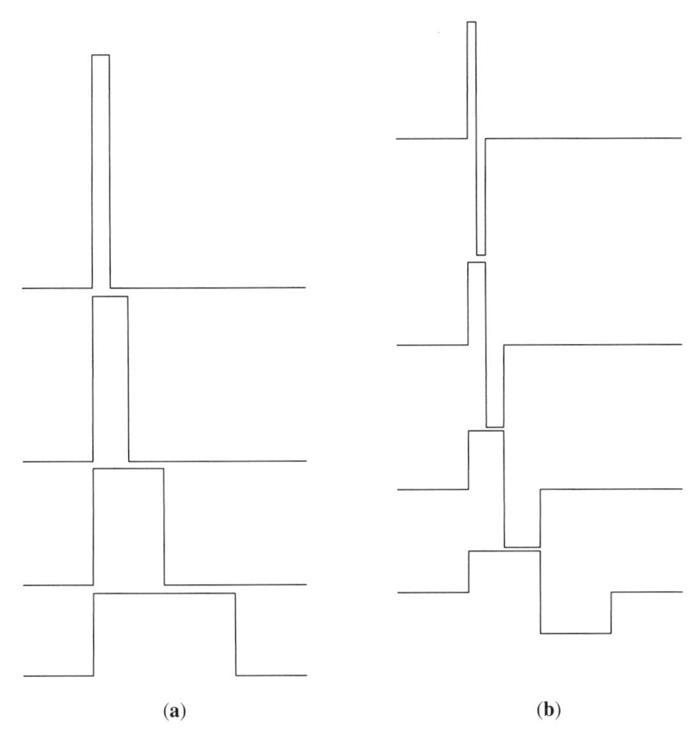

Figure 3.11 Scaling and wavelet function for the Haar wavelet on scales $j = 0, -1, -2, -3$: (a) $\phi(x)$, $\phi_{-1,0}(x)$, $\phi_{-2,0}(x)$, and $\phi_{-3,0}(x)$; (b) $\psi(x)$, $\psi_{-1,0}(x)$, $\psi_{-2,0}(x)$, and $\psi_{-3,0}(x)$.

It is often convenient to represent $f(x) \in L^2(\mathbb{R})$ in terms of a scaling function at one resolution with finer details represented in wavelet components. For example, one may represent $f(x)$ on $V_0 \oplus \cup_{j=-\infty}^{0} W_j = L^2(\mathbb{R})$ as

$$f(x) = \sum_{n\in\mathbb{Z}} f_{a0}\phi_{0,n}(x) + \sum_{j=-\infty}^{0} \sum_{n\in\mathbb{Z}} f_{j,n}\psi_{j,n}(x) \tag{3.116}$$

The coefficients f_{a0} are called the "averages," and the terms $f_{j,n}$ are called the "differences" in this expansion. The averages provide a coarse representation of $f(x)$ on the scaling function basis; this representation is nonredundantly refined by adding differences (wavelet coefficients) in each successive order.

Expansion coefficients for the function of Fig. 3.9 the spaces V_0, W_0 through V_{-3}, W_{-3} are plotted in Fig. 3.12. The upper trace in each plot is the expansion coefficients on V_j (shifted up by 1 to separate the averages and differences). Since $V_{j-1} = V_j + W_j$, coefficients on V_{j-1} can be calculated from the coefficients on V_j and W_j. Accordingly, the coefficients on V_0 in Fig. 3.12 would be saved in a decomposition of $f(x)$ as the averages and the wavelet coefficients at each level would be saved as the differences.

In this case the range of the function over $(-8, 8)$ produces 16 average values and 16 difference values on the zeroth order, 32 -1 order differences, 64 -2 order

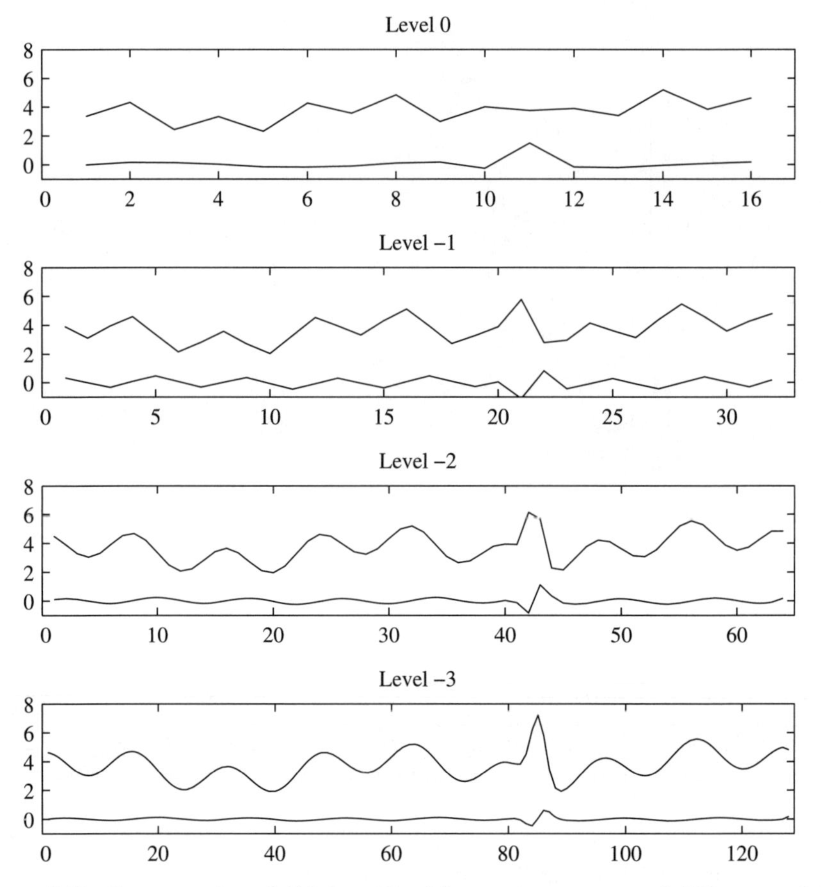

Figure 3.12 Representation of $f(x)$ from Fig. 3.9 onto the averages and differences of the Haar scaling function and wavelets to order −3.

differences, and 128 −3 order differences. These values may be combined to produce 256 average values in a −3 order estimate as shown in Fig. 3.9. Multiscale sampling is particularly useful in communication systems where one may like to send coarse signal representations first and fill in finer data as further signals arrive.

Wavelet analysis may be extended to multidimensional signal analysis by various mechanisms. The simplest approach uses spatially separable 2D functions based on 1D scaling functions and wavelets. The scaling function $\phi(x, y) = \phi(x)\phi(y)$ generates a vector space $V_0^2 \in L^2(\mathbb{R}^2)$. The wavelet space $W_0^2 \in V_{-1}^2$ for the 2D function is generated from three wavelets:

$$\begin{aligned} \psi^1(x, y) &= \phi(x)\psi(y) \\ \psi^2(x, y) &= \psi(x)\phi(y) \\ \psi^3(x, y) &= \psi(x)\psi(y) \end{aligned} \tag{3.117}$$

The wavelet family

$$\psi^k_{j,nm}(x) = \frac{1}{\sqrt{2^j}} \psi^k \left(\frac{x}{2^j} - n, \frac{y}{2^j} - m \right) \tag{3.118}$$

for n, $m \in$ *integers* is an orthonormal basis of W_j^2 and

$$\left\{ \psi^1_{j,nm}(x, y), \psi^2_{j,nm}(x, y), \psi^3_{j,nm}(x, y) \right\}_{(j,n,m) \in \mathbb{Z}^3} \tag{3.119}$$

is an orthonormal basis of $L^2(\mathbb{R})$ [164]. The 2D analog of Eqn. (3.116) is

$$f(x) = \sum_{n,m \in \mathbb{Z}} f_{a0} \phi_{0,n}(x) \phi_{0,m}(y) + \sum_{k=1}^{3} \sum_{j=-\infty}^{0} \sum_{n \in \mathbb{R}} f_{j,nm} \psi^k_{j,nm}(x, y) \tag{3.120}$$

Wavelet decomposition may operate from either coarse to fine, as in Eqn. (3.120), or from fine to coarse. One might assume, for example, that a given image consists of wavelet coefficients on V_0^2 and project onto spaces V_j^2 and W_j^2 for $j > 0$. Figure 3.13(b) shows the averages on V_3^2 for the V_0 image shown in Fig. 3.13(a). For the Haar wavelet, the averages image is a simple lower resolution version of the original image. Figure 3.14 shows the wavelet coefficients on levels 1, 2, and 3. The original image in this case consists of 1024 × 768 coefficients. The level 3

Figure 3.13 (a) Original image and (b) wavelet averages on V_3^2.

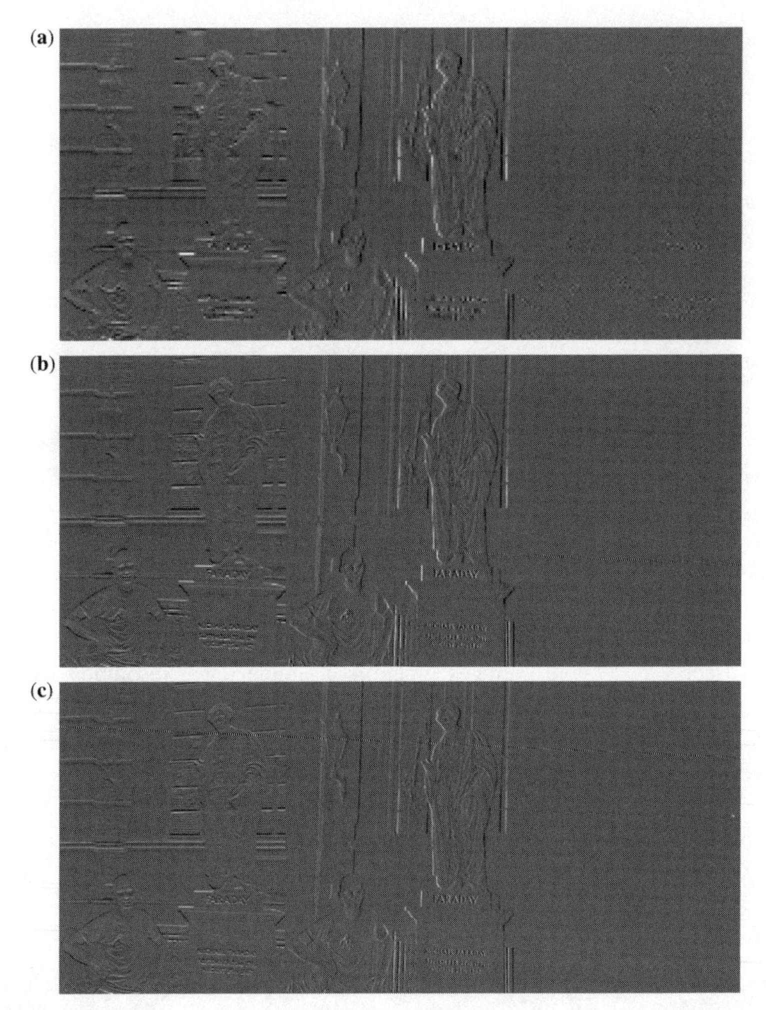

Figure 3.14 Wavelet coefficients for the Fig. 3.13(a) image on (a) W_3^2, (b) W_2^2, and (c) W_1^2.

averages image is reduced by $2^3 = 8$ in each dimension to a 128×96 array. The level 3 wavelet coefficients consist of three arrays of *horizontal*, *vertical*, and *diagonal* differences corresponding to $\psi^1_{j,nm}$, $\psi^2_{j,nm}$, $\psi^3_{j,nm}$. Figure 3.14(a) conjoins the difference arrays in a single 128×288 array. The level 2 difference arrays shown in Fig. 3.14(b) consist of 3 256×192 arrays, and the level 1 wavelet coefficients in Fig. 3.14(c) consist of 3 512×384 arrays. Since the total number of coefficients in the arrays shown in Figs. 3.13(b) and 3.14 is $4 \times 128 \times 96 + 3 \times 256 \times 192 + 3 \times 512 \times 384 = 1024 * 768$, a communication system that sent averages in Fig. 3.13(b) first followed by the differences in Fig. 3.14 would send a coarse image much faster than a system that simply transmitted a raster version of Fig. 3.13(a). In practice, compression and communication algorithms extend this concept by identifying and discarding low value differences.

There are many families of scaling functions in addition to the Haar function that produce orthogonal wavelets. The sinc function, in particular, is an appropriate orthogonal scaling function and generates a corresponding *Shannon* wavelet. Using the Shannon wavelet, the space of bandlimited functions may be considered as V_0 and Eqn. (3.92) may be extended to represent nonbandlimited functions by including wavelet expansions of higher order. The Shannon basis is little used in practice however, because practical sampling systems do not directly measure on this basis and because of challenges in using the sinc as a scaling function.

3.9 B-SPLINES

The Haar scaling function is the lowest order of the set of polynomial basis functions known as *B-splines*. B-spline functions of higher order are recursive convolutions of $\beta^0(x)$ defined according to

$$\beta^m(x) = \beta^0\left(x + \frac{1-(-1)^m}{2}\right) * \beta^{m-1}(x) \tag{3.121}$$

where $m \geq 1$ [236, 237]. Spatial shifts are included in the β^0 functions to ensure that the B-splines of even order are centered at $x = \frac{1}{2}$ and those of odd order are centered at $x = 0$. The first six orders of the B-splines are shown in Fig. 3.15. Note that while each order covers finite support, the size of the support grows by 1 from each order to the next. The support of the zeroth-order B-spline is (0, 1), the first order is B-spline (−1, 1), and the second-order B-spline is (−1, 2). The support for the nth order is of magnitude $n + 1$. The mth-order spline traces out a piecewise continuous polynomial curve of order m.

In contrast to the Haar scaling function, B-splines of higher order do not generate an orthogonal basis for unit shifts on a regular grid. It is nevertheless useful to represent $f(x)$ discretely on these functions. Each order of B-splines does generate an independent basis for unit shifts and the B-splines appear frequently in physical models of optical sensor systems. For example, the sampling function $\hat{h}_{i,j}(\theta_x, \theta_y)$ derived in Chapter 2 for coded aperture imaging is the separable product of B-splines

$$\hat{h}_{i,j}(\theta_x, \theta_y) = \beta^1\left(\theta_x - i\frac{\Delta}{z_o}\right)\beta^1\left(\theta_y - j\frac{\Delta}{z_o}\right) \tag{3.122}$$

This sampling function is depicted in Fig. 2.27. In Chapter 2 we considered a discrete array of inner products of this sampling function with the continuous field as the "image" of the field. This section describes a method for generating an estimate of the continuous field consistent with the discrete measurements generated from the sampling function.

In a typical measurement system, discrete measurements of the form $f_n = \langle \beta^m(x - n)| f(x)\rangle$ are recorded. This discretization is described by Eqn. (2.38)

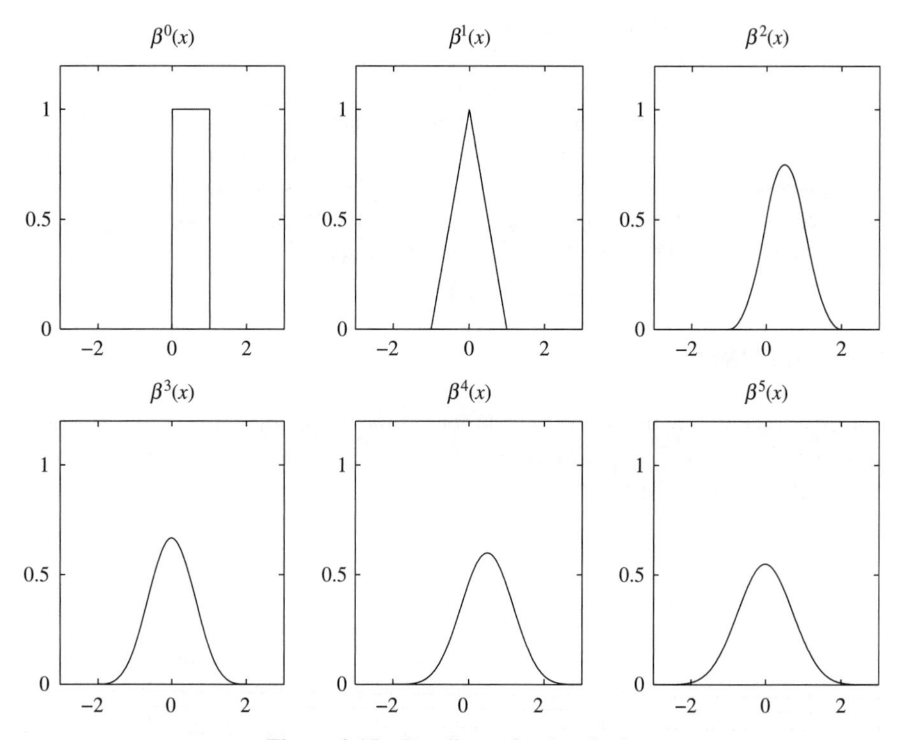

Figure 3.15 B-splines of orders 0–5.

in the coded aperture system. In the following we represent $\beta^m(x-n)$ by the generic sampling function $\phi_n(x)$. Our goal in this section is to describe how to estimate $f(x)$ given the discrete measurements. Since at this point we know nothing about the nature of $f(x)$ besides the measurements, all continuous signals that would result in the same measurements are equally likely.

In general, there exist infinitely many continuous functions that are consistent with the measured data. Choosing which of these functions to use as *the* estimated signal is challenging, but as a general rule we might like to start by choosing an estimate that is at least consistent with the measurements. Consistency determines the part of the estimated signal, $f_{\text{est}}(x)$, that is, in the Hilbert space $V(\phi)$ generated by $\phi(x)$. For each subspace $V(\phi) \subset L^2(\mathbb{R})$ there exists a complementary subspace $V_\perp(\phi)$ such that

$$\begin{aligned} V_\perp(\phi) &\subset L^2(\mathbb{R}) \\ V_\perp(\phi) \cap V(\phi) &= \{\mathbf{0}\} \\ V_\perp(\phi) \cup V(\phi) &= L^2(\mathbb{R}) \end{aligned} \tag{3.123}$$

A function $f(x)$ may be divided into components $f_\phi(x) \in V(\phi)$ and $f_\perp(x) \in V_\perp(\phi)$ such that $f(x) = f_\phi(x) + f_\perp(x)$. The discrete measurements f_n completely determine $f_\phi(x)$ but tell us nothing about $f_\perp(x)$.

A function $f_\phi(x) \in V(\phi)$ may be represented as

$$f_\phi(x) = \sum_{n\in\mathbb{Z}} c_n \phi_n(x) \tag{3.124}$$

In contrast with the sampling theorem and with the Haar wavelet expansion, the expansion coefficients are not samples of f_ϕ or inner products between f_ϕ and the basis vectors. For the B-splines it turns out that we can derive complementary functions $\check{\phi}_n(x)$ for each $\phi_n(x) = \beta^m(x - n)$ such that $\langle \check{\phi}_n | \phi_{n'} \rangle = \delta_{nn'}$. The complementary functions can be used to produce a continuous estimate for $f(x)$ that is completely consistent with the discrete measurements. This interpolated function is

$$f_{\text{est}}(x) = \sum_{n\in\mathbb{Z}} \langle \phi_n | f \rangle \check{\phi}_n(x) \tag{3.125}$$

Given the orthogonality relationship between the sampling functions and the complementary functions, f_{est} is by design consistent with the measurements. We can further state that $f_{\text{est}} = f_\phi$ if the complementary functions are such that $\check{\phi} \in V(\phi)$, in which case there exist discrete coefficients $p(k)$ such that

$$\check{\phi}(x) = \sum_{k\in\mathbb{Z}} p(k)\phi(x - k) \tag{3.126}$$

Using the convolution theorem, the Fourier transform of $\check{\phi}(x)$ is

$$\hat{\check{\phi}}(u) = \hat{\phi}(u)\left[\sum_{k\in\mathbb{Z}} p(k)e^{-i2\pi ku}\right] \tag{3.127}$$

The orthogonality between the dual bases may be expressed as

$$\begin{aligned} \langle \check{\phi}_n | \phi_{n'} \rangle &= \delta_{nn'} \\ &= \sum_{k\in\mathbb{Z}} p(k)a(n' - k - n) \end{aligned} \tag{3.128}$$

where $a(n) = \langle \phi(x) | \phi(x - n) \rangle$. Without loss of generality, we set $n = 0$ and sum both sides of Eqn. (128) against the discrete kernel $e^{-i2\pi n'u}$ to obtain

$$\begin{aligned} \sum_{n'\in\mathbb{Z}} \delta_{0n'} e^{i2\pi n'u} &= 1 \\ &= \sum_{k\in\mathbb{Z}} \sum_{n'\in\mathbb{Z}} p(k)a(n' - k)e^{-i2\pi n'u} \\ &= \left[\sum_{k\in\mathbb{Z}} p(k)e^{-i2\pi ku}\right]\left[\sum_{n''\in\mathbb{Z}} a(n'')e^{-i2\pi n''u}\right] \end{aligned} \tag{3.129}$$

where we use the substitution of variables $n'' = n' - k$.

Poisson's summation formula is helpful in analyzing the sums in Eqn. (3.129). The summation formula states that for $g(x) \in L^1(\mathbb{R})$

$$\sum_{n\in\mathbb{Z}} g(n)e^{-i2\pi nu} = \sum_{k\in\mathbb{Z}} \hat{g}(u+k) \tag{3.130}$$

where $\hat{g}(u)$ is the Fourier transform of $g(x)$. To prove the summation formula, note that

$$h(u) = \sum_{k\in\mathbb{Z}} \hat{g}(u+k) \tag{3.131}$$

is periodic in u with period 1. The Fourier series coefficients for $h(u)$ are

$$\begin{aligned}\dot{h}_n &= \int_0^1 h(u)e^{2\pi inu}\,du \\ &= \sum_{k\in\mathbb{Z}} \int_0^1 \hat{g}(u+k)e^{2\pi inu}\,du \\ &= \sum_{k\in\mathbb{Z}} \int_k^{k+1} \hat{g}(u)e^{2\pi inu}\,du \\ &= \int_{-\infty}^{\infty} \hat{g}(u)e^{2\pi inu}\,du \\ &= g(n)\end{aligned} \tag{3.132}$$

Since $a(x)$ is the autocorrelation of ϕ, its Fourier transform is $|\hat{\phi}(u)|^2$. Thus by the Poisson summation formula

$$\sum_{n\in\mathbb{Z}} a(n)e^{-i2\pi nu} = \sum_{k\in\mathbb{Z}} |\hat{\phi}(u+k)|^2 \tag{3.133}$$

Reconsidering Eqn. (3.129), we find

$$\begin{aligned}\sum_{k\in\mathbb{Z}} p(k)e^{-i2\pi ku} &= \frac{1}{\sum_{n\in\mathbb{Z}} a(n)e^{-i2\pi nu}} \\ &= \frac{1}{\sum_{k\in\mathbb{Z}} |\hat{\phi}(u+k)|^2}\end{aligned} \tag{3.134}$$

Substitution in Eqn. (3.127) yields

$$\hat{\hat{\phi}}(u) = \frac{\hat{\phi}(u)}{\sum\limits_{k \in \mathbb{Z}} |\hat{\phi}(u+k)|^2} \tag{3.135}$$

We can evaluate Eqn. (3.135) to determine $\hat{\hat{\phi}}(u)$ and $\check{\phi}(x)$ if $\sum_{k \in \mathbb{Z}} |\hat{\phi}(u+k)|^2$ is finite. The requirement that there exist positive constants A and B such that

$$A \leq \sum_{k \in \mathbb{Z}} |\hat{\phi}(u+k)|^2 \leq B \tag{3.136}$$

is the defining feature of a *Riesz basis*. A Riesz basis may be considered as a generalization of an orthonormal basis. In the case that $\sum_{k \in \mathbb{Z}} |\hat{\phi}(u+k)|^2 = 1$, Eqn. (135) reduces to $\hat{\hat{\phi}}(u) = \hat{\phi}(u)$ and an orthonormal basis may be obtained.

The Fourier transform of the mth-order B-spline is

$$\begin{aligned}\hat{\beta}^m(u) &= [\text{sinc}(u)]^{(m+1)} e^{-i\pi\xi u} \\ &= \left[\hat{\beta}^0(u)\right]^{(m+1)} e^{i\pi(m+1-\xi)u}\end{aligned} \tag{3.137}$$

where $\xi = 0$ if m is odd and $\xi = 1$ if m is even. For the B-spline basis, we obtain

$$Q_m(u) = \sum_{k \in \mathbb{Z}} |\hat{\phi}(u+k)|^2 = \sum_{k \in \mathbb{Z}} |\text{sinc}(u+k)|^{2(m+1)} \tag{3.138}$$

Since the zeroth-order B-spline produces an orthogonal basis, we know that $Q_0(u) = 1$. For higher orders we note that $|\text{sinc}(u+k)|^{2(m+1)} \leq |\text{sinc}(u+k)|^2$, meaning that $Q_m(u) \leq Q_o(u)$. Thus, $0 < Q_m(u) < 1$ and the B-spline functions of all orders satisfy the Riesz basis condition.

In contrast with the B-splines themselves, the complementary functions $\check{\phi}(x)$ do not have finite support. It is possible, nevertheless, to estimate $\check{\phi}(x)$ over a finite interval for each B-spline order by numerical methods. Estimation of $Q_m(u)$ from Eqn. (3.138) is the first step in numerical analysis. This objective is relatively easily achieved because $Q_m(u)$ is periodic with period 1 in u. Evaluation of the sum over the first several thousand orders for closely spaced values of $0 \geq u \leq 1$ takes a few seconds on a digital computer.

Given $Q_m(u)$, we may estimate $\check{\phi}(x)$ by using a numerical inverse Fourier transform of Eqn. (3.135) or by calculating $p(k)$ from Eqn. (3.134). Since $p(k)$ must be real and since $Q_m(u)$ is periodic, we obtain

$$p(k) = \int_0^1 \frac{\cos(2\pi ku)}{Q_m(u)} du \tag{3.139}$$

Estimation of $p(k)$ was the approach taken to calculate $\check{\phi}(x)$ for Fig. 3.16.

Given $\phi(x) = \beta^m(x - n)$ and $\check{\phi}(x)$, we can calculate $f_\phi(x)$ for target functions. For example, Fig. 3.17 shows the signals of Figs. 3.8 and 3.9 projected onto the $V(\phi)$ subspaces for B-splines of orders 0–3. Higher-order splines smoothly represent signals with higher-order local polynomial curvature. Note that higher-order splines are not more localized than the lower-order functions, however, and thus do not immediately translate into higher signal resolution. Notice also the errors at the edges of the signal windows in Fig. 3.17. These arise from the *boundary conditions* used to truncate the infinite time signal $f(x)$. In the case of these figures, $f(x)$ was assumed to be periodic in the window width, such that sampling and interpolation functions extending beyond the window could be wrapped around the window.

The interpolated signals plotted in Fig. 3.17 are the projections $f_\phi(x) \in V(\phi)$ of $f(x)$ onto the corresponding subspaces $V(\phi)$. The consistency requirement designed into the interpolation strategy means that these functions, despite their obvious discrepancies relative to the actual signals, would yield the same sample projections. Corrections that map the interpolated signals back onto the actual signal lie in $V_\perp(\phi)$. Strategies for sampling and interpolation to take advantage of known constraints on $f(x)$ to so as to infer correction components $f_\perp(x)$ are discussed in Chapter 7.

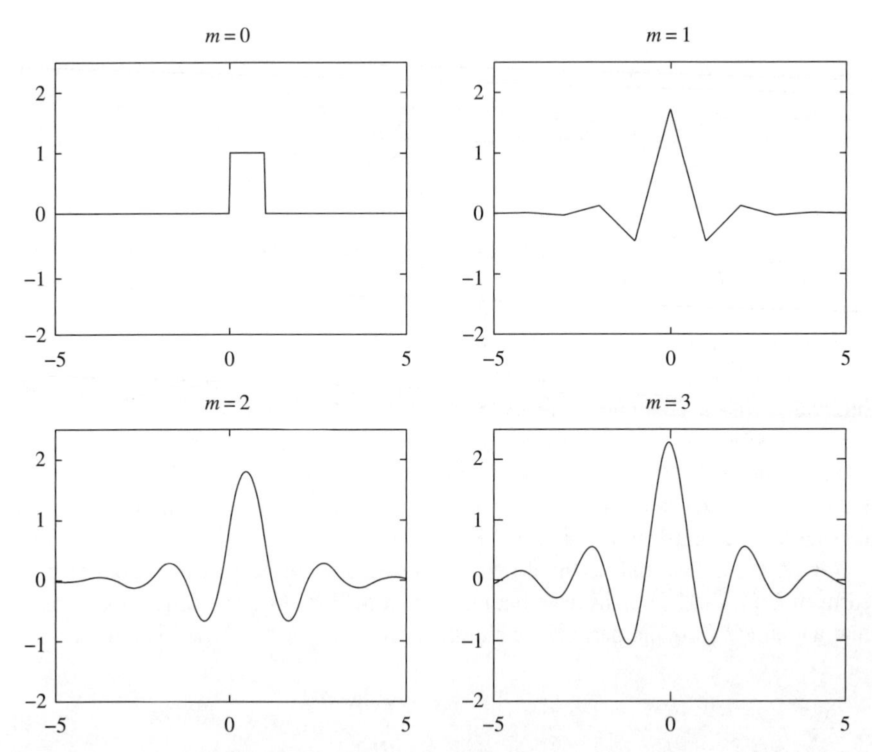

Figure 3.16 Complementary interpolation functions $\check{\phi}(x)$ for the B-splines of orders 0–3. The zeroth-order B-spline is orthonormal such that $\check{\phi}(x) = \beta^0(x)$.

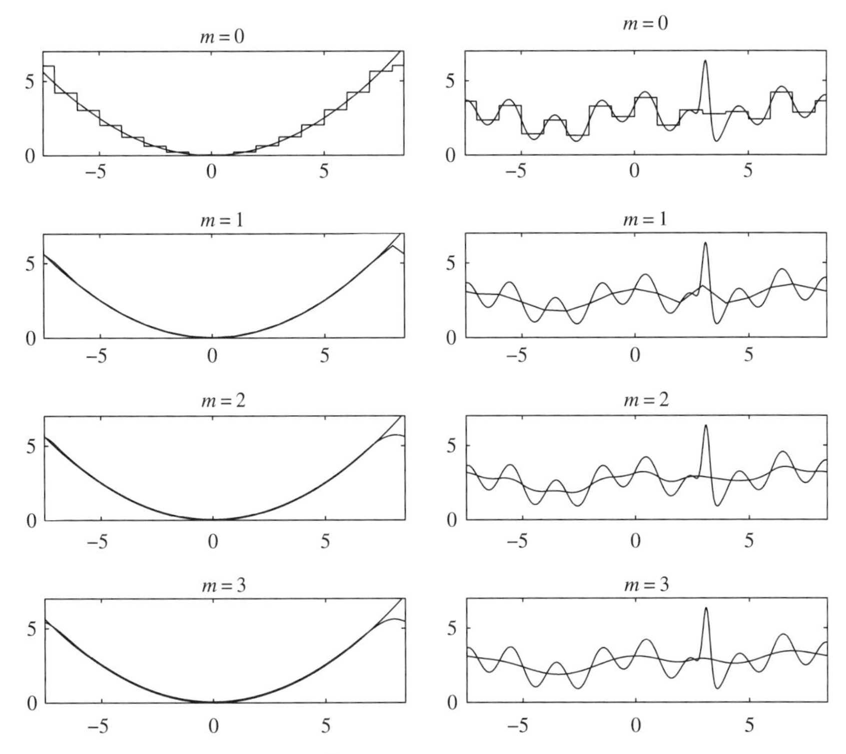

Figure 3.17 Projection of $f(x) = x^2/10$ and the signal of Fig. 3.9 onto the $V(\phi)$ subspace for B-splines of orders 0–3.

Use of Eqn. (3.125) to estimate $f(x)$ is somewhat unfortunate given that $\check{\phi}_n(x)$ does not have finite support. A primary objection to the use of the original sampling theorem [Eqn. (3.92)] for signal estimation is that sinc(x) has infinite support and decays relatively slowly in amplitude. While $\phi(x)$ is better behaved for low-order B-splines, it is is still true that accurate estimation of $f(x)$ may be computationally expensive if a large window is used for the support of $\check{\phi}$. As the order of the B-spline tends to infinity, $\check{\phi}(x)$ converges on sinc(x) [235]. If we remove the requirement that $\check{\phi}(x) \in V(\phi)$, it is possible to generate a *biorthogonal* dual basis for $\beta^m(x)$ with compact support [49]. The compactly supported biorthogonal wavelets in this case introduce a complementary subspace $\check{V}$ spanned by $\check{\phi}(x)$.

The goal of the current section has been to consider how one might use a set of discrete B-spline inner products to estimate a continuous signal. This problem is central to imaging and optical signal analysis. We have already encountered it in the coded aperture and tomographic systems considered in Chapter 2, and we will encounter it again in the remaining chapters of the text. We leave this problem for now, however, to consider the use of sampling functions and multiscale representations in signal and system analysis. One may increase the resolution and fidelity of the reconstructions in Fig. 3.17 by increasing the resolution of the sampling function in a manner similar to the wavelet approach taken in Section 3.8.

3.10 WAVELETS

As predicted in the Section 3.1, this chapter has developed three distinct classes of mathematics: transformation tools, sampling tools, and analysis tools. In the first several sections we considered fields and field transformations. We have just completed three sections focusing on sampling. Section 3.9 describes a method for representing a function $f(x)$ on the space $V(\phi)$ spanned by the scaling function $\phi(x) = \beta^m(x)$. This section extends our consideration of B-splines to wavelets, similar to our extension of Haar analysis in Section 3.8. We have already considered mathematical bases suitable for field analysis in terms of the Fourier transform and Hermite–Gaussian functions. In fact, many functional families could be used to analyze fields. The choice of which family to use depends on which family arises naturally in the physical specification of the problem (e.g., Laguerre–Gaussian functions arise naturally in the specification of cylindrically symmetric fields), which family arises at sampling interfaces, and which family enables the most computationally efficient and robust analysis of field transformations.

Wavelet theory is a broad and powerful branch of mathematics, and the student is well advised to consult standard courses and texts for deeper understanding [53,164]. Wavelets often describe images and other natural signals well. The intuitive match between wavelets and images arises from the assumption that "features" in natural signals tend to cluster, meaning that higher resolution is desirable in the vicinity of a feature than elsewhere in the signal. Multiscale clustering enables wavelet representations to estimate signals with fewer samples than might be used with uniform regular sampling. Under the Whittaker–Shannon sampling strategy, functional samples are distributed uniformly in space even in regions with no significant image features. Wavelets enable samples to be dynamically assigned to regions with interesting features. This dynamic resource allocation is the basis of natural signal compression.

B-splines may be used to generate semiorthogonal bases as in Section 3.9, biorthogonal spaces and orthogonal wavelet bases. As before, we imagine a hierarchy of spaces

$$\{0\} \subset \cdots \subset V_2 \subset V_1 \subset V_o \subset V_{-1} \subset V_{-2} \subset \cdots \subset L^2(\mathbb{R}) \tag{3.140}$$

Semiorthogonal bases are spanned by sets of functions that are not themselves orthogonal but are orthogonal to a complementary set of functions. Biorthogonal bases generate complementary spaces spanned by complementary sets of functions. Orthogonal bases generate a single hierarchy of spaces spanned by a single set of orthogonal functions. We have already encountered an orthogonal wavelet basis in the form of the Haar wavelets of Section 3.8. In this section we extend the Haar analysis to orthogonal bases based on higher-order B-splines.

The orthonormal basis for spaces spanned by discretely shifted B-splines was introduced by Battle [15] and Lemarie [150]. For the Battle–Lemarie basis, $\phi(x)$ is a scaling function on the space $V(\beta^m(x))$ spanned by the mth-order B-spline. Since $\phi(x) \in V(\beta^m(x))$ there exist expansion coefficients $p[n]$ such that

$$\phi(x) = \sum p[n]\beta^m(x - n) \tag{3.141}$$

The Fourier transform of Eqn. (3.141) yields

$$\hat{\phi}(u) = \hat{p}(u)\hat{\beta}^m(u) \tag{3.142}$$

Our goal is to select $\phi(x)$ to be an orthonormal scaling function such that

$$\begin{aligned} \langle \phi(x - n), \phi(x - m) \rangle &= \int_{-\infty}^{\infty} \phi^*(x - n)\phi(x - m)dx \\ &= \delta_{nm} \end{aligned} \tag{3.143}$$

We may apply the Poisson summation formula as in Section 3.9 to derive a simple identity from Eqn. (3.143). Again letting $a(x) = \langle \phi(x'), \phi(x' - x) \rangle$, we note from Eqn. (3.130) that

$$\sum_{n \in \mathbb{Z}} a(n)e^{-i2\pi nu} = \sum_{k \in \mathbb{Z}} \hat{a}(u + k) \tag{3.144}$$

For an orthonormal scaling function, however, $\sum_{n \in \mathbb{Z}} a(n)e^{-i2\pi nu} = 1$ and $\hat{a}(u) = |\hat{\phi}(u)|^2$, which yields the identity for orthonormal scaling functions

$$\sum_k |\hat{\phi}(u + k)|^2 = 1 \tag{3.145}$$

Referring to Eqn. (142), we see that $\hat{\phi}(u)$ satisfies Eqn. (145) if we select

$$\hat{p}(u) = \frac{1}{\sqrt{\sum_k |\hat{\beta}^m(u + k)|^2}} \tag{3.146}$$

where $\hat{p}(u)$ is finite and well defined because the B-splines form a Riesz basis, as discussed in Section 3.9. Since $\hat{p}(u)$ is periodic with period 1 in u, it generates a discrete series $p[n]$ for use in Eqn. (3.141). Substituting $\hat{\beta}^m(u)$ from Eqn. (3.137) in Eqns. (3.146) and (3.142) yields

$$\hat{\phi}(u) = \frac{e^{-i\pi\xi u}}{u^{m+1}\sqrt{S_{2m+2}(u)}} \tag{3.147}$$

where

$$S_n(u) = \sum_{k \in \mathbb{Z}} \frac{1}{(u+k)^n} \tag{3.148}$$

We know that the $m = 0$ spline produces the Haar scaling function

$$\hat{\phi}^0(u) = e^{-i\pi u} \frac{\sin(\pi u)}{\pi u} \tag{3.149}$$

Comparing Eqns. (3.147) and (3.149), we see that

$$S_2(u) = \frac{\pi^2}{\sin^2(\pi u)} \tag{3.150}$$

Higher orders of $S_n(u)$ are obtained by noting that $S_{n+1}(u) = -S_n'(u)/n$. This yields

$$S_4(u) = \frac{\pi^4(2 + \cos(2\pi u))}{6\,\sin^4(\pi u)} \tag{3.151}$$

$$S_6(u) = \frac{\pi^6(33 + 26\,\cos(2\,\pi u) + \cos(4\pi u))}{180\,\sin^6(\pi u)} \tag{3.152}$$

and

$$S_8(u) = \frac{\pi^8(1208 + 1191\,\cos(2\pi u) + 120\,\cos(4\pi u) + \cos(6\pi u))}{10{,}080\,\sin^8(\pi u)} \tag{3.153}$$

To satisfy the requirement that $V_j \subset V_{j-1}$, we require that $\phi_{j,n}(x) \in V_{j-1}$, which means that there exist expansion coefficients $h[n]$ such that

$$\frac{1}{\sqrt{2^j}} \phi\left(\frac{x}{2^j} - n\right) = \sum_{n'} \frac{1}{\sqrt{2^{j-1}}} h[n' - n] \phi\left(\frac{x}{2^{j-1}} - n'\right) \tag{3.154}$$

Equation (3.154) reduces without loss of generality to

$$\frac{1}{\sqrt{2}} \phi \frac{x}{2} = \sum_n h[n]\phi(x - n) \tag{3.155}$$

The Fourier transform of Eqn. (155) yields

$$\sqrt{2}\hat{\phi}(2u) = \hat{h}(u)\hat{\phi}(u) \tag{3.156}$$

where

$$\hat{h}(u) = \sum_{n \in \mathbb{Z}} h[n] e^{-2\pi i n u} \tag{3.157}$$

For the Battle–Lemarie scaling functions

$$\begin{aligned} \hat{h}(u) &= \sqrt{2}\frac{\hat{\phi}(2u)}{\hat{\phi}(u)} \\ &= e^{-i\pi\xi u}\sqrt{\frac{S_{2m+2}(u)}{2^{2m+1}S_{2m+2}(2u)}} \end{aligned} \tag{3.158}$$

As with the Haar scaling function, we are interested in obtaining orthogonal wavelets spanning the spaces W_j such that $V_{j-1} = V_j \oplus W_j$. Such wavelets are immediately obtained using the *conjugate mirror filter* $\hat{h}(u)$. The wavelet corresponding to the scaling function $\phi(x)$ has the Fourier transform

$$\hat{\psi}(u) = \frac{1}{\sqrt{2}} e^{-i\pi u} \hat{h}^*\left(\frac{u+1}{2}\right) \hat{\phi}\left(\frac{u}{2}\right) \tag{3.159}$$

The Battle–Lemarie scaling function and wavelet can be reconstructed by inverse Fourier transforming Eqns. (3.147) and (3.159). These functions satisfy the same orthogonality and scaling rules as the Haar wavelets discussed earlier; specifically

$$\phi_{j,n}(x) = \frac{1}{\sqrt{2^j}} \phi\left(\frac{x}{2^j} - n\right) \tag{3.160}$$

$$\psi_{j,n}(x) = \frac{1}{\sqrt{2^j}} \psi\left(\frac{x}{2^j} - n\right) \tag{3.161}$$

$$\langle \phi_{j,n} | \phi_{j',n'} \rangle = \delta_{jj'} \delta_{nn'} \tag{3.162}$$

$$\langle \psi_{j,n} | \psi_{j',n'} \rangle = \delta_{jj'} \delta_{nn'} \tag{3.163}$$

$$\langle \psi_{j,n} | \phi_{j',n'} \rangle = 0 \tag{3.164}$$

As with the Haar wavelets, the Battle–Lemarie functions span $L^2(\mathbb{R}^2)$ in the hierarchy of spaces described by Eqn. (3.140). The Battle–Lemarie wavelets are presented here to provide an accessible introduction to wavelet theory. Many other wavelet families have been developed [164]; the selection of which family to use for a particular class of signals is application-specific. Some wavelets are attractive because they have compact support, which the Shannon wavelet famously does not. Other wavelets, such as the Haar and B-splines, arise naturally from physical or system design considerations. In still other cases, a particular basis may prove more amenable to compact support of a particular signal class.

PROBLEMS

3.1 *Fourier Uncertainty*. Show that for $f(x) = ae^{-b(x-x_o)^2}$

$$\sigma_f^2 \sigma_{\hat{f}}^2 = \frac{1}{16\pi^2} \tag{3.165}$$

3.2 *Fourier Rotation*. Derive Eqn. (3.39).

3.3 *Fresnel Identities*:

(a) Derive Eqn. (3.63).

(b) Derive Eqn. (3.64).

3.4 *Hermite–Gaussian Eigenfunctions*. The Hermite polynomial $H_n(x)$ is defined as

$$H_n(x) = (-1)^n e^{x^2} \frac{d^n}{dx^n} e^{-x^2} \tag{3.166}$$

Defining

$$\phi_n(x) = e^{-\pi x^2} H_n(\sqrt{2\pi}x) \tag{3.167}$$

show that for $n > 0$

$$\sqrt{2\pi}\phi_n(x) = 2\pi x \phi_{n-1}(x) - \frac{d}{dx}\phi_{n-1}(x) \tag{3.168}$$

Combine this relationship with Eqns. (3.13) and (3.57) to show by recursion that

$$\mathcal{F}\{\phi_n(x)\} = i^n \phi_n(u) \tag{3.169}$$

3.5 *One-dimensional Numerical Analysis*:

(a) Plot $\sin(2\pi ux)$ on [0, 1] using 1024 uniformly spaced samples for $u = 16,32,64,128,256$. At what point does aliasing become significant? Can you describe the structure of the aliased signal?

(b) Plot the discrete Fourier transform of $\sin(2\pi ux)$ on [0, 1] using 1024 uniformly spaced samples for $u = 16, 32, 64$. Label the plot in frequency units. What is the width of the Fourier features that you observe? What causes this width?

(c) Plot the discrete Fourier transform of $\beta^0(x)\sin(2\pi ux)$ on [−1.5, 2.5] using 4096 uniformly spaced samples for $u = 16, 32, 64$. Label the plot in frequency units and explain the plot.

3.6 *Fourier Analysis of a Hermite–Gaussian*:

(a) Plot the Hermite–Gaussian $\phi_5(x)$ over the range of the function.

(b) Plot $\tilde{\phi}_{5_\tau}(x)$ over a representative range of τ.

3.7 *Transformations*:

(a) Plot the Fourier transform of the function $e^{-2(x-.1)^2}\,\mathrm{circ}(x/0.3)\cos(20\pi x)$. Mark units on your plot.

(b) Plot the Fresnel transform for $\tau = 1, 3, 10$ for the function from part (**a**).

3.8 *Fresnel Transformation of the Hermite–Gaussian Functions*. Prove Eqn. (3.76).

3.9 *Fresnel Transformation of the Laguerre–Gaussian Functions*. Use the convolution theorem and the fast Fourier transformation to numerically calculate the Fresnel transformation of the Laguerre–Gaussian modes for m, n equal to 0,0, 1,0, 1,1, 2,0, 2,1, and 4,3 for $\tau = 0.5$, $\tau = 1$, and $\tau = 2$. Use your computational result and the analytic result given by Eqn. (3.83) to plot the absolute value and phase of the mode distribution at $\tau = 0$ and for the transform values of τ in each case. Submit your code, plots, and comments regarding features of the modes or discrepancies between the computational methods.

3.10 *Haar Analysis*:

(a) Generate and plot a function of similar complexity to $f(x)$ in Fig. 3.9.

(b) Replicate Figs. 3.9, 3.12, and 3.17 for your function.

3.11 *2D Wavelet Analysis*. Replicate Figs. 3.13 and 3.14 for an image of your choosing.

3.12 *Spline Interpolation*:

(a) Show that a one-dimensional pinhole imaging system produces measurements

$$g_n = f_n = \int f(x)\beta^1\left(\frac{x - n\Delta}{\Delta}\right) dx \tag{3.170}$$

(b) Plot the values of f_n for

$$f(x) = \cos\left(2\pi\frac{x}{5\Delta}\right)e^{-(x/30\Delta)^2} \tag{3.171}$$

(c) Use Eqn. (3.125) to estimate $f(x)$ from f_n. Plot $f(x)$ at a sampling period of $\Delta/10$. Compare your plots.

3.13 *Wavelets*. For the Battle–Lemarie bases, plot the scaling function $\phi(x)$ and wavelet $\psi(x)$ for orders 0, 1, 2, and 3.

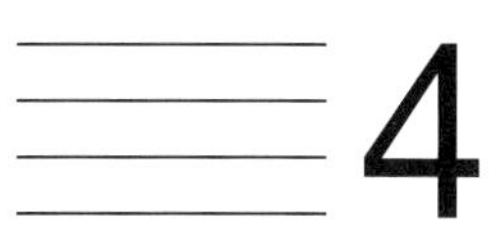

WAVE IMAGING

The physical phenomenon called *diffraction* is of the utmost importance in the theory of optical imaging systems.

—J. W. Goodman [100]

4.1 WAVES AND FIELDS

The optical field is an electromagnetic field. The physical nature of the field is determined by the laws of electromagnetic propagation and by quantum mechanical and thermal laws describing the interaction between the field and materials. In the design and analysis of optical systems we consider

- How the field is generated. Common mechanisms include
  - Thermal radiation generated, for example, by the Sun, a flame, or an incandescent lightbulb
  - Electrical discharge by gases such as neon or mercury vapor
  - Fluorescence
  - Electrical recombination in semiconductors

While we do not consider light generation in detail in this text, differences in the *coherence* properties of the source are central to our discussion. Coherence theory, which relates the electromagnetic nature of the field to statistical properties of quantum (e.g., photonic) processes, is the focus of Chapter 6.

- How the field is detected. The field may be detected by optically induced chemical, physical, thermal, and electronic effects. Optoelectronic detection interfaces for imaging and spectroscopy are the focus of Chapter 5.

Optical Imaging and Spectroscopy. By David J. Brady

- How the field propagates and how propagating fields are modulated by materials. Field propagation is described by the Maxwell equations for electromagnetic waves, and field–matter interactions are described by materials equations. The electromagnetic description of optical waves and optical interactions is the focus of this chapter.

In view of the peculiarly quantum mechanical nature of optical field generation and detection, it is important to understand that the conventional electromagnetic field of the Maxwell equations is not a sufficient description of optical fields. The description derived in this chapter provides a basis for optical analysis, but complete understanding of optical field propagation and field properties must incorporate the detection and coherence processes discussed in Chapters 5 and 6. In short, the student must understand the next three chapters as a group to have a vision for the peculiar and beautiful nature of optical fields.

4.2 WAVE MODEL FOR OPTICAL FIELDS

The Maxwell equations for electromagnetic propagation are

$$\nabla \times \mathbf{E} = -\frac{\partial}{\partial t}\mathbf{B} \tag{4.1}$$

$$\nabla \times \mathbf{H} = \mathbf{J} + \frac{\partial}{\partial t}\mathbf{D} \tag{4.2}$$

$$\nabla \cdot \mathbf{D} = \rho \tag{4.3}$$

$$\nabla \cdot \mathbf{B} = 0 \tag{4.4}$$

where $\mathbf{E}$ is the *electric field*, $\mathbf{D}$ is the *electric displacement*, $\mathbf{B}$ is the *magnetic induction*, $\mathbf{H}$ is the *magnetic field*, $\mathbf{J}$ is the *current density*, and ρ is the *charge density*. Equation (4.1), which expresses the tendency of a moving magnet to generate an electromotive force, is called *Faraday's law*. Equation (4.2), which expresses the tendency of an electric current to generate a magnetic flux, is called *Ampere's law*. The electric displacement current $\partial\mathbf{D}/\partial t$ in Ampere's law was added by Maxwell as a means of explaining electrodynamics. Equation (4.3), which expresses the Coulomb attraction of electromagnetic charge, is called *Gauss' law*. Equation (4.4), is called *Gauss' law for magnetism* and expresses the absence of magnetic monopoles.

The fields are further related by the material equations

$$\mathbf{D} = \varepsilon_0\mathbf{E} + \mathbf{P} \tag{4.5}$$

$$\mathbf{B} = \mu_0\mathbf{H} + \mathbf{M} \tag{4.6}$$

where $\mathbf{P}$ is the *polarization* of the material and $\mathbf{M}$ is the *magnetization*. In most optical materials, $\mathbf{M} = 0$ and $\mathbf{P}$ is a function of $\mathbf{E}$. The simplest and most common case is the linear dielectric relationship

$$\mathbf{P} = \varepsilon_0 \chi_e \mathbf{E} \tag{4.7}$$

such that

$$\mathbf{D} = \varepsilon \mathbf{E} \tag{4.8}$$

where

$$\varepsilon = (1 + \chi_e)\varepsilon_0 \tag{4.9}$$

Since charge dynamics at optical frequencies are described by quantum mechanical processes that cannot be accurately analyzed by continuous models, the space charge density ρ and the current density $\mathbf{J}$ are generally neglected in optical analysis. A nonzero current density is sometimes applied to formally account for optical absorption. We also note that $\mathbf{E}$ and $\mathbf{D}$ need not be collinear, meaning that ε is in general tensor-valued. Materials in which ε is a scalar are called *isotropic*. Optical glasses are isotropic, but optical crystals are often anisotropic. While most of the optical systems discussed in this text utilize isotropic materials, we consider the application of anisotropic materials to tunable filters in Section 9.7.

Using the material relations to eliminate $\mathbf{B}$ and $\mathbf{D}$ from the Faraday and Ampere relationships yields

$$\nabla \times \mathbf{E} = -\mu_0 \frac{\partial}{\partial t} \mathbf{H} \tag{4.10}$$

$$\nabla \times \mathbf{H} = \varepsilon \frac{\partial}{\partial t} \mathbf{E} \tag{4.11}$$

Operating on Eqns. (4.10) and (4.11) with the curl yields the wave equations

$$\nabla \times \nabla \times \mathbf{E} = -\mu_0 \varepsilon \frac{\partial^2}{\partial t^2} \mathbf{E} \tag{4.12}$$

$$\nabla \times \nabla \times \mathbf{H} = -\mu_0 \varepsilon \frac{\partial^2}{\partial t^2} \mathbf{H} \tag{4.13}$$

The equations are reduced to a simpler form by the vector identity

$$\nabla \times \nabla \times \mathbf{A} = \nabla(\nabla \cdot \mathbf{A}) - \nabla^2 \mathbf{A} \tag{4.14}$$

From Gauss' law we know that

$$\nabla \cdot (\varepsilon \mathbf{E}) = \mathbf{E} \cdot \nabla \varepsilon + \varepsilon \nabla \cdot \mathbf{E} = 0 \tag{4.15}$$

where we have assumed for the moment that ε is a scalar. We can reexpress Eqn. (4.15) as

$$\nabla \cdot \mathbf{E} = -\mathbf{E} \cdot \nabla \log(\varepsilon) \tag{4.16}$$

Substituting Eqn. (4.16) in the wave equation yields

$$\nabla^2\mathbf{E} - \nabla(\mathbf{E}\cdot\nabla\log(\varepsilon)) = \mu_0\varepsilon\frac{\partial^2}{\partial t^2}\mathbf{E} \tag{4.17}$$

A medium in which $\nabla\varepsilon = 0$ is termed *homogeneous*. Most optical dielectrics (i.e., transparent glasses and crystals and liquids) are homogeneous. Optically interesting inhomogeneous media include photonic crystals, optical fiber, graded-index lenses, and volume holograms.

In isotropic homogeneous media, the wave equations reduce to

$$\nabla^2\mathbf{E} - \mu_0\varepsilon\frac{\partial^2}{\partial t^2}\mathbf{E} = 0 \tag{4.18}$$

$$\nabla^2\mathbf{H} - \mu_0\varepsilon\frac{\partial^2}{\partial t^2}\mathbf{H} = 0 \tag{4.19}$$

4.3 WAVE PROPAGATION

Solutions to Eqn. (4.12) take many forms, but the linearity of the equations and the natural harmonic nature of optical sources make harmonic solutions particularly attractive. The basic harmonic solution is the plane wave described by

$$\mathbf{E} = \mathbf{E_0}e^{i2\pi(\nu t - \mathbf{u}\cdot\mathbf{r})} \tag{4.20}$$

Substituting this solution into Eqn. (4.13) yields

$$\mathbf{u}\times\mathbf{u}\times\mathbf{E_0} = -\mu_0\varepsilon\nu^2\mathbf{E_0} \tag{4.21}$$

If ε is a scalar, this equation has solutions $\mathbf{E_0}$ only if $|\mathbf{u}|^2 = \mu_0\varepsilon\nu^2$. Allowing for the possibility that ε is a tensor, solutions correspond to values of $\mathbf{u}$ such that

$$\left|\mathbf{u}\times\mathbf{u}\times + \mu_0\varepsilon\nu^2\right| = 0 \tag{4.22}$$

Equation (4.22) reduces the range of $\mathbf{u}$ from three dimensions to two. The surface defined by this equation is called the *wave normal surface*. In isotropic materials, the wave normal surface is a sphere in $\mathbf{u}$ space of radius $\nu\sqrt{\mu_0\varepsilon}$, as sketched in Fig. 4.1. In anisotropic materials (crystals), the wave normal surface splits into two sheets, so that there are two solutions for $\mathbf{u}$ in almost every direction. The relationship between $\mathbf{u}$ and ν (or radial coordinates $\mathbf{k} = 2\pi\mathbf{u}$ and $\omega = 2\pi\nu$) expressed by the wave normal surface is called the *dispersion relationship*, which reduces the four-dimensional $\mathbf{u}$, ν space to a three-dimensional manifold of allowed solutions for wave propagation. Each solution for $\mathbf{u}$ corresponds to an eigenvector $\mathbf{E_0}$. The direction of $\mathbf{E_0}$ is the polarization. For the isotropic case, two possible polarizations exist for each $\mathbf{u}$. In the general case, each eigenvector corresponds to a different value of $\mathbf{u}$.

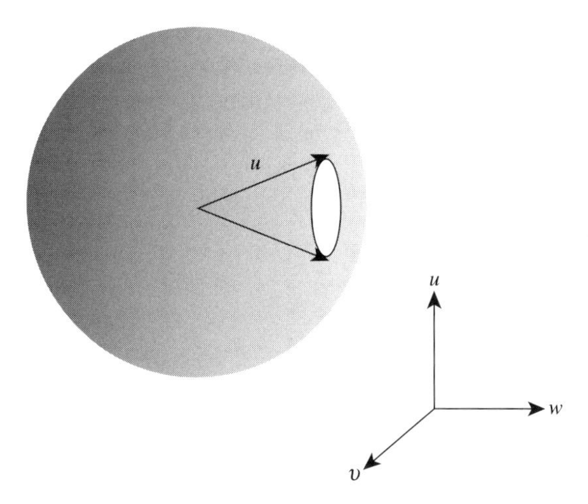

Figure 4.1 The wave normal surface in free space.

The primary attraction of plane wave analysis is that more general solutions can be expressed as superpositions of plane waves. In the remainder of this chapter we restrict our attention to isotropic materials, in which case ε is a scalar and $|\mathbf{u}|^2 = \mu_0 \varepsilon \nu^2$. A general solution to Eqn. (4.18) in this case is

$$\mathbf{E}(\mathbf{r}, t) = \iiint F(u, v, \nu)\mathbf{p}(u, v, \nu)e^{-i2\pi\left(\nu t - ux - vy - \sqrt{\mu\varepsilon\nu^2 - u^2 - v^2}z\right)} du\, dv\, d\nu \tag{4.23}$$

where $(u, v, w = \sqrt{\mu\varepsilon\nu^2 - u^2 - v^2})$ corresponds to $\mathbf{u}, \mathbf{p}(u, v, \nu) \cdot \mathbf{u} = 0$ and $|\mathbf{p}(u, v, \nu)| = 1$.

As discussed in some detail in subsequent chapters, the space–time field $\mathbf{E}(\mathbf{r},t)$ is not generally measurable at optical frequencies. It will take us a while to introduce functions that are measurable over the optical band; for the present purposes we prefer to analyze the field using the temporal Fourier transform of $\mathbf{E}(\mathbf{r}, t)$:

$$\mathbf{E}(\mathbf{r}, \nu) = \int \mathbf{E}(\mathbf{r}, t)e^{i2\pi\nu t} dt \tag{4.24}$$

In this chapter ν is treated as an implicit variable in the function $\mathbf{E}(\mathbf{r}) = \mathbf{E}(\mathbf{r}, \nu)$. We according drop the harmonic time dependence $e^{-i2\pi\nu t}$ from Eqn. (4.23) and describe spatial distribution of the field amplitude according to

$$\mathbf{E}(\mathbf{r}) = \iint F(u, v)\mathbf{p}(u, v)e^{i2\pi\left(ux + vy + \sqrt{\mu\varepsilon\nu^2 - u^2 - v^2}z\right)} du\, dv \tag{4.25}$$

In the remainder of this chapter we also assume that the field propagates *paraxially*. The paraxial approximation consists of the assumption that values of u and v for which $|F(u, v)|$ is nonzero lie on a compact window on the wave normal sphere centered on the w axis, as illustrated in Fig. 4.1. This window is centered on the z axis, such that $w \gg u, v$ over the full spatial bandwidth. This means that the polarization vector $\mathbf{p}(u, v)$ is nearly parallel to the (x, y) plane over the entire spatial bandwidth.

In an isotropic material $\mathbf{p}(u, v)$ may be represented on any basis orthogonal to $\mathbf{u}$. We select as an example a basis in which one of the polarization vectors is also orthogonal to the y axis. The resulting orthonormal basis for $\mathbf{p}(u, v)$ is

$$
\begin{aligned}
\mathbf{p_x} &= -\kappa\left(u\mathbf{i_x} + v\mathbf{i_y} + w\mathbf{i_z}\right) \times \mathbf{i_y} \\
&= \frac{\lambda w\mathbf{i_x} - \lambda u\mathbf{i_z}}{\sqrt{1-\lambda^2 v^2}} \\
\mathbf{p_y} &= -\left(\lambda u\mathbf{i_x} + \lambda v\mathbf{i_y} + \lambda w\mathbf{i_z}\right) \times \mathbf{p_x} \\
&= \frac{(1-\lambda^2 v^2)\mathbf{i_y} + \lambda^2 uv\mathbf{i_x} + \lambda^2\mathbf{i_z}}{\sqrt{1-\lambda^2 v^2}}
\end{aligned}
\tag{4.26}
$$

where κ is a normalization constant and $\lambda = 1/\nu\sqrt{\mu\varepsilon}$. Substituting the polarization into Eqn. (4.25) separates the field into polarized components

$$
\begin{aligned}
\mathbf{f}_x(\mathbf{r}) &= \iint F_x(u, v)\mathbf{p}_x e^{i2\pi\left(ux+vy+\sqrt{1/\lambda^2-u^2-v^2}z\right)}du\,dv \\
\mathbf{f}_y(\mathbf{r}) &= \iint F_y(u, v)\mathbf{p}_y e^{i2\pi\left(ux+vy+\sqrt{1/\lambda^2-u^2-v^2}z\right)}du\,dv
\end{aligned}
\tag{4.27}
$$

Equation (4.27) is an exact vector model relating the Fourier distribution of the field in linear polarizations to the spatial field distribution in three dimensions. The inverse relationship is

$$
\begin{aligned}
F_x(u, v)\mathbf{p_x} e^{i2\pi\sqrt{1/\lambda^2-u^2-v^2}z} &= \iint \mathbf{f}_x(x, y, z)e^{-i2\pi(ux+vy)}dx\,dy \\
F_y(u, v)\mathbf{p_y} e^{i2\pi\sqrt{1/\lambda^2-u^2-v^2}z} &= \iint \mathbf{f}_y(x, y, z)e^{-i2\pi(ux+vy)}dx\,dy
\end{aligned}
\tag{4.28}
$$

From Eqn. (4.28) we see that knowledge of $\mathbf{f}_x(x, y, z)$ and $\mathbf{f}_y(x, y, z)$ as functions of (x, y) for any specific value of z is sufficient to calculate $F_x(u, v)$ and $F_y(u, v)$. In particular, if we know $f(x, y, z = 0)$, we can then calculate

$$
\begin{aligned}
F_x(u, v)\mathbf{p_x} &= \iint \mathbf{f}_x(x, y, z = 0)e^{-i2\pi(ux+vy)}dx\,dy \\
F_y(u, v)\mathbf{p_y} &= \iint \mathbf{f}_y(x, y, z = 0)e^{-i2\pi(ux+vy)}dx\,dy
\end{aligned}
\tag{4.29}
$$

Once we have determined $F_x(u, v)$ and $F_y(u, v)$, the field at all (x, y, z) may be calculated from Eqn. (4.27). Specification of the field on a surface, such as the plane $z = 0$, is called a *boundary condition* and the evolution of the field distribution from one boundary to another is called *diffraction*. Equations (4.27) and (4.28) enable us to computationally model diffraction in homogeneous media.

4.4 DIFFRACTION

Diffraction is the process of wave propagation from one boundary to another. A canonical example of optical diffraction, propagation of a monochromatic field from the plane $(x, y, z = 0)$ to the plane $(x', y', z = d)$, is illustrated in Fig. 4.2. Given the electric field distribution on the input plane, we seek to estimate the field distribution on the output plane. Viewed as a transformation between a function $f(x, y)$ over the input plane and a function $g(x', y')$ over the output plane, diffraction is linear and shift-invariant. Our goal in this section is to describe the transfer function and impulse response corresponding to diffraction from one plane to another.

An arbitrary vector field $\mathbf{f}(x, y)$ in the plane $z = 0$ corresponds to the Fourier space distribution

$$\mathbf{F}(u, v) = \iint \mathbf{f}(x, y) e^{-i2\pi(ux+vy)} dx\, dy \tag{4.30}$$

where $\mathbf{F}(u, v)$ may be separated into x, y and z components $F_x(u, v) = \mathbf{F}(u, v) \cdot \mathbf{p}_x(u, v)$, $F_y(u, v) = \mathbf{F}(u, v) \cdot \mathbf{p}_y(u, v)$, and $F_z(u, v) = \mathbf{F}(u, v) \cdot \lambda\mathbf{u}$. The F_z component does not produce a propagating field.

Let $G_x(u, v)$ and $G_y(u, v)$ be the Fourier distributions of the field in the x', y' plane at $z = d$. From Eqn. (4.28) we see that

$$\begin{aligned} G_x(u, v) &= e^{i2\pi\sqrt{1/\lambda^2 - u^2 - v^2}d} F_x(u, v) \\ G_y(u, v) &= e^{i2\pi\sqrt{1/\lambda^2 - u^2 - v^2}d} F_y(u, v) \end{aligned} \tag{4.31}$$

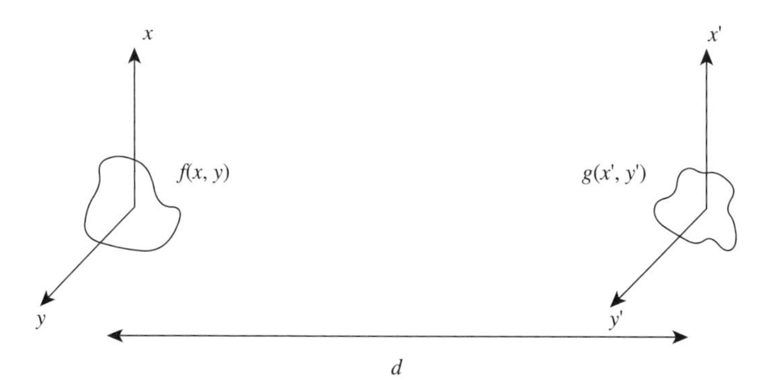

Figure 4.2 Diffraction between two planes.

the factor $T(u, v) = e^{i2\pi\sqrt{1/\lambda^2 - u^2 - v^2}d}$ is the *transfer function* for diffraction from the $z = 0$ plane to the $z = d$ plane.

Nominally, the *impulse response* for diffraction is the inverse Fourier transform of the transfer function. We continue along this line with care, however, by briefly accounting for the vector nature of the field. Using the transfer function and Eqn. (4.27), we obtain

$$\mathbf{g}(x', y') = \int\int \left[\mathbf{F}(u, v) \cdot \mathbf{p_x p_x} + \mathbf{F}(u, v) \cdot \mathbf{p_y p_y}\right] \times e^{i2\pi\left(ux' + vy' + \sqrt{1/\lambda^2 - u^2 - v^2}z\right)} du\, dv \tag{4.32}$$

If there are no longitudinal (nonpropagating) field components on the input boundary then $\mathbf{F}(u, v) \cdot \lambda\mathbf{u} = 0$, and

$$\mathbf{F}(u, v) \cdot \mathbf{p_x p_x} + \mathbf{F}(u, v) \cdot \mathbf{p_y p_y} = \mathbf{F}(u, v) \tag{4.33}$$

In this case Eqn. (4.32) simplifies considerably to yield

$$\begin{aligned} \mathbf{g}(x', y') &= \int\int \mathbf{F}(u, v) e^{i2\pi\left(ux' + vy' + \sqrt{1/\lambda^2 - u^2 - v^2}z\right)} du\, dv \\ &= \int\int \mathbf{f}(x, y) h(x' - x, y' - y) dx\, dy \end{aligned} \tag{4.34}$$

where

$$h(x, y) = \int\int e^{i2\pi\left(ux + vy + \sqrt{1/\lambda^2 - u^2 - v^2}z\right)} du\, dv \tag{4.35}$$

Equation (4.35) integrates in closed form to yield [21]

$$h(x, y) = \frac{d}{\lambda}\left[\frac{-i}{(d^2 + x^2 + y^2)} + \frac{\lambda}{2\pi(d^2 + x^2 + y^2)^{3/2}}\right] e^{i(2\pi/\lambda)\sqrt{d^2 + x^2 + y^2}} \tag{4.36}$$

Given $\mathbf{f}(x, y)$, it is not difficult to numerically apply Eqns. (4.30) and (4.34) to calculate the diffracted field. In paraxial systems such that $u, v \ll w$ over the spatial bandwidth of the field, we assume that $\mathbf{p}_x \approx \mathbf{i}_x$ and $\mathbf{p}_y \approx \mathbf{i}_y$. Under this assumption the polarization components are independent of u and v and Eqn. (4.34) reduces to independent scalar equations for each polarization. Accordingly, we base our model for diffraction in the remainder of the text on the scalar transformation

$$g(x', y') = \int\int f(x, y) h(x' - x, y' - y) dx\, dy \tag{4.37}$$

Integration of Eqn. (4.37) using $h(x, y)$ as given by Eqn. (4.36) is a bit tricky, but one can numerically model diffraction by applying the transfer function using the methods described in Section 3.7. In analytic work, however, one generally chooses to work with a simplified approximate impulse response. As an example, $\lambda \ll d$ in essentially all optical systems, meaning that the $1/\lambda d$ term in Eqn. (4.36) dominates the $1/d^2$ term. In imaging system analysis, the impulse response is often simplified by the *Fresnel* (near-field) approximation or the more restrictive *Fraunhofer* (far-field) approximation. Both approximations are *paraxial*, meaning that we restrict our attention to field distributions over the space close to the axis of optical propagation (the z axis in Fig. 4.2).

The Fresnel approximation is just the paraxial approximation that $d \gg |x - x'|$, $|y - y'|$ for all x, y and x', y' of interest. In this case

$$h(x, y) \approx \frac{1}{i\lambda d} e^{i(2\pi d/\lambda)} e^{i(\pi/\lambda d)(x^2+y^2)} \tag{4.38}$$

Under the Fresnel approximation, diffraction in homogeneous isotropic space is described by a 2D version of the Fresnel transform discussed in Section 3.5 evaluated at $\tau = \sqrt{\lambda d}$. Noting from Fig. 3.7 that the Fresnel transformation produces significant blurring for Gaussian features of width Δ when $\tau/\Delta > 1$, one might expect features of size Δ to blur on propagation at distances greater than $d = \Delta^2/\lambda$. This suggests that wavelength scale features will blur quite rapidly on diffraction. Features with an initial scale of 10 wavelengths blur in 100 wavelengths, while features on a scale of 100 wavelengths blur in 10,000 wavelengths. This effect is illustrated in Fig. 4.3, which shows diffraction of Gaussian spots of various sizes. Notice, however, that high-frequency features reappear at 10 mm as a result of interference between the diffracting spots. Such interference appears in the diffraction of coherent laser fields, but is not observed in the diffraction of incoherent fields.

Figure 4.3 was generated using numerical analysis in Matlab. The figure used a 2×2-mm spatial window sampled with 1024×1024 pixels. The Fresnel transfer function multiplied the DFT of the input field and an inverse DFT was used to generate the diffracted field. Of course, numerical analysis is not necessary for analysis of diffraction of these particular sources because, as discussed in Section 3.5, Hermite–Gaussian distributions are eigenfunctions of the Fresnel transform. According to Eqns. (3.76) and (4.38), if the input field $f(x, y) = \phi_n(x/w_0)\phi_m(y/w_0)$ for real w_0, then the diffracted field is

$$g(x, y) = \frac{e^{i(\pi/4)} e^{i(2\pi d/\lambda)} e^{-i(n+m+1)\arctan\left(\lambda d/w_0^2\right)}}{\sqrt{w_0^4 + \lambda^2 d^2}} e^{i\pi[(x^2+y^2)\lambda d/(w_0^4+\lambda^2 d^2)]}$$
$$\times \phi_n\left(\frac{xw_0}{\sqrt{w_o^4 + \lambda^2 d^2}}\right) \phi_m\left(\frac{xw_0}{\sqrt{w_0^4 + \lambda^2 d^2}}\right) \tag{4.39}$$

With specific reference to the Gaussian input spots of Fig. 4.3, the diffracted field for the input distribution $f(x, y) = \exp(-\pi(x^2 + y^2)/w_0^2)$ is

$$g(x, y) = \frac{e^{-i(\pi/4)}}{w_0^2 + i\lambda d} e^{i(2\pi d/\lambda)} e^{-\pi[(x^2+y^2)/(w_0^2+i\lambda d)]} \tag{4.40}$$

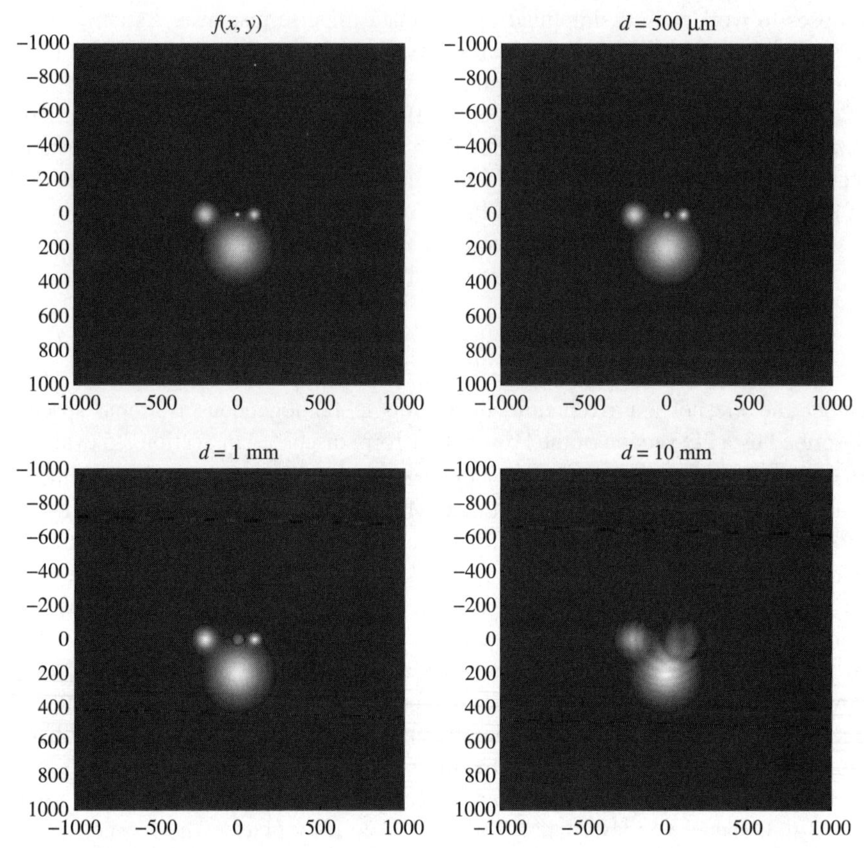

Figure 4.3 Absolute values of the diffracted field for the Gaussian spots $e^{-\pi[(x^2+y^2)/20^2]}$, $e^{-\pi\{[(x-100)^2+y^2]/50^2\}}$, $e^{-\pi\{[(x+200)^2+y^2]/100^2\}}$, and $e^{-\pi\{[x^2+(y-200)^2]/250^2\}}$ under the Fresnel approximation for various diffraction distances. All units are in microns and $\lambda = 1$ μm.

Additional interesting Fresnel diffraction effects are observed in Figs. 4.4 and 4.5. These figures were generated from the same spatial window and sampling as above, but the images are zoomed to focus on features of interest. Figure 4.4 is a harmonic field modulated by a Gaussian envelope. Note that the harmonic features do not blur (features with the same frequency are present at all diffraction lengths). At $d = 10$ mm, the diffracted field has begun to separate horizontally into multiple images of the Gaussian envelope and harmonic modulation is observed only in the interference between the separating spots, not within individual spots.

Figure 4.5 is a chirped harmonic field modulated by a Gaussian envelope. For this input, the diffracting field sharpens to a focus at $d = 2$ mm rather than blurring on propagation. After the focus, the field blurs. In both Figs. 4.4 and 4.5 it is interesting to note that blur is not a fundamental process of diffraction for coherent fields. In fact, a diffracting coherent field maintains its spatial frequency bandwidth on propagation.

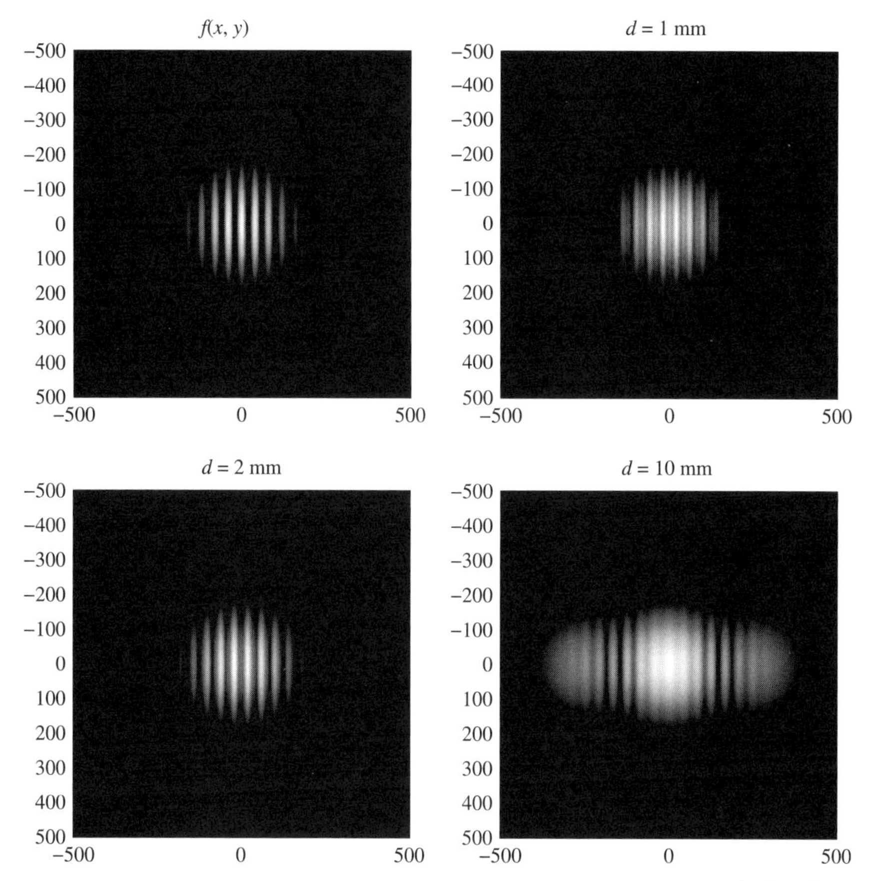

Figure 4.4 Absolute value of the diffracted field for $f(x, y) = e^{-\pi[(x^2+y^2)/250^2]}[1 + \cos(0.05\pi x)]$. All units are in micrometers; $\lambda = 1$ μm.

Blur in the normally observed sense of optical fields is a property of partially coherent or incoherent fields, as discussed in Chapter 6.

As illustrated in Fig. 4.4, however, Fourier components of diffracting fields tend to separate on propagation. This effect is easily explained in the context of Fraunhofer diffraction theory. Fraunhofer diffraction is most easily derived from the integral form of Fresnel diffraction

$$
\begin{aligned}
g(x', y') &= \frac{e^{i(2\pi d/\lambda)}}{i\lambda d} \iint e^{i(\pi/\lambda d)}[(x - x')^2 + (y - y')^2] f(x, y) dx\, dy \\
&= \frac{e^{i(2\pi d/\lambda)} e^{i(\pi/\lambda d)}(x'^2 + y'^2)}{i\lambda d} \iint \exp\left(-i2\pi \frac{xx' + yy'}{\lambda d}\right) \\
&\quad \times \exp\left[i \frac{\pi}{\lambda d}(x^2 + y^2)\right] f(x, y) dx\, dy
\end{aligned}
\tag{4.41}
$$

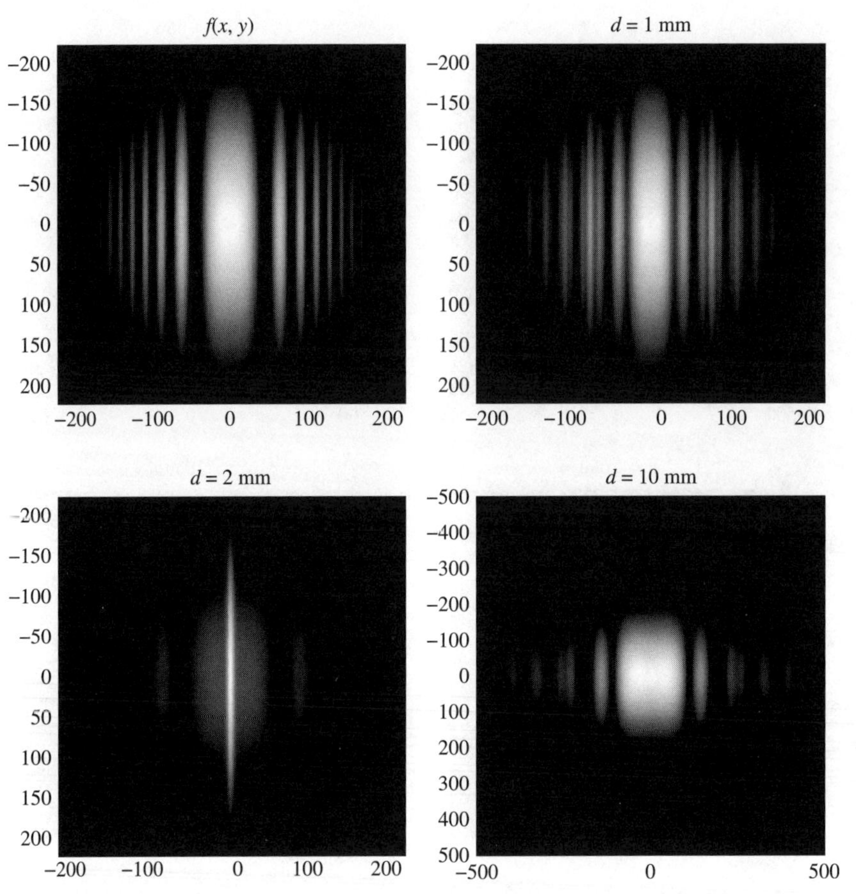

Figure 4.5 Absolute value of the diffracted field for $f(x, y) = e^{-\pi[(x^2+y^2)/250^2]}$ $[1 + \cos(5\pi 10^{-4}x^2)]$. All units are in micrometers; $\lambda = 1$ μm.

Assuming that $x^2 \ll \lambda d$ and $y^2 \ll \lambda d$ over the support of $f(x, y)$, we may drop the second exponential in the integrand to obtain

$$g(x', y') \approx \frac{e^{i(2\pi d/\lambda)} e^{i(\pi/\lambda d)(x'^2+y'^2)}}{i\lambda d} \hat{f}\left(u = \frac{x'}{\lambda d}, v = \frac{y'}{\lambda d}\right) \tag{4.42}$$

meaning that the diffracted field is proportional to the Fourier transform of the input field. Note that the Fraunhofer assumption is quite restrictive, however. For example, a 100 wavelength scale input must diffract for well over 10,000 wavelengths to reach the Fraunhofer regime and a 1000 wavelength feature, for well over 1,000,000 wavelengths. Fraunhofer diffraction is nevertheless often useful in determining the rough size and spatial frequency structure of objects. The Fraunhofer assumption is commonly applied at opposite ends of the electromagnetic imaging frequency scale, such as in X-ray crystallography and radio astronomy.

4.5 WAVE ANALYSIS OF OPTICAL ELEMENTS

Chapter 2 considered the use of optical elements to shape the mutual visibility of source points and detection points. The visibility function, renamed the *impulse response* or *point spread function*, remains of central interest under the wave model. The goal of optical sensor design is to use optical elements to program the impulse response, within physical constraints, to usefully encode target object features into detected data.

This section presents wave models for the optical elements that we described using geometric models in Section 2.2. In addition, we consider *diffractive optical elements*, which cannot be described by ray models. As in Section 2.2, analysis of refraction and reflection at dielectric interfaces is a good starting point for optical element analysis. In analogy with Fig. 2.4, the effect of a dielectric interface on a plane wave is illustrated Fig. 4.6. A plane wave is incident on the interface in a medium of index of refraction n_1. The incident field is $E_i(\mathbf{r}) = E_i \exp(2\pi i \mathbf{u_i} \cdot \mathbf{r})$. The incident wave is refracted at the interface into the second medium of index of refraction n_2, and a reflected wave is returned into the first medium. The refracted and reflected fields are $E_t(\mathbf{r}) = E_t \exp(2\pi i \mathbf{u_t} \cdot \mathbf{r})$ and $E_r(\mathbf{r}) = E_r \exp(2\pi i \mathbf{u_r} \cdot \mathbf{r})$.

Boundary conditions derived from the Maxwell equations determine the relative amplitudes of these waves. The boundary conditions may be stated as follows:

- Vector components of **E** and **H** that lie in the plane of the interface are continuous.
- Vector components of **D** and **B** normal to the plane of the interface are continuous.

In both cases we assume that there are no surface charges or currents, which is always the case at optical frequencies. These boundary conditions are used in standard texts

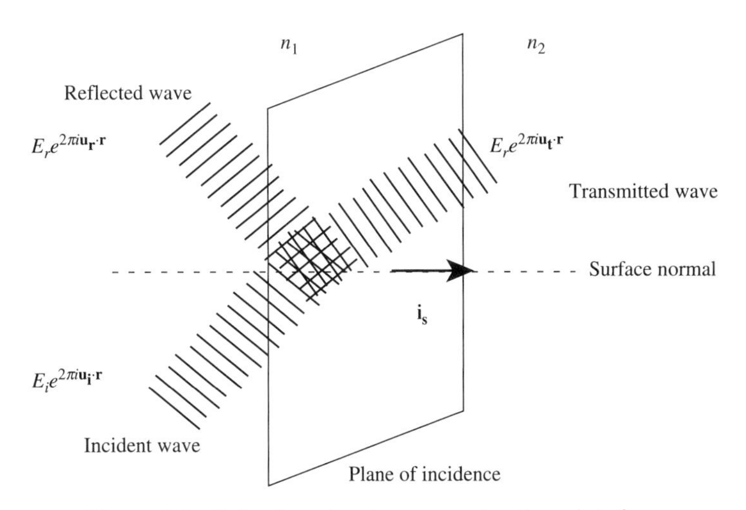

Figure 4.6 Refraction of a plane wave at a planar interface.

on optics and electromagnetics to relate the amplitudes of the refracted and reflected waves to the amplitude of the incident wave. Typically, the power in the reflected wave at a dielectric interface is a few percent of the incident power, and most of the power is transmitted. Thin film layers are often used to encode the impedance at the interface to suppress or enhance reflection.

It is not necessary to model reflection and refraction in detail to understand the functional utility of optical elements in shaping the impulse response. The most important features from a wave perspective are obtained simply by noting that the functional form of the wave distribution must be maintained on both sides of the interface for the boundary conditions to be satisfied. To satisfy the boundary conditions in the plane of the interface, we require that $[\mathbf{E_i}(\rho) + \mathbf{E_r}(\rho)] \times \mathbf{i_s} = \mathbf{E_t}(\rho) \times \mathbf{i_s}$, for all ρ on the interface. $\mathbf{i_s}$ is the surface normal for the interface. To satisfy this condition, one must require that

$$\mathbf{u_i} - \mathbf{u_i} \cdot \mathbf{i_s}\mathbf{i_s} = \mathbf{u_t} - \mathbf{u_t} \cdot \mathbf{i_s}\mathbf{i_s} \tag{4.43}$$

In combination with the requirement that $|\mathbf{u_t}| = n_2/\lambda$, we find that

$$\mathbf{u_t} = \mathbf{u_i} - \mathbf{u_i} \cdot \mathbf{i_s}\mathbf{i_s} + \mathbf{i_s}\sqrt{\frac{n_2^2}{\lambda^2} - \frac{n_1^2}{\lambda^2} + (\mathbf{u_i} \cdot \mathbf{i_s})^2} \tag{4.44}$$

If we use angles relative to the surface normal to decompose $\mathbf{u_i}$ and $\mathbf{u_t}$ into transverse and longitudinal components, Eqn. (4.44) immediately reduces to Snell's law [Eqn. (2.5)]. In the the paraxial case $\mathbf{i_s}$ and $\mathbf{u_i}$ are nearly collinear and Eqns. (4.44) can be approximated by

$$\mathbf{u_t} \approx \mathbf{u_i} + \mathbf{i_s}\frac{\Delta n \bar{n}}{\lambda^2 \mathbf{u_i} \cdot \mathbf{i_s}} \tag{4.45}$$

where $\Delta n = n_2 - n_1$ and $\bar{n} = (n_1 + n_2)/2$.

As in Section 2.2, we first apply Snell's law to the analysis of prism refraction. As illustrated in Fig. 4.7, a prism consists of a series of two tilted planar interfaces. The prism of Fig. 4.7 consists of a dielectric of index n_2 embedded in a dielectric of index n_1. If $\mathbf{i}_1$ is the surface normal at the first interface of a prism and $\mathbf{i}_2$ the surface normal at the second interface, recursive application Eqn. (4.45) produces an estimate of the output wavevector $\mathbf{u}_3$

$$\begin{aligned} \mathbf{u}_3 &\approx \mathbf{u}_2 - \mathbf{i}_2\frac{\Delta n \bar{n}}{\lambda^2 \mathbf{u}_2 \cdot \mathbf{i}_2} \\ &\approx \mathbf{u}_1 + \frac{\Delta n \bar{n}}{\lambda}(\mathbf{i}_1 - \mathbf{i}_2) \end{aligned} \tag{4.46}$$

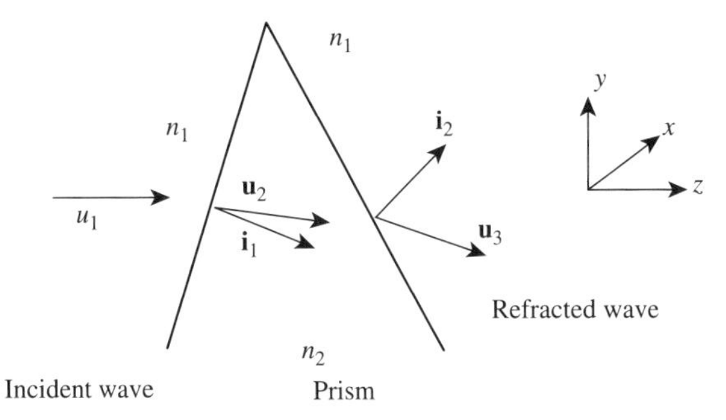

Figure 4.7 Refraction of a plane wave by a prism. For $n_2 \gg n_1$, the incident wavevector refracts to move $\mathbf{u}_2$ closer to the surface normal $\mathbf{i}_1$ than the incident wavevector $\mathbf{u}_1$. Refraction at the output interface moves $\mathbf{u}_3$ away from the surface normal $\mathbf{i}_2$ in comparison with $\mathbf{u}_2$.

We find, therefore, that the plane wave $E_i(\mathbf{r}) = E_i \exp(2\pi i \mathbf{u_i} \cdot \mathbf{r})$ incident on a prism is refracted under the paraxial approximation into the plane wave

$$E_t(\mathbf{r}) = E_i \exp(2\pi i \mathbf{u_i} \cdot \mathbf{r}) e^{i\phi_p} \times \exp\left(2\pi i \frac{\Delta n(n_1 + n_2)}{\lambda} [\mathbf{i}_1 - \mathbf{i}_2] \cdot \mathbf{r}\right) \tag{4.47}$$

where the phase ϕ_p is the phase shift the wave experiences in propagating through the prism.

Equation (4.47) has the form

$$E_t(\mathbf{r}) = t(\mathbf{r}) E_i(\mathbf{r}) \tag{4.48}$$

Because $t(x, y)$ is independent of $\mathbf{u_i}$ in the plane $z = 0$, one can imagine taking an inverse Fourier transform of Eqn. (4.47) with respect to x and y to show that Eqn. (4.48) holds for any input wave, not just plane waves. $t(x, y)$ is the "transmittance" of the prism. Transmittance functions are commonly used to approximate the action of thin optical elements, such as prisms, lenses, aperture stops, gratings, and mirrors. We made use of the transmittance concept in Section 2.5 in the context of coded aperture imaging and again apply this concept in Chapter 9 in considering coded aperture spectroscopy. In the present chapter, we find transmittance extremely useful in describing prisms, lenses, gratings, and holograms.

The basic idea of a transmittance function is illustrated in Fig. 4.8. The wave field, $E_i\,(x'', y'')$, is incident on an optical element. The field to the immediate right of the element is $t(x'', y'')E_i\,(x'', y'')$. One models a system involving the optical element by first propagating the field $f(x, y)$ to the input of the element using the Fourier methods described in Section 4.4, then modulating the field by the transmittance, and finally propagatating the modulated field to the output plane to determine $g(x', y')$.

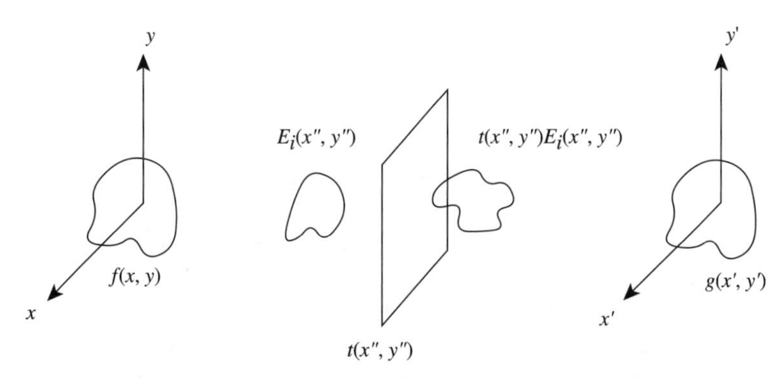

Figure 4.8 Transmittance of a thin optical element.

The transmittance for the prism of Fig. 4.7 is

$$t(\rho) = e^{i\phi_p} e^{2\pi i((n^2-1)/\lambda)[\mathbf{i}_1 - \mathbf{i}_2]\cdot\rho} \tag{4.49}$$

where we assume no reflective loss, $n_1 = 1$ and $n_2 = n$. The effect of the prism is to introduce a linear phase shift in the modulation plane. This linear phase shift is consistent with the idea that the variable thickness of a wedge will shift the phase of the field in proportion to the local thickness of the wedge. In the case $\mathbf{i}_1 - \mathbf{i}_2 = 2\sin\psi\,\mathbf{i_x}$, the transmittance of a prism is

$$t(x) = e^{i\phi_0} e^{2\pi i(n^2-1)(x/\lambda)\sin\psi} \tag{4.50}$$

If this prism is illuminated by a plane wave with wavevector

$$\mathbf{u_i} = \frac{1}{\lambda}[\sin\theta\,\mathbf{i_x} + \cos\theta\,\mathbf{i_z}] \tag{4.51}$$

The refracted wave vector is approximately

$$\mathbf{u_s} = \frac{1}{\lambda}[\sin\theta'\mathbf{i_x} + \cos\theta'\mathbf{i_z}] \tag{4.52}$$

where $\sin\theta' = \sin\theta + (n^2 - 1)\sin\psi$. Note that θ' is independent of λ. In practice, of course, spectral dispersion is observed in prism refraction. As discussed in Section 2.2, this dispersion is due to the wavelength dependence of n.

Transmittance functions may also be used to model diffractive optical elements. Consider, for example, the diffraction grating sketched in Fig. 4.9. The surfaces are curved such that $\mathbf{i}_1 - \mathbf{i}_2$ varies harmonically. If we assume, for example, that $\mathbf{i}_1 - \mathbf{i}_2 = \alpha\sin(Kx)\mathbf{i_x}$, then, for some constant α, we find in analogy with Eqn. (4.49) that transmittance is

$$t(x, y) = e^{i\phi_0} e^{2\pi i(\gamma/\lambda)\sin(Kx)} \tag{4.53}$$

where we have absorbed materials constants into the grating amplitude γ and ϕ_0 is a phase constant. We neglect the constant ϕ_0 phase factor in the following analysis.

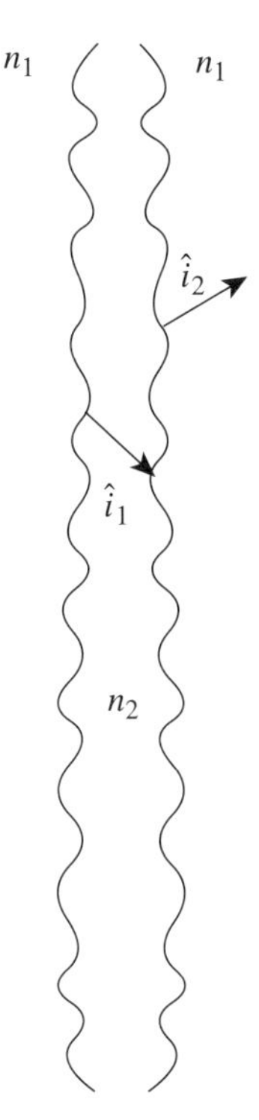

Figure 4.9 A phase modulating diffraction grating.

Since the transmittance of the grating is periodic, it may be expanded in a Fourier series. In the case of Eqn. (4.53) the series is obtained from the Jacobi–Anger expansion [246]

$$e^{im\sin(x)} = \sum_{q=-\infty}^{\infty} J_q(m)e^{iqx} \tag{4.54}$$

where $J_q(m)$ is a Bessel function of the first kind. Substituting in Eqn. (4.53) yields

$$t(x, y) = \sum_{q=-\infty}^{\infty} J_q\left(\frac{2\pi\alpha}{\lambda}\right)e^{iqKx} \tag{4.55}$$

If the grating described by Eqn. (4.55) is illuminated by the incident plane wave $E_i(\mathbf{r}) = E_i \exp(2\pi i \mathbf{u_i} \cdot \mathbf{r})$, then the field in the plane immediately after the grating is

$$E_d(x, y) = E_p \sum_{q=-\infty}^{\infty} E_i J_q\left(\frac{2\pi\alpha}{\lambda}\right)e^{iqKx}e^{2\pi i u_{ix}x}e^{2\pi i u_{iy}y} \tag{4.56}$$

This field consists of an infinite series of harmonic components, or *diffraction orders*. The qth component is of amplitude $E_i J_q(2\pi\gamma\lambda)$ and produces a plane wave with wavevector

$$\mathbf{u_q} = (u_{ix} + qK)\mathbf{i_x} + u_{iy}\mathbf{i_y} + \mathbf{i_z}\sqrt{\frac{1}{\lambda^2} - (u_{ix} + qK)^2 + u_{iy}^2} \tag{4.57}$$

The effect of both the grating and the prism is to redirect an incident wave. Neglecting reflections, however, a prism produces only one refracted order while a grating tends to produce many diffraction orders. An even more fundamental difference between the effect of prisms and gratings is observed in the spectral domain. For a prism, $\mathbf{u_s} - \mathbf{u_i}$ is inversely proportional to λ and, as expressed in Eqn. (4.52), the shift in angle between the incident and refracted beams depends on wavelength only through the materials dispersion. For a grating, in contrast, the shift in the incident wavevector, $(\mathbf{u_q} - \mathbf{u_i}) \cdot \mathbf{i_x} = qK$, is independent of λ. A wavelength independent shift in the wavevector produces strong wavelength dependence (dispersion) in the direction of the scattered field. In the paraxial approximation, a plane wave with angle of incidence θ is diffracted in the qth order to the transmission angle

$$\sin\theta' = \sin\theta + q\frac{\lambda}{\Lambda} \tag{4.58}$$

where $\Lambda = 2\pi/K$ is the period of the grating.

The optical element illustrated in Fig. 4.9 is called a *surface relief grating*. Since the optical element is nonabsorbing, it is called a *phase element*. It is also possible to produce phase elements by spatially varying the index of refraction rather than the surface relief. This approach produces a transmittance function substantially similar in functional form and diffraction properties to the surface relief grating. It is also possible to produce diffraction gratings by modulating absorption properties of a material or both phase and absorption. The transmittance of an absorption grating might be $t(x, y) = 1/(1 + m)[1 + m\cos(2\pi x/\Lambda)]$, where m is a constant between zero and one. For the normally incident plane $U(\mathbf{r}) = A_0 e^{jk_0 z}$, the field transmitted by the absorption grating is $f(x, y) = A_0/(1 + m)[1 + m\cos(2\pi x/\Lambda)]$. This field

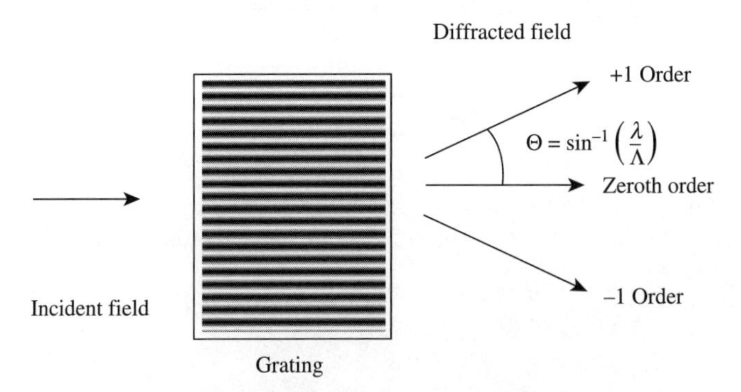

Figure 4.10 Diffraction by an amplitude grating.

corresponds to a plane wave of amplitude $A_0/(1+m)$ propagating along the z axis, a plane wave of amplitude $mA_0/2(1+m)$ propagating with wavevector $K\mathbf{i_x}+\sqrt{k_0^2-K^2}\mathbf{i_z}$, and a plane wave of amplitude $mA_0/2(1+m)$ propagating with wavevector $-K\mathbf{i_x}+\sqrt{k_0^2-K^2}\mathbf{i_z}$. As illustrated in Fig. 4.10, the angular spacing between diffraction orders is λ/Λ.

4.6 WAVE PROPAGATION THROUGH THIN LENSES

As illustrated in Fig. 2.9, a lens may be considered as an array of prisms with the direction of the surface normal varying across the transverse plane. The difference in the surface normal directions varies linearly with radial position on a lens such that

$$\mathbf{i}_1 - \mathbf{i}_2 = -\alpha\rho\mathbf{i}_\rho \tag{4.59}$$

where ρ is the radial coordinate in the plane of the lens and α is a constant. Substituting Eqn. (4.59) in Eqn. (4.49), we obtain the transmittance of a lens as

$$t(\rho) = e^{i\phi_0}e^{-\pi i(\Delta n\alpha\rho^2/\lambda)} \tag{4.60}$$

To understand the action of the lens, recall that a lens transforms light radiated by a point source at the front focal point into a plane wave. Figure 4.11 illustrates a point source illuminating a lens from the front focal point. The field striking the lens is the impulse response for propagating over a distance F:

$$h(x, y) = \frac{e^{ik_0z}}{\lambda z}\exp\left(i\pi\frac{x^2+y^2}{\lambda F}\right)$$

The lens transforms this field into a plane wave by modulating it by a transmittance function $t(x, y)$. Since the field of a plane wave propagating along the z axis is

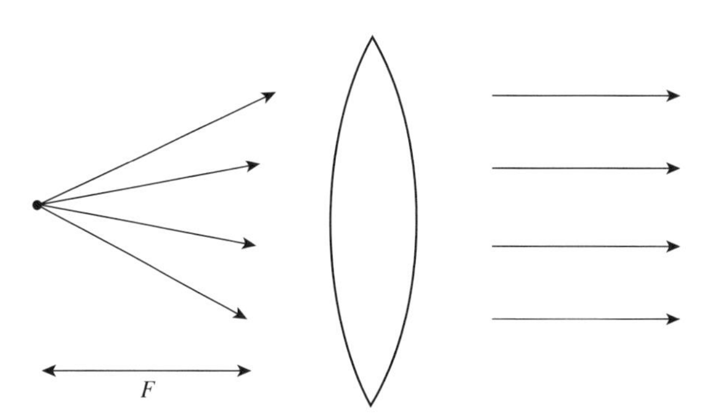

Figure 4.11 Collimation by a thin lens.

constant across the transverse plane, the lens transmittance must be

$$t(x, y) = \exp\left(-i\pi\frac{x^2 + y^2}{\lambda F}\right)$$

We find, therefore, that the constant α in Eqn. (4.59) is related to the focal length according to $F = 1/\Delta n\alpha$. Note that, neglecting material dispersion, F is independent of λ.

If the incident point source is not on the optical axis, but is rather at point (x_0, y_0) in the plane a distance F in front of the lens, the field striking the lens is

$$h(x, y) = \frac{e^{ik_0 z}}{\lambda z}\exp\left(i\pi\frac{(x - x_0)^2 + (y - y_0)^2}{\lambda F}\right)$$

Modulation of this field by the lens transmittance produces the field

$$t(x, y)h(x, y) = \exp\left(i2\pi\frac{xx_0 + yy_0}{\lambda F}\right)\exp\left(i\pi\frac{x_0^2 + y_0^2}{\lambda F}\right) \tag{4.61}$$

which corresponds to a plane wave with wavevector

$$\frac{x_0}{\lambda F}\mathbf{i_x} + \frac{y_0}{\lambda F}\mathbf{i_y} + \sqrt{k_0^2 - 4\pi^2\frac{x_0^2 + y_0^2}{\lambda^2 F^2}}\,\mathbf{i_z}$$

An alternative derivation of the transmittance considers the phase modulation placed on the field by propagation through the lens. Since the lens is nonabsorbing, the transmittance is of the form $t(x, y) = \exp[i\phi(x, y)]$, where $\phi(x, y)$ is the phase delay that the field encounters in passing through the lens at (x, y). This phase delay is $\phi(x, y) = (2\pi/\lambda)(\Delta_0 - \Delta(x, y)) + (2\pi n/\lambda)\Delta(x, y)$, where $\Delta(x, y)$ is the thickness of the lens at (x, y), $\Delta_0 = \Delta(0, 0)$ and n is the index of refraction of the lens.

As sketched in Fig. 4.12, a lens consists of a dielectric with curved surfaces. The front surface of the lens is a section of a sphere of radius R_1, and the back surface is a section of a sphere of radius R_2. The thickness of the lens at its center is Δ_0. The front surface is described by the equation $x^2 + y^2 + [z - (R_1 - \Delta_0/2)]^2 = R_1^2$. The back surface is described by surface is described by the equation $x^2 + y^2 + [z - (R_2 + \Delta_0/2)]^2 = R_2^2$. The thickness of the lens is the difference between z on the front surface and z on the back surface at (x, y):

$$\begin{aligned}\Delta(x, y) &= \Delta_0 - R_1 + R_2 + \sqrt{R_1^2 - x^2 - y^2} - \sqrt{R_2^2 - x^2 - y^2} \\ &\approx \Delta_0 - \frac{x^2 + y^2}{R_1} + \frac{x^2 + y^2}{R_2}\end{aligned}$$

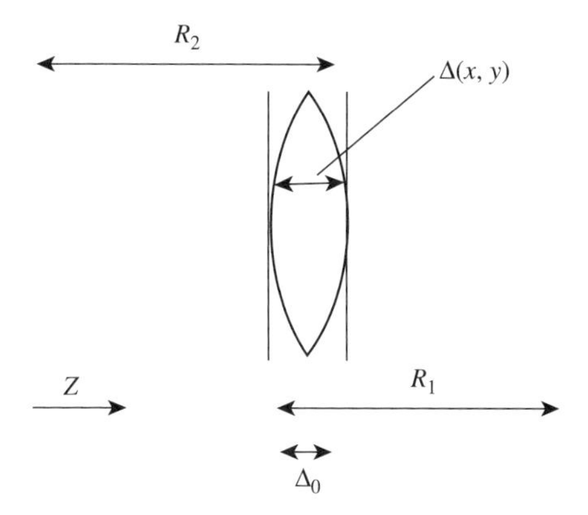

Figure 4.12 Lens geometry.

The phase delay propagating through the lens is thus approximately

$$\phi(x, y) = \frac{2\pi}{\lambda}\Delta_0 - \frac{2\pi(n-1)}{\lambda}\left(\frac{1}{R_1} - \frac{1}{R_2}\right)(x^2 + y^2)$$

Comparing with the transmittance derived above from focal collimation, we see that $1/F = (n-1)[(1/R_1) - (1/R_2)]$, which is the *lensmaker's equation* that we previously encountered in Eqn. (2.16). R_1 and R_2 are accounted negative if the center of curvature is to the right of the lens and positive if it is to the left. Thus for a lens convex on both surfaces, R_1 is positive and R_2 is negative. Depending on the values of R_1 and R_2, F may be positive or negative. A negative focal length converts an incident plane wave into a diverging wave.

Since the front and back surfaces of convex lenses meet at edges of zero thickness in essentially all our lens drawings, we are, of course, aware that the lens transmittance described by Eqn. (4.60) cannot hold over an infinite aperture. We account for the finite transverse aperture and lens shape aberrations of real lenses by introducing the pupil function, $P(x, y)$. With the pupil function the lens transmittance is modeled as

$$t(x, y) = \exp\left(-i\pi\frac{x^2 + y^2}{\lambda F}\right)P(x, y) \tag{4.62}$$

For a round lens of diameter D, for example,

$$P(x, y) = \mathrm{circ}\left(\frac{\sqrt{x^2 + y^2}}{D}\right)$$

While we have introduced the pupil function with an innocuous comment, $P(x, y)$ is central to our discussion for much of the remainder of the text. We have, in fact, gone to some lengths to introduce it. The small-angle approximations that we used to derive the transmittance of the prism and lens are, in fact, routinely violated in optical systems. We find it more convenient, however, to account for these violations in aberration and coding terms in secondary analysis. As we see in the next section, the simple elegance of the pupil function is intoxicating.

4.7 FOURIER ANALYSIS OF WAVE IMAGING

Consider the system sketched in Fig. 4.13. The field at the (x, y) plane diffracts along the $\mathbf{i}_z$ axis a distance z_1 to the (x', y') plane, which contains a lens. The field at this plane is modulated by the transmittance of the lens before diffracting along the $\mathbf{i}_z$ axis a distance z_2 to the (x'', y'') plane. Our goal is to determine the field $g(x'', y'')$ on the plane (x'', y'') given the field $f(x, y)$ on the (x, y) plane. Because all transformations in this process are linear, the overall process may be viewed as a linear transformation and characterized by an impulse response. An impulse at (x_0, y_0) in the (x, y) plane generates the field

$$U(x', y') = \frac{e^{ik_0 z_1}}{\lambda z_1} \exp\left(i\pi \frac{(x' - x_0)^2 + (y' - y_0)^2}{\lambda z_1} \right) \tag{4.63}$$

on the left side of the lens. After passing through the lens, the modulated field is

$$\begin{aligned} t(x', y')U(x', y') = {} & \frac{e^{ik_0 z_1}}{\lambda z_1} \exp\left(-i\pi \frac{x'^2 + y'^2}{\lambda F} \right) \\ & \times \exp\left(i\pi \frac{(x' - x_0)^2 + (y' - y_0)^2}{\lambda z_1} \right) P(x', y') \end{aligned} \tag{4.64}$$

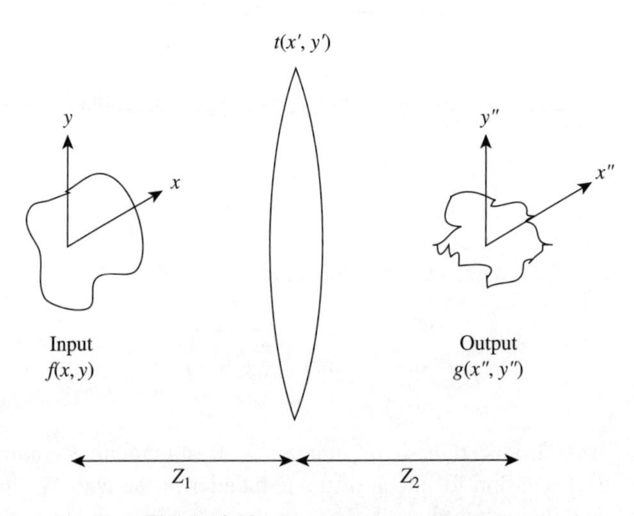

Figure 4.13 Lens system geometry.

We find the impulse response $h(x'', y'', x_0, y_0)$ by convolving the field on the right side of the lens with the Fresnel kernel. This yields

$$h(x'', y'', x_o, y_o) = \frac{e^{ik_0(z_1+z_2)}}{\lambda^2 z_1 z_2} \iint \exp\left(-i\pi \frac{x'^2 + y'^2}{\lambda F}\right) P(x', y') \times \exp\left(i\pi \frac{(x' - x_0)^2 + (y' - y_0)^2}{\lambda z_1}\right) \times \exp\left(i\pi \frac{(x'' - x')^2 + (y'' - y')^2}{\lambda z_2}\right) dx'\, dy' \tag{4.65}$$

or

$$h(x'', y'', x_0, y_0) = \frac{e^{ik_0(z_1+z_2)}}{\lambda^2 z_1 z_2} \exp\left(i\pi \frac{x''^2 + y''^2}{\lambda z_2}\right) \exp\left(i\pi \frac{x_0{}^2 + y_0{}^2}{\lambda z_1}\right) \times \iint P(x', y') \exp\left[i\frac{\pi}{\lambda}\left(x'^2 + y'^2\right)\left(\frac{1}{z_1} + \frac{1}{z_2} - \frac{1}{F}\right)\right] \times \exp\left\{-i\frac{2\pi}{\lambda}\left[x'\left(\frac{x_o}{z_1} + \frac{x''}{z_2}\right)\right] + \left[y'\left(\frac{y_0}{z_1} + \frac{y''}{z_2}\right)\right]\right\} dx'\, dy' \tag{4.66}$$

We consider two cases. Suppose first that $z_1 = z_2 = F$. Then

$$h(x'', y'', x_0, y_0) = \frac{e^{i(4\pi F/\lambda)}}{\lambda^2 z_1 z_2} \exp\left(i\pi \frac{x''^2 + y''^2}{\lambda z_2}\right) \exp\left(i\pi \frac{x_0^2 + y_0^2}{\lambda z_1}\right) \times \iint P(x', y') \exp\left[-i\frac{\pi}{\lambda F}\left(x'^2 + y'^2\right)\right] \times \exp\left\{i\frac{2\pi}{\lambda}\left[x'\left(\frac{x_0}{z_1} + \frac{x''}{z_2}\right)\right] + \left[y'\left(\frac{y_0}{z_1} + \frac{y''}{z_2}\right)\right]\right\} dx'\, dy' \tag{4.67}$$

If we neglect the effect of the aperture, for instance, if we assume that $P(x, y) = 1$, then this integral is the Fourier transform of the complex Gaussian, which is also a complex Gaussian. Applying the similarity theorem, we find

$$h(x'', y'', x_0, y_0) = \frac{e^{i2\,k_0 F}}{i\lambda F} \exp\left(-i\pi \frac{x''^2 + y''^2}{\lambda F}\right) \exp\left(-i\pi \frac{x_0{}^2 + y_0{}^2}{\lambda F}\right) \times \exp\left[-i\pi\lambda F\left(\frac{x_0}{\lambda F} + \frac{x''}{\lambda F}\right)^2 + \left(\frac{y_0}{\lambda F} + \frac{y''}{F}\right)^2\right] = \frac{e^{i2\,k_0 F}}{i\lambda F} \exp\left[-i\frac{2\pi}{\lambda F}(x_0 x'' + y_0 y'')\right] \tag{4.68}$$

where $h(x'', y'', x_0, y_0)$ is the impulse response for propagation from the front focal plane of a lens to the back focal plane. Putting this impulse response into the transformation for an arbitrary input field, we find the transformation from the front focal plane to the back focal plane to be

$$g(x'', y'') = \frac{e^{i2k_0F}}{i\lambda F}\int f(x, y)\exp\left[-i\frac{2\pi}{\lambda F}(xx'' + yy'')\right]dx\,dy \tag{4.69}$$

The output field is the Fourier transform of the incident field evaluated at $u = x''/\lambda F$, $v = y''/\lambda F$. Note that diffraction from the front focal plane to the back focal plane has the same form as Fraunhofer diffraction. In this sense, the field distribution at the back focal plane of a lens is the far-field diffraction pattern of an object in the front focal plane.

Equation (4.67) can be evaluated for a finite aperture using *the method of stationary phase* [23], which yields

$$g(x'', y'') = \frac{e^{i2k_0F}}{i\lambda F}\int f(x, y)\exp\left[-i\frac{2\pi}{\lambda F}(xx'' + yy'')\right]P(x + x'', y + y'')dx\,dy \tag{4.70}$$

$g(x'', y'')$ is proportional to the Fourier transform of $f(x, y)$ windowed (vignetted) by $P(x + x'', y + y'')$.

As a second example, suppose that $(1/z_1) + (1/z_2) = (1/F)$, which is the familiar thin lens imaging rule. For consistency with Chapter 2, let $d_i = z_2$ be the *image distance* and $d_o = z_1$ be the *object distance*. The imaging condition eliminates quadratic terms in Eqn. (4.66) so that the impulse response becomes

$$h(x'', y'', x_0, y_0) = \frac{e^{2\pi i[(d_i+d_o)/\lambda]}}{\lambda^2 d_i d_o}\exp\left(i\pi\frac{x''^2 + y''^2}{\lambda d_i}\right)\exp\left(i\pi\frac{x_0^2 + y_0^2}{\lambda d_o}\right)$$
$$\iint P(x', y')\exp\left\{-i\frac{2\pi}{\lambda}\left[x'\left(\frac{x_0}{d_o} + \frac{x''}{d_i}\right)\right] + \left[y'\left(\frac{y_0}{d_o} + \frac{y''}{d_i}\right)\right]\right\}dx'dy' \tag{4.71}$$

Equation (4.71) is simplified with the substitution $x_r = Mx$, $y_r = My$ where, as in Chapter 2, the magnification is $M = -d_i/d_o$. This yields

$$h(x'', y'', x_r, y_r) = \frac{|M|}{\lambda^2 d_i^2}e^{2\pi i[(d_i+d_o)/\lambda]}\exp\left(i\pi\frac{x''^2 + y''^2}{\lambda d_i}\right)\exp\left(i\pi\frac{{x_r}^2 + {y_r}^2}{\lambda d_i}\right)$$
$$\times\iint P(x', y')\exp\left\{-i\frac{2\pi}{\lambda d_i}[x'(x'' - x_r)] + [y'(y'' - y_r)]\right\}dx'\,dy'$$
$$= |M|e^{2\pi i[(d_i+d_o)/\lambda]}\exp\left(i\pi\frac{x''^2 + y''^2}{\lambda d_i}\right)$$
$$\times\exp\left(i\pi\frac{{x_r}^2 + {y_r}^2}{\lambda d_i}\right)h_r(x'' - x_r, y'' - y_r) \tag{4.72}$$

where the shift-invariant component of the impulse response $h_r(x, y)$ is

$$h_r(x, y) = \frac{1}{\lambda^2 d_i^2} \int\int P(x', y') e^{-i(2\pi/\lambda d_i)(x'x + y'y)} dx' \, dy' \tag{4.73}$$

Ideally, the imaging impulse response would consist of a Dirac delta function centered on $x'' = Mx_0$ and $y'' = My_0$. We find in practice that the impulse response is proportional to the Fourier transform of the pupil function. $h_r(x, y)$ would be a delta function if the pupil function were constant and of infinite extent; the finite extent of the pupil function thus acts as a bandpass filter on the imaged field. $P(x, y)$ must have finite support to be consistent with the paraxial and Fresnel approximations used to derive Eqn. (4.74).

For the most common case of a circular lens aperture of diameter A, $P(x, y) = \text{circ}(\sqrt{x^2 + y^2}/A)$ and

$$h_r(x, y) = \frac{A^2}{\lambda^2 d_i^2} \text{jinc}\left(\frac{A}{\lambda d_i}\sqrt{x^2 + y^2}\right) \tag{4.74}$$

The term $\text{jinc}[(A/\lambda d_i)\sqrt{x^2 + y^2}]$ is a delta-like distribution around the point $x = 0$, $y = 0$. As we saw in Chapter 2, the first zero occurs at $\sqrt{x^2 + y^2} \approx 1.22\lambda d_i/A$. For a system imaging an object at infinity, $d_i = F$. The ratio $f/\# = F/A$ is the *f number* of the imaging system. In discussing the f number, $f/\#$ is treated as a unified symbol, with # replaced by numerical values in reference to specific systems. To avoid severe violations of the paraxial approximation, the f number must be greater than 1. An $f/2$ system, for example, produces a *diffraction-limited* impulse response that is 4.88 wavelengths in diameter from first zero to first zero.

The phase terms in Equation (4.72) are troubling both as a phase modulation in the output x'', y'' space and as a distortion in the input space. The fact that the shift invariant component of the impulse response is highly localized at $x'' = x_r$ and $y'' = y_r$ leads us to wonder whether we might approximate $\exp\{i\pi[(x_r{}^2 + y_r{}^2)/\lambda d_i]\}$ as $\exp\{i\pi[(x''^2 + y''^2)/\lambda d_i]\}$. Approximating the support of $\text{jinc}((1/(\lambda f/\#))\sqrt{x^2 + y^2})$ by $\lambda f/\#$, the maximum difference between x^2 and x''^2 over the support is approximately $x\lambda f/\#$. This corresponds to a maximum phase difference between $\exp\{i\pi[(x_r^2 + y_r^2)/\lambda d_i]\}$ and $\exp\{i\pi[(x''^2 + y''^2)/\lambda d_i]\}$ of approximately $\pi f/\# x/d_i$, meaning that the phase of the input quadratic changes by π across the focal spot at the edge of the field with linearly reduced phase distortion toward the center of the field.

The phase terms have significant impact on the imaging of coherent fields, but are eliminated in the consideration of incoherent systems. As a prelude to later discussion of the incoherent impulse response, it is helpful to consider the effect of dropping the quadratic phase terms. This assumption is reasonable near the center of the field where $x/d_i \ll 1$. Under this assumption, the mapping from object field to the image field is

$$g(x', y') = \frac{1}{|M|} \int\int f\left(\frac{x}{M}, \frac{y}{M}\right) h_r(x' - x, y' - y) dx\, dy \tag{4.75}$$

The imaging transformation is a shift-invariant linear transformation of the magnified input distribution. The transfer function is the Fourier transformation of $h_r(x, y)$:

$$\hat{h}(u, v) = P(-\lambda d_i u, -\lambda d_i v) \tag{4.76}$$

The imaging transformation is expressed in Fourier space as the product

$$\hat{g}(u, v) = |M| P(-\lambda d_i u, -\lambda d_i v) \hat{f}(Mu, Mv) \tag{4.77}$$

As illustrated in Fig. 4.14, $P(x, y)$ is typically constant and centered on the origin of the (x, y) plane. $P(x, y)$ acts as a lowpass filter on most imaging systems, passing image frequencies such that $|u| \leq A/(2\lambda d_i)$ and blocking higher frequencies. On the basis of our analysis of bandlimited sampling in Chapter 3, this suggests that the

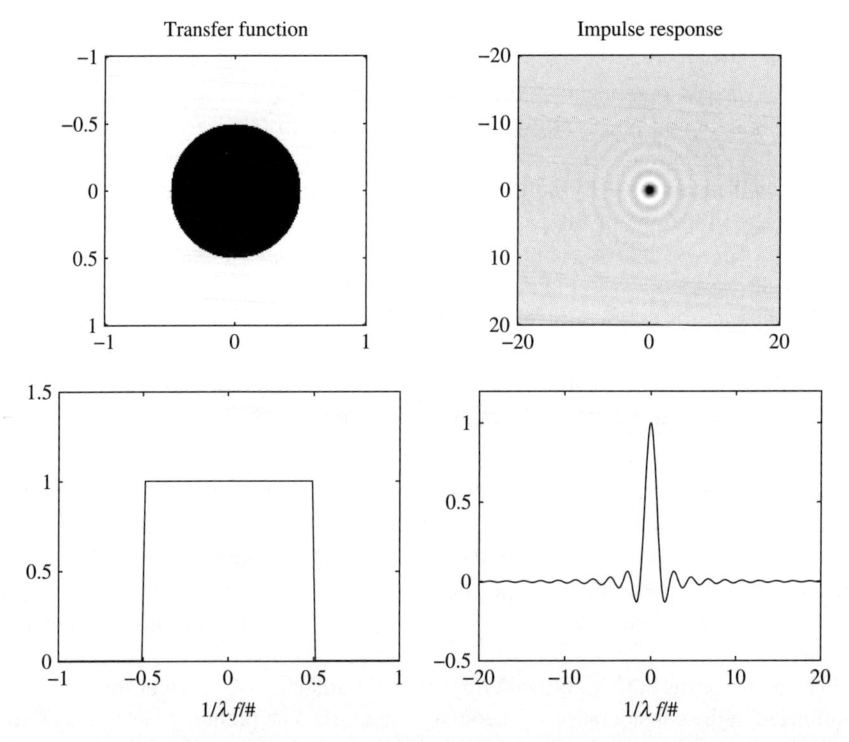

Figure 4.14 Transfer function and impulse response for an $f/1$ optical system imaging an object at infinity. The distance between the first two zeros of the impulse response is 2.44 wavelengths, the full-width half-maximum is 1.4 wavelengths.

image should be sampled with a sampling period less than $\lambda f/\#$, where we have substituted $f/\#$ for d_i/A. Of course, the coherent field is complex-valued, so sampling is nontrivial. Measurement and sampling of coherent fields requires holography or interferometery, which are the subject of the next section. We discuss image sampling rates in more detail in Chapter 7.

One may select pupil transmittance functions other than the circular aperture. For example, for coherent imaging systems one may block the center of the pupil and create a "highpass" imaging system. An example pupil–impulse response for an annular aperture appropriate to highpass imaging is shown in Fig. 4.15. The effect of imaging through the lowpass system of Fig. 4.14 and the highpass system of Fig. 4.15 is illustrated in Fig. 4.16. It is important to note that very different imaging behavior is observed for these imaging systems under incoherent illumination. In this regard, compare Fig. 4.16 with Fig. 6.19 Alternative pupil functions, useful for even incoherent systems, are discussed in Chapter 10. For example, Section 10.2 discusses the use of deliberate phase modulation in the pupil function to extend the imaging system depth of field. Nonuniform pupil functions may also be used to describe aberrations and other artifacts of optical systems.

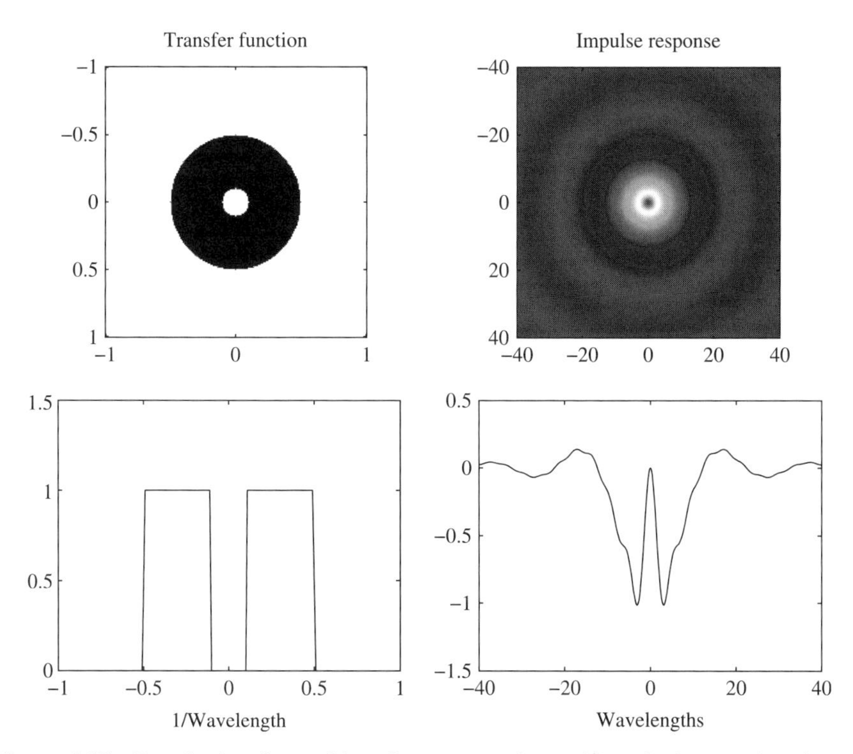

Figure 4.15 Transfer function and impulse response for an $f/1$ optical system imaging an object at infinity with an annular pupil. The radius of the blocked center disk is 20% of the radius of the full aperture.

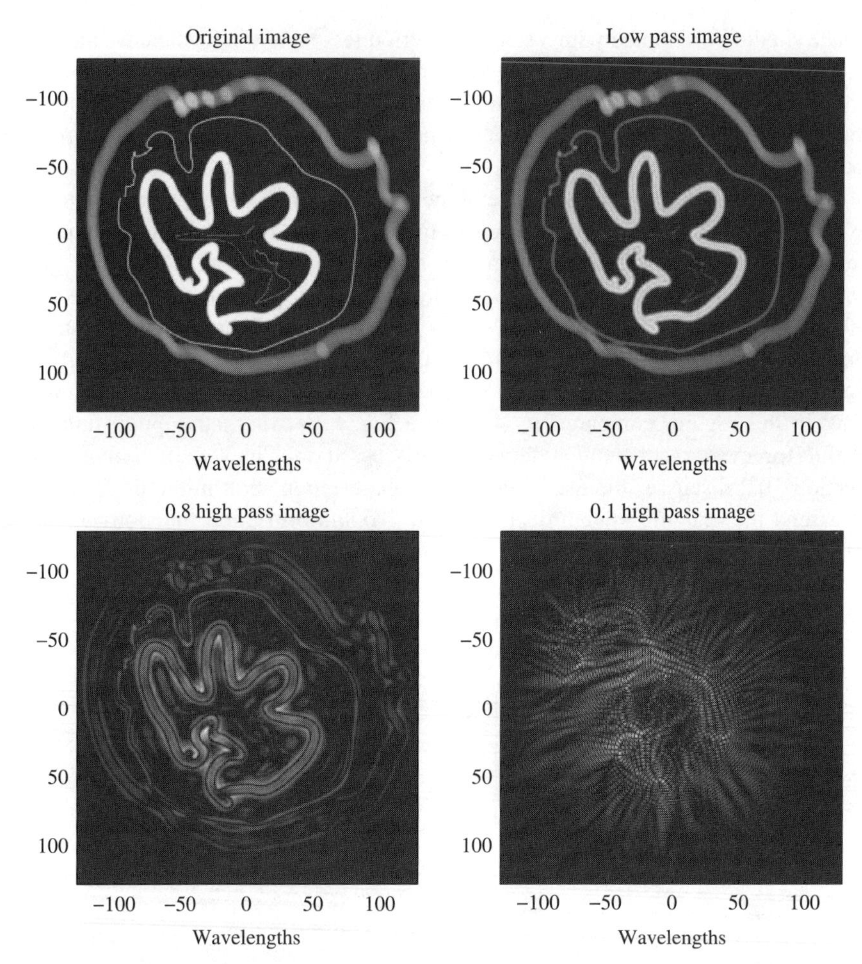

Figure 4.16 Effect of pupil filtering on in the imaging system corresponding to Figs. 4.14 and 4.15. The lower left image is filtered by exactly the transfer function of Fig. 4.15, which corresponds to a lens with the center 0.2 radius component obscured. The lower right image is filtered by a lens with the center 0.9 radius component obscured. Knowledge of the $f/\#$ and the spatial scale of the image is sufficient to accurately model the system scaled in wavelengths.

4.8 HOLOGRAPHY

Following the present section, the remainder of this text focuses exclusively on sensing of naturally occuring "incoherent" fields (with the notable exception of our discussion of optical coherence tomography in Section 6.5). Prior to turning our attention away from coherent fields, however, we briefly turn our attention to holography. *Holography* is a form of optical interferometry invented by Gabor in 1948 [83] and substantially extended by many investigators after the invention of the laser. Interferometric imaging based on the van Cittert–Zernike theorem, as discussed in Section 6.4.2, predates Gabor's work, but holography is fundamentally different from classical interferometry in that it provides a mechanism for imaging the coherent field itself, rather than just the object irradiance or spectral density.

Given the revolutionary nature of holography, as evidenced by Gabor's Nobel prize and the many associated Nobel prizes in laser technology, nonlinear optics, and optical interferometry, the reader may be surprised that holography was not included among the revolutions discussed in Chapter 1. The author's response is to note that while the invention of the laser may be the most revolutionary event in the history of optical science, the impact of coherent light and holography on optical sensing to date is relatively modest. The vast majority of images and spectra recorded are generated by incoherent processes, although in the case of spectroscopy these processes are often driven by laser excitation. I believe, however, that the full impact of coherent excitation and interferometric detection are yet to come. As noted in Section 1.4, a fourth revolution is emerging in interferometric optical processing and coherence detection. Although it is now 60 years old, holography may be regarded as the first salvo in this fourth revolution.

While hope for mass market applications of holographic displays and memories continues, the principal modern applications of holography are spatiospectral filters for liquid crystal displays, dispersive spectrometers, and laser line stabilization. Holograms are also used as transmittance filters in imaging and in illumination and optical interconnection systems. *Analog* holograms, which are recorded using laser illumination and photochemical materials, are used for most display and filter applications. *Digital* holograms, which use optical lithography to create mathematically derived transmission functions, are used in imaging and interconnection applications.

This section covers three useful aspects of modern holography:

1. We review the basic nature of *off-axis* analog holography. A basic understanding of how holography can be used to record and reconstruct a coherent field is intrinsically interesting and is illuminating in considering the spatial band structure of images.
2. We describe volume holography, which is essential both to explaining how diffraction gratings for spectroscopic and filtering applications achieve 80–90% diffraction efficiencies and how static display holograms function with white-light illumination.
3. We discuss modal analysis of volume holograms, which is helpful in understanding the band structure of photonic and electronic crystals.

An analog hologram is formed when a coherent field is used to produce an optical element with transmittance proportional to the product of the field and a reference wave. The recording signal field is then recovered by illuminating the holographically recorded transmittance with a reference field. A typical recording geometry is illustrated in Fig. 4.17. The hologram is recorded on a plate or film coated with a photochemical layer. Optical properties of the photochemical layer are changed on

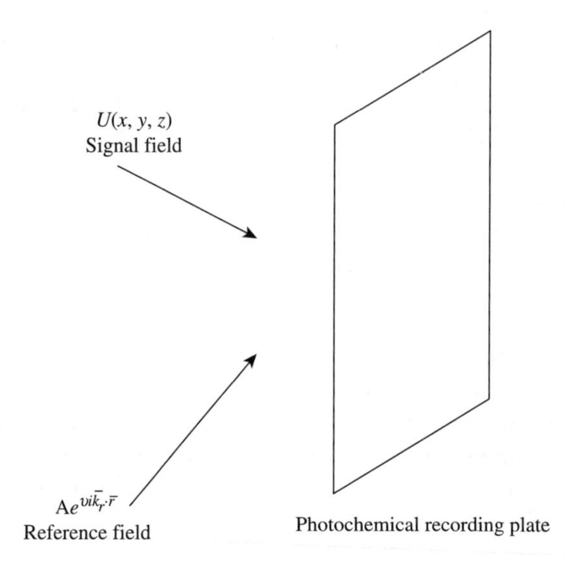

Figure 4.17 Hologram recording geometry.

absorption of light. The creation of grains of metallic silver from silver halide microcrystals is the classical photographic process. The metal particles darken the film to modulate the optical transmission. Absorption modulation visible to the human eye is desirable for photographic processes, but phase modulation by varying the thickness, surface relief, or index of refraction of the developed film is more popular for holography. Phase modulation is commonly achieved by photoinitiated polymerization.

A hologram is recorded through interference of a signal field $U(x, y, z)$ and a reference field. To reconstruct the signal field with high fidelity, the reference field must have uniform intensity over the exposure plane. The simplest field satisfying this constraint is the plane wave $Ae^{i\mathbf{k}\cdot\mathbf{r}}$. As discussed in Chapter 5, optical absorption is proportional to the *irradiance* I. The irradiance is proportional to the square of the electromagnetic field. Supposing that the signal field and the reference field record a hologram in the plane $z = 0$, the recording irradiance is

$$I(x, y) = |U(x, y, 0) + Ae^{ik_x x}|^2 \tag{4.78}$$

A photochemical process records a transmittance feature in proportion to the recording irradiance. For simplicity, we initially assume here that the recording irradiance modulates the real transmission such that $t(x, y) \propto I(x, y)$. In this case

$$t(x, y) \propto |A|^2 + |U(x, y, 0)|^2 + U(x, y, 0)A^*e^{-ik_x x} + U^*(x, y, 0)Ae^{ik_x x} \tag{4.79}$$

As illustrated in Fig. 4.18, a hologram is reconstructed by illuminating it with the original recording field. Under illumination by the original reference plane, the field

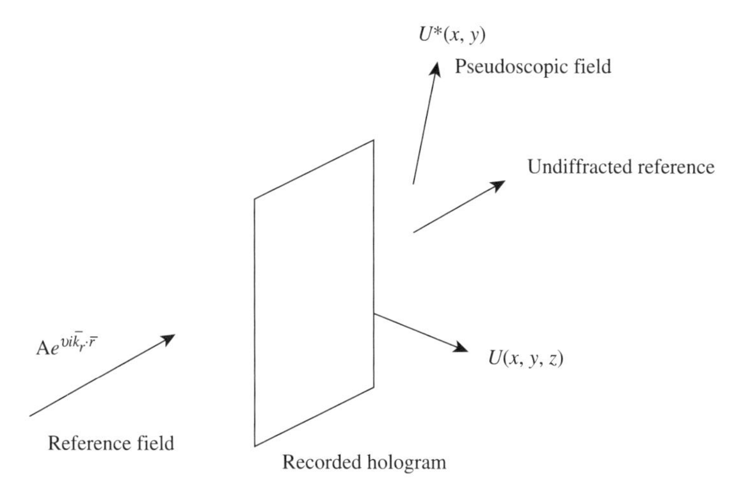

Figure 4.18 Hologram reconstruction geometry.

after modulation by the developed hologram is

$$t(x, y)Ae^{ik_x x} \propto |A|^2 Ae^{ik_x x} + |U(x, y, 0)|^2 Ae^{ik_x x} + U(x, y, 0)|A|^2 + U^*(x, y, 0)A^2 e^{i2k_x x} \tag{4.80}$$

The reconstructed field is a linear superposition of four field components. The term $|A|^2 Ae^{ik_x x}$ is the zeroth-order or undiffracted reference field. The term $|U(x, y, 0)|^2 Ae^{ik_x x}$ propagates along the same optical axis as the undiffracted field. The term $U^*(x, y, 0)A^2 e^{i2k_x x}$ is called the *pseudoscopic* field and propagates in some ways like the object field projected back on itself (e.g., if U is a diverging spherical wave, U^* is a converging wave).

The component $U(x, y, 0)|A|^2$ is proportional to the original signal field and diffracts exactly as though the original object were present. An observer of this diffracting component sees the object as if the object were present. It is interesting to note at this point that holography is not a multidimensional imaging system in the same sense as projection tomography. A tomographic imaging system estimates the density of an object at every point in a volume. A hologram records the 2D boundary conditions necessary to describe the field scattered off the object. Monochromatic holographic data cannot be inverted to reconstruct a 3D image, but polychromatic or multiangle holograms can be computationally inverted to form volume images (as can polychromatic and multiangle photographs). A hologram provides greater functionality than does a conventional photograph in that the hologram is essentially a window through which one can observe the object. In contrast with a normal photograph, one can look through a holographic window from any direction and see different perspectives on the object.

Since a hologram can be used to reconstruct the original signal, one may say that holography provides a mechanism for measuring the electromagnetic field using materials that can measure only the irradiance. The signal field may, in fact, be estimated by digital analysis of the recorded holographic pattern. This strategy is particularly effective if the hologram is recorded on an electronic detector array, but for reasons discussed momentarily, holographic recording systems generally require substantially higher spatial resolution than do those obtained by normal photography. Electronic detector arrays with resolution and pixel count consistent with holographic recording are just now becoming available.

The holographic recording strategy described here differs from Gabor's original proposal in that the reference is on a *carrier frequency* of wavenumber k_x. This approach is called *off-axis* or *Leith–Upatnieks holography* [149]. The carrier frequency is essential in isolating the signal field from the undiffracted and pseudoscopic components. It was not possible to generate a reasonable intensity reference wave for off-axis holography at the time of Gabor's original invention, but the intervening invention of the laser made this approach straightforward.

The utility of off-axis holography is illustrated by considering the Fourier transform of the reconstructed field:

$$|A|^2 A\delta\left(u - \frac{k_x}{2\pi}\right) + \hat{U}^*(u, v) * \hat{U}\left(u - \frac{k_x}{2\pi}, v\right)A + \hat{U}(u, v)|A|^2 + \hat{U}^*\left(u - \frac{k_x}{\pi}, v\right)A^2 \tag{4.81}$$

A cross section of this spatial spectrum along the u axis is sketched in Fig. 4.19. If $U(x, y)$ is bandlimited such that $|\hat{U}(u, v)| = 0$ for $u > B$, then the cross-correlation $\hat{U}^*(u, v) * \hat{U}(u, v)$ will have bandwidth $2B$. The signal $U(x, y, z)$ can be spatially filtered from the reconstructed hologram if the various terms cover distinct regions of the u axis. Refering again to Fig. 4.19, we see that spatial spectrum of the reconstructed signal may be separated from other components if $k_x > 6\pi B$. Interpreting this

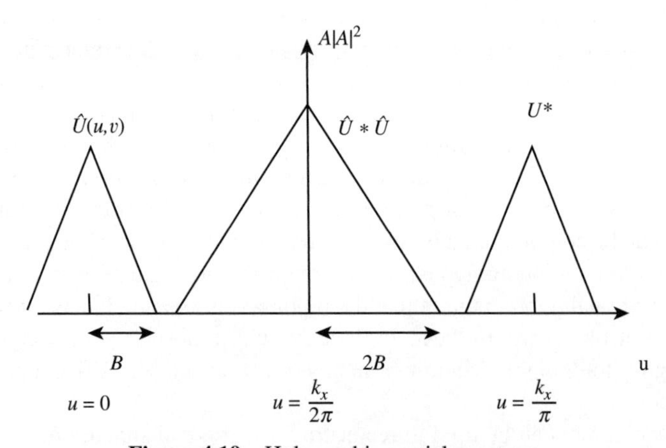

Figure 4.19 Holographic spatial spectrum.

result, one sees that off axis holography uses a high-spatial-frequency "carrier" to separate the holographic signal from background terms. Since the carrier frequency must be a factor of 3 greater than the maximum frequency in the holographically recorded image, much of the spatial bandwidth available in an off-axis holographic recording system is dedicated to separating components rather than the holographic signal. This carrier frequency explains the need for higher resolution in holographic media as compared to photographic media. In practice, especially for recording on electronic focal planes, signal disambiguation strategies other than spatial filtering may be considered and may yield substantially improved bandpass utilization.

A hologram recorded at one wavelength or orientation may be reconstructed using a reference wave at a different wavelength or angle of incidence. Changing the angle of incidence of the reference wave redirects the reconstructed hologram, changing the reconstruction wavelength changes the scale of the reconstruction. A hologram reconstructed at at a longer wavelength than the recording wavelength magnifies the object field. Gabor's original proposal focused on the potential of holography to magnify an object. Holograms may also be reconstructed by more complex probe fields; the use of holograms to correlate a coherent probe and a fixed signal is a core technique of optical signal processing [240].

To this point we have focused on "thin" and "transmission" holograms. The simple model of a multiplicative transmittance applies to such holograms. Several potential drawbacks must be considered for this technique, however. First, the signal conversion efficiency from the reference field to the reconstructed signal field is limited for thin holograms to at best 25% of the reference signal power. Also, thin holograms are not visible under white-light illumination. Since all angles and colors are diffracted by a thin hologram, white light remains white and no clear diffraction pattern emerges.

These drawbacks are resolved in volume holograms. For display holograms, the primary advantage of volume holograms are that they spatially and spectrally filter the reconstruction beam. Thus, a volume display hologram illuminated by white light produces a color image. The reconstruction color is not the natural color of the object; rather, it is determined by the recording and reconstruction geometries and wavelengths for the hologram. Display holograms generally use reflection geometries to maximize spectral filtering. We focus here on properties of transmission volume holograms, however, in anticipation of our discussion of spectroscopy in Chapter 9. The advantages of transmission holograms in spectroscopy are that near 100% diffraction efficiency can be achieved with very high spectral dispersion rates and that holograms may be used as spectral filters.

We limit our discussion to volume holograms recorded between two plane waves as illustrated in Fig. 4.20. Recording beam 1 is described by the plane wave $A\exp[i(-Kx/2 + k_z z)]$ and recording beam 2 by the plane wave $A\exp[i(Kx/2 + k_z z)]$. The recording irradiance in the holographic emulsion is

$$I(x) = |A|^2[1 + \cos(Kx)] \tag{4.82}$$

We assume that the emulsion is of infinite extent in x and y and of thickness d along the z axis.

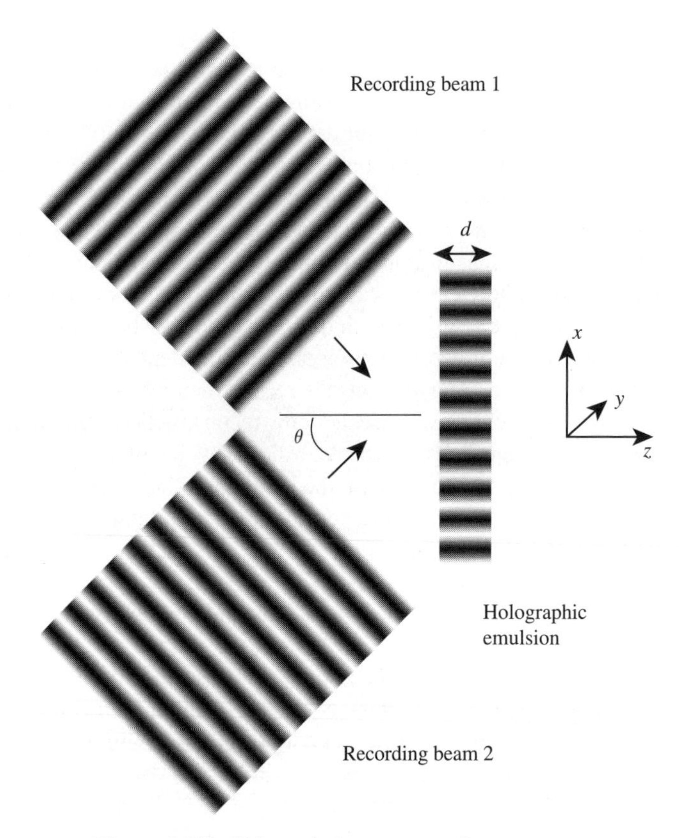

Figure 4.20 Volume hologram recording geometry.

Volume holograms are generally recorded in phase modulating materials, such as photopolymers, gelatins, or photorefractives. We assume that the permittivity of the recording material is modulated in proportion to the recording field, that is, that

$$\varepsilon = \varepsilon_m + \alpha I(x) \tag{4.83}$$

where ε_m is the material permittivity prior to holographic modulation. The holographic change in permittivity is typically very weak, ranging from a factor of 10^{-5} to 10^{-1} of the unperturbed value.

Since the medium is of finite thickness, analysis of volume holographic reconstruction is a wave propogation problem. We begin by considering the wave equation in the hologram. For simplicity, we consider the "transverse electric field" solution such that $\mathbf{E} \cdot \nabla \log(\varepsilon) = 0$, which allows us to neglect the corresponding term in Eqn. (4.17). The wave equation with this term included is considered in Problem 4.11. For a scalar field $U(x, y, z)$, the wave equation for a monochromatic field in a hologram recorded with the irradiance of Eqn. (4.82) takes the form

$$\nabla^2 U + \mu\omega^2[\varepsilon + \Delta\varepsilon \cos(Kx)]U = 0 \tag{4.84}$$

This equation is often considered using "coupled wave" analysis [137,179]. We assume that a plane wave $R\exp[i(k_{rx}x + k_{rz}z)]$ is incident on the hologram. Both k_{rx} and k_{rz} may be changed relative to the recording beams, and the reconstruction wavelength may also be different from the recording wavelength. Scattering from the hologram generates a signal plane wave $S\exp[i(k_{sx}x + k_{sz}z)]$. The transfer of light from the reconstruction wave to the signal wave is often modeled using the *slowly varying envelope approximation*, under which the amplitudes R and S are assumed to be slowly varying functions of z. "Slow" in this case means that $S'' \ll k_{sz}^2 S$.

Substituting $U = R\exp[i(k_{rx}x + k_{rz}z)] + S\exp[i(k_{sx}x + k_{sz}z)]$ in Eqn. (4.84), we note first that consistency with respect to x requires that $k_{sx} = k_{rx} - K$. Separating terms of similar spatial frequency with respect to z produces the coupled wave equations

$$ik_{rz}\frac{dR}{dz} + \frac{k^2}{2}\frac{\Delta\varepsilon}{\varepsilon}Se^{i(k_{sz}-k_{rz})z} = 0 \tag{4.85}$$

$$ik_{sz}\frac{dS}{dz} + \frac{k^2}{2}\frac{\Delta\varepsilon}{\varepsilon}Re^{i(k_{rz}-k_{sz})z} = 0 \tag{4.86}$$

where $k_{sz} = \sqrt{k^2 - (k_{rz} - K)^2}$, $k = \sqrt{\mu\varepsilon}\omega$, and we neglect terms in S'' and R''. Equation (4.85) is simplified by defining $\tilde{S} = S(z)e^{i(k_{sz}-k_{rz})z}$ such that

$$\frac{dS}{dz} = e^{i(k_{rz}-k_{sz})z}\frac{d\tilde{S}}{dz} + i(k_{rz} - k_{sz})e^{i(k_{rz}-k_{sz})z}\tilde{S} \tag{4.87}$$

Substituting in Eqns. (4.85) and (4.86) yields

$$ik_{rz}\frac{dR}{dz} + \frac{k^2}{2}\frac{\Delta\varepsilon}{\varepsilon}\tilde{S} = 0 \tag{4.88}$$

$$ik_{sz}\frac{d\tilde{S}}{dz} - k_{sz}(k_{rz} - k_{sz})\tilde{S} + \frac{k^2}{2}\frac{\Delta\varepsilon}{\varepsilon}R = 0 \tag{4.89}$$

Assuming that the input plane of the hologram is $z = 0$ and the output plane is $z = d$, the solution to Eqn. (4.88) consistent with the boundary conditions that $S(0) = 0$ and $R(0) = R_0$ is

$$\begin{aligned} R(z) &= e^{i(\Delta k_z z/2)}R_0\left[\cos(\gamma z) - i\frac{\Delta k_z}{2\gamma}\sin(\gamma z)\right] \\ S(z) &= \frac{i}{2}\frac{k^2}{k_{sz}\gamma}\frac{\Delta\varepsilon}{\varepsilon}e^{-i(\Delta k_z z/2)}R_0\sin(\gamma z) \end{aligned} \tag{4.90}$$

where $\Delta k_z = k_{rz} - k_{sz}$ and $\gamma = \frac{1}{2}\sqrt{\Delta k_z^2 + k^4\Delta\varepsilon^2/(k_{rz}k_{sz}\varepsilon^2)}$.

Equations (4.90) simplify enormously under the condition that $k_{rz} = k_{sz}$, in which case $\Delta k_z = 0$. In this case

$$\begin{aligned} R(z) &= R_0 \cos(\gamma z) \\ S(z) &= i R_0 \sin(\gamma z) \end{aligned} \tag{4.91}$$

Defining the diffraction efficiency of a hologram to be the ratio of the diffracted signal irradiance to the incident reconstruction irradiance, for example

$$\eta = \frac{|S|^2}{|R_0|^2} \tag{4.92}$$

we see from Eqns. (4.91) that the diffraction efficiency reaches 1 at $z = \pi/2\gamma = \lambda \cos(\theta)\varepsilon/(2\Delta\varepsilon)$, where θ is the angle between the reconstruction and signal wavevectors and the z axis. As an example, $\varepsilon/\Delta\varepsilon = 100$ achieves 100% diffraction efficiency in a hologram that is approximately 50 wavelengths thick.

The condition that $k_{rz} = k_{sz}$ is known as the *Bragg condition*, in honor of pioneering work on x-ray scattering from crystals by W. H. Bragg and W. L. Bragg [32]. The condition is most easily understood by returning to the wave normal surface of Fig. 4.1. A harmonic holographic modulation at spatial frequency $\mathbf{K}$ probed by a reconstruction plane wave with spatial frequency $\mathbf{k_r}$ is Bragg-matched for scattering if either of the two waves with spatial frequency $\mathbf{k_r} + \mathbf{K}$ or $\mathbf{k_r} - \mathbf{K}$ lies on the wave normal surface in the holographic material. The basic geometry for Bragg matching is illustrated in Fig. 4.21, which shows probe and reconstruction wavevectors. As illustrated in the figure, Bragg matching requires that the reconstruction wavevector $\mathbf{k_r}$ and the signal wavevector $\mathbf{k_s} = \mathbf{k_r} \pm \mathbf{K}$ lie on the the wave normal sphere. The

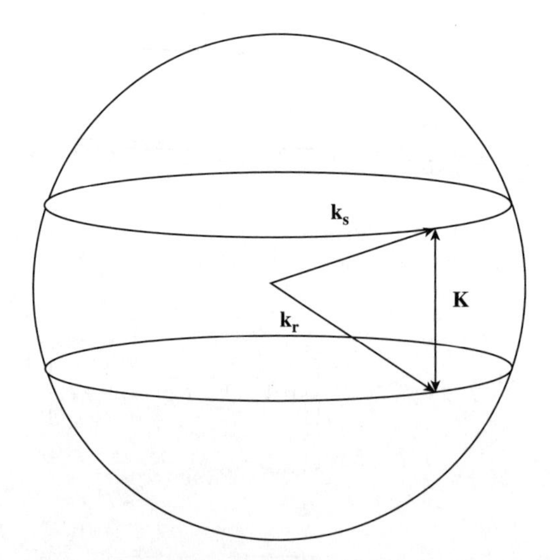

Figure 4.21 Bragg matching condition on the wave normal sphere.

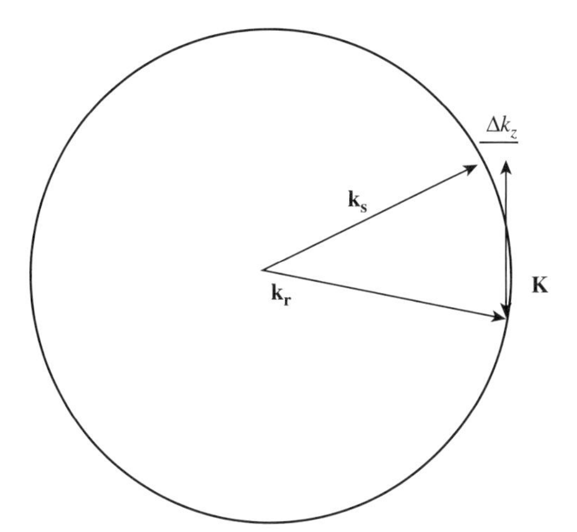

Figure 4.22 Reconstruction with a mismatched probe beam.

circles illustrated on the wave normal surface illustrate the degeneracy of the Bragg condition. Given **K**, any matched pair of probe and signal waves on the degeneracy curves will be Bragg-matched.

As illustrated in Fig. 4.22, a Bragg mismatch occurs when the probe beam is incident at an angle such that $\mathbf{k_r} + \mathbf{K}$ does not lie on the wave normal surface. The mismatch parameter Δk_z from Eqn. (4.90) is also illustrated in the figure. Under mismatch conditions, the maximum power transfer efficiency from the probe to the signal is

$$\frac{|S|^2_{\max}}{|R_0|^2} = \frac{k^4}{k_{sz}^2 \gamma^2} \frac{\Delta\varepsilon^2}{\varepsilon^2} \tag{4.93}$$

The peak diffraction efficiency as a function of angular mismatch of the probe beam for an example geometry is illustrated in Fig. 4.23. For the particular geometry chosen, the angular bandwidth of the hologram is approximately 1.7°. As discussed in Section 9.6, Bragg limitations on the angular and spectral sensitivity of volume holograms are important in spectrograph design.

As a final comment on holographic systems, we briefly consider rigorous scalar solutions of Eqn. (4.84) [35]. We assume solutions of the form $U(x, y, z) = e^{ik_y y} e^{ik_z z} \psi(x)$, which reduces Eqn. (4.84) to the Mathieu equation [183]

$$\frac{d^2\psi}{dx^2} + [a + b\cos(Kx)]\psi = 0 \tag{4.94}$$

where $a = k^2 - k_y^2 - k_z^2$ and $b = k^2 \Delta\varepsilon/\varepsilon$. Equation (4.94) has solutions of the form

$$\psi(x) = e^{iqx} \sum_{n=-\infty}^{\infty} \alpha_n e^{inKx} \tag{4.95}$$

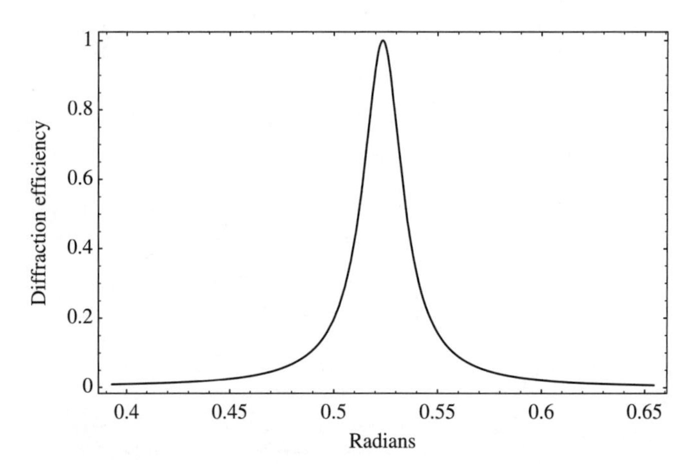

Figure 4.23 Maximum diffraction efficiency as a function angular mismatch for $K = k_0 \sin(\pi/6)$ and $\Delta\varepsilon/\varepsilon = 10^{-2}$.

Substitution of Eqn. (4.95) in Eqn. (4.94) yields a recursion relationship:

$$(a - (q + nK)^2)\alpha_n + \frac{b}{2}\alpha_{n+1} + \frac{b}{2}\alpha_{n-1} = 0 \tag{4.96}$$

The determinant of this infinite-order relationship can be transformed into the *Hill determinant* and evaluated in closed form [182]. The determinant produces an eigenvalue relationship for q in terms of k_z and k_y.

Without the holographic modulation, $q = \sqrt{k^2 - k_z^2 - k_y^2}$ with $\alpha_0 = 1$ and $\alpha_n = 0$ for $n \neq 0$. However, the holographic grating produces the multiharmonic solution of Eqn. (4.95). Figure 4.24 is a plot of the eigenvalue q as a function of k_z for $k_y = 0$. In

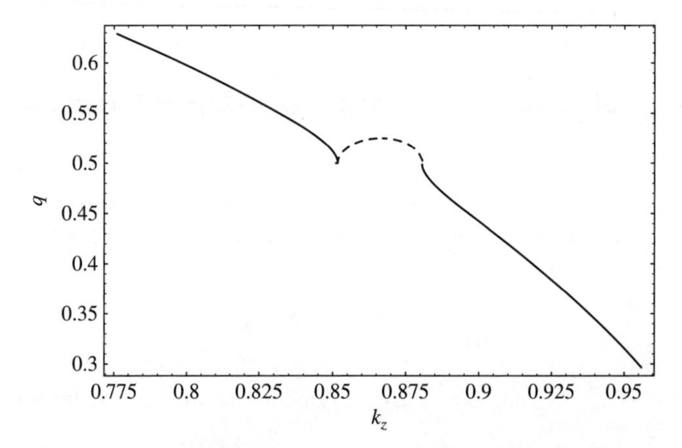

Figure 4.24 Dispersion relationship q versus k_z for $K = 2k\sin(\pi/6)$ in Eqn. (4.94). The plot is in units of k along both axes. In this example $\Delta\varepsilon = 0.05\varepsilon$. The imaginary component of q is shown in a dashed line within the stopband.

this figure, $K = 2k\sin(\pi/6)$ and the Bragg resonance occurs at $k_z = \cos(\pi/6)$. In the vicinity of the resonance all eigensolutions for q are complex, meaning that the mode described by Eqn. (4.95) is evanescent in x. The region in coordinates k_z, k_y, ω over which q is evanescent is termed the *stopband* of the grating.

We find then that within the Bragg region energy is transfered from the reconstruction wave to the signal wave because the reconstruction wave does not couple to a propagating wave in the hologram. Oscillations and localization of energy around the entrance aperture occur because neither coupled wave propagates in x. The stopband is effectively a gap in the wave normal sphere for propagating modes.

The stopband is a one-dimensional representation of a more general phenomenon in 2D and 3D periodic structures, termed *photonic crystals*. In higher-dimensional structures one may observe *bandgaps* in which $\mathbf{k}$ is complex along all axes for certain values of ω. In such structures modes may be localized in multiple dimensions [128].

Our main purpose in discussing holography in this text is to facilitate discussion of dispersive components in spectroscopic systems. Ultimately, photonic crystal structures and complex diffractive devices hold great promise for integrated dispersive and imaging components. For the present, however, the main use of our analysis of band structure is to facilitate discussion of electronic bands in Chapter 5.

PROBLEMS

4.1 *Bessel Beams:*

(a) Verify that the field

$$E(\mathbf{r}, t) = \frac{e^{i(\omega t - \beta z)}}{2\pi} \int_0^{2\pi} e^{-i\alpha(x\cos\phi + y\sin\phi)} d\phi \tag{4.97}$$

where $\beta^2 + \alpha^2 = 4\pi^2/\lambda^2$ is a solution to the wave equation (Eqn (4.18).

(b) Equation (4.97) is called a *Bessel beam* in view of the identity

$$2\pi J_0(\alpha\rho') = \int_0^{2\pi} e^{i\alpha\rho' \cos(\phi - \phi')} d\phi \tag{4.98}$$

where $J_0(x)$ is the zeroth-order Bessel function of the first kind. Plot the magnitude of the Bessel beam as a function of x and z for $\alpha = 0.2\pi/\lambda$. Explain why the Bessel beam is called a *diffraction-free* or *propagation-invariant* beam.

(c) Using a lens and an aperture mask, design a system to generate a Bessel beam.

(d) The Bessel beam is no longer propagation-invariant when the support of the beam is limited to a finite aperture. As an example, use the numerical Fresnel transformation to analyze diffraction of the input field distribution

$$\mathrm{circ}\left(\frac{\rho}{N\lambda}\right)J_0\left(\frac{\pi\rho}{10\lambda}\right) \tag{4.99}$$

for $N = 500$ and $N = 1000$ over a diffraction range from $z = 0$ to $z = 10{,}000\lambda$.

4.2 *Laguerre–Gaussian Modes:*

(a) Derive an expression similar to Eqn. (4.39) for $g(\rho, \phi)$ as a function of d for the case $f(\rho, \phi) = \psi_{mn}(\rho/w_0, \phi)$, where ψ_{mn} is the Laguerre–Gaussian function of Eqn. (3.82).

(b) Plot the amplitude and phase of $g(\rho, \phi)$ at $d = 0, w_0^2/\lambda, 10w_0^2/\lambda$ for $f(\rho, \phi) = \psi_{97}(\rho/w_0, \phi)$.

(c) Plot the amplitude and phase of $g(\rho, \phi)$ at $d = 0, 0.5w_0^2/\lambda, w_0^2/\lambda, 10w_0^2/\lambda$ for $f(\rho, \phi) = \psi_{97}(\rho/w_0, \phi) - 10\psi_{75}(\rho/w_0, \phi)$.

4.3 *The Talbot Effect.* According to the Talbot effect, coherent fields periodic in the transverse coordinates of an input aperture are "self-imaging," meaning that the original input field reappears at various planes in the z direction.

(a) Assuming a period of Λ in the x and y directions, derive an expression for the ranges at which the original field reappears.

(b) Assume that the input field is a 5×5 grid of circles. The circles are 5 wavelengths in diameter and are spaced on 15 wavelength centers. Assume that the field is zero outside the circles and uniform with constant phase and amplitude within each circle. Use Matlab to calculate the field diffracted from this input aperture at 5 self-imaging and at 5 non-self-imaging ranges.

4.4 *Fraunhofer Diffraction:*

(a) Design an experiment to use Fraunhofer diffraction of a $\lambda = 633$ nm laser beam to determine the size of a small circular pinhole. Plot the diffraction pattern observed and describe quantities one might measure to characterize the pinhole.

(b) Design an experiment to use Fraunhofer diffraction of a $\lambda = 633$ nm laser beam to determine the size of a human hair strand. Plot the diffraction pattern observed and describe quantities that one might measure to characterize the pinhole.

4.5 *Diffraction Patterns.* Generate a 1-mm-scale letter E and a 1-mm-scale letter O. Calculate the 2D Fourier transform of each in Matlab. Calculate the diffraction pattern when each is normally illuminated by a plane wave with 1 μm wavelength light. Find the diffraction pattern at ranges of 0 m, 10 cm, 1 m, and 10 m. Be sure to mark distance scales on your plots.

4.6 *The Grating Equation.* Equation (4.58) is called the *grating equation.* With reference to this equation

(a) Given Λ and $q \neq 0$, what is the longest wavelength that diffracts off a grating into a propagating mode? What is the angle of incidence for which diffraction occurs?

(b) Given λ and Λ, what is the largest value of q corresponding to a propagating mode? For what range of θ is this diffraction order observed?

(c) Plot θ' versus θ for all propagating modes and orders for $\lambda/\Lambda = 1$.

4.7 *The Coherent Impulse Response.* Consider a 2-cm-aperture lens with a 5-cm focal length illuminated by light with a wavelength of 1 μm. Use Matlab to calculate the impulse response for imaging from 10 cm in front of the lens to approximately 10 cm behind the lens. Plot the coherent impulse response over a defocus range of ± 0.5 cm (i.e., from 9 to 11 cm behind the lens).

4.8 *Fresnel Zone Plates.* A cylindrically symmetric mask with amplitude transmittance

$$t(\rho) = \tfrac{1}{2}[1 + \cos(\alpha\rho^2)] \tag{4.100}$$

is called a Fresnel zone plate. It acts as a lens with multiple focal lengths.

(a) Plot $t(\rho)$ for $\alpha = 50\ \text{cm}^{-2}$.

(b) What are the focal lengths associated with the zone plate?

(c) What fraction of incident irradiance is mapped into the field associated with each focal component?

4.9 *Highpass Spatial Filtering.* Consider a lens with a square aperture. The center of the lens is blocked by a square of side length 1 cm. The outer aperture is defined by an enclosing square of side length 1.01 cm. The image distance is 10 cm.

(a) Plot the coherent optical transfer function.

(b) Simulate a macroscopic image, such as a letter, imaged through this system.

4.10 *Absorption Holograms.* Prove that the maximum diffraction efficiency for a thin absorption hologram is 0.25.

4.11 *Floquet–Bloch Modes:*

(a) We neglected wave equation terms in $\nabla\varepsilon$ in deriving Eqn. (4.84). Explain why this is a valid approximation for reflection holograms.

(b) Solutions of the Floquet–Bloch form [Eqn. (4.95)] may still be found for the holographically modulated wave equation even if we retain the $\nabla\varepsilon$ term. Derive the recursion relationship replacing Eqn. (4.96) for this case.

(c) Is there a geometry (e.g., polarization and grating orientation) in which the $\nabla\varepsilon$ term significantly influences wave dynamics?

4.12 *Volume Holography:*

(a) Plot the maximum diffraction efficiency of a volume hologram as a function of reconstruction beam angle of incidence assuming that $\Delta\varepsilon/\varepsilon = 10^{-3}$ and that $K = \sqrt{2}\,k_0$.

(b) A volume hologram is recorded with $\lambda = 532$ nm light. The half-angle between the recording beams in free space is 20°. The surface normal of the holographic plate is along the bisector of the recording beams. The index of refraction of the recording material is 1.5. What is the period of grating recorded? Plot the maximum diffraction efficiency at the recording Bragg angle of the hologram as a function of reconstruction wavelength.

4.13 *Computer-Generated Holograms.* A computer-generated hologram (CGH) is formed by lithographically recording a pattern that reconstructs a desired field when illuminated using a reference wave. The CGH is constrained by details of the lithographic process. For example CGHs formed by etching glass are *phase-only* holograms. Multilevel phase CGHs are formed using multiple step etch processes. Amplitude-only CGHs may be formed using digital printers or semiconductor lithography masks. The challenge for any CGH recording technology is how best to encode the target hologram given the physical nature of the recording process. This problem considers a particular rudimentary encoding scheme as an example.

(a) Let the target signal image be the letter E function from Problem 4.5. Model a CGH on the basis of the following transmittance function

$$t(x, y) = \begin{cases} 1 & \text{if } \arg\left(\mathcal{F}\{E\}|_{u=\frac{x}{\lambda d}, v=\frac{y}{\lambda d}}\right) > 0 \\ 0 & \text{otherwise} \end{cases} \quad (4.101)$$

where λ is the intended reconstruction wavelength and $d \gg x$ is the intended observation range. $\mathcal{F}\{E\}$ is the Fourier transform of your letter E function. Numerically calculate the Fraunhofer diffraction pattern at range d when this transmittance function is illuminated by a plane wave.

(b) A more advanced transmittance function may be formed according to the following algorithm:

$$t(x, y) = \begin{cases} 1 & \text{if } \arg\left(e^{0.2\pi i[(x+y)/\lambda]}\mathcal{F}\{E\}|_{u=(x/\lambda d), v=(y/\lambda d)}\right) > 0 \\ 0 & \text{otherwise} \end{cases} \quad (4.102)$$

Numerically calculate the Fraunhofer diffraction pattern at range d when this transmittance function is illuminated by a plane wave. It is helpful when displaying these diffraction patterns to suppress low-frequency scattering components (which are much stronger than the holographic scattering).

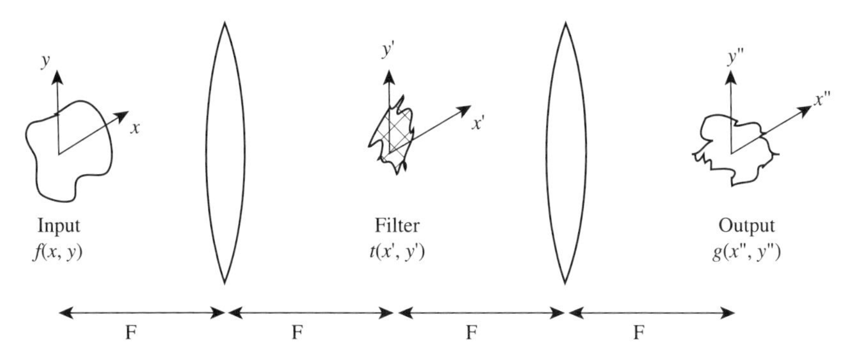

Figure 4.25 A Vanderlught correlator.

(c) A still more advanced transmittance function may be formed by multiplying the letter E function by a high frequency random phase function prior to taking its Fourier transform. Numerically calculate the Fraunhofer diffraction pattern for a transmission mask formed according to

$$t(x, y) = \begin{cases} 1 & \text{if } \arg\left(e^{0.2\pi i[(x+y)/\lambda)}\mathcal{F}\{e^{\phi(x,y)}E\}|_{u=(x/\lambda d),v=(y/\lambda d)}\right) > 0 \\ 0 & \text{otherwise} \end{cases} \tag{4.103}$$

where $\phi(x, y)$ is a random function with a spatial coherence length much greater than λ.

(d) If all goes well, the Fraunhofer diffraction pattern under the last approach should contain a letter E. Explain why this is so. Explain the function of each component of the CGH encoding algorithm.

4.14 *Vanderlught Correlators.* A Vanderlught correlator consists of the 4F optical system sketched in Fig. 4.25.

(a) Show that the transmittance of the intermediate focal plane acts as a shift-invariant linear filter in the transformation between the input and output planes.

(b) Describe how a Vanderlught correlator might be combined with a holographic transmission mask to optically correlate signals $f_1(x, y)$ and $f_2(x, y)$. How would one create the transmission mask?

(c) What advantages or disadvantages does one encounter by filtering with a 4F system as compared to simple pupil plane filtering?

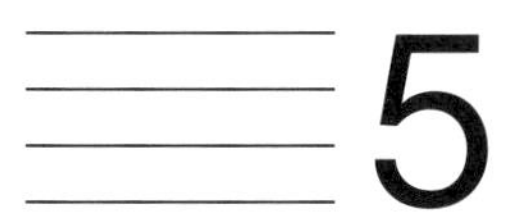

5

DETECTION

Despite the wide variety of applications, all digital electronic cameras have the same basic functions:

1. Optical collection of photons (i.e., a lens)
2. Wavelength discrimination of photons (i.e., filters)
3. A detector for conversion of photons to electrons (e.g., a photodiode)
4. A method to read out the detectors [e.g., a charge-coupled device (CCD)]
5. Timing, control, and drive electronics for the sensor
6. Signal processing electronics for correlated double sampling, color processing, and so on
7. Analog-to-digital conversion
8. Interface electronics

—E. R. Fossum [78]

5.1 THE OPTOELECTRONIC INTERFACE

This text focuses on just the first two of the digital electronic camera components named by Professor Fossum. Given that we are starting Chapter 5 and have several chapters yet to go, we might want to expand optical systems in more than two levels. In an image processing text, on the other hand, the list might be (1) optics, (2) optoelectronics, and (3–8) detailing signal conditioning and estimation steps. Whatever one's bias, however, it helps for optical, electronic, and signal processing engineers to be aware of the critical issues of each major system component. This chapter accordingly explores electronic transduction of optical signals.

We are, unfortunately, able to consider only components 3 and 4 of Professor Fossum's list before referring the interested reader to specialized literature. The specific objectives of this chapter are to

- Motivate and explain the need to augment the electromagnetic field theory of Chapter 4 with the more sophisticated coherence field theory of Chapter 6 and to clarify the nature of optical signal detection
- Introduce noise models for optical detection systems
- Describe the space–time geometry of sampling on electronic focal planes

Pursuit of these goals leads us through diverse topics ranging from the fundamental quantum mechanics of photon–matter interaction to practical pixel readout strategies. The first third of the chapter discusses the quantum mechanical nature of optical signal detection. The middle third considers performance metrics and noise characteristics of optoelectronic detectors. The final third overviews specific detector arrays. Ultimately, we need the results of this chapter to develop mathematical models for optoelectronic image detection. We delay detailed consideration of such models until Chapter 7, however, because we also need the coherence field models introduced in Chapter 6.

5.2 QUANTUM MECHANICS OF OPTICAL DETECTION

We introduce increasingly sophisticated models of the optical field and optical signals over the course of this text. The geometric visibility model of Chapter 2 is sufficient to explain simple isomorphic imaging systems and projection tomography, but is not capable of describing the state of optical fields at arbitrary points in space. The wave model of Chapter 4 describes the field as a distribution over all space but does not accurately account for natural processes of information encoding in optical sources and detectors. Detection and analysis of natural optical fields is the focus of this chapter and Chapter 6.

Electromagnetic field theory and quantum mechanical dynamics must both be applied to understand optical signal generation, propagation, and detection. The postulates of quantum mechanics and the Maxwell equations reflect empirical features of optical fields and field–matter interactions that must be accounted for in optical system design and analysis. Given the foundational significance of these theories, it is perhaps surprising that we abstract what we need for system design from just one section explicitly covering the Maxwell equations (Section 4.2) and one section explicitly covering the Schrödinger equation (the present section). After Section 4.2, everything that we need to know about the Maxwell fields is contained in the fact that propagation consists of a Fresnel transformation. After the present section, everything we need to know about quantum dynamics is contained in the fact that charge is generated in proportion to the local irradiance.

Quantum mechanics arose as an explanation for three observations from optical spectroscopy:

1. A hot object emits electromagnetic radiation. The energy density per unit wavelength (e.g., the spectral density) of light emitted by a thermal source has a

temperature-dependent maximum. (A source may be red-hot or white-hot.) The spectral density decays exponentially as wavenumber increases beyond the emission peak.

2. The spectral density excited by electronic discharge through atomic and simple molecular gases shows sharp discrete lines. The discrete spectra of gases are very different from the smooth thermal spectra emitted by solids.
3. Optical absorption can result in *cathode rays*, which are charged particles ejected from the surface of a metal. A minimum wavenumber is required to create a cathode ray. Optical signals below this wavenumber, no matter how intense, cannot generate a cathode ray.

These three puzzles of nineteenth-century spectroscopy are resolved by the postulate that materials radiate and absorb electromagnetic energy in discrete quanta. A quantum of electromagnetic energy is called a *photon*. The energy of a photon is proportional to the frequency ν with which the photon is associated. The constant of proportionality is Planck's constant h, such that $E = h\nu$. Quantization of electromagnetic energy in combination with basic statistical mechanics solves the first observation via the Planck radiation formula for thermal radiation. The second observation is explained by quantization of the energy states of atoms and molecules, which primarily decay in single photon emission events. The third observation is the basis of Einstein's "workfunction" and is explained by the existence of structured bands of electronic energy states in solids.

The formal theory of quantum mechanics rests on the following axioms:

1. A quantum mechanical system is described by a state function $|\boldsymbol{\Psi}\rangle$.
2. Every physical observable a is associated with an operator $\mathbf{A}$. The operator acts on the state $\boldsymbol{\Psi}$ such that the expected value of a measurement is $\langle\boldsymbol{\Psi}|\mathbf{A}|\boldsymbol{\Psi}\rangle$.
3. Measurements are quantized such that an actual measurement of a must produce an eigenvalue of $\mathbf{A}$.
4. The quantum state evolves according to the Schrödinger equation

$$\mathbf{H}\boldsymbol{\Psi} = i\hbar \frac{\partial \boldsymbol{\Psi}}{\partial t} \tag{5.1}$$

where $\mathbf{H}$ is the Hamiltonian operator.

The first three postulates describe perspectives unique to quantum mechanics, the fourth postulate links quantum analysis to classical mechanics through Hamiltonian dynamics.

There are deep associations between quantum theory and the functional spaces and sampling theories discussed in Chapter 3: $\boldsymbol{\Psi}$ is a point in a Hilbert space V, and V is spanned by orthonormal state vectors $\{\Psi_n\}$. The simplest observable operator is the state projector $\mathbf{P_n} = |\Psi_n\rangle\langle\Psi_n|$. The eigenvector of $\mathbf{P_n}$ is, of course, Ψ_n. If $\mathbf{P_n}\boldsymbol{\Psi} = \mathbf{0}$, then the system is not in state Ψ_n. If $\mathbf{P_n}\boldsymbol{\Psi} = \boldsymbol{\Psi}$, then the system is definitely in state Ψ_n. In the general case, we interpret $|\langle\Psi_n|\boldsymbol{\Psi}\rangle|^2$ as the probability that the system is in state Ψ_n.

For a static system, the eigenvalue of the Hamiltonian operator is the total system energy. For the Hamiltonian eigenstate Ψ_n, we have

$$\mathbf{H}|\Psi_n\rangle = E_n|\Psi_n\rangle \tag{5.2}$$

This eigenstate produces a simple solution to the Schrödinger equation in the form

$$\Psi(t) = e^{-i(E_n t/\hbar)}|\Psi_n\rangle \tag{5.3}$$

Having established the basic concepts of quantum mechanics, we turn to the quantum description of optical detection. Detection occurs when a material system, such as photographic film, a semiconductor, or a thermal detector interacts with the optical field. We assume that the Hamiltonian of the isolated material system is $\mathbf{H}_0$ and that the system is initially in a ground eigenstate Ψ_g corresponding to energy value E_g. Interaction between charge in the material system and the electromagnetic field of the incident optical signal perturbs the system Hamiltonian. Let $\mathbf{H}_1$ represent the energy operator for this perturbation. The system Hamiltonian including the perturbation is $\mathbf{H} = \mathbf{H}_0 + \mathbf{H}_1$.

The perturbation to the system Hamiltonian raises the possibility that the state of the system may change. When this occurs, a photon is absorbed from the optical signal, meaning that the energy state of the field drops by one quantum and the energy state of the material system increases by one quantum. Let $|\Psi_e\rangle$ represent the excited state of the material system. We may attempt a solution to the Schrödinger equation using a superposition of the ground state and the excited state:

$$|\Psi(t)\rangle = a(t)e^{-i(E_g t/\hbar)}|\Psi_g\rangle + b(t)e^{-i(E_e t/\hbar)}\Psi_e\rangle \tag{5.4}$$

The transition between the ground and excited states is mediated by the perturbation $\mathbf{H}_1$. $\mathbf{H}_1$ is an operator corresponding to the classical potential energy induced in the material system by the incident field. Since the spatial scale of the quantum system is typically just a few angstroms or nanometers, we may safely assume that the field is spatially constant over the range of the interaction potential. The field varies as a function of time, however. Suppose that the field has the form $\mathbf{A}e^{i2\pi\nu t}$. The interaction potential is typically linear in the field, as in

$$\mathbf{H}_1 = \mathbf{p}\cdot\mathbf{A}e^{i2\pi\nu t} + \text{c.c.} \tag{5.5}$$

where $\mathbf{p}\cdot\mathbf{A}$ is an operator and c.c. refers to the complex conjugate and $\mathbf{p}$ is typically related to the dipole moment induced in the material. In the following we substitute $\mathbf{f} = \mathbf{p}\cdot\mathbf{A}$.

Substituting $\Psi(t)$ in the Schrödinger equation produces

$$\begin{aligned}
\mathbf{H\Psi} &= aE_g e^{-i(E_g t/\hbar)}|\Psi_g\rangle + a\mathbf{H}_1 e^{-i(E_g t/\hbar)}|\Psi_g\rangle \\
&\quad + bE_e e^{-i(E_e t/\hbar)}|\Psi_e\rangle + b\mathbf{H}_1 e^{-i(E_e t/\hbar)}|\Psi_e\rangle \\
&= i\hbar \frac{\partial \mathbf{\Psi}}{\partial t} \\
&= aE_g e^{-i(E_g t/\hbar)}|\Psi_g\rangle + ia' e^{-i(E_g t/\hbar)}\hbar\,|\Psi_g\rangle \\
&\quad + bE_e e^{-i(E_e t/\hbar)}|\Psi_e\rangle + ib' e^{-i(E_e t/\hbar)}\hbar\,|\Psi_e\rangle
\end{aligned} \tag{5.6}$$

where $a' = da/dt$ and $b' = db/dt$.

With elimination of redundant terms and operating from the left with the orthogonal states $\langle\Psi_g|$ and $\langle\Psi_e|$, Eqn. (5.6) produces the coupled equations

$$\begin{aligned}
a'(t) &= \frac{a}{i\hbar} e^{i2\pi\nu t}\langle\Psi_g|\mathbf{f}|\Psi_g\rangle + \frac{b}{i\hbar}\exp^{\left(i\left[\frac{(E_g-E_e)}{\hbar}+2\pi\nu\right]t\right)}\langle\Psi_g|\mathbf{f}|\Psi_e\rangle \\
b'(t) &= \frac{b}{i\hbar} e^{i2\pi\nu t}\langle\Psi_e|\mathbf{f}|\Psi_e\rangle + \frac{a}{i\hbar}\exp^{\left(i\left[\frac{(E_e-E_g)}{\hbar}-2\pi\nu\right]t\right)}\langle\Psi_e|\mathbf{f}^*|\Psi_g\rangle
\end{aligned} \tag{5.7}$$

where we have dropped terms oscillating at high frequencies $(E_e - E_g)/\hbar + 2\pi\nu$. Assuming that the system is initially in the ground state with $a = 1$ and $b = 0$

$$\frac{1}{i\hbar}\exp^{\left(i\left[\frac{(E_e-E_g)}{\hbar}-2\pi\nu\right]t\right)}\langle\Psi_e|\mathbf{f}^*|\Psi_g\rangle \tag{5.8}$$

is the rate at which the excited-state amplitude increases. The probability that the system is in the excited state as a function of time, for small values of $t = \Delta t$, is

$$\left|\int_o^{\Delta t} b'(t)dt\right|^2 = \frac{\Delta t^2}{4\hbar^2}|\langle\Psi_g|\mathbf{f}|\Psi_e\rangle|^2 \text{sinc}^2\left(\left[\frac{(E_e - E_g)}{\hbar} - 2\pi\nu\right]\frac{\Delta t}{2}\right) \tag{5.9}$$

We learn three critical facts from Eqn. (5.9):

1. The transition probability from the ground state to the excited state is vanishingly small unless the energy difference between the states, $E_e - E_g$, is equal to $h\nu$. This characteristic is reflected in strong spectral dependence in photodetection systems. At energies for which there are no quantum transitions, materials are transparent, no matter how intense the radiation. At energies for which there are transitions, materials are absorbing.
2. The transition probability is proportional to $|\mathbf{f}|^2$, where $\mathbf{f}$ is proportional to the amplitude of the electromagnetic field.

3. The transition from the ground state to the excited state adds a quantum of energy $(E_e - E_g)$ to the material system and removes a quantum of energy $h\nu = (E_e - E_g)$ from the electromagnetic field. While a broader theory detailing quantum states of the field is necessary to develop the concept of the *photon number operator*, the basic idea of absorption as an exchange of quanta between the field and the material system is established by Eqn. (5.9).

Practical detectors consist of very large ensembles of quantum systems. Photoexcited states rapidly decohere in such systems as the excited-state energy is transferred from the excited state through electrical, chemical, or thermal processes. Replacing the transition time Δt by a quantum coherence time t_c the signal generated in such systems is

$$i = \kappa \int |E(\nu)|^2 \mu_{eg}(\nu) g(\nu) d\nu \tag{5.10}$$

where κ is a constant and the *oscillator strength* $\mu_{eg}(\nu)$ is proportional to the square of the coherence time and of the quantum transition probability $\langle \Psi_e | \mathbf{f} | \Psi_g \rangle$. Removing the electric field amplitude from the quantum operation is a *semiclassical approximation* in that we consider quantum materials states but do not quantize states of the electromagnetic field. $g(\nu)$ is the *density of states* of the material system at frequency ν. While Eqn. (5.9) predicts that state transitions occur only at the quantum resonance frequency, large ensembles of detection states are spectrally broadened by *homogeneous* effects such as environmental coupling [which decreases the coherence time and broadens the sinc function in Eqn. (5.9)] and by *inhomogeneous* effects corresponding to the integration of signals from physically distinguishable quantum systems.

The power flux of an electromagnetic field, in watts per square meter $(\mathrm{W/m^2})$ is represented by the *Poynting vector*

$$\mathbf{S} = \mathbf{E} \times \mathbf{H} \tag{5.11}$$

For a harmonic field, one may use the Maxwell equations to eliminate $\mathbf{H}$ and show that the amplitude of the Poynting vector is $\sqrt{\varepsilon/\mu}|\mathbf{E}|^2$. This relationship is derived for the field as a function of time, but one may, of course, use Plancherel's theorem [Eqn. (3.19)] to associate $\sqrt{\varepsilon/\mu}|E(\nu)|^2$ with the *power spectral density* $S(\nu)$. We present a careful derivation of the power spectral density with the field considered as a random process in Chapter 6; for the present purposes it is sufficient to note that our basic model for photodetection is

$$i = \kappa \int S(\nu) h(\nu) d\nu \tag{5.12}$$

where $S(\nu) \sim |E(\nu)|^2$ is the power per unit area per unit frequency in the field and $h(\nu)$ describes the spectral response of the detector on the basis of quantum, geometric, and readout effects.

Despite our efforts to sweep all the complexity of optical signal transduction into the simple relationship of Eqn. (5.12), idiosyncracies of the quantum process still affect the final signal. The transition probability of Eqn. (5.9) reflects a process under which the material system changes state when a photon of energy equal to $h\nu$ is extracted from the field. At energy fluxes typical of optical systems the number of quanta in a single measurement varies from a few thousand to a million or more. As discussed in Section 5.5, measurements of a few thousand quanta produce noise statistics typical of counting processes.

The difference in scales between the quantum coherence time and the readout rate of the photodetector is also significant. The detected signal is proportional to the time average of the $|\mathbf{f}|^2$ over some macroscopic observation time. Since temporal fluctuations in the readout signal are many orders of magnitude slower than the oscillation frequency of the field, the detected signal is "rectified" and the temporal structure of the field is lost in noninterferometric systems.

To be useful as an optical detector, the state transition from the ground state to the excited state must produce an observable effect in the absorbing material. Photographic and holographic films rely on photochemical effects. In analog photography absorption converts silver salt into metallic silver and catalyzes further conversion through a chemical development process. This change is observed in light transmitted or reflected from the film. Since phase modulation based on variations in the density and surface structure of a material is preferred in holography, holographic films tend to use photoinitiated polymerization. Bolometers and pyroelectric detectors rely on physical phenomena, specifically thermal modulation of resistivity or electric potential. For digital imaging and spectroscopy, we are most interested in detectors that directly induce electronic potentials or currents. Mechanisms by which state transitions in these detectors induce signals are discussed in Section 5.3.

5.3 OPTOELECTRONIC DETECTORS

Optical signals are transduced into electronic signals by (1) *photoconductive effects*, under which optical absorption changes the conductivity of a device or junction; or (2) *photovoltaic effects*, under which optical absorption creates an electromotive force and drives current through a circuit. Photoconductive devices may be based on direct optical modulation of the conductivity of a semiconductor or on indirect effects such as photoemission or bolometry. Photovoltaic effects occur at junctions between photoconducting materials. Depending on the operating regime and the detection circuit, a photovoltaic device may produce a current proportional to the optical flux or may produce a voltage with a more complex relationship to the optical signal. This section reviews photoconductive and photovoltaic effects in semiconductors. We briefly overview photoconductive thermal sensors in Section 5.8.

5.3.1 Photoconductive Detectors

Solid-state materials are classified as metals (conductors), dielectrics (insulators), or semiconductors according to their optical properties. Metals are reflective. Dielectrics are transparent. Semiconductors are nominally transparent, but become highly absorbant beyond a critical optical frequency associated with the *bandgap energy*. The optical properties of semiconductors are sensitive to material composition and can be changed by doping with ionizable materials as well as by compounding and interface structure. On the basis of dopant, interface, and electrical parameters, semiconductors may be switched between conducting and dieletric states.

The optical properties of materials may be accounted for using a complex valued index of refraction $n' = n - i\kappa$. The field for a propagating wave in an absorbing material is

$$\mathbf{E} = \mathbf{E}_0 e^{i2\pi\nu t} e^{-i2\pi(n-i\kappa)(z/\lambda)} \tag{5.13}$$

The wave decays exponentially with propagation. Typically, one characterizes the loss of field amplitude by monitoring the irradiance $I \propto |\mathbf{E}|^2$. The decay of the irradiance is described by $I = I_o e^{-\alpha z}$, where $\alpha = 4\pi\kappa/\lambda$. The range over which the irradiance decays by $1/e$, $\delta = 1/\alpha$, is called the *skin depth*. The skin depth of metals is generally much less than one free-space wavelength. The vast majority of light incident on a metal is reflected, however, due to the large impedance mismatch at the dielectric–metal interface. The skin depth near the band edge in semiconductors may be 10–100 wavelengths. The real part of the index is near 1 for a metal. It is typically 3–4 near the band edge of a semiconductor.

The properties of conductors, insulators, and semiconductors are explained through quantum mechanical analysis. A solid-state material contains approximately 10^{25} quanta of negative charge and 10^{25} quanta of positive charge per cubic centimeter (cm^3). These quanta, electrons, and protons mixed with uncharged neutrons exist in a quantum mechanical state satisfying the Schrödinger equation. Description of the quantum state is particularly straightforward in crystalline materials, where the periodicity of the atomic arrangement produces bands of allowed and evanescent electron wavenumbers.

The Schrödinger equation for charge in a crystal lattice is a wave equation balancing the potential energy of charge displaced relative to the lattice against kinetic energy

$$\frac{\hbar^2}{2m}\nabla^2\Psi + V(\mathbf{r})\Psi = E\Psi \tag{5.14}$$

where $V(\mathbf{r})$ is the potential energy field and E is the energy eigenvalue for the state Ψ. In a crystal the potential energy distribution is periodic in three dimensions. The basic behavior of the states can be understood by considering the one-dimensional potential $V(x) = V_o \cos(Kx)$. For this case, Eqn. (5.14) is identical to Eqn. (4.95) and a similar dispersion relationship results. $\Psi(\mathbf{r})$ is a charged particle wavefunction and the Fourier "**k** space" corresponds to the charge momentum, but the structure of the momentum dispersion is the same as in Fig. 4.24. Just as we saw for Bragg diffraction, a *stopband* in which there are no allowed momentum states is created in the semiconductor crystal.

Since crystals are periodic in three dimensions, the solution of Eqn. (5.14) in natural crystals results in a 3D wavenormal surface with *bandgaps* of momentum space in which no energy eigenstates exist. The band structures of three particularly important semiconductor materials are shown in Fig. 5.1. The band diagrams show energy eigenvalues as a function of the momentum value k for the *Floquet modes*, which are 3D versions of Eqn. (4.96).

A significant difference between photonic and electronic band structure arises from the *Pauli exclusion principle*, which states that no two identical fermions may simultaneously occupy the same quantum state. The exclusion principle is a critical component in explaining the structure of atomic nulcei, atoms, molecules, and crystals. In addition to charge and mass, quantized values of angular momentum, or *spin* are associated with fundamental particles. *Fermions* are particles with spin states that preclude multiple quanta occupying the same quantum state. Electrons,

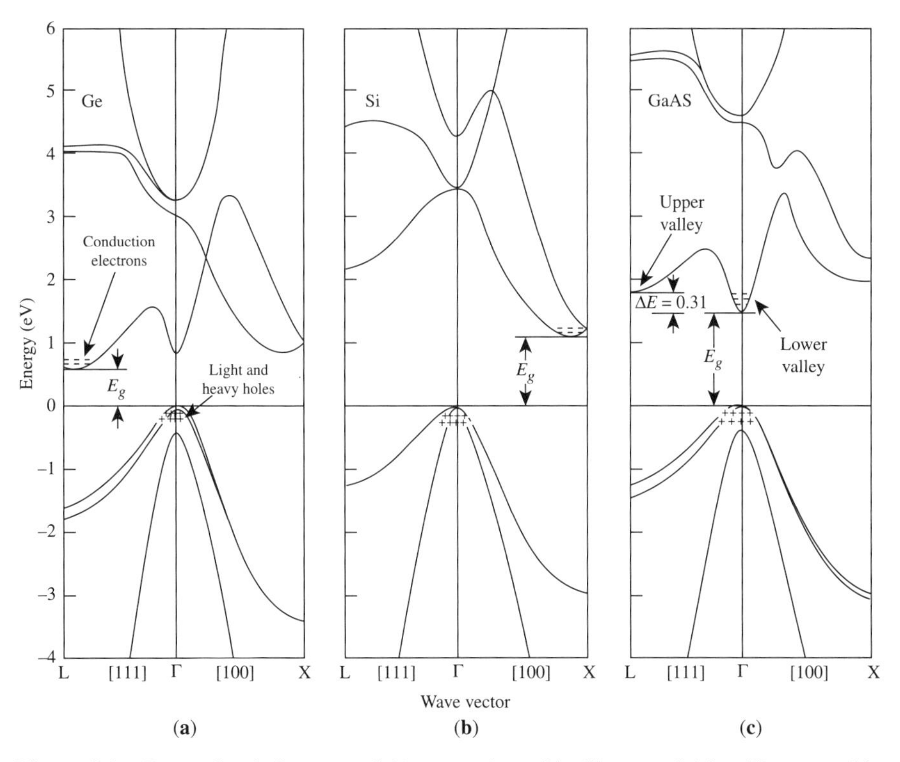

Figure 5.1 Energy band diagrams of (a) germanium, (b) silicon, and (c) gallium arsenide. The diagrams show k versus E for eigensolutions of the schrödinger equation. The k axis corresponds to critical directions with respect to the underlying crystal structure and the lines in k space between these points. (From Sze, *Physics of Semiconductor Devices* © 1981. Reprinted with permission of John Wiley & Sons, Inc.)

protons, and neutrons are fermions. *Bosons* are particles that allow multiple occupation of the same state. *Photons* are bosons. As an example, laser action occurs when many photons occupy the same state. Similar highly populated states are not available to fermions.

Just as an atom or molecule has ground and excited states, the energy eigenstates of a crystal corresponding to the eigenvalues shown in the band diagram may be occupied or unoccupied in the ground state. For the semiconductor materials of Fig. 5.1, the fully occupied ground state fills a continuous range of k values. The filled range is called the *valence band*. The next available excited states correspond to k values in the *conduction band*. For metals, the conduction band is partially filled even in the ground state and for dielectrics the gap between the ground state and the excited state is so large that crystal binding is disrupted by the excitation energy.

Charge transport occurs in semiconductors via conduction electrons and holes. *Conduction electrons* are charges excited thermally, electrically, or optically from the valance band to the conduction band. Depending on the material, these excited charges persist for some time and diffuse or move in response to an applied voltage. Similarly, the positively charged valence band states created by the excitation, the *holes*, can move through the crystal prior to recombination.

The energy difference between the top of the valance band and the bottom of the conduction band is E_g, the bandgap. E_g is indicated in Fig. 5.1. The E_g values for various materials are indicated in Table 5.1. Table 5.1 also lists a *cutoff wavelength* for each material. Because there are no vacant energy states between the top of the valance band and the bottom of the conduction band, absorption does not occur in semiconductors if the frequency of incident radiation is less than $\nu_c = E_g/h$. With E_g in electron volts and λ in micrometers, this corresponds to a cutoff wavelength $\lambda_c = 1.24/E_g$.

TABLE 5.1 Bandgap Energies, Cutoff Wavelengths, and Electron and Hole Mobilities of Several Semiconductors at Room Temperature

Material	E_g (eV)	λ_c (μm)	μ_n [cm^2/($V \cdot s$)]	μ_p [cm^2/($V \cdot s$)]
CdS	2.42	0.51	400	—
CdSe	1.74	0.71	650	—
GaAs	1.35	0.92	9,000	500
GaP	2.24	0.55	300	150
Ge	0.67	1.85	3,800	1,820
HgTe	0.15	8.27	25,000	350
InAs	0.33	3.76	33,000	460
InP	1.27	0.98	5,000	200
InSb	0.17	7.29	78,000	750
PbS	0.37	3.35	800	1,000
PbSe	0.26	4.77	1,500	1,500
Si	1.1	1.13	1,900	500
ZnS	3.54	0.35	180	—

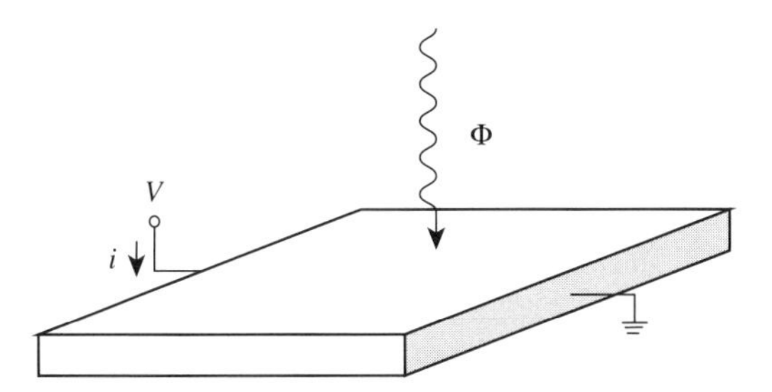

Figure 5.2 Photoconductive detector structure.

The basic structure of an optical detector based on photoconduction in semiconductors is sketched in Fig. 5.2. Light incident on the detector generates electron–hole pairs. The photogenerated charge migrates under the influence of an applied voltage V. The current in the circuit is

$$i = \eta\Phi eG \tag{5.15}$$

where η is the *quantum efficiency*, Φ is the photon flux, e is the electron charge, and G is the *photoconductive gain*; η is a number between zero and one indicating the fraction of incident photons that generate an electron–hole pair, and Φ is the ratio of the incident optical power P to the photon energy $h\nu$. (For a polychromatic source Φ is the average over the spectral range of the number of photons per second striking the detector.)

A conduction electron accelerated in a semiconductor by the electromotive force $E = V/l$ acquires a drift velocity

$$v_d = -\mu_n \frac{V}{l} \tag{5.16}$$

where μ_n is the electron mobility. If τ is the mean lifetime of the excited charge, the displacement of the charge during excitation is $v_d\tau$. The maximum displacement in a photodetector is l, but to maintain charge neutrality electrons absorbed by the positive electrode are replaced in the circuit by electrons entering the semiconductor from the negative electrode. Thus, a single excited charge may flow through the detector $G = |v_d|\tau/l$ times. Accounting for both electrons and holes, we find

$$G = \frac{(\mu_n + \mu_p)\tau V}{l^2} \tag{5.17}$$

While Eqn. (5.17) indicates that the photoconductive gain increases linearly in the bias voltage, gain saturation arises in practice through dependence of the carrier lifetime and mobilities on V. The mobilities may change as a result of thermal heating due to photo- and dark currents, but even before heating becomes significant surface and contact recombination effects significantly reduce the value of τ. The effective carrier lifetime $\tau_{\rm eff}$ is the harmonic mean

$$\frac{1}{\tau_{\rm eff}} = \frac{1}{\tau_0} + \frac{1}{\tau_c} + \frac{1}{\tau_s} \tag{5.18}$$

where τ_0 is the carrier lifetime in the bulk semiconductor, τ_c is the contact recombination lifetime, and τ_s is the lifetime for recombination with surface states. Mobility values for various semiconductors are given in Table 5.1. τ_0 ranges from a few milliseconds in indirect semiconductors like germanium and silicon to microseconds in direct materials such as GaAs, but in most detectors of interest for imaging and spectroscopy $\tau_{\rm eff}$ will be dominated by the contact recombination lifetime. Assuming that the diffusion length of the minority carrier is greater than the device thickness

$$\tau_c \approx \frac{1}{12}\frac{l^2}{D} \tag{5.19}$$

where l is the device thickness and D is the minority carrier diffusion constant [17]. The diffusion constant is linearly proportional to the carrier mobility according to $D = \mu kT/e$.

In summary, a photoconductive detector is a current source with current proportional to the incident photon flux and gain determined by the applied voltage and the diffusion length.

Spectral sensitivity is the most important aspect of photoconductive materials for applications in imaging and spectroscopy. The most common material, silicon, absorbs light from the near ultraviolet through the near infrared (roughly 300–1100 nm). As illustrated in Fig. 5.3, the skin depth of silicon, and thus the quantum efficiency of silicon devices, varies considerably over this range. Shorter wavelengths are absorbed more strongly; longer wavelengths tend to penetrate farther. On the micrometer scale,

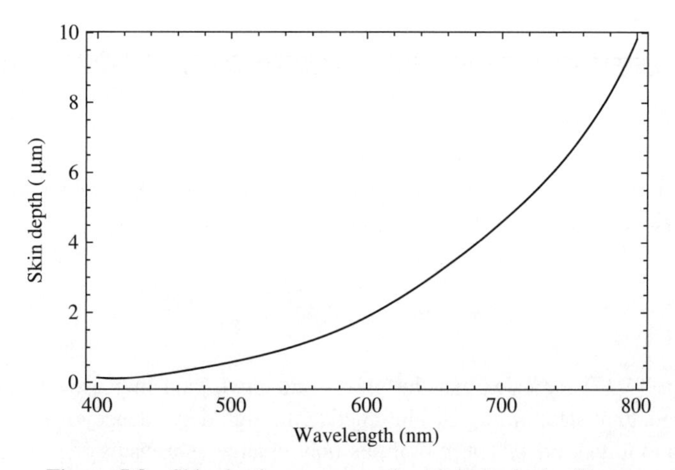

Figure 5.3 Skin depth versus wavelength in intrinsic silicon.

skin depth is a critical feature in determining minimum device size because devices smaller than an absorption length on the surface of detector tend to induce crosstalk.

It is possible to tailor the spectral response of a semiconductor by

- *Doping*. A pure semiconductor is called an *intrinsic* material and behaves as discussed thus far in this section. A material doped with donor or acceptor species (an *extrinsic material*) may support a population of conduction electrons or holes at thermal equilibrium. The impurity atoms create quantum states within the bandgap that may be ionized by optical radiation.
- *Compounding*. Table 5.1 lists elemental and compound semiconductors. The band structure of compound semicoductors is tuned in mixtures like $Ga_{1-x}Al_xAs$ and $Hg_{1-x}Cd_xTe$.
- *Quantum Confinement*. Nanometer-scale spatial structure in semiconductor materials creates artificial electronic resonances based on electron wavefunction cavity effects. Quantum wells and quantum dots are nanostructured devices designed to shape the absorption and conductivity properties of materials.

5.3.2 Photodiodes

Semiconductor circuits and devices are built from complex structures combining metal, dielectric, and semiconductor interfaces. The simplest semiconductor device, a *p–n junction diode*, is formed at the interface between p-type and n-type extrinsic photoconductors. A *p-type* material is doped with acceptor impurities such that valence band electrons are bound to impurity sites and unbound holes are produced in the valence band. An *n-type* material is doped with donor sites that contribute unbound conduction electrons.

Unbound charge diffuses across an interface between a p-type and an n-type material, meaning that holes from the p-type region enter the n-type material and electrons from the n-type material enter the p-type material. Charge diffusion creates a space charge region at the interface. The space charge region is negatively charged in the p-type material and positively charged in the n-type material. The space charge distribution creates an electromagnetic field across the junction that inhibits further charge diffusion.

The charge density, electric field, and electric potential across a p–n junction is illustrated in Fig. 5.4. The p-type interface is negatively charged at the acceptor density N_A, corresponding to compensation of ionized acceptor sites by donor electrons. The n-type material is similarly positively charged with peak density N_D. Compensation of the acceptor and donor sites creates a *depletion region* in which no free charge is present and conduction is inhibited.

Electric current is induced through a p–n junction diode by either an applied potential across the junction or by electron–hole pair generation. The *diffusion* current due to an applied field across the diode is

$$i_d = i_{\text{sat}}\left(e^{eV/kT} - 1\right) \tag{5.20}$$

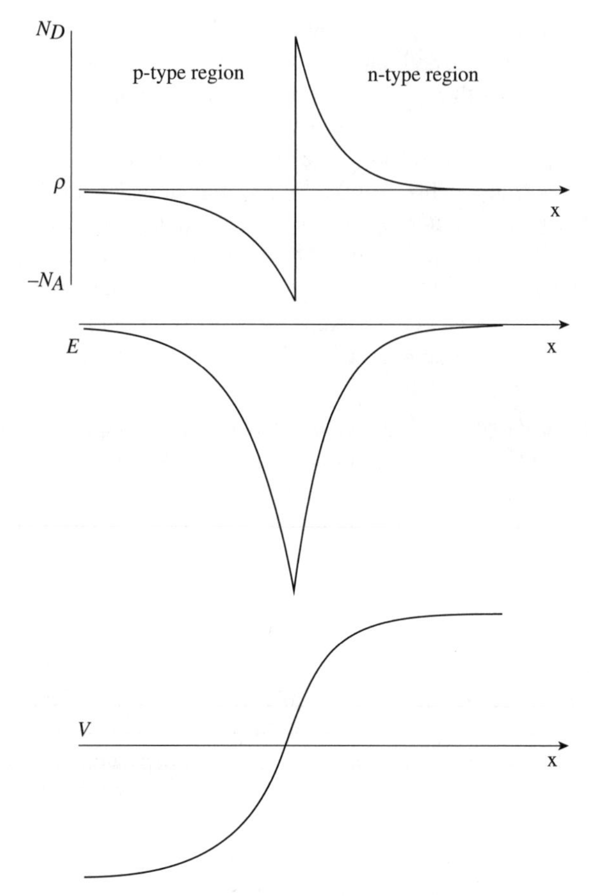

Figure 5.4 Charge density, electric field, and electric potential across the space charge region of a p–n junction. The peak charge density is equal to the donor density N_D in the n-type region and $-N_A$ in the p-type region.

where T is temperature, k is the Boltzmann constant, e is the electron charge, and V is the applied voltage. The exponential form of Eqn. (5.20) arises from the Boltzmann distribution of charge in thermal equilibrium [136]. The depletion region creates a barrier to charge flow across the diode. The Boltzmann distribution predicts the charge density above the barrier potential. At room temperature, kT/e is approximately 25 mV; i_{sat} depends on the geometry of the junction and materials properties.

The diffusion current produces the characteristic diode I–V curve illustrated in Fig. 5.5. The reverse-biased ($V < 0$) current saturates at $-i_{\text{sat}}$, which is typically very small. In forward bias ($V > 0$) the diode is highly conducting. The turn-on voltage is a multiple of the 25 mV thermal voltage in practical materials, for example in Si the turn-on voltage is approximately 0.7 V. A strong reverse bias produces "breakdown" in the diode and results in low junction resistance. Breakdown

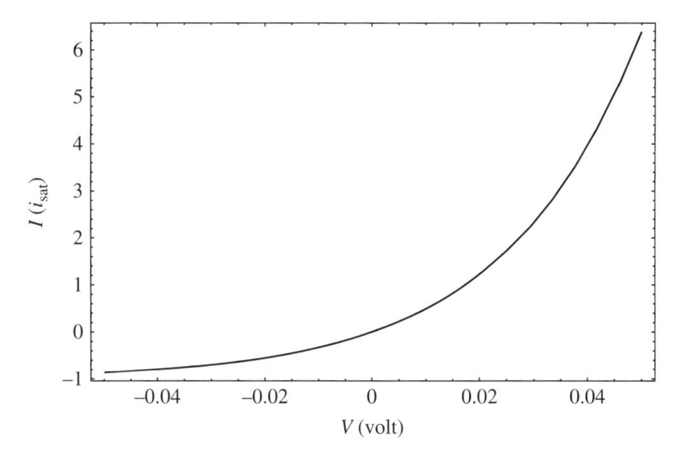

Figure 5.5 Current I versus voltage V for an ideal p–n diode.

is associated with electron–hole pair generation in the depletion region through the acceleration of charge with sufficient energy to ionize bound charge.

A photon flux Φ incident on the depletion region generates a current $-\eta e\Phi$ in the diode, where η is the quantum efficiency for electron–hole pair generation at the frequency of the incident light. Combining photogenerated and diffusion components, the total current across the diode is

$$i = -\eta e\Phi + i_{\text{sat}}\left(e^{eV/kT} - 1\right) \tag{5.21}$$

The optical signal absorbed by the diode may be characterized by measuring either the voltage generated across the diode or the current generated through the diode. A voltage is generated across the diode even in an open circuit. Setting $i = 0$ and solving for the photogenerated voltage in Eqn. (5.21) yields

$$V = \frac{kT}{e}\log\left(1 + \frac{\eta e\Phi}{i_{\text{sat}}}\right) \tag{5.22}$$

Equation (5.21) indicates that photovoltaic measurements of a diode might be an effective temperature gauge, but the nonlinear response with respect to Φ is generally unattractive.

While the total current through the diode depends on the applied potential, the change in the photocurrent due to light is linear in Φ. Photocurrent detection using an operational amplifier, as illustrated in Fig. 5.6, provides a simple mechanism for amplifying the diode current into a readout voltage linear in Φ; in this case the voltage is

$$V = e\eta R_f\Phi \tag{5.23}$$

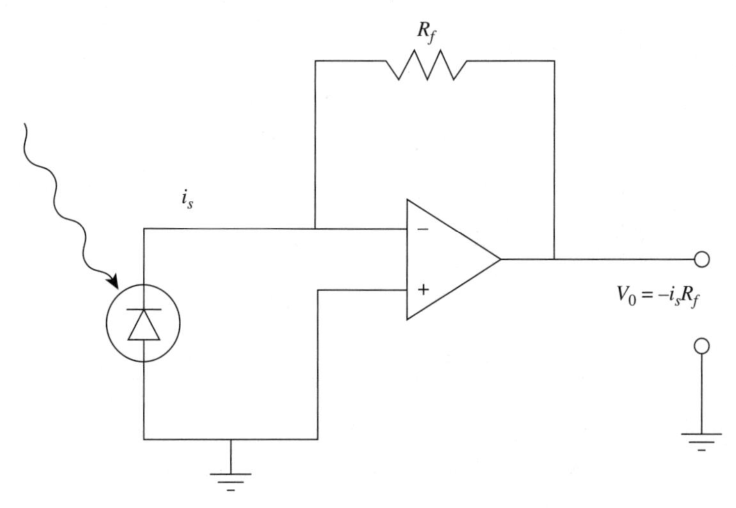

Figure 5.6 Diode photocurrent detection using an operational amplifier.

In summary, photodiodes act as current sources similar to photoconductive devices but without photoconductive gain. Gain is often provided by an operational amplifier, which also converts the photocurrent into a voltage proportional to the photon flux.

Photodiode geometry and circuits are extended in many ways for functional photodetectors. For example, the *p–i–n diode* adds an intrinsic absorption layer between the p- and n-type materials. The intrinsic layer affords a uniform depletion region with a constant acceleration field. *Avalanche photodiodes* are strongly reverse biased such that photogenerated charge induces a cascade of electron–hole pair ionizations, amplifying the current response by factors of 1000 or more. Finally, Schottky barrier photodiodes rely on the contact potential between a metal and an insulator to produce the space charge region.

5.4 PHYSICAL CHARACTERISTICS OF OPTICAL DETECTORS

Detectors are described according to

- Geometric characteristics, such as the spatial, spectral, and temporal structure of measurements and sampling rates
- Noise and statistical characteristics
- Physical characteristics, such as spectral and polarization sensitivities, linearity, dynamic range, and responsivity

Detector geometry defines sampling structure, which influences detector peformance so profoundly that it arises throughout the text as well as absorbing the entirety of

Chapter 7. Noise is considered in detail in Section 5.5 The present section briefly overviews essential physical characteristics of optical sensors.

Physical characteristics begin with the transduction mechanism. Common detectors rely on

- Photoinduced chemical or physical changes, as in photographic and holographic plates and films
- Photoemission, as in vacuum tubes
- Photoconduction and photovoltaic effects in semiconductors, as discussed in Section 5.3
- Thermal effects

We focus on large electrically addressable arrays of photodetectors, which rely on semiconducting or thermal detectors. The primary difference between semiconductor and thermal detectors is that the former is sensitive to the incident photon flux, while the latter is sensitive to the total absorbed optical energy. Assuming uniform quantum efficiency with respect to wavelength, the response of a photon detector at 1 W of power at $\lambda = 500\,\text{nm}$ is half the response of the same detector to 1 W of power at $\lambda = 1\,\mu\text{m}$. The distinction, of course, is that the photon flux at 500 nm is

$$\Phi = \frac{P}{h\nu} = \frac{P\lambda}{hc} = 2.5 \times 10^{18}\,\text{quanta/s} \tag{5.24}$$

while at 1 μm $\Phi = 5 \times 10^{18}$ quanta/s (quanta per second). For a thermal detector, the change in temperature is linearly proportional to the power deposited. Assuming uniform spectral absorption, a thermal detector produces the same response for 1 W at 500 nm as for 1 W at 1 μm.

Thermal detector arrays are commonly used in infrared spectral ranges where semiconductor detectors are expensive and difficult to fabricate and where relatively broad and uniform spectral response is desirable. Noise characteristics of thermal detectors are not generally as attractive as photon detectors, however, so most high-performance systems use photon detectors.

Responsivity, defined as

$$\mathcal{R} = \frac{\text{output signal}}{\text{input power}} \tag{5.25}$$

is a commonly referenced detector metric. The responsivity may be reported in volts per watt or amperes per watt, depending on the units in which the output signal is measured. For the photoconductive detector of Eqn. (5.15) the responsivity is

$$\mathcal{R} = \frac{I}{h\nu\Phi} = \frac{\eta\lambda e G}{hc} \tag{5.26}$$

which yields 0.8 $\eta G\lambda$ A/W for λ in micrometers. Similarly, the responsivity of the amplified photodiode of Eqn. (5.23) is

$$\mathcal{R} \approx 0.8\, \eta\lambda R_f \text{ V/W} \tag{5.27}$$

with R_f in ohms (Ω). Responsivity in linear proportion to the quantum efficiency η and λ is characteristic of photon detectors. The responsivity is also a function of λ through spectral variation in the quantum efficiency. The responsivity of a thermal detector, in contrast, is linearly proportional to the absorption efficiency but is insensitive to wavelength.

Response time is a measure of the minimum temporal variation in the irradiance a detector can resolve. While the spectral response is, in most cases, determined by the materials composition of the detector, temporal and temporal frequency responses of the detector system are determined primarily by details of the readout circuit. Capacitive effects produce exponential decay impulses in the temporal response, which yields $1/f$ decay in the frequency response of the detector.

Linearity describes the relationship between the input irradiance and the output signal. We have implicitly assumed in our definition of the responsivity that the output signal is linearly proportional to the input power. In practice, detectors respond linearly over a limited range. Beyond this range, saturation effects limit the detector response. Potential saturation effects are clear in our previous discussion of photodiodes, we discuss saturation in CCD detectors in Section 5.6.

The *signal-to-noise ratio* (SNR) is a commonly quoted detector characteristic. In electrical systems, the signal-to-noise ratio (SNR) is the ratio of the power in an electrical system to the noise power. The definition of SNR sometimes varies in applications to imaging systems, where "signal power" is not as easy to define as one might think. An image is ultimately a digital object; imaging scientists sometimes define SNR to mean the ratio between the mean or peak digital signal value and the noise-based standard deviation of the signal. In many cases, the digital signal values are proportional to the optical power, although they may be proportional to the magnitude of an electronic current or voltage.

This text uses the definition

$$\text{SNR} = \frac{\bar{s}}{\sigma_n} \tag{5.28}$$

where $\bar{s}$ is the mean reconstructed signal value and σ_n is the noise standard deviation. SNR is often quoted in decibels (e.g., $\text{SNR} = 10\log_{10} \bar{s}/\sigma_n$ dB). Depending on the definition and units, our definition of SNR may vary by a factor of 2 in dB or by a square in relation to alternative definitions.

Dynamic range refers to the number of distinct detector states that can be translated into digital values. Dynamic range is most often quoted in terms of decibels or bits; for example, a 16-bit detector produces 2^{16} values. The dynamic range may be more or less equivalent to the peak SNR for a linear system, but more commonly, the quoted value refers to the number of bits in the digitization circuitry, without

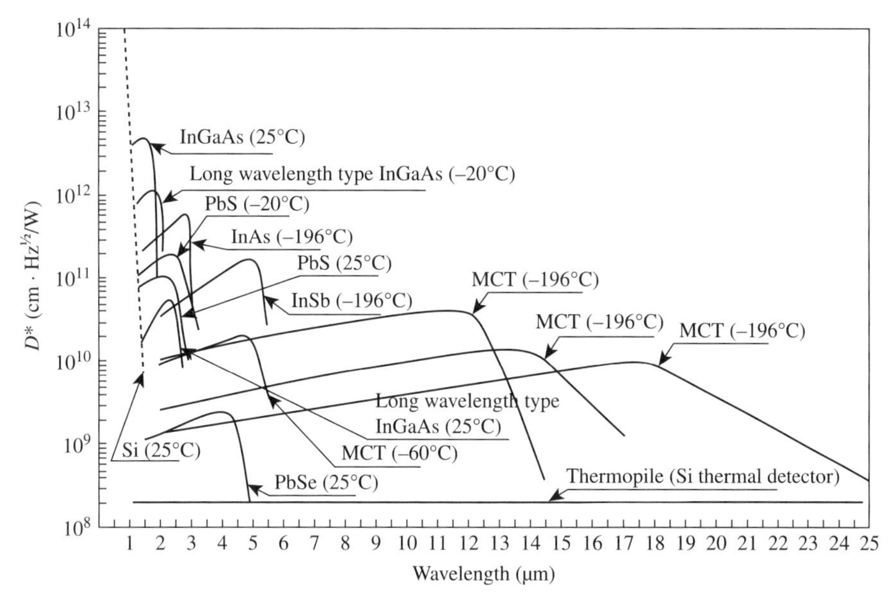

Figure 5.7 D^* as a function of wavelength for various detector materials (courtesy of Hamamatsu Corporation).

necessarily ensuring that the detector itself can meaningfully produce all 2^{16} values or that the mapping is linear. The process of analog to digital conversion produces *digitization noise* associated with the mapping from an continuous value to digital number, in most optical systems; however, this source of uncertainty insignificant.

Noise equivalent power (NEP) is the optical power inducing a signal-to-noise ratio of 1. It is often appropriate to assume that the noise power is proportional to the square root of the the detector bandwidth Δf and the detector area A. In this case, the detectivity D^*

$$D^* = \frac{\sqrt{A\Delta f}}{\text{NEP}} \tag{5.29}$$

is used as a detector metric. The common unit of detectivity is 1 "jones" = 1 cm · $\text{Hz}^{1/2}$/W. Values of D^* under similar operating conditions for various detector materials are plotted as a function of wavelength in Fig. 5.7. A value of D^* of 10^{10} jones, for example, implies that a 1-cm^2 detector operating with 1 Hz of bandwidth produces an SNR of 1 for 0.1 nW of incident power.

5.5 NOISE

An optical measurement returns a distribution of different values when repeatedly sampled under identical circumstances. Variation in recorded values is due to quantum mechanical uncertainty, unaccounted background processes or sampling and digitization structure. If different detectors are used to make the same measurement, variation due to differences in detector characteristics are also observed. One cannot eliminate variation in measurements, but by understanding the physical processes that create it and by modeling its statistics, one can account for its impact on images. Knowledge of probability density functions for measurement noise enables us to choose the most likely estimate of signal parameters and to bound estimation errors.

The optical signal incident on a detector is a *random process*. Optical detectors convert a stream of photons into a photocurrent, thus transforming one random process into another. Differences in implementing this transformation account for some of the controversy in defining SNR mentioned in Section 5.4. Systems with just a few detection channels digitize the current produced by a photodetector as a time series. On large detector arrays, such as those used in imaging systems, the photocurrent cannot be continuously observed for all detector elements. Instead, the photocurrent is integrated to create a charge. This charge is read at discrete intervals to produce discrete voltage signals. Noise arising in this transformation is usually modeled using a probability density function for the discrete signal. This section accordingly focuses on the statistics of discrete measurements rather than those of random processes.

Noise in imaging and spectroscopy differs from noise in optical communication, data processing, and data storage systems because temporal signal modulation and read frequencies tend to be substantially lower while spatial parallelism tends to be much higher. A typical imaging system operates at frame rates of 1–100 Hz with pixel read frequencies in the 1–10 MHz range. Modern optical imaging relies on very large detector arrays. Noise on arrays arises from pixel-level signal fluctuations (temporal noise) and pixel-to-pixel detector variations (fixed pattern noise). Temporal noise is due to quantum and thermal fluctuations; fixed pattern noise is caused by variations in the physical characteristics of detectors. Fixed pattern noise may in principle be ameliorated by characterizing the response of each pixel, but accurate characterization is impractical for large arrays.

Quantum photon and charge fluctuations produce *shot noise*. Shot noise is characteristic of processes that count discrete events, such as the detection of individual quanta of light and electricity. To understand shot noise, suppose that an extremely stable source generates an average of $\bar{n}$ photons in a time window of duration T. Dividing the time window into N subwindows each of duration T/N, the probability of exactly n counts in a particular time window is

$$p(n) = \frac{N!}{n!(N-n)!}\left(\frac{\bar{n}}{N}\right)^{n}\left(1-\frac{\bar{n}}{N}\right)^{N-n} \tag{5.30}$$

where the factorial term accounts for the number of permutations for assigning n counts to N slots and the exponential terms describe the probability of a specific realization of n slots with one count and $N-n$ slots with no counts. We assume that T/N is sufficiently short that the probability of two counts in any one subwindow vanishes.

In the limit $N \to \infty$, approximation of $N!$ and $(N-n)!$ using Stirling's approximation $N! \approx \sqrt{2\pi}N^{N+(1/2)}e^{-N}$ and the use of the limit $e^{\alpha} = \lim_{N\to\infty}(1-\alpha/N)^N$ yields

$$p(n) = \frac{\bar{n}^n}{n!}e^{-\bar{n}} \tag{5.31}$$

$p(n)$ is the *Poisson distribution*. Properties of the distribution include

$$\begin{aligned}
\sum_{n=0}^{\infty} p(n) &= 1 \\
\langle n \rangle &= \sum_{n=0}^{\infty} np(n) = \bar{n} \\
\langle n^2 \rangle &= \sum_{n=0}^{\infty} n^2 p(n) = \bar{n}^2 + \bar{n}
\end{aligned} \tag{5.32}$$

The variance of the distribution is thus $\sigma^2 = \langle n^2 \rangle - \langle n \rangle^2 = \bar{n}$.

Even though the energy per pixel is quite small, $\bar{n}$ is generally large in optical systems. For example, a typical image might correspond to 1–10 nW of power on a detector array. Integration for 30 ms deposits 10–100 nJ of energy. Over a megapixel sensor, this corresponds to 10–100 fJ per sensor element. Assuming 1 eV photons, this corresponds to between 10^5 and 10^6 photons per pixel. With such large photon counts, the probability of any specific number of photons is small, for example, for $\bar{n} = 10^5$, $p(\bar{n}) = 0.0013$. The standard deviation of the Poisson distribution, $\sigma = \sqrt{\bar{n}}$ grows as with the signal amplitude, but the signal grows at a faster rate and the signal-to-noise ratio increases in proportion to $\sqrt{\bar{n}}$.

The mean number of quanta detected for optical power P over an integration time of T is $\bar{n} = \eta TP/h\nu$. The shot-noise-limited SNR for this process is

$$\text{SNR} = \frac{\bar{n}}{\sigma_n} = \sqrt{\frac{\eta P}{h\nu\,\Delta f}} \tag{5.33}$$

where the bandwidth Δf is associated with the integration time by $T = 1/\Delta f$. Setting SNR = 1, we find that the quantum-limited noise equivalent power is

$$\text{NEP} = \frac{h\nu\Delta f}{\eta} \tag{5.34}$$

For 1 eV photons integrated at 30 Hertz with 50% quantum efficiency, this corresponds to NEP=10 attojoules (aJ).

While visible optical fields are generated by quantum processes far from thermal equilibrium, near-equilibrium processes in detector circuits lead to thermal, or *Johnson*, noise. Johnson noise arises from random fluctuations of thermally excited charge. These fluctuations may augment or detract from photogenerated currents. The structure of Johnson noise is derived by considering "modes" of the detector circuit. A mode at frequency f is populated according to the Boltzmann distribution for thermal blackbody radiators. The overall noise energy density in the circuit at energy hf is

$$E = \frac{2hf}{e^{hf/kT} - 1} \tag{5.35}$$

At room temperature, $kT/h = 6.25$ THz. At kHz–MHz frequencies of interest to imaging systems, one may safely assume that $E \approx kT$. Integrating over the active spectral range of the detector, the mean-square power of the thermal noise is $P_n = 2kT\Delta f$.

Assuming a zero-bias detector resistance R, Johnson noise results in mean-square current $\bar{i}^2 = 4kT\Delta f/R$. The SNR for Johnson noise is

$$\mathrm{SNR} = \left(\frac{e\eta P}{h\nu}\right)\sqrt{\frac{R}{4kT\Delta f}} \tag{5.36}$$

The NEP is

$$\mathrm{NEP} = \left(\frac{h\nu}{e\eta}\right)\sqrt{\frac{4kT\Delta f}{R}} \tag{5.37}$$

and the detectivity is

$$D^* = \frac{e\eta}{h\nu}\sqrt{\frac{RA}{4kT}} \tag{5.38}$$

To first order, R is inversely proportional to A and the value RA is independent of detector area. At room temperature for $\lambda = 1\ \mu$m, $RA = 50\ \mathrm{M\Omega \cdot cm^2}$ and $\eta = 0.9$, $D^* = 4.2 \times 10^9$ Jones. Shot noise– and Johnson noise–dominated processes both correspond to detectivity that increases linearly in wavelength and in the quantum efficiency η. Of course, η is itself a function of λ. We observe a monotonic increase in detectivity as a function of λ in Fig. 5.7 up to a critical point, where η quickly goes to zero and the detectivity collapses.

In addition to shot noise and Johnson noise, noise arises from sampling and device response characteristics ($1/f$ noise) and from readout, amplification, and digitization electronics. An individual optical measurement consists of a sum of signal and noise components of the form

$$i = s + \sum_i n_i \tag{5.39}$$

Assuming that the components are independent, each noise component is distributed according to its own statistics. Combining the joint probabilities allows us to develop a distribution for the overall noise. If components n_1 and n_2, for example, have distributions $p_1(n)$ and $p_2(n)$, then the distribution of $n = n_1 + n_2$ is

$$p(n) = \int p_1(n_1 = n - n_2) p_2(n_2) dn_2 \tag{5.40}$$

If n_1 and n_2 are normally distributed with zero mean and deviations σ_1 and σ_2, then

$$p_i(n) = \frac{1}{\sigma_i \sqrt{2\pi}} e^{-n^2/\sigma_i^2} \tag{5.41}$$

and

$$\begin{aligned} p(n) &= \frac{1}{\sigma_1 \sigma_2 2\pi} \int \exp\left(\frac{-(n - n_2)}{\sigma_1^2}\right) \exp\left(\frac{-n_2^2}{\sigma_2^2}\right) dn_2 \\ &= \frac{1}{\sqrt{\sigma_1^2 + \sigma_2^2}\sqrt{2\pi}} \exp\left(\frac{-n^2}{\sigma_1^2 + \sigma_2^2}\right) \end{aligned} \tag{5.42}$$

Thus, the distribution of a sum of normally distributed noise components is normally distributed with variance equal to the sum of the variances of each individual component.

Although their origins are quite different, most noise components (with the notable exception of shot noise) may be assumed to be normally distributed. Fixed pattern noise, for example, arises from a distribution of manufacturing parameters in a sensor array. In principle these parameters might be characterized to enable noise-free calibrated measurement, but in practice the complexity of precise characterization over a large array and a complete range of temperatures and operating conditions is impossible. Some aspects of fixed pattern noise are eliminated by *correlated double sampling*, which measures the difference between a pixel value immediately after reset and again after signal integration. Because of the multiple sources of fixed pattern noise (device size, temperature, and materials variation), it is often safe to assume a normal distribution.

The overall noise distribution is the convolution of the shot noise and additive noise components, for example, the convolution of the Poisson or compound Poisson distribution with a normal distribution for additive components. Optical detection may involve a compound Poisson process due to both photon and dark-current variations. In this case, the shot noise distribution is a correlation of

Poisson processes and the overall noise distribution is a convolution of the compound Poisson distribution and a normal distribution.

Independent of the structure of the distribution, if noise may be assumed to be independent from each source and between each pixel, the variance of the noise at each pixel is the sum of the variances due to all sources at that pixel. The pixel SNR is

$$
\begin{aligned}
\text{SNR} &= 10\,\log_{10}\frac{P}{\sqrt{\sum_i \sigma_i^2}}\,d\text{B} \\
&= 10\,\log_{10}\frac{P}{\sqrt{\kappa_p P + \sigma_r^2}}\,d\text{B}
\end{aligned}
\tag{5.43}
$$

where κ_p is a constant and σ_r^2 is the "read noise" variance, including components from Johnson, $1/f$, dark-current and fixed pattern noise. For low signal values, read noise dominates and the SNR grows 10 dB per decade of signal power. At higher signal values, shot noise dominates and the SNR grows 5 dB per decade of signal power.

5.6 CHARGE-COUPLED DEVICES

We turn finally to Professor Fossum's point 4, detector readout. In imaging and spectroscopy our interest is in massively parallel arrays consisting of millions of photodetectors. From the origins of electronic imaging through the 1980s, image transduction was implemented using cathode ray tubes. Incident light generated charge on a photoconductive service, and the charge density on this surface was scanned using an electron beam in a vacuum tube. Such systems are bulky and have poor quantum efficiency. The development of all solid-state *focal plane arrays* (FPAs) enabled the development of much more compact, efficient, and robust imaging systems.

A solid-state FPA performs the same two basic functions as a vacuum tube system: (1) transducing incident light into electronic charge and (2) transforming the 2D array of photodetector signals into a 1D temporal signal for readout into a digital processor. We understand how to implement step 1 through our discussion of photodiodes. Step 2 requires us to venture a bit further into discussion of electronic circuit design.

As suggested by steps 5–8 of Professor Fossum's list, the optoelectronic interface is somewhat more complex than simply detecting the light and reading it out. The interface transforms the physical distribution of an image light field into a mathematical data array. This transformation consists of diverse electronic amplification, noise reduction, signal conditioning, and analog-to-digital operations. FPA design is separated into systems that implement some of these operations at individual photodetection sites (active pixel sensors) and systems that implement most of these operations after serializing the image data stream (charge-coupled devices) (CCD).

The CCD, invented in 1969 at Bell Laboratories [25], consists of an array of *gates*. As illustrated in Fig. 5.8, a single photogate consists of a metal–oxide–semiconductor

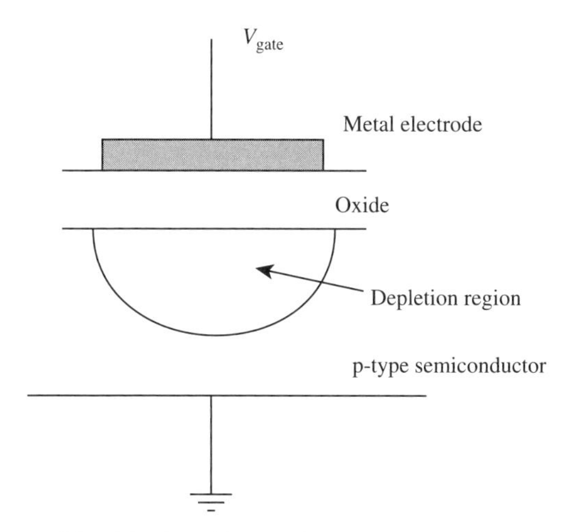

Figure 5.8 Metal–oxide–semiconductor (MOS) photogate.

capacitor. The metal is typically a transparent material such as indium tin oxide or very thin and heavily doped polycrystalline silicon. Light passes through the metal and oxide layers and is absorbed in the semiconductor. Alternatively, on may use a "back-illuminated" geometry in which the semiconductor is thinned to allow light penetration through the wafer. Although current does not flow through the oxide layer, a positive voltage applied to the gate creates a space charge field and moves mobile positive charge (assuming a p-type semiconductor) away from the gate electrode and toward the ground electrode. This creates a charge depletion region around the gate electrode. *Buried-channel* CCDs include an n-type layer between the oxide and the p-type layer to move the depletion region away from the oxide–semiconductor interface, which tends to be contaminated by surface states. Buried-channel designs to isolate the depletion region from noise due to the surface states are used in all modern CCD and active pixel sensors.

Photon absorption creates electron–hole pairs. The positively charged hole is repelled by the gate electrode, while negatively charged electrons accumulate in the depletion region. The accumulated charge reduces the voltage across the gate by an $\Delta V = ne/C$, where n is the number of photogenerated electrons and C is the capacitance. The number of charges that can be accumulated, the *well capacity*, depends on the applied voltage and the capacitance, which depends in turn on the oxide layer thickness and the electrode area. Well capacities of 50,000–500,000 electronics are typical.

Charge-coupled device gates implement both photocharge accumulation and charge transfer functions. The logical architecture of a CCD pixel consisting of three gates is illustrated in Fig. 5.9. The central gate is positively biased with voltage V, while the adjacent gates are negatively biased. Photoelectrons generated anywhere in the pixel accumulate in the potential energy well under the central gate. The actual physical architecture is somewhat more complicated because the

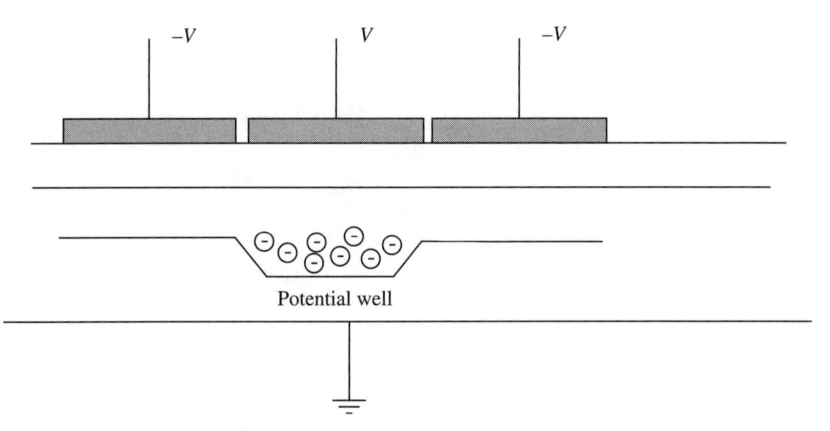

Figure 5.9 Three phase CCD pixel.

conducting regions of the gates are interleaved with oxide layers to create overlapping gates and the photoelectron accumulation gate geometry is not quite symmetric with the transfer gates.

The voltage applied to the gates is cycled in time to shift the accumulated charge from one gate to the next, as illustrated in Fig. 5.10. In the first timestep photoelectrons accumulate under the positive voltage (V_2) gates. The V_2 and V_3 lines are set at equal positive voltage in the second timestep, broadening the potential well to include both gates. When the V_2 line is negatively biased in the third timestep, all of the charge has been transferred to the V_3 gates. Repeating this process 2 more times with the voltages shown shifts the photogenerated charge by one pixel. The scheme described in Fig. 5.10 is *three-phase* charge transfer; variations using two or four phases or using physically asymmetric gate structures are used in some designs.

The "bucket brigade" strategy of charge transfer illustrated in Fig. 5.10 is effective in reading a linear array of pixels. Most commonly, however, one is interested in

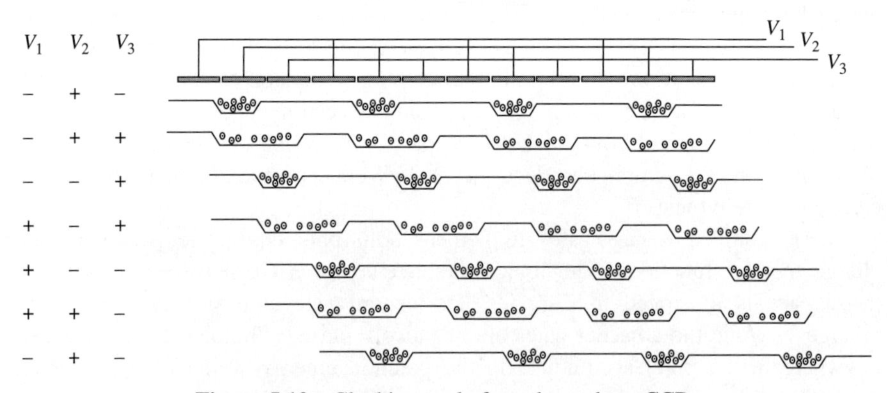

Figure 5.10 Clocking cycle for a three-phase CCD.

reading pixels distributed over a plane. 2D readout can be implemented by crossing CCD systems as illustrated in Fig. 5.11. A parallel array of 1D charge transfer gates dumps charge into an orthogonal serial array. The readout array transfers charge into an amplification circuit that converts each charge packet into an output voltage. The serial array transfer rate is set such that one line can be transfered out in the period of one pixel shift on the parallel array.

To get an idea of the electronic bandwidths associated with readout, consider a 1-megapixel (1-Mpixel) CCD read at 30 frames per second (fps). With no compression, the serial data stream must encode 30 Mpixels/s. With a readout dynamic range of 16 bits, this would correspond to 480 megabits per second (Mbits/s). One might compare this to the video bandwidth of the National Television System Committee (NTSC) standard for analog service in the United States, which is 4.2 MHz, the

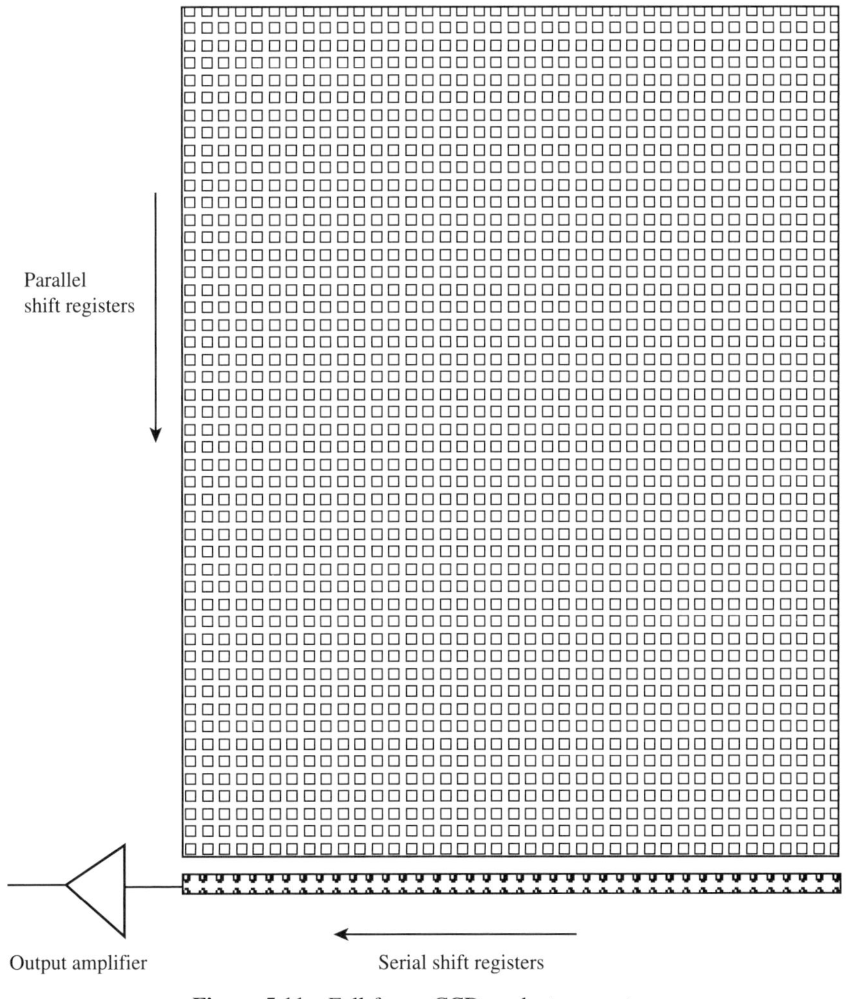

Figure 5.11 Full-frame CCD readout geometry.

difference in frequencies is due to the fact that NTSC resolution is around 0.1 Mpixel and the dynamic range is much less than 16 bits. The parallel shift register bandwidth necessary to read a column of 1000 pixels at 30 fps is $\frac{1}{1000}$ of the serial bandwidth, but the serial shift register must operate 1000 times faster to stream out the parallel array data. Of course, these calculations do not allow for any photointegration time. If one wishes to integrate light for just half of the available time, then the readout burst bandwidth must be doubled. The challenges associated with these readout bandwidths maintain a divide between low-resolution video and high-resolution still imaging.

The geometry shown in Fig. 5.11 is "full-frame readout." To register the charge detected at a specific pixel, one must inhibit new photocharge generation once the transfer process has begun. In full-frame systems this is achieved using a shutter to shield in the CCD during readout. The shutter may be mechanical in still cameras (mechanical shutter lifetimes are usually just a few million frames) or liquid crystal–based. Alternatively, one may use digital processing to remove or take advantage of the motion blur associated with charge generation during readout. Full-frame readout is not used in video systems. Full frame is used in astronomical and other low-light scientific applications where one may wish to integrate for several seconds before reading at a slower rate. The advantage of full-frame systems is that the *fill factor*, the fraction of the FPA surface over which light is usefully detected, is nearly 100%. The overall quantum efficiency of an FPA is the product of the fill factor and the photodetector quantum efficiency.

Alternative readout schemes when one cannot or does not wish to use a shutter include "frame transfer" and "interline" readout. In a frame transfer system half of the parallel shift register is permanently masked. For example, one might build a 1024×512 array. A 512×512 section is open for light detection, and a 512×512 section is masked. Once every frame period the open section is rapidly shifted in parallel into the masked section. The masked section is read into a serial transfer register during the next accumulation period.

Interline transfer, which is commonly in consumer imagers, is illustrated in Fig. 5.12. Alternating CCD columns are permanently masked from the light field.

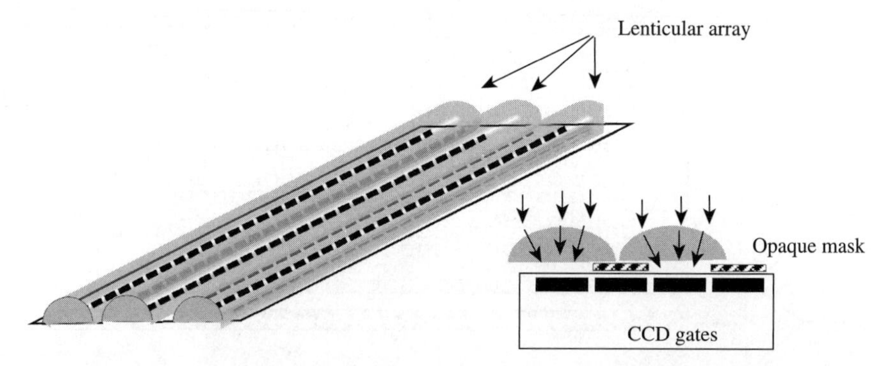

Figure 5.12 Interline CCD readout geometry. Alternate columns of the CCD are blocked from the light field. A microlenticular array with aperture equal to two column widths sits over the open columns to focus the incident light and increase the apparent fill factor.

Photodiodes in the open columns transfer charge into the masked CCD pixels, which can then be streamed out periodically without further masking or readout blur. Both this approach and frame transfer readout reduce the net fill factor by 50%, but the frame transfer strategy has 100% fill factor over the image surface. Interline transfer has nominally 50% fill, although the photosensitive columns may in practice be somewhat larger than the transfer columns. A lenticular array (a set of microscale cylindrical lenses) is often used to recover the lost fill factor by focusing the incident beam on the read pixels. The effectiveness of microlenses on the FPA is a fascinating topic; we will need to develop our system analysis toolbox somewhat before returning to this topic in Chapter 7. For present purposes we simply note that the microlenses are likely to significantly complicate our sensor response model. Various other optical components, such as spectral and polarization filters, may also be integrated on the FPA.

Charge transfer efficiency (CTE), such as the fraction of charge contained in one pixel that is transferred to the next pixel in a clock cycle, is a critical parameter for a CCD FPA. Charge may be lost during a transfer cycle if it is caught in a local trapping site or if it drifts into the wrong potential well. Incomplete transfer is a particular problem if the clock time is too short to allow effective diffusion from one well to the next. On an $N \times M$ array the most remote pixel is transfered $N + M$ times before readout. The fraction of the photogenerated charge at the readout circuit is thus $(\mathrm{CTE})^{N+M}$. For $N + M = 1000$, a worst-case readout efficiency of 0.9 requires a charge transfer efficiency of 0.999895. If we increase $N + M$ to 10,000, this charge transfer efficiency would deliver only a third of the original charge in the correct pixel. CTE limits CCD size to the megapixel size, although larger arrays may be implemented by a mosaic of multiple CCDs on a single substrate.

The spectral response of a CCD is determined primarily by the bandgap of the semiconductor substrate, although the gate electrode electronics often also plays a role. Polysilicon and indium tin oxide electrodes preferentially absorb blue and ultraviolet ranges. For this reason, back-illuminated geometries are preferred at shorter wavelengths. Variation in the skin depth as a function of wavelength in silicon is illustrated in Fig. 5.3. The power $1/e$ range is half of the skin depth. The skin depth affects several issues in CCD design. Making a back-illuminated device thin enough that charge generated within the blue absorption range is effectively collected in a pixel can substantially reduce the red and infrared quantum efficiency. As discussed in Chapter 7, one may ultimately like to make pixels of subwavelength scale. If the skin depth is longer than the pixel pitch, however, light striking the CCD photodepletion barrier at one point may actually generate charge one or more pixels away.

As a final comment, we note that CCDs are susceptible to "blooming," which occurs when the irradiance at a pixel or set of pixels exceeds well capacity. Excess charge from the saturated gates bleeds into adjacent pixels, causing error and saturation over the affected region. This problem may be regarded as an example of the general issue of dynamic range management on focal planes. If a scene consists of only bright or only faint sources, one may adjust exposure to capture them, but if a scene contains both bright and faint sources, one cannot reduce exposure to prevent saturation of the bright target without loosing sensitivity to the faint targets. The first-order response in CCD design is to include antiblooming gates, which remove overflow and prevent interpixel crosstalk. A second-order response might be to consider pixels with dynamically addressable individual gain and well capacity. Such innovations are the focus of active pixel sensor design, which is the subject of the next section.

5.7 ACTIVE PIXEL SENSORS

Active pixel sensors (APSs) integrate diverse optoelectronic signal transduction and conditioning functions at the pixel level. Active pixel sensors are implemented in "complementary" metal–oxide–semiconductor (CMOS) silicon technology. CMOS is the dominant integration technology for microprocessors, memories, and application-specific circuits. In addition to active pixel arrays, CMOS readout integrated circuits (ROICs) are used as backplanes for detector arrays fabricated using bolometers or photodiodes in diverse material systems.

The simplest active pixel approach integrates amplification and charge–voltage conversion at the pixel level. Image sensors with integrated per pixel amplifiers were developed almost simultaneously with CCD technology and have long been used in infrared cameras. At first, however, the low-noise and high-fill-factor characteristics of CCDs enabled them to dominate visible and near-infrared (NIR) focal planes. Numerous fundamental disadvantages of the CCD approach and the radical improvement of CMOS integration technologies in support of successive microprocessor and memory generations eventually enabled active pixel designs to become competitive.

The basic design of an active pixel sensor is illustrated in Fig. 5.13. Each pixel is addressed individually by a parallel row bus and a serial column bus. Each pixel consists of a photodiode and a multiple transistor buffer, amplification, and readout circuit. The photodiode acts as a current source, generating charge at a rate proportional to the incident photon flux. This charge determines the gate voltage on a source follower transistor, which is coupled to a row select transitor to enable nondestructive readout. A reset transistor periodically clears the accumulated photocharge.

Fixed pattern noise results from variations in the responsivity of individual elements in an array. Variation may be due to differences in the pixel offset potential, the intrinsic photodiode response, or pixel dark current. Offset variation is particularly troublesome in active pixel sensors. This problem is substantially reduced by *correlated double sampling*, which augments the photodiode with a gate to enable the output voltage to be sampled twice: once immediately after reset and again after photocharge accumulation. A signal value consisting of the difference between the final and reset potentials removes sensitivity to the reset value.

Active pixel sensors operate with much lower power than a CCD array. A typical CCD camera may draw several watts of power where the corresponding active pixel array draws just a few milliwatts. Of course, if postdetection image processing is

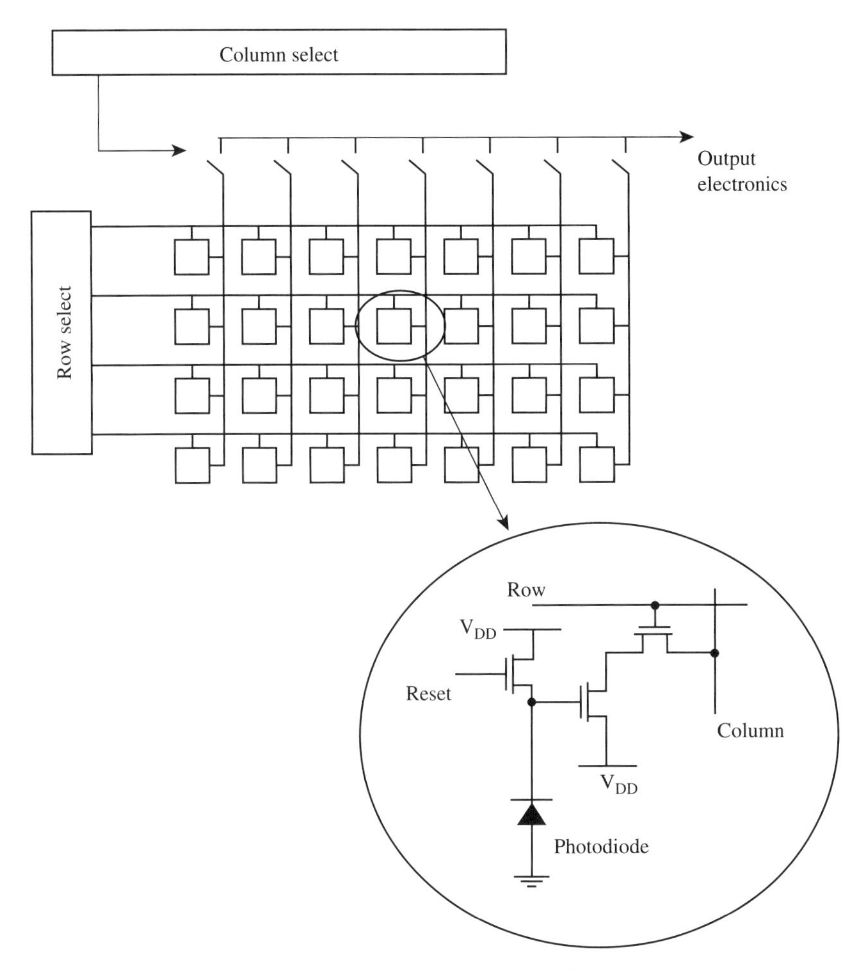

Figure 5.13 Active pixel sensor layout.

implemented in the camera, the relative advantage of the APS approach may be degraded, but in general the power advantage leads to a substantial advantage for CMOS in mobile devices.

Since a portion of the pixel area is dedicated to amplification and readout circuits, the fill factor for active pixel devices is substantially worse than for CCDs. The CCD fill factor for frame transfer devices is essentially 100%, active pixel fill factors typically range within 30–50%. The effective fill factor is often increased using microlens arrays, but as discussed in Section 7.4, the utility of this approach is somewhat dubious. In any case, the fill factor of interline transfer CCDs is just comparable to CMOS devices.

In addition to the power advantage, the potential to implement generalized sampling strategies at the pixel level is the primary long-term advantage of active pixel sensors. Examples of generalized sampling include advanced dynamic range management, such as circuitry to nonlinearly adjust the pixel response during charge generation or to implement time-dependent sampling based on the current read voltage, as well as space–time readout strategies for compressive or multiscale analysis.

Active pixel sensors have been demonstrated with foveated or other non-Cartesian readout geometries, with shift-variant and adaptive spatial impulse responses and with nonsequential readout strategies [181]. Over time, one expects that on-chip sampling and processing with active pixel sensors, in combination with optical multiplexing and prefiltering and the generalized sampling theories discussed in Section 7.5, will revolutionize camera design.

5.8 INFRARED FOCAL PLANE ARRAYS

The CCD and CMOS focal planes of Sections 5.6 and 5.7 are utilized in spectral ranges accessible to silicon photodiodes, including the visible and the NIR ($300\,\text{nm} < \lambda < 1100\,\text{nm}$). Silicon is a particularly attractive materials system for focal plane integration because silicon technology for electrical signal transfer and processing is well developed. In view of the general attractiveness of silicon integrated circuits, all-silicon focal planes have been developed for X-ray, UV, and NIR imaging by processing silicon layers to accentuate the responsivity in the targeted spectral range. More commonly, however, imaging arrays for applications outside the visible and NIR ranges rely on different materials systems. Common materials for shortwave infrared (SWIR, $1\mu < \lambda < 2.5\mu$), midwave infrared (MWIR, $3\mu < \lambda < 7\mu$), longwave infrared (LWIR, $8\mu < \lambda < 30\mu$) are indicated in Fig. 5.7.

In principle, one could create readout electronics in these materials systems. For example, some success has been achieved with SWIR InGaAs and MWIR HgCdTe CCD arrays. Much greater success has been achieved by heterogeneous integration of detector materials on silicon circuits or by bonding nonsilicon diode arrays to silicon backplanes. The most common approach uses indium bump bonds to transfer signals directly from discrete detector junctions into CMOS readout circuits, as illustrated in Fig. 5.14(a). Alternatively, one may use the "loophole" technique illustrated in Fig. 5.14(b). In the loophole system detector material is adhesively bonded to the silicon readout and thinned. Vias etched through the detector layer are metallized to contact pads on the readout circuit.

While the diverse array of materials systems and nanostructures used in photon counting infrared sensors each have their idiosyncrasies, at the level of detail of this chapter we may consider them functionally similar to photon counting visible devices. We refer the reader to the specialized literature for surveys of IR materials [57,211]. The common use of thermal detectors in infrared systems has no parallel in visible focal planes, however, so we offer a brief account of thermal arrays here.

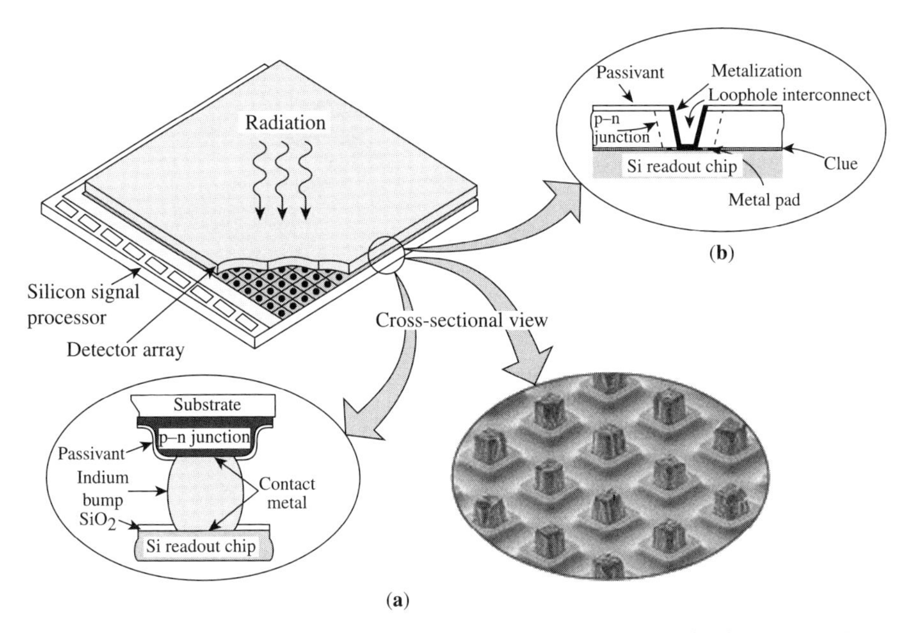

Figure 5.14 Hybrid chip bonding strategies to join a silicon readout circuit with an infrared focal plane array: (a) illustration of the indium bump technique and (b) scheme for the loophole technique. (From A. Rogalski [210], © 2003 Elsevier Limited. Reprinted with permission.)

A thermal detector measures radiation based on an optically induced change in temperature. The temperature is measured via

- Photoconductive effects in *bolometers*, in which a thermally induced change in electrical resistance changes the current in a circuit
- Photovoltaic effects in *thermocouples*, in which a change in temperature produces a change in junction voltage
- Pyroelectric effects, in which a change in temperature produces a change in electrical polarization

Bolometers are the most commonly used thermal detectors in imaging arrays.

Thermal arrays are attractive in some infrared imaging applications because they use simpler materials systems (silicon or vanadium oxide) than do photon counting devices, which makes manufacturing less expensive and more reliable, because they respond uniformly over broad spectral ranges and do not require active cooling. The last point illustrates a critical difference between the infrared and visible spectral ranges; objects at typical environmental temperatures are active sources of infrared radiation, whereas most environmental visible light arises from scattered sunlight or active fluorescence. To avoid detector saturation with infrared photon counting devices, one must use a "cold shield" to isolate the detector array from ambient light. Thermal arrays, ironically, have sufficiently poor detectivity to

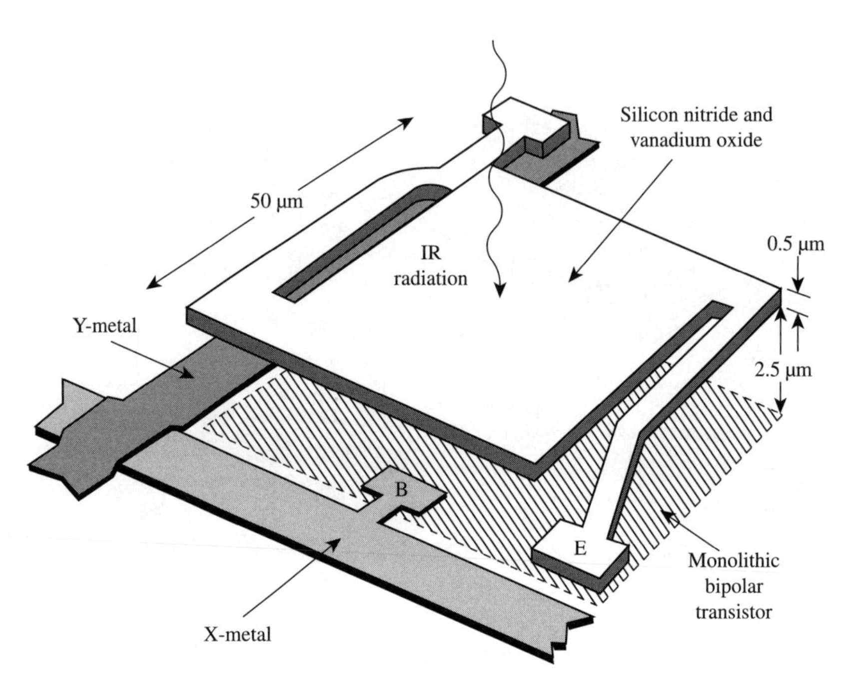

Figure 5.15 Microbolometer pixel structure. (From A. Rogalski [211], © 2003 Elsevier Limited. Reprinted with permission.)

measure the temperature difference between image objects and the ambient scene without background saturation.

A typical microbolometer structure is illustrated in Fig. 5.15 [253]. The device consists of a silicon nitride microstructure created by sacrificial layer selective etching. A vanadium oxide layer serves as a temperature dependent resistor. The bridge structure isolates the bolometer from the underlying silicon readout circuit. The entire package is hermetically sealed in vacuum to thermally isolate the bolometer.

Our analysis of microbolometer photodetection follows Kruse [141]. The thermal balance of the bolometer is described by

$$C\frac{d\Delta T}{dt} + G(\Delta T) = \eta P \tag{5.44}$$

where C is the heat capacity of the bolometer, ΔT is the difference between bolometer temperature and the temperature of the surrounding environment, η is the quantum efficiency for optical absorption, and P is the incident radiant power. The thermal relaxation constant G describes the cooling of the bolometer due to radiation and the thermal conductance of the bridge support. The thermal conductance of the support is the dominant cooling channel.

The Fourier transform of Eqn. (5.44) yields

$$\Delta\hat{T}(\nu) = \frac{\eta}{G + 2\pi i\nu C}\hat{P} \tag{5.45}$$

Defining the thermal response time $\tau = C/G$, the transfer function mapping temporal variations in the irradiance into thermal fluctuations is $\hat{h}(\nu) = \exp(i\phi(\nu))$ $\eta/G\sqrt{1+4\pi^2\nu^2\tau^2}$. The thermal impulse response of the bolometer is

$$h(t) = \begin{cases} \frac{\eta}{G}e^{-t/\tau} & t > 0 \\ 0 & t < 0 \end{cases} \tag{5.46}$$

The transfer function and impulse response are plotted in Fig. 5.16. The impulse response in this case is the temperature change in the bolometer due to a temporal spike in the illuminating radiation.

The goals of microbolometer design are to make the thermal response as large as possible and the response time as short as possible. One maximizes the response by making η big and G small. One maximizes η by coating the bolometer surface with a strongly absorbing film. One minimizes G by creating good thermal isolation in the microstructure. A problem arises, of course, in that as one reduces G one also increases the response time τ. One would like to minimize τ to enable high-frame-rate imaging. In contrast with the circuit dominated response times of photon counting detectors, the thermal relaxation time is the dominant factor in determining the response time of bolometers. One minimizes τ in a microbolometer structure by making C as small as possible. The value of C is determined by the mass of the resistive material. The area of the resistive layer is determined by the pixel size, which is matched to the wavelength and the optical system. To minimize C, therefore, one reduces thickness of the resistive layer to obtain the desired response time. Layer thicknesses of less than 1 μm are common.

A bolometer translates the temperature change into a change in electrical resistance R. One generally operates in a linear regime such that

$$\Delta R = \alpha R \, \Delta T \tag{5.47}$$

where α is the *thermal coefficient of resistance* (TCR). α may be positive or negative; typical values are 0.002/K in metals, -0.02/K in semiconductors, and 2/K in

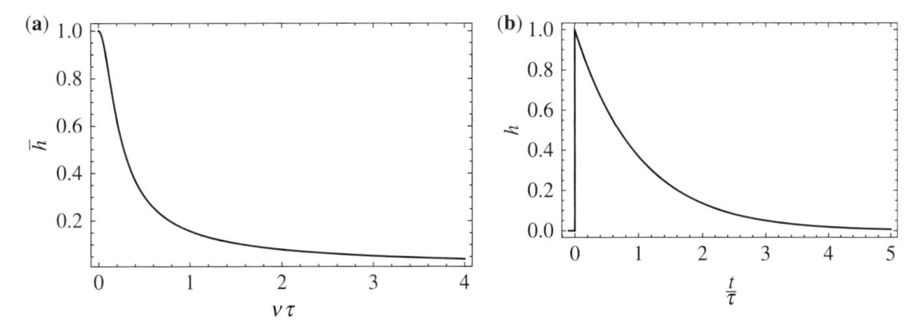

Figure 5.16 (a) Absolute value of the temporal transfer function $\hat{h}(\nu)$ of a microbolometer and (b) the impulse response $h(t)$. The vertical axis is in units of η/G.

superconducting films. Assuming that the bolometer is probed by a bias current i_b to produce signal voltage $V_s = i_b \Delta R$, the responsivity $R(\nu) = \hat{V}_s / \hat{P}$ is

$$R = \frac{i_b \alpha R \eta}{G\sqrt{1 + 4\pi^2 \nu^2 \tau^2}} \tag{5.48}$$

The maximum value of i_b is determined by the need to avoid resistive heating of the bolometer. Defining the bias voltage $V_b = i_b R$, we see that the bolometer operating at frequencies $\nu < 1/\tau$ amplifies the bias voltage by a factor $\alpha\eta/G$. Since relatively little can be done to change α or η, minimization of G is at the heart of microbolometer design. A value of $G = 2 \times 10^{-7}$ W/K with a semiconducting resistive layer yields $\mathcal{R} = 10^5\, V_b$ V/W.

In contrast with the responsivity of photon counting detectors described by Eqns. (5.26) and (5.27), we note that the responsivity of a thermal detector is not proportional to wavelength. As discussed in Sections 5.4 and 5.5, responsivity and detectivity define the operating limits of photodetectors. The detectivity of a thermal detector is typically several orders of magnitude worse than for a photon detector. An example is seen in the difference between MCT and thermopile detectors in Fig. 5.7. Note that as with the responsivity, the detectivity of the thermal detector is constant as a function of wavelength.

Uncooled microbolometer array performance is most often evaluated in terms of the *noise equivalent temperature difference* (NETD) and *minimum resolvable temperature difference* (MRTD) rather than in terms of the noise equivalent power or detectivity. Since microbolometers operate in environments with substantial background illumination the goal is not to measure the total infrared flux as much as to image sources radiating powers above the thermal background. Noise arises from both (1) the thermal cycle of heat exchange between the bolometer and the environment and (2) fluctuations in the background radiance. The NETD is the temperature difference relative to background at which a large blackbody imaged on an array produces a signal equal to the noise variance. The NETD is the temperature difference corresponding to SNR = 1.

The definition of NETD incorporates imaging system properties, specifically the light collection efficiency of the optical system influences the power at the focal plane and thus the response to a change in object temperature. To understand this relationship, one must consider the relationship between the power density radiated by the blackbody and the power density incident on the focal plane. The blackbody is usually modeled as a *Lambertian* source, meaning that it radiates most strongly in the direction normal to the object surface and that the power per unit area per unit solid angle (the radiance) falls with the squared cosine of the angle between radiant direction and the surface normal [61]. A patch on the surface of a Lambertian blackbody of area x_o^2 produces power density $P_o x_o^2 / \pi z_o^2$ at range z_o from the surface. P_o is the irradiance on the surface of the blackbody. A lens of aperture A collects power $A^2 P_o x_o^2 / 4 z_o^2$ from the patch on the blackbody. The object patch is mapped to a patch of area $M^2 x_o^2$ on the focal plane. The image power density P_i on the focal plane is thus

$$P_i = \frac{A^2}{4M^2 z_o^2} P_o = \frac{P_o}{4(f/\#)^2} \tag{5.49}$$

The signal generated by a microbolometer for blackbody temperature change ΔT is

$$V = \frac{\tau_0 D^2 \mathcal{R}}{4(f/\#)^2} \frac{\Delta P_o}{\Delta T} \tag{5.50}$$

where τ_o is the transmittance of the optics, D^2 is the detector area, and $\Delta P_o/\Delta T$ is the change in blackbody irradiance with respect to temperature. Values of $\Delta P_o/\Delta T$ for sources near human body temperature range from $10^{-5}\,\mathrm{W/(cm^2 \cdot K)}$ in the 3–5 μm wavelength range to $10^{-4}\,\mathrm{W/cm^2 \cdot K}$ in the 8–14 μm wavelength range. The NETD of an imaging system is thus

$$\text{NETD} = \frac{4(f/\#)^2 V_N}{\tau_o D^2 \mathcal{R}(\Delta P/\Delta T)} \tag{5.51}$$

where V_N is the standard deviation of the detector measurement. The impact of the imaging system on NETD is entirely encapsulated in the $f/\#$, absent other issues (such as aliasing and aberrations); the system designer seeks to make the $f/\#$ as small as possible.

The minimum resolvable temperature difference is the temperature change apparent to a human observer as a function of spatial frequency. It is typically evaluated by display of bar chart images. The MRTD is related to the NETD by

$$\text{MRTD}(u) = K(u) \frac{\text{NETD}}{\text{MTF}(u)} \tag{5.52}$$

where $K(u)$ is the spatial frequency transfer function for the human observer and $\text{MTF}(u)$ is the modulation transfer function for the optical system.

PROBLEMS

5.1 *Photoconductive Devices.* Consider a 1-μm-thick silicon photoconductor at $T = 300\text{K}$. The material is p-type with background doping of $N_d = 10^{15}\ \mathrm{cm^{-3}}$, an electron mobility of $\mu_n = 1300\ \mathrm{cm^2/(V \cdot s)}$, and a hole mobility of $\mu_p = 400\ \mathrm{cm^2/(V \cdot s)}$. The bulk electron and hole minority carrier lifetimes are $t = 1.0$ ms. A bias of $V = 5$ V is applied to the detector. When illuminated with wavelength $\lambda = 600$ nm, the doped silicon has an absorption coefficient of $\alpha = 0.5\ \mu\mathrm{m}^{-1}$ and a surface reflectivity of $R = 0.30$.

(a) Compute the photoconductive gain assuming that the carrier lifetime is dominated by the bulk lifetime. Comment on whether this value seems reasonable. Then compute the photoconductive gain assuming that the contact lifetime of Eqn. (5.19) dominates and use this value for the rest of the problem.

(b) By what factor would the gain be increased by cooling the detector to $T = 77$ K? Under the assumptions leading to Eqn. (5.19), show that the photoconductive gain is proportional to the ratio of the kinetic energy gained by a carrier crossing the detector and the thermal energy of a free carrier.

(c) Compute the responsivity of the detector.

5.2 *Photoconductive and Photovoltaic Devices*. Compare the responsivity of photoconductive detectors and photodiodes as described by Eqns. (5.26) and (5.27). Explain why focal plane arrays use photodiode gates rather than photoconductive detectors.

5.3 *D^**. The spectral dependence of the quantum efficiency for a certain detector is described by

$$\eta = \frac{\eta_0}{2}\left(1 - \tanh\left(\frac{\lambda - 1.5}{0.1}\right)\right) \tag{5.53}$$

for λ in micrometers and $\eta_0 = 0.9$. Assuming that $RA = 50\,\mathrm{M\Omega \cdot cm^2}$, plot the Johnson noise–limited detectivity for this detector over the range $\lambda = 0.5$–$2\ \mu$m at temperatures $T = 77$ and $T = 270$K. Include units for your plots. Compare your results with Fig. 5.7.

5.4 *Focal Plane Arrays*. Datasheets for CCD and active pixel sensors manufactured by diverse suppliers are available online. A *scatterplot* considers each device as a data point and plots correlations between performance parameters.

(a) Make a scatterplot of pixel size versus saturation signal for full-frame CCDs found at supplier websites. Identify the model numbers on your plot.

(b) Make a scatterplot of pixel size versus dynamic range.

(c) Counting each product datasheet as one data point, plot histograms of quantum efficiency at 550 nm for full-frame, interline, and CMOS focal plane arrays.

(d) Plot the quantum efficiency versus wavelength for a selection of front- and back-illuminated CCDs. How is η optimized for near-infrared focal planes?

(e) CMOS focal planes are characterized by diverse fill factor, dynamic range management, and pixel complexity factors. Briefly describe and classify the range of available CMOS devices. How are CMOS sensors from various leading suppliers differentiated?

5.5 *Photon Flux*. Estimate the saturation irradiance in $\mathrm{W/m^2}$ for a CCD sensor under illumination by green light. Assume 5 μm pixel pitch, a well capacity of 100,000 electrons, and a quantum efficiency of 40% read at 30 fps.

5.6 *Microbolometers.* Consider a microbolometer array with 50 μm square pixels operating at 30 frames per second with $G = 10^{-7}$ W/K and a resistive material specific heat of 1 J/(cm$^3 \cdot$ K).

(a) As illustrated in Fig. 5.16(a), the frequency response of a microbolometer has a peak at $\nu = 0$. Find τ such that the thermal transfer function at the frame rate is 70% of the direct current (DC) response.

(b) How thin must the resistive layer of the bolometer be to obtain the target value of τ?

(c) Suppose that this array images a 1 cm^2 area blackbody radiating 100 W at a range of 20 m using $f/1$ optics with a focal length of 20 cm. Assuming a responsivity 10^5 V/W, estimate the voltage change induced. Assuming a semiconducting resistive layer, estimate the temperature change induced.

(d) Assuming that the variance of the voltage signal is 0.1% of the detected value in (**c**), estimate NETD for this array assuming operation in the 3–5 range and (separately) in the 8–14 μm spectral range.

6

COHERENCE IMAGING

> ...subjects that often appear to be well understood and perhaps even a little old-fashioned have frequently some surprises in store for us.
>
> —E. Wolf [249]

6.1 COHERENCE AND SPECTRAL FIELDS

The *mutual coherence* of the optical field is

$$\Gamma(\mathbf{r}_1, \mathbf{r}_2, t_1, t_2) = \langle E^*(\mathbf{r}_1, t_1) E(\mathbf{r}_2, t_2) \rangle \tag{6.1}$$

where $E(\mathbf{r}, t)$ is the electric field at spatial position $\mathbf{r}$ and time t. For simplicity, we ignore the polarization of the field, which could be accounted for by a tensor-valued mutual coherence. The angular brackets $\langle\ \rangle$ signify the expected value of the terms contained over an ensemble of identical physical systems. The mutual coherence and related functions described in this section are of interest because

1. Mutual coherence, like the irradiance but unlike the electric field, is observable. The mutual coherence can be completely described by measuring the irradiance at a suitable range of sampling points.
2. Mutual coherence, like the electric field but unlike the irradiance, can be calculated over a volume given its value on a boundary. The mutual coherence at the input to an optical system uniquely determines the mutual coherence at the output. In Chapter 4 we derived input/output transformations for the electric field in imaging systems. In this chapter, we show that impulse responses derived for the electric field can be immediately applied to describe the propagation of coherence functions.

Coherence functions completely determine optical fields observed via irradiance detectors, meaning that if one knows the coherence function on a boundary, one can

Optical Imaging and Spectroscopy. By David J. Brady

predict the irradiance that would be measured at any point and, conversely, knowing the irradiance that would be measured at all points is equivalent to knowing the mutual coherence. There are situations in ultrafast and nonlinear optics where the coherence functions described here are insufficient to fully characterize the field, but coherence functions are the most fundamental tool for analysis of irradiance-based imaging and spectroscopy (which is to say, essentially all optical imaging and spectroscopy).

The enormous disparity between the oscillation frequency of the optical field and the temporal sampling rate of optoelectronic detectors is as important to the nature of optical detection as the restriction to irradiance measurements. The optical frequency is a few hundred terahertz. While the fastest optoelectronic point detectors sample at terahertz rates, detectors used in imaging and spectroscopy operate in the kilohertz–megahertz range. Even at terahertz rates detectors average over hundreds of optical cycles, a detector in a focal plane array averages billions of cycles of the field. Under these conditions, each optical measurement may safely be regarded as a good statistical sample of the state of the field.

The statistical nature of optical measurement is enshrined in two assumptions. First, we assume that $\Gamma(\mathbf{r}_1, \mathbf{r}_2, t_1, t_2)$ is *stationary* with respect to time. A random process is stationary if its statistics are independent of the origin of the temporal axis. Formally, the mutual coherence is stationary with respect to time if $\Gamma(\mathbf{r}_1, \mathbf{r}_2, t_1, t_2) = \Gamma(\mathbf{r}_1, \mathbf{r}_2, \tau)$, where $\tau = t_1 - t_2$. The optical field is not generally stationary on long timescales. For example, the statistics of sunlight are different between day and night. However, the difference in timescales between the optical period and sample times on the one hand and such macroscopic events on the other is enormous. So long as sampling is much faster than rate of macroscopic variation in the irradiance, it is safe to assume that the mutual coherence is stationary with respect to the time axis. Note that we do not assume that mutual coherence is stationary with respect to $\mathbf{r}$.

Second, we assume that the field is *ergodic*. A random process is ergodic if the time average of the signal is equal to the statistical mean. In the case of the mutual coherence the ergodic assumption is

$$\langle E^*(\mathbf{r}_1, t_1)E(\mathbf{r}_2, t_2)\rangle = \lim_{T\to\infty} \frac{1}{T} \int_{-(T/2)}^{T/2} E^*(\mathbf{r}_1, t_1)E(\mathbf{r}_2, t_2)dt \tag{6.2}$$

Noting that practical optical measurements average over a very large number of cycles, each measurement may be regarded under the ergodic assumption as an ensemble average.

Assuming ergodicity and stationarity, relationships between the mutual coherence and the irradiance are easily derived. We saw in Eqn. (5.12) that photodetectors measure

$$\begin{aligned} I(\mathbf{r}) &= \lim_{T\to\infty} \frac{1}{T} \int_{-(T/2)}^{T/2} |E(\mathbf{r}, t)|^2 dt \\ &= \Gamma(\mathbf{r}, \mathbf{r}, \tau = 0) \end{aligned} \tag{6.3}$$

where we assume uniform spectral response and we set constants equal to 1. Conversely, if the fields $E_1 = E(\mathbf{r}_1, t_1)$ and $E_2 = \mathrm{E}(\mathbf{r}_2, t_2)$ are superimposed by an optical system, the irradiance is

$$I(\mathbf{r}) = \left\langle |E_1 + e^{i\phi} E_2|^2 \right\rangle = \Gamma_{11} + \Gamma_{22} + e^{i\phi}\Gamma_{12} + e^{-i\phi}\Gamma_{21} \tag{6.4}$$

where $\Gamma_{12} = \langle E_1^* E_2 \rangle$ and ϕ is an optically induced phase difference between the fields. By varying ϕ over several irradiance measurements, one may generate a nondegerate dataset for algebraic estimation of Γ_{12}.

Recalling Eqn. (5.12) again, we note that photocurrents are most correctly modeled as projections of the power spectrum of the field. The *power spectral density* $S(\mathbf{r}, \nu)$ is the distribution of irradiance per unit spectral range such that the total irradiance is

$$I(\mathbf{r}) = \int S(\mathbf{r}, \nu)\, d\nu \tag{6.5}$$

Nominally, we might define the power spectral density in terms of the ensemble average power spectrum of the electric field as

$$S(\mathbf{r}, \nu) = \left\langle \left| \widehat{E}(\mathbf{r}, \nu) \right|^2 \right\rangle \tag{6.6}$$

but this definition must be treated with care because, as this is a stationary random process, we cannot assume that $|E(\mathbf{r}, t)|$ tends to zero as $t \to \infty$. This means that the field is not square integrable and does not therefore have a well-defined Fourier transform. It is nevertheless possible for us to consider expectation values for the mutual coherence and spectral density. For example, we relate the ensemble average $\langle \widehat{E}^*(\mathbf{r}_1, \nu)\widehat{E}(\mathbf{r}_2, \nu') \rangle$ to the mutual coherence as

$$\begin{aligned}
\left\langle \widehat{E}^*(\mathbf{r}_1, \nu)\widehat{E}(\mathbf{r}_2, \nu') \right\rangle &= \iint \langle E^*(\mathbf{r}_1, t_1) E^*(\mathbf{r}_2, t_2) \rangle e^{-2\pi i \nu t_1} e^{2\pi i \nu' t_2} dt_1\, dt_2 \\
&= \iint \Gamma(\mathbf{r}_1, \mathbf{r}_2, t_1 - t_2) e^{-2\pi i \nu t_1} e^{2\pi i \nu' t_2} dt_1\, dt_2 \\
&= \int e^{2\pi i (\nu' - \nu) t} dt \int \Gamma(\mathbf{r}_1, \mathbf{r}_2, \tau) e^{-2\pi i \nu \tau} dt\, d\tau \\
&= \delta(\nu' - \nu) \int \Gamma(\mathbf{r}_1, \mathbf{r}_2, \tau) e^{-2\pi i \nu \tau} d\tau
\end{aligned} \tag{6.7}$$

Defining

$$W(\mathbf{r}_1, \mathbf{r}_2, \nu) = \lim_{\Delta\nu \to 0} \int_{\nu - \Delta\nu}^{\nu + \Delta\nu} \left\langle \widehat{E}^*(\mathbf{r}_1, \nu)\widehat{E}(\mathbf{r}_2, \nu') \right\rangle d\nu' \tag{6.8}$$

TABLE 6.1 Coherence Functions

$\Gamma(\mathbf{r}_1, \mathbf{r}_2, t_1, t_2) = \langle E^*(\mathbf{r}_1, t_1)E(\mathbf{r}_2, t_2)\rangle$	Mutual coherence
$W(\mathbf{r}_1, \mathbf{r}_2, \nu) = \lim_{\Delta\nu\to 0} \int_{\nu-\Delta\nu}^{\nu+\Delta\nu} \left\langle \widehat{E}^*(\mathbf{r}_1, \nu)\widehat{E}(\mathbf{r}_2, \nu')\right\rangle d\nu'$	Cross-spectral density
$S(\mathbf{r}, \nu) = W(\mathbf{r}, \mathbf{r}, \nu)$	Spectral density
$J(\mathbf{r}_1, \mathbf{r}_2) = \Gamma(\mathbf{r}_1, \mathbf{r}_2, \tau = 0)$	Mutual intensity
$I(\mathbf{r}) = J(\mathbf{r}, \mathbf{r})$	Irradiance

we find from Eqn. (6.7) that

$$W(\mathbf{r}_1, \mathbf{r}_2, \nu) = \int \Gamma(\mathbf{r}_1, \mathbf{r}_2, \tau)e^{-2\pi i\nu\tau}d\tau \tag{6.9}$$

where $W(\mathbf{r}_1, \mathbf{r}_2, \nu)$ is the cross-spectral density. The Fourier transform relationship between $\Gamma(\mathbf{r}_1, \mathbf{r}_2, \tau)$ and $W(\mathbf{r}_1, \mathbf{r}_2, \nu)$ is a version of the *Wiener–Khintchine theorem* and may be regarded as an extension of Plancherel's theorem [(Eqn. (3.19)] to stationary processes. The power spectral density is related to the cross-spectral density as $S(\mathbf{r}, \nu) = W(\mathbf{r}, \mathbf{r}, \nu)$ and the Wiener–Khintchine relationship between $S(\mathbf{r}, \nu)$ and the mutual coherence is simply

$$S(\mathbf{r}, \nu) = \int \Gamma(\mathbf{r}, \mathbf{r}, \tau)e^{-2\pi i\nu\tau}\, d\tau \tag{6.10}$$

The inverse Fourier relationship

$$\Gamma(\mathbf{r}, \mathbf{r}, \tau) = \int S(\mathbf{r}, \nu)e^{2\pi i\nu\tau}\, d\nu \tag{6.11}$$

immediately yields Eqn. (6.5) for $\tau = 0$.

Finally, we complete our definitions of coherence functions by noting that Γ evaluated at $\tau = 0$ is sufficiently useful to deserve a name, the *mutual intensity* $J(\mathbf{r}_1, \mathbf{r}_2)$ such that

$$J(\mathbf{r}_1, \mathbf{r}_2) = \Gamma(\mathbf{r}_1, \mathbf{r}_2, \tau = 0). \tag{6.12}$$

Table 6.1 summarizes the optical coherence functions.

6.2 COHERENCE PROPAGATION

In Chapter 4 we derived various impulse responses for propagation of the electromagnetic field from one boundary to the next through free space and optical systems. The primary utility of the E-field impulse response is that it can be trivially extended to model the impulse response for coherence propagation. The transformation of an electric field at temporal frequency ν on the input (x, y) plane to the electric field on the output (x', y') plane in Chapter 4 takes the form

$$g(x', y', \nu) = \int\int f(x, y, \nu) h_c(x, x', y, y', \nu)\, dx\, dy \tag{6.13}$$

We refer to $h_c(x, x', y, y', \nu)$ as the *coherent impulse response* of the optical system.

This section uses the coherent impulse response to derive an impulse response for propagation of coherence functions from an input interface to an output interface. We refer to the simple diffractive system sketched in Fig. 6.1. Just as we found it most convenient to work in temporal Fourier space in Chapter 4, we find it more convenient to propagate the spectral density than the mutual coherence. Our basic problem differs somewhat from the diffraction problem of Fig. 4.2 in that the cross-spectral density is defined over 4D correlation spaces at the input and output rather than just over the input and output planes. As illustrated by the point pairs in Fig. 6.1, the cross-spectral density is a defined between each pair of points on the input plane and each pair of points on the output plane. Given the input cross-spectral density $W(x_1, y_1, x_2, y_2, \nu)$, we must determine the output cross-spectral density $W(x'_1, y'_1, x'_2, y'_2, \nu')$.

To determine the impulse response appropriate for transformation of the cross-spectral density from the coherent impulse response, we note from Eqn. (6.13) that

$$\begin{aligned} \hat{g}^*(x'_1, y'_1, \nu)\hat{g}(x'_2, y'_2, \nu') = \int\int\int\int & \hat{f}^*(x_1, y_1, \nu)\hat{f}(x_2, y_2, \nu') \\ & \times h_c^*(x_1, x'_1, y_1, y'_1, \nu) \\ & \times h_c(x_2, x'_2, y_2, y'_2, \nu')\, dx_1\, dy_1\, dx_2\, dy_2 \end{aligned} \tag{6.14}$$

Noting that

$$W(x_1, y_1, x_2, y_2, \nu) = \lim_{\Delta\nu \to 0} \int_{\nu-\Delta\nu}^{\nu+\Delta\nu} \left\langle \hat{f}^*(x_1, y_1, \nu)\hat{f}(x_2, y_2, \nu') \right\rangle d\nu' \tag{6.15}$$

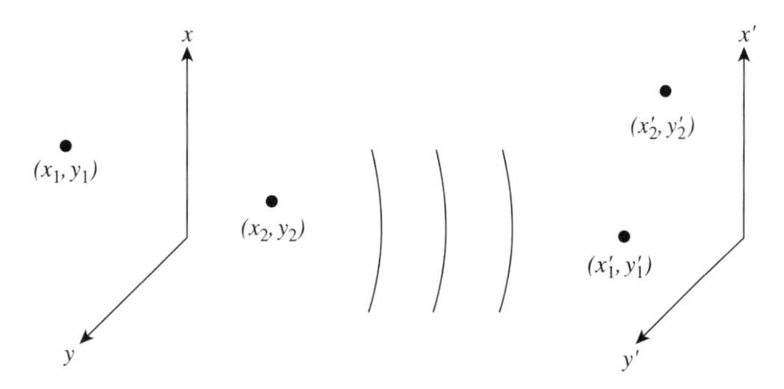

Figure 6.1 Input and output boundaries for propagation of coherence.

and

$$W(x_1', y_1', x_2', y_2', \nu) = \lim_{\Delta\nu \to 0} \int_{\nu-\Delta\nu}^{\nu+\Delta\nu} \langle \hat{g}^*(x_1, y_1, \nu)\hat{g}(x_2, y_2, \nu') \rangle \, d\nu' \tag{6.16}$$

we operate on the left and right sides of Eqn. (6.14) with $\lim_{\Delta\nu \to 0} \int_{\nu-\Delta\nu}^{\nu+\Delta\nu} \langle \cdot \rangle$ to obtain

$$\begin{aligned} W(x_1', y_1', x_2', y_2', \nu) = \iiiint & W(x_1, y_1, x_2, y_2, \nu) \\ & \times h_c^*(x_1, x_1', y_1, y_1', \nu) \\ & \times h_c(x_2, x_2', y_2, y_2', \nu) \, dx_1 \, dy_1 \, dx_2 \, dy_2 \end{aligned} \tag{6.17}$$

The impulse response for the 4D transformation of the cross-spectral density from the input plane to the output plane is thus

$$h_W(x_1, y_1, x_2, y_2, x_1', y_1', x_2', y_2', \nu) = h_c^*(x_1, x_1', y_1, y_1', \nu) h_c(x_2, x_2', y_2, y_2', \nu) \tag{6.18}$$

Equation (6.18) provides a very general basis for propagation of coherence functions in optical analysis. Once one knows the coherent impulse response, one can easily apply this principle to find the cross-spectral density response. The remainder of this section applies Eqn. (6.17) in analysis of three examples:

1. The propagation W from a 2D *spatially incoherent* primary source to an observation plane
2. The propagation of W from a remote source through an intermediate modulation plane (e.g., an aperture stops or an optical distortion) to an observation plane
3. The propagation of W from an object illuminated by partially coherent light to an observation plane

Turning to example 1, we note that most nonlaser optical radiators, such as the Sun and the stars, fluorescent and incandescent lightbulbs, and photochemical reactions, are well-modeled as sources of *spatially incoherent light*. Formally, the optical field on a plane is said to be spatially incoherent if

$$W(x_1, y_1, x_2, y_2, \nu) = l^2 S(x_1, y_1, \nu)\delta(x_1 - x_2)\delta(y_1 - y_2) \tag{6.19}$$

where l is a finite measure of the spatial coherence cross section. *Spatial incoherence* means that the light from any two distinct points on the plane is uncorrelated. While as a practical matter any physically realized field has a finite coherence cross section, we nevertheless find the incoherent model of Eqn. (6.19) quite useful as a first approximation to natural sources. Substituting the incoherent cross-spectral density in Eqn. (6.17) yields

$$W(x_1', y_1', x_2', y_2', \nu) = l^2 \int\int S(x, y, \nu) h_c^*(x, x_1', y, y_1', \nu) h_c(x, x_2', y, y_2', \nu)\, dx\, dy \tag{6.20}$$

In the case of free-space diffraction, h_c is given by Eqn. (4.38) under the Fresnel approximation and

$$\begin{aligned} W(x_1', y_1', x_2', y_2', \nu) &= \frac{l^2}{\lambda^2 z^2} \int\int S(x, y, \nu) e^{-i(\pi\nu/cz)[(x-x_1')^2+(y-y_1')^2]} e^{i(\pi\nu/cz)[(x-x_2')^2+(y-y_2')^2]}\, dx\, dy \\ W(\Delta x, \Delta y, q, \nu) &= \frac{l^2}{\lambda^2 z^2} e^{-i(2\pi\nu q/cz)} \int\int S(x, y, z, \nu) e^{i(2\pi\nu/cz)(x\Delta x + y\Delta y)}\, dx\, dy \end{aligned} \tag{6.21}$$

where $q = \bar{x}\Delta x + \bar{y}\Delta y$, $\bar{x} = (x_1' + x_2')/2$, and $\Delta x = x_1' - x_2'$. We find, therefore, that the cross-spectral density radiated by an incoherent source is proportional to the Fourier transform of the spatial distribution of the source and that it is quasistationary with respect to space (only the phase factor q depends on absolute spatial position).

Most importantly, note that the cross-spectral density of the radiated field no longer describes a spatially incoherent field. For sources of finite extent, the field "gains coherence" on propagation. If the spatial support of the source is A, one expects by Fourier uncertainty that the spatial support (*the coherence cross section*) of the coherence function will be approximately $\Delta x_{\max} \approx \lambda z/A = \lambda/\Delta\theta$, where $\Delta\theta = z/A$ is the angular extent of the source viewed from the output plane. As an example, if we assume that the Sun is a circular disk described by the spectral density $S(\rho, \nu) = S(\nu)\text{circ}(\rho/A)$, then the cross-spectral density of sunlight on Earth is

$$W(\Delta x, \Delta y, \nu) = \kappa S(\nu) \text{jinc}\left(\frac{\nu \Delta\theta \sqrt{\Delta x^2 + \Delta y^2}}{c} \right) \tag{6.22}$$

where we drop the q phase term because the maximum phase change that it produces is proportional to the ratio of the range of our measurement space to diameter of the Sun. The angular extent of the Sun viewed from Earth is 8.6 milliradians (mrad), meaning that the diameter of the cross-spectral density is approximately 284 wavelengths. The coherence cross section defined as the maximum value of Δx or Δy such that $|W(\Delta x, \Delta y, \nu)|$ is nonvanishing is an important measure of the coherence of the field. The related concepts of coherence length and coherence time are discussed in Section 6.3.1.

The spectral density $S(\nu) = W(\Delta x = 0, \Delta y = 0, q = 0, \nu)$ is uniform at all points in the Fresnel diffraction plane for an incoherent source. If the source is homogeneous such that the input spectral density separates into $f(x, y)S(\nu)$ (as in our solar model), then the diffracted spectrum is equal to the source spectrum. In general, the diffracted spectral density is equal to the mean of the spectral density over the input plane. Note that both the coherence and the spectrum of the field evolve on propagation.

It is strange, of course, that the intensity and spectral density of the source should be uniform in the Fresnel domain. We are used to the idea that the intensity of the source blurs slowly as it diffracts. The key to this mystery is our assumption that the source is incoherent, which essentially means that the input field has very high spatial frequencies that diffract rapidly. In the near field of actual sources a more advanced coherence model is necessary, but the incoherent model is satisfactory for most imaging applications.

As an important final comment on example 1, consider the power spectral density in the output plane under the propagation transformation Eqn. (6.20)

$$\begin{aligned} S(x', y', \nu) &= W(x', y', x', y', \nu) \\ &= l^2 \iint S(x, y, \nu) h_{ic}(x, y, x', y', \nu)\, dx\, dy \end{aligned} \tag{6.23}$$

where

$$h_{ic}(x, y, x', y', \nu) = |h_c(x, x', y, y', \nu)|^2 \tag{6.24}$$

Equation (6.23) expresses the general rule that *the impulse response describing the transformation of the incoherent power spectral density by an optical system is equal to the squared magnitude of the coherent impulse response.* We find this rule extremely useful in considering imaging of incoherent sources in Section 6.4.

Examples 2 and 3 consider imaging systems that sense objects using scattered light or sense primary sources through intervening apertures, systems, and media. Full 3D analysis of these systems is quite challenging; as a first approximation we consider the planar modulation system sketched in Fig. 6.2. Light from a remote primary source illuminates a 2D transmission mask in the input (x, y) plane. If $W_{in}(x_1, y_1, x_2, y_2, \nu)$ is the cross-spectral density of the light illuminating the mask, the cross-spectral density

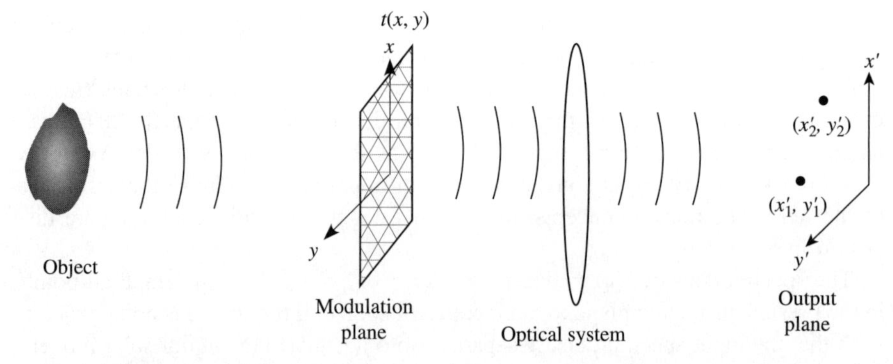

Figure 6.2 Coherence propagation through a 2D transmittance mask.

immediately after the mask is

$$W_{\text{out}}(x_1, y_1, x_2, y_2, \nu) = t^*(x_1, y_1, \nu)t(x_2, y_2, \nu)W_{\text{in}}(x_1, y_1, x_2, y_2, \nu) \tag{6.25}$$

where we allow for the possibility of spectral dependence in the mask transmittance. The mask transmittance is in general complex, reflecting phase and amplitude modulation.

In example 2 one images a primary source through an intervening modulation. The impulse response of the optical system in Fig. 6.2 to the right of the mask is a Fourier kernel. The optical system may consist, for example, of a Fourier transform lens with coherent impulse response given by Eqn. (4.68). The coherence transformation from the input to the mask to the output plane is

$$\begin{aligned} W(x_1', y_1', x_2', y_2', \nu) = & \iiiint W(x_1 - x_2, y_1 - y_2, \nu)t^*(x_1, y_1)t(x_2, y_2) \\ & \times \exp\left(2\pi i \frac{x_1 x_1' + y_1 y_1'}{\lambda F}\right) \exp\left(-2\pi i \frac{x_2 x_2' + y_2 y_2'}{\lambda F}\right) dx_1\, dy_1\, dx_2\, dy_2 \end{aligned} \tag{6.26}$$

where we assume the spatially stationary cross-spectral density of an incoherent source neglecting the q term under the assumption that $q/\lambda z \ll 1$. We consider imaging systems in which the q term is not negligible in Section 6.4.2.

Substituting the cross-spectral density from Eqn. (6.21) yields

$$\begin{aligned} W(x_1', y_1', x_2', y_2', \nu) = \kappa & \iint S(x, y, \nu) \\ & \times \hat{t}^*\left(\frac{x}{\lambda z} - \frac{x_1'}{\lambda F}, \frac{y}{\lambda z} - \frac{y_1}{\lambda F}\right) \\ & \times \hat{t}\left(\frac{x_2'}{\lambda F} - \frac{x}{\lambda z}, \frac{y_2'}{\lambda F} - \frac{y}{\lambda z}\right) dx\, dy \end{aligned} \tag{6.27}$$

where, as always, $\hat{t}$ is the Fourier transform of t. Since t is acting as the effective pupil stop for an imaging system, Eqn. (6.27) is hardly surprising. Recognizing that this is an imaging system, one immediately recognizes that the coherent impulse response is the Fourier transform of the pupil, which one can insert in Eqn. (6.20) to get Eqn. (6.27).

It is, however, worth emphasizing a couple of details with respect to Eqn. (6.27). First, note that even though the input source is incoherent, its image is partially coherent. The Fourier transform of Eqn. (6.27) with respect to all spatial variables yields

$$\begin{aligned} \hat{W}(u_1, v_1, u_2, v_2, \nu) = & \kappa \hat{S}\left(\frac{(u_1 + u_2)z}{F}, \frac{(v_1 + v_2)z}{F}, \nu\right) \\ & \times t^*(-\lambda F u_1, -\lambda F v_1)t(\lambda F u_2, \lambda F v_2) \end{aligned} \tag{6.28}$$

Since t must have finite support, W is bandlimited in all four spatial variables. By the Fourier localization relationship of Eqn. (3.25) one expects that if the support of t is A, such that the support of $\hat{W}$ is $A/\lambda F$, then the coherence cross section of the image will be approximately $\lambda F/A$. This postulate is trivially confirmed if the input object is a point source. A more interesting case considers the spectrally homogeneous spatially uniform source corresponding to $S(x, y, \nu) = S(\nu)$, in which case

$$W(\Delta x, \Delta y, \nu) = \kappa S(\nu) \iint \hat{t}^* \left(\frac{x}{\lambda f}, \frac{y}{\lambda F} \right) \hat{t} \left(\frac{x - \Delta x}{\lambda F}, \frac{y - \Delta y}{\lambda F} \right) dx\, dy \tag{6.29}$$

If t is a circular aperture of diameter A, $\hat{t}$ is the jinc function of Eqn. (4.75). Since $\text{jinc}(\rho)$ is invariant under autoconvolution, the cross-spectral density of the diffraction limited image of a uniform incoherent source is

$$W(\Delta x, \Delta y, \nu) = \kappa S(\nu) \text{jinc} \left(\frac{A}{\lambda F} \sqrt{\Delta x^2 + \Delta y^2} \right) \tag{6.30}$$

Our second comment with respect to Eqn. (6.27) is that the cross-spectral density, even when imaging an incoherent source, may contain information that is otherwise difficult to extract from irradiance or spectral density measurements. The mapping between the input and output power spectra represented by Eqn. (6.23) discards phase and cross-correlation data from the transfer function $t(x, y)$ that may be used to image through distortions or turbulence. Investigators commonly use a diversity of pupil modulations or use "wavefront sensors" to overcome this problem.

Example 3 images the mask in Fig. 6.2 onto the output plane. We again assume that the mask is illuminated by a random field that is stationary in both space and time, such that the output cross-spectral density is

$$\begin{aligned} W(x_1', y_1', x_2', y_2', \nu) = \iiiint & W(\Delta x, \Delta y, \nu) t^*(x_1, y_1) t(x_2, y_2) \\ & \times h^*(x_1' - x_1, y_1' - y_1, \nu) \\ & \times h(x_2' - x_2, y_2' - y_2, \nu)\, dx_1\, dy_1\, dx_2\, dy_2 \end{aligned} \tag{6.31}$$

where h is an imaging kernel as described by Eqn. (4.73). The goal in this case is to image the scattering object, $t(x, y)$. One may usually assume that the illuminating cross-spectral density and the imaging kernel are known a priori.

To illustrate the significance of Eqn. (6.31), consider an object consisting of two points, For example, $t(x, y) = \delta(x - a, y) + e^{i\phi} \delta(x + a, y)$, such that

$$\begin{aligned} W(x_1', y_1', x_2', y_2', \nu) = & W(0, 0, \nu)[h^*(x_1' - a, y_1', \nu) h(x_2' - a, y_2', \nu) \\ & + h^*(x_1' + a, y_1', \nu) h(x_2' + a, y_2', \nu)] \\ & + e^{i\phi} W(2a, 0, \nu) h^*(x_1' - a, y_1', \nu) h(x_2' + a, y_2', \nu) \\ & + e^{-i\phi} W(-2a, 0, \nu) h^*(x_1' + a, y_1', \nu) h(x_2' - a, y_2', \nu) \end{aligned} \tag{6.32}$$

If the two object points are within a spatial coherence cross section on the target, then there is interference between their impulse responses in the image. The relative phase of the scattering objects ϕ may potentially be abstracted from this interference or may appear as an image artifact if no attempt is made to measure it. As we saw with sunlight, even large incoherent sources may illuminate a scene with sufficient coherence that such interference effects play a role.

It is particularly interesting to consider the interference of two point scatterers that cannot be resolved by the imaging system. In this case $h(a, y, \nu) \approx h(-a, y, \nu)$ and the power spectral density at $x_1', y_1' = 0$ is

$$S(0, 0, \nu) = 2W(0, 0, \nu) + 2|W(2a, 0, \nu)|\cos(\phi + \phi_a) \tag{6.33}$$

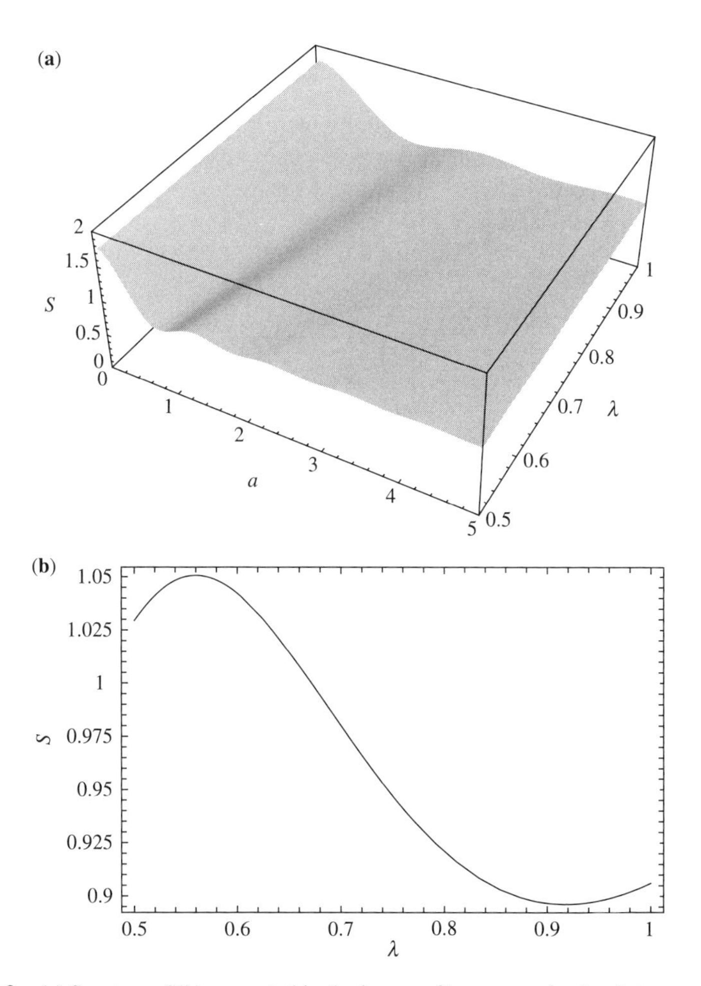

Figure 6.3 (a) Spectrum $S(\lambda)$ generated in the image of two unresolved point sources illuminated by a wave with cross-spectral density $\text{jinc}(\Delta\rho\,\Delta\theta/\lambda)$, where a is in units of $\lambda/\Delta\theta$ (we assume one octave of uniform spectral density); (b) plots the spectrum for $a = 1.5\lambda/\Delta\theta$.

where $\phi_a = \arg[W(2a, 0, \nu)]$. Note that the relative position of the two point sources affects the spectrum of the image field even though the points are unresolved. The scattered spectrum observed on the optical axis as a function of a and wavelength for a jinc distributed cross-spectral density is illustrated in Fig. 6.3. We assume that the spectrum of the illuminating source is uniform across the observed range. The scattered spectrum is constant if the two points are in the same position or if the two points are widely separated. The scattered power is doubled if the two points are at the same point as a result of constructive interference. If the two points are separated in the transverse plane by 1–2 wavelengths, the spectrum is weakly modulated, as illustrated Fig. 6.3(b). The spectral modulation is much greater if the sources are displaced longitudinally or if the scattered light is observed from an off-axis perspective. This example is considered in Problem 6.3; more general discussion of spectral modulation by secondary scattering is presented in Sections 6.5 and 10.3.1.

While the three examples that we have discussed have various implications for imaging and spectroscopy, our primary goal has been to introduce the reader to analysis of cross-spectral density transformations and diffraction. Equation (6.20) is quite general and may be applied to many optical systems. Now that we know how to propagate the cross-spectral density from input to output, we turn to the more challenging topic of how to measure it.

6.3 MEASURING COHERENCE

We saw in Section 6.2 that given the cross-spectral density (or equivalently the mutual coherence) on a boundary, the cross-spectral density can be calculated over all space. But how do we characterize the coherence function on a boundary? We have often noted that optical detectors measure only the irradiance $I(x, y, t)$ over points x, y, and t in space and time. Coherence functions must be inferred from such irradiance measurements. The goal of optical sensor design is to lay out physical structures such that desired projections of coherence fields are revealed in irradiance data.

Sensor performance metrics are complex and task-specific, but it is useful to start with the assumption that one wishes simply to measure natural cross-spectral densities or mutual coherence functions with high fidelity. We explore this approach in simple Michelson and Young interferometers before moving on to discuss coherence measurements of increasing sophistication based on parallel and indirect methods.

6.3.1 Measuring Temporal Coherence

The temporal coherence of the field at a point r may be characterized using a Michelson interferometer, as sketched in Fig. 6.4. Input light from pinhole is collimated and split into two paths. Both paths are retroreflected on to a detector using mirrors. One of the mirrors is on a translation stage such that its longitudinal position

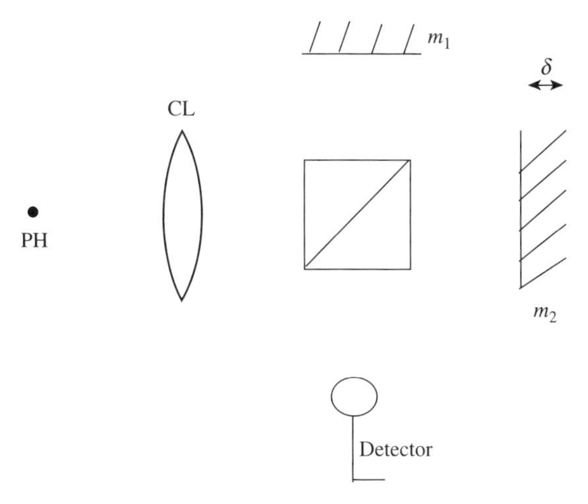

Figure 6.4 Measurement of the mutual coherence using a Michelson interferometer. Light from an input pinhole or fiber is collimated into a plane wave by lens CL and split by a beamsplitter. Mirror M2 may be spatially shifted by an amount δ along the optical axis, producing a relative temporal delay $2\delta/c$ for light propagating along the two arms. Light reflected from M1 interferes with light from M2 on the detector.

may be varied by an amount δ. If the input field is $E(t)$, the irradiance striking the detector is

$$\begin{aligned} I(\delta) &= \frac{1}{4}\left\langle \left| E(t) + E\left(t + \frac{2\delta}{c}\right)\right|^2 \right\rangle \\ &= \frac{\Gamma(0)}{2} + \frac{1}{4}\Gamma\left(\frac{2\delta}{c}\right) + \frac{1}{4}\Gamma\left(-\frac{2\delta}{c}\right) \end{aligned} \tag{6.34}$$

where we have abbreviated the single-point mutual coherence $\Gamma(\mathbf{r}, \mathbf{r}, \tau)$ with $\Gamma(\tau)$. $\Gamma(\tau)$ is isolated from $\Gamma(0)$ and $\Gamma(-\tau)$ in Eqn. (6.34) by Fourier filtering. The Fourier transform of $I(\delta)$ is

$$\hat{I}(u) = \frac{\Gamma(0)}{2}\delta(u) + \frac{c}{8}S\left(\nu = \frac{uc}{2}\right) + \frac{c}{8}S\left(\nu = -\frac{uc}{2}\right) \tag{6.35}$$

$S(\nu)$ is the positive frequency component of $\hat{I}(u)$, and $\Gamma(\tau)$ is the inverse Fourier transform of $S(\nu)$.

The Fourier transform pairing between the power spectrum and the mutual coherence corresponds to a relationship between spectral bandwidth and coherence time through the Fourier uncertainty relationship. The bandwidth σ_ν measures the support of $S(\nu)$, and the coherence time $\tau_c \propto 1/\sigma_n u$ measures the support of $\Gamma(\tau)$. Various precise definitions for each may be given; the variance of Eqn. (3.22) may be the best measure. For present purposes it most useful to consider the relationship in the context of common spectral lines, as listed in Table 6.2.

TABLE 6.2 Spectral Density and Mutual Coherence

Lineshape	$S(\nu)$	$\Gamma(\tau)$
Monochromatic	$\delta(\nu - \nu_0)$	$e^{2\pi i\nu_0\tau}$
Gaussian	$(1/\sigma_\nu)e^{-\pi[(\nu-\nu_0)^2/\sigma_\nu^2]}$	$e^{2\pi i\nu_0\tau}e^{-\pi\sigma_\nu^2\tau^2}$
Lorentzian	$\sigma_\nu/[(\nu - \nu_0)^2 + \sigma_\nu^2]$	$2\pi e^{2\pi i\nu_0\tau}e^{-2\pi\sigma_\nu\|\tau\|}$

The Gaussian and Lorentzian spectra are plotted in Fig. 6.5. A common characteristic is that the spectrum is peaked at a center frequency ν_0 and has a characteristic width σ_ν. The mutual coherence function oscillates rapidly as a function of τ with period ν_0. The mutual coherence peaks at $\tau = 0$ and has characteristic width $1/\sigma_\nu$.

Mechanical accuracy and stability must be precise to measure coherence using a Michelson interferometer. The output irradiance $I(\delta)$ oscillates with period $\lambda_0/2$, where $\lambda_0 = c/\nu_0$. Nyquist sampling of $I(\delta)$ therefore requires a sampling period of less than $\lambda_0/4$, which corresponds to 100–200 nm at optical wavelengths. Fine sampling rates on this scale are achievable using piezoelectric actuators to translate

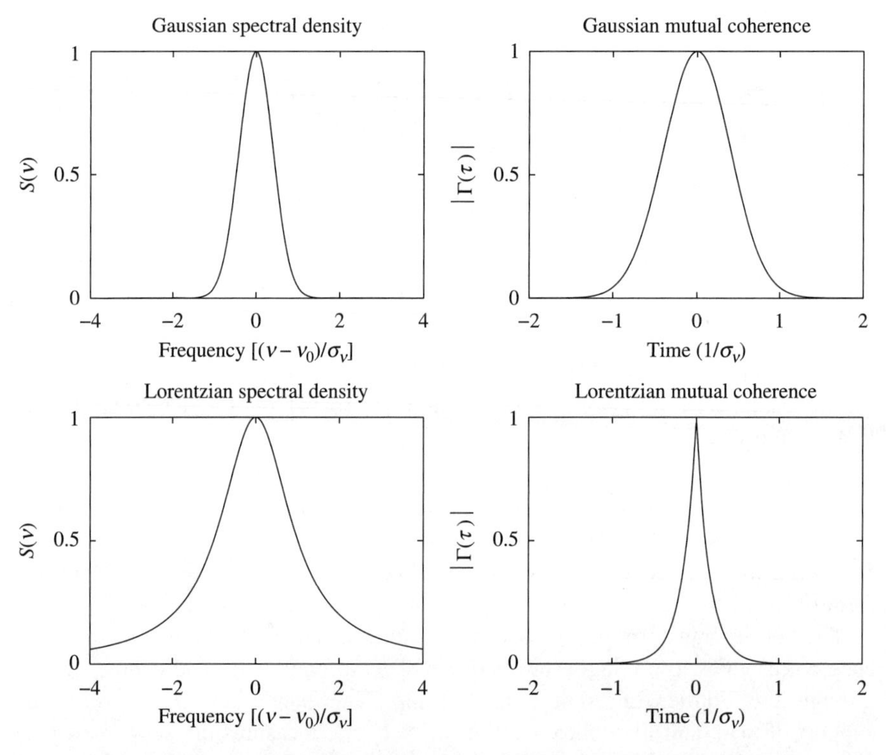

Figure 6.5 Spectral densities and mutual coherence of Gaussian and Lorentzian spectra. The mutual coherence is modulated by the phasor $e^{2\pi i\nu_0\tau}$; the magnitude of the mutual coherence is plotted here.

the mirror M2. Ideally, the range over which one samples should span the coherence time τ_c. This corresponds to a sampling range $D = c/2\tau_c$.

The Michelson interferometer is used in this way is a *Fourier transform spectrometer* (there are many other interferometer geometries that also produce FT spectra). The Michelson is the first encounter in this text with a true spectrometer. While we begin to mention spectral degrees of freedom more frequently, we delay most of our discussion of Fourier instruments until Chapter 9. For the present purposes it is useful to note that the FT instrument is particularly useful when one wants to measure a spectrum using only one detector. FT instruments are favored for spectral ranges where detectors are noisy and expensive, such as the infrared (IR) range covering 2–20 μm. Instruments in this range are sufficiently popular that the acronym *FTIR* covers a major branch of spectroscopy.

6.3.2 Spatial Interferometry

One must create interference between light from multiple points to characterize spatial coherence. The most direct way to measure $W(x_1, y_1, x_2, y_2, \nu)$ samples the interference of every pair of points as illustrated in Fig. 6.6. Pinholes at points P_1 and P_2 transmit the fields $E(P_1, \nu)$ and $E(P_2, \nu)$. Letting $h(\mathbf{r}, P, \nu)$ represent the impulse response for propagation from point P on the pinhole plane to point $\mathbf{r}$ to the detector plane, the irradiance at the detector array is

$$\begin{aligned} I(\mathbf{r}) &= \int \left\langle |E(P_1, \nu)h(\mathbf{r}, P_1, \nu) + E(P_2, \nu)h(\mathbf{r}, P_2, \nu)|^2 \right\rangle d\nu \\ &= I(P_1) + I(P_2) + \int W(P_1, P_2, \nu)h^*(\mathbf{r}, P_1, \nu)h(\mathbf{r}, P_2, \nu)\, d\nu \\ &\quad + \int W(P_2, P_1, \nu)h^*(\mathbf{r}, P_2, \nu)h(\mathbf{r}, P_1, \nu)\, d\nu \end{aligned} \tag{6.36}$$

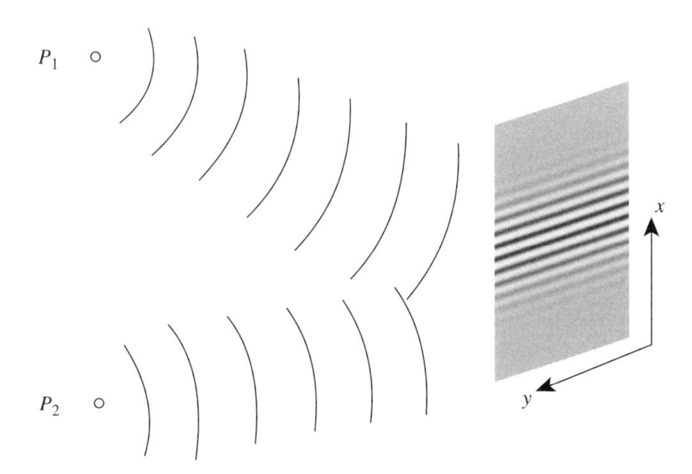

Figure 6.6 Interference between fields from points P_1 and P_2.

Approximating h with the Fresnel kernel models the irradiance at point (x, y) on the measurement plane as

$$\begin{aligned} I(x, y) = & I(P_1) + I(P_2) \\ & + \int W(x_1, y_1, x_2, y_2, \nu) \\ & \times \exp\left(2\pi i\nu \frac{x\Delta x + y\Delta y}{cd}\right) \exp\left(-2\pi i\nu \frac{q}{cd}\, d\nu\right) \\ & + \int W(x_2, y_2, x_1, y_1, \nu) \\ & \times \exp\left(-2\pi i\nu \frac{x\Delta x + y\Delta y}{cd}\right) \exp\left(2\pi i\nu \frac{q}{cd}\, d\nu\right) \end{aligned} \tag{6.37}$$

where d is the distance from the pinhole plane to the measurement plane and as before $\Delta x = x_1 - x_2$ and $q = \bar{x}\Delta x + \bar{y}\Delta y$.

With Fresnel diffraction, the interference pattern produced by a pair of pinholes varies along the axis joining the pinholes and is constant along the perpendicular bisector, as illustrated in Fig. 6.6. We isolate the 1D interference pattern mathematically by rotating variables in the x, y plane such that $\tilde{x} = (x\Delta x + y\Delta y)/\sqrt{\Delta x^2 + \Delta y^2}$ and $\tilde{y} = (x\Delta x - y\Delta y)/\sqrt{\Delta x^2 + \Delta y^2}$. In the rotated coordinate system the interference term in the two-pinhole diffraction pattern becomes

$$\int W(x_1, y_1, x_2, y_2, \nu) \exp\left(2\pi i\nu \frac{\tilde{x}\sqrt{\Delta x^2 + \Delta y^2}}{cd}\right) \exp\left(2\pi i\nu \frac{q}{cd}\, d\nu\right), \tag{6.38}$$

which is independent of $\tilde{y}$.

The interference term is the inverse Fourier transform of the cross-spectral density with respect to ν, which means by the Wiener–Khintchine theorem that the interference is proportional to the mutual coherence. Specifically

$$\begin{aligned} I(\tilde{x}) = & I(P_1) + I(P_2) \\ & + \Gamma\left(x_1, y_1, x_2, y_2, \tau = \frac{q - \tilde{x}\sqrt{\Delta x^2 + \Delta y^2}}{cd}\right) \\ & + \Gamma\left(x_2, y_2, x_1, y_1, \tau = \frac{\tilde{x}\sqrt{\Delta x^2 + \Delta y^2 - q}}{cd}\right) \end{aligned} \tag{6.39}$$

Like the Michelson interferometer, the two-pinhole interferometer measures the mutual coherence. In this case, however, samples are distributed at a single moment in time along a spatial sampling grid.

Sampling for the pinhole system is somewhat complicated by the uneven scaling of the sampling rate. Ideally, one would sample τ over the range $(0, \tau_c)$ at resolution $1/2\nu_{\max}$, where $\Delta\nu$ is the bandwidth of the field and $\nu_{\max}$ is the maximum temporal frequency. This corresponds to a spatial sampling range $X = c\tau_c d/\Delta x$ at sampling rate $cd/2\nu_{\max}$. If the pixel pitch for sampling the interference pattern is 10λ, which may be typical of current visible focal planes, one would need to ensure that $d/\Delta x > 20$. In this case $\tau_c = 100\,\text{fs}$ would correspond to $X = 0.6\,\text{mm}$.

As with the Michelson interferometer, one isolates the cross-spectral density from $I(\tilde{x})$ by Fourier analysis. The Fourier transform of Eqn. (6.39) with respect to $\tilde{x}$ yields the following term in the range $u > 0$:

$$\hat{I}(u > 0) = W\left(x_1, y_1, x_2, y_2, \nu = \frac{cdu}{\sqrt{\Delta x^2 + \Delta y^2}}\right)\exp\left(-2\pi i \frac{qu}{\sqrt{\Delta x^2 + \Delta y^2}}\right) \tag{6.40}$$

Thus we are able to isolate the complex coherence function by Fourier filtering. In the current example we use an entire plane to characterize $W(x_1, y_1, x_2, y_2, \nu)$ as a function of ν with (x_1, y_1, x_2, y_2) held constant.

As an example, suppose that a primary source consisting of a point radiator with a spectral radiance $S(\nu)$ illuminates the pinholes. Assuming that the point source is located at $(x_0, y_0, z = 0)$, Eqn. (6.21) immediately yields the cross-spectral density at planes $z \neq 0$

$$\begin{aligned} W(\Delta x, \Delta y, q, \nu) = {} & \frac{l^4}{\lambda^2 z^2} S(\nu) \\ & \times \exp\left(-i2\pi\nu \frac{(x_0\Delta x + y_0\Delta y)}{cz}\right)\exp\left(-i2\pi\nu\frac{q}{cz}\right) \end{aligned} \tag{6.41}$$

and, for the two-pinhole system of Fig. 6.6

$$I(x, y) = 2I_0 + \Gamma[\tau(x, y)] + \Gamma[-\tau(x, y)] \tag{6.42}$$

where $\Gamma(\tau)$ is the mutual coherence and the inverse Fourier transform of $S(\nu)$ and

$$\begin{aligned} \tau(x, y) &= \frac{(x_0 - x)\Delta x + (y_0 - y)\Delta y + 2q}{cd} \\ &= \frac{x_0\Delta x + y_0\Delta y - 2\tilde{x}\sqrt{\Delta x^2 + \Delta y^2} + 2q}{cd} \end{aligned} \tag{6.43}$$

A plot of $I(\tilde{x})$ for $q = 0$ and for $\Delta x/d = 0.1$ is shown in Fig. 6.7 for a Gaussian spectrum of width 10 nm with a central wavelength of 600 nm. The interference pattern produced has a period of $\lambda_0 d/\Delta x$, which is 6 μm in this case. Thus, one would hope to spatially sample at 3 μm resolution over the 1.5 mm range to capture this interference pattern.

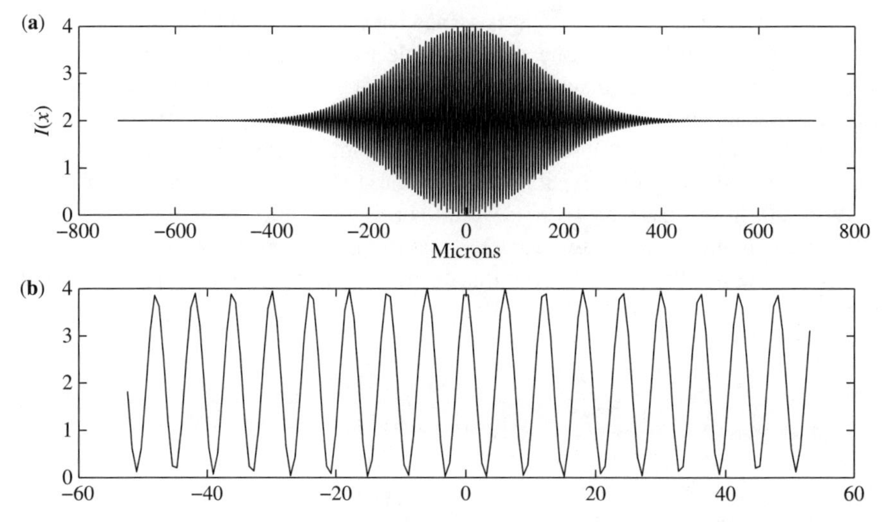

Figure 6.7 Irradiance pattern $I(x)$ produced by a 10 nm spectral bandwidth source centered on 600 nm observed through a two-pinhole interferometer with $\Delta x/d = 0.1$. Plot (a) details the center region of plot (b).

Each configuration of the pinholes enables us to characterize $W(\Delta x, \Delta y, \bar{x}, \bar{y}, \nu)$ as a function of ν for a particular value of $(\Delta x, \Delta y, \bar{x}, \bar{y})$. One can imagine moving the pinholes around the plane to fully sample the cross-spectral density, but the two-pinhole approach is not a very efficient sampling mechanism and faces severe challenges with respect to sampling rate and range for large or small values of Δx. The two-pinhole approach is nevertheless the basic strategy underlying the *Michelson stellar interferometer* [58]. The sampling efficiency can be improved by using lens combinations to reduce the spatial pattern due to one pair of pinholes to a line, thus enabling "two slit" characterization of distinct values of Δx and $\bar{x}$ in parallel. Dual-slit sampling enables full utilization of a 2D measurement plane for independent measurements, but the mechanical complexity and limited throughput of this approach pose challenges.

6.3.3 Rotational Shear Interferometry

The cross-spectral density on a plane is a five-dimensional function of four spatial dimensions and temporal frequency. A rotational shear interferometer (RSI) characterizes this space from nondegenerate measurements on a 2D plane. The basic structure of an RSI is sketched in Fig. 6.8. Figure 6.9 is a photograph of an RSI.

The structure is the same as for a Michelson interferometer, but the flat mirrors have been replaced by wavefront folding mirrors. A wavefront folding mirror is a right angle assembly of two reflecting surfaces. A light beam entering such an interferometer is inverted across the fold axis, as described below. In the RSI of Fig. 6.6 the fold mirrors consist of right-angle prisms. The "fold axis" is the right-angle edge

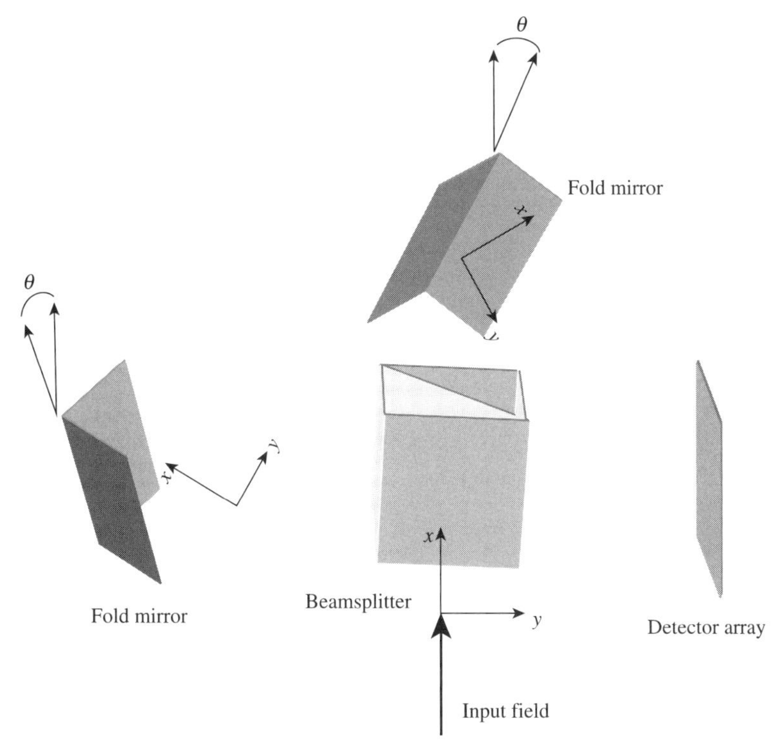

Figure 6.8 System layout of a rotational shear interferometer.

Figure 6.9 Photograph of a rotational shear interferometer. The fold mirrors consist of right-angle prisms, one of the prisms is mounted in a computer controlled rotation stage to adjust the longitudinal displacement and shear angle.

of the prism. As illustrated in Fig. 6.8, the fold axes of the mirrors are displaced from the vertical (x) axis by angle θ on one arm and by $-\theta$ on the other arm.

The effect of angular displacement of the fold axes is to produce a field distribution from each arm rotated in the x, y plane with respect to the field from the other arm. Let $E(x, y)$ be the electromagnetic field that would be produced on the detection plane of an RSI after reflection from a flat mirror. If this same field is reflected by a fold mirror with fold axis is parallel to y, the resulting reflected field is $E(-x, y)$. If the fold axis is parallel to x, the resulting field is $E(x, -y)$. If the fold axis lies at an arbitrary angle θ with respect to the x axis in the xy plane, the resulting field is $E[x\cos(2\theta) + y\sin(2\theta), x\sin(2\theta) - y\cos(2\theta)]$. With the fold axes of the mirrors on the two reflecting arms of the RSI counter rotated by θ and $-\theta$, the electromagnetic field on the detection plane is

$$\begin{aligned} &E[x\cos(2\theta) + y\sin(2\theta), x\sin(2\theta) - y\cos(2\theta)] \\ &\quad + E[x\cos(2\theta) - y\sin(2\theta), -x\sin(2\theta) - y\cos(2\theta)]\exp(i\phi) \end{aligned} \tag{6.44}$$

where, as with a Michelson interferometer, $\phi = 4\pi\nu\delta/c$ is the phase difference between the two arms produced by a relative longitudinal displacement δ between mirrors on the two arms.

The spectral density on the detection plane is found by taking appropriate expectation values of the square of Eqn. (6.44), which yields

$$\begin{aligned} S_{\text{rsi}}(x, y, \nu) = {} & S[x\cos(2\theta) + y\sin(2\theta), x\sin(2\theta) - y\cos(2\theta), \nu] \\ & + S[x\cos(2\theta) - y\sin(2\theta), -x\sin(2\theta) - y\cos(2\theta), \nu] \\ & + e^{4\pi i(\nu\delta/c)} W[\Delta x = 2y\sin(2\theta), \Delta y = 2x\sin(2\theta), \\ & \quad \bar{x} = 2x\cos(2\theta), \bar{y} = -2y\cos(2\theta), \nu] \\ & + e^{-4\pi i(\nu\delta/c)} W[\Delta x = -2y\sin(2\theta), \Delta y = -2x\sin(2\theta), \\ & \quad \bar{x} = 2x\cos(2\theta), \bar{y} = -2y\cos(2\theta), \nu] \end{aligned} \tag{6.45}$$

where $S(x, y, \nu)$ and $W(\Delta x, \Delta y, \bar{x}, \bar{y}, \nu)$ are the spectral densities that would appear on the detection plane if the fold mirrors were replaced by flat mirrors.

As an example, suppose that an RSI is illuminated by a remote point source with spectral density $S(\nu)$. The cross-spectral density incident on the RSI measurement plane for this case is given by Eqn. (6.41). Substituting in Eqn. (6.45) and ignoring constant factors, the spectral density observed on the RSI measurement plane is

$$S_{\text{rsi}}(x, y, \nu) = S(\nu)\left(1 + \cos\left\{4\pi\frac{\nu}{c}[\theta_x y\sin(2\theta) + \theta_y x\sin(2\theta) + \delta)]\right\}\right) \tag{6.46}$$

where $\theta_x = x/z$ and $\theta_y = y/z$ are the angular positions of the point source as observed at the RSI. Figure 6.10 shows interference patterns detected by an RSI observing a remote point illuminated at two wavelengths. In this

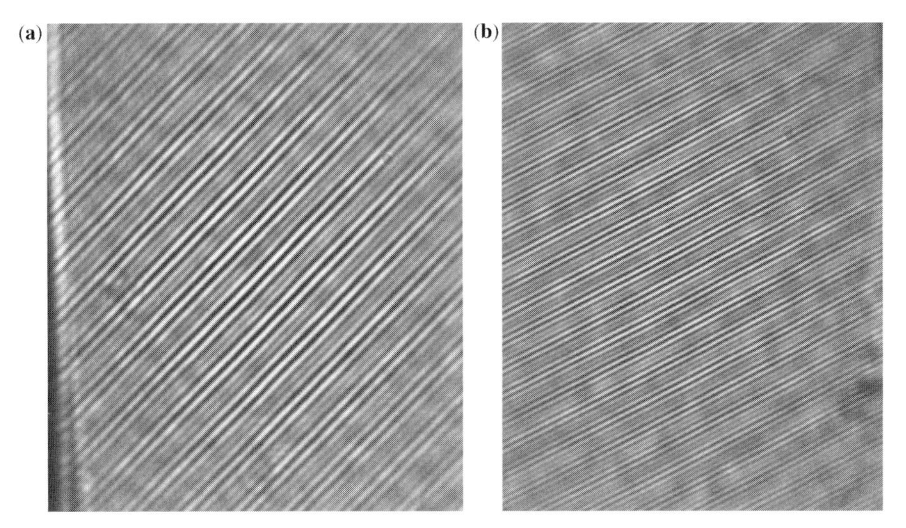

Figure 6.10 RSI raw data image for the two-color point source of Eqn. (6.47): (a) $\sqrt{\theta_x^2 + \theta_y^2} = 1.34^\circ$; (b) $\sqrt{\theta_x^2 + \theta_y^2} = 3^\circ$ ($\theta = 2^\circ$ in both cases).

case $S(\nu) = I_1 \delta(\nu - \nu_1) + I_2 \delta(\nu - \nu_2)$, and the irradiance on the detector is

$$I(x, y) = I_1 + I_2 + I_1 \cos\left[\frac{4\pi\nu_1 \sin(2\theta)}{c}(\theta_x y + \theta_y x)\right] + I_2 \cos\left[\frac{4\pi\nu_2 \sin(2\theta)}{c}(\theta_x y + \theta_y x)\right] \quad (6.47)$$

Consistent with Eqn. (6.47), the images in Fig. 6.10 show beating between two harmonics, as confirmed in Fig. 6.11, which shows the 2D FFT of the irradiance patterns with DC frequencies suppressed. The FFT produces images of the illuminating point source at $[u = 2\theta_y \sin(2\theta)/\lambda,\ v = 2\theta_x \sin(2\theta)/\lambda]$. The point image further from the origin thus corresponds to the image of the source at the bluer illuminating wavelength.

The figures show interference patterns for two different angular displacements of the point source from the optical axis. As expected, the fringe frequency increases as the angle increases. The dark vertical lines at the left edge of Fig. 6.10(a) are shadows of the fold edge of the wavefront folding mirrors. The total angular displacement the fold mirrors is 4°, meaning $\theta = 2^\circ$.

Note from Eqn. (6.46) that the fringe frequency is proportional to $\sin(2\theta)$, so θ may be set to match the fringe pattern to the sampling rate on the detector plane. The fringe frequency is also proportional to ν and the angular position. If θ is fixed, $\nu\theta_x$ and $\nu\theta_y$ may be determined from Fourier analysis of Eqn. (6.46). ν and θ_x, θ_y may be disambiguated by varying δ or the orientation of the RSI relative to the scene.

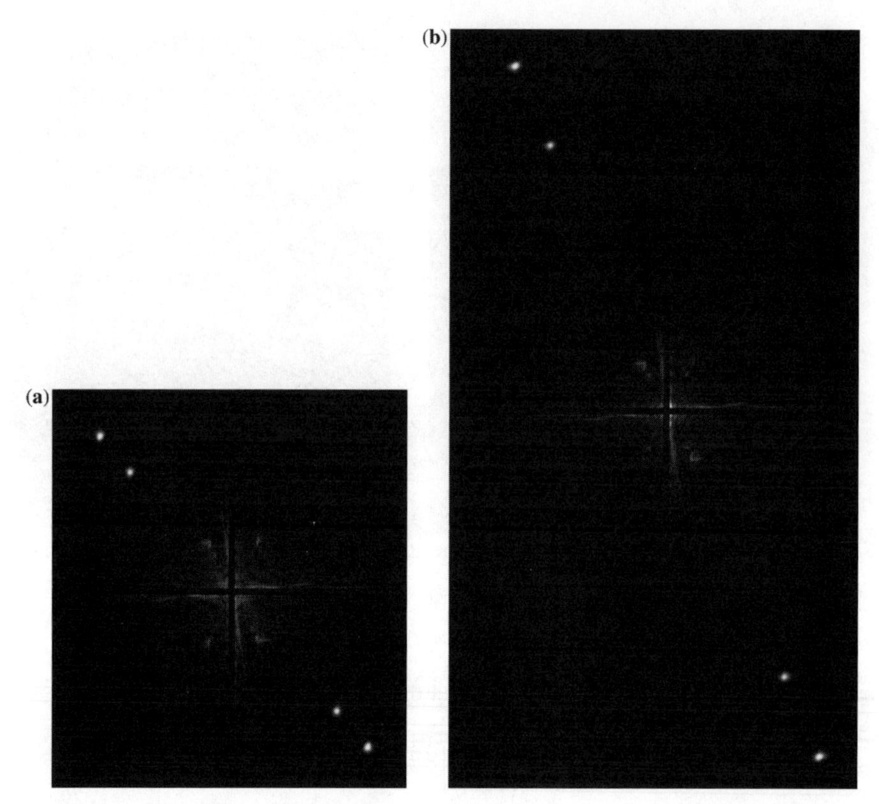

Figure 6.11 FFT of Figs. 6.10(a) and (b). The plot scale is the same in both cases; (a) the higher-frequency fringes of (b) correspond to a source at greater angular displacement.

Since Eqn. (6.47) is the impulse response for incoherent imaging, the RSI irradiance created by an arbitrary 3D incoherent primary source is

$$I(x, y) = \int S(x, y, z, \nu)\, dx\, dy\, dz\, d\nu + \int S(\theta_x, \theta_y, \nu) \times \cos\left[\frac{4\pi\nu\sin(2\theta)}{c}\left(\theta_x y + \theta_y x\right) + \frac{4\pi\nu\delta}{c}\right] d\theta_x\, d\theta_y\, d\nu \tag{6.48}$$

where, as in Eqn. (2.31), we obtain

$$S(\theta_x, \theta_y, \nu) \int S(x = z\theta_x, y = z\theta_y, z, \nu)\, dz \tag{6.49}$$

The second term in Eqn. (6.48), the 3D cosine transform of $S(\theta_x, \theta_y, \nu)$, is invertible given the real and nonnegative nature of the power spectral density. Thus, the RSI can function as an infinite depth of the field imaging system [170]. Unfortunately, however, noise from the DC background tends to dominate image reconstruction from Eqn. (6.48). For shot noise–dominated imagers, for example, the pixel SNR in reconstructing $S(\theta_x, \theta_y, \nu)$ using linear estimators is

$$\mathrm{SNR}_{ij} = \frac{P_{ij}}{\sqrt{2P}} \tag{6.50}$$

where P_{ij} is the expected photon count from pixel ij and P is the total number of photons detected by the RSI [14]. If, for example, the image consists of N pixels of approximately equal intensity, the SNR is $\sqrt{P_{ij}/2N}$. This compares with an SNR of $\sqrt{P_{ij}}$ for a conventional focal image, although the comparison is not quite fair given the RSI's infinite depth of field. We discuss image depth of field in detail in Section 10.2.

Measurement of the full 5D cross-spectral density using an RSI is most easily described on the $\Delta x = (x_1 - x_2)$, $\bar{x} = (x_1 + x_2)/2$ basis. We see from Eqn. (6.45) that each point in the RSI plane measures $W(\Delta x, \Delta y, \nu)$ for a unique value of Δx, Δy, and that the mean positions $\bar{x}, \bar{y}$ vary linearly across the RSI plane. The process of cross-spectral density measurement is illustrated in experimental data in Fig. 6.12. The first step is to gather a data cube of RSI measurements for displacements δ covering the spectral coherence length. Each pixel of this data cube is Fourier-transformed along the δ axis to transform from the mutual coherence to the cross-spectral density. Slices of the transformed data cube in the transverse plane correspond to a plane of Δx, Δy data tilted with respect to the $\bar{x}, \bar{y}$ axes. Slices of W at specific frequencies and may be transformed to image an incoherent source as shown in the figure. One samples a full range of mean positions using relative lateral translation of the RSI and object.

The RSI presents an efficient and powerful direct method for measuring the cross-spectral density. As we have seen, however, the method provides poor SNR and requires a sophisticated positioning and scanning system. It is clear from Section 6.2 that a sensor to measure the true cross-spectral density is a boon to optical imaging, but direct two-beam interferometry is not the only means of measuring W. We turn to subtler methods in the next section.

6.3.4 Focal Interferometry

The vast majority of optical measurements use lens systems rather than pointwise interferometry. A focal system is also an interferometer; the magical transformation from diffuse light to well-focused image arises from wave interference. Focal interference, however, is based on global transformations of coherence functions rather than two-beam correlations.

Transformation of coherence functions in focal systems is the most basic tool of optical system analysis. One may use coherence functions to analyze the action of focal imaging systems on optical fields, which approach we adopt in

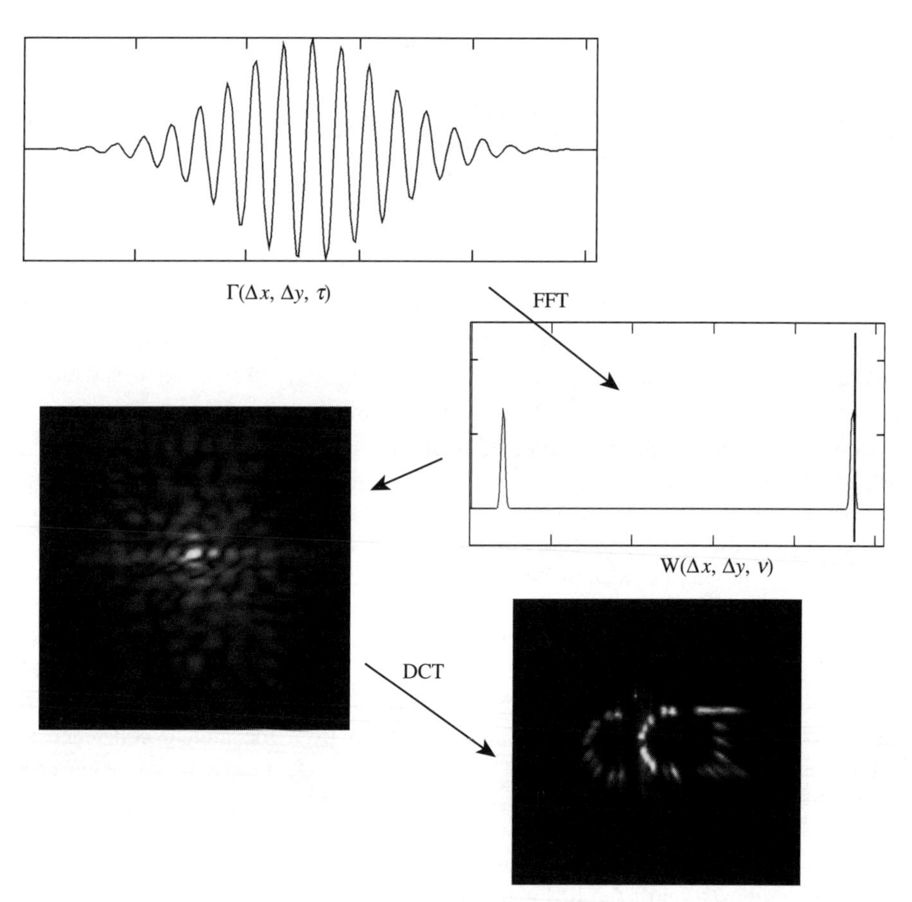

Figure 6.12 Measurement of $W(\Delta x, \Delta y, \nu)$ with an RSI. Plotted at upper left is the irradiance measured by a single pixel as a function of the longitudinal delay δ. The absolute value of the FFT of this trace is shown below to the right with the DC terms suppressed. A single complex value corresponding to the cross-spectral density at a particular wavelength is selected from this trace. The particular frequency selected is marked with a vertical line in the FFT trace. The image at lower left shows the magnitude of the cross-spectral density at this frequency at each pixel on the RSI. The image at lower right is the inverse discrete cosine transform of this map, which for an incoherent source produces an image. The object is a "LiteBrite" toy with red pegs stuck in paper in front of an incandescent lightbulb. The letters CCI denote the shortlived Center for Computational Imaging.

Section 6.4, or one may use focal systems to analyze coherence functions, which approach we take in the present section.

We specifically consider the transformation between the cross-spectral density on the input aperture of a lens and the spectral density in the focal volume, as illustrated in Fig. 6.13. Modeling diffraction by the Fresnel approximation [Eqn. (4.38)], and the lens transmittance by thin parabolic phase modulation [Eqn. (4.62)], the spectral

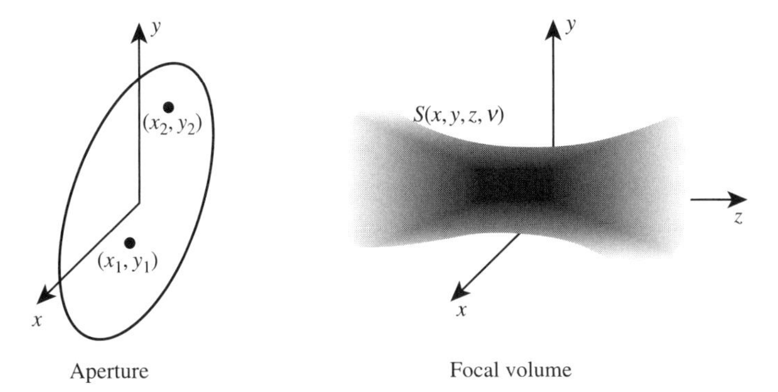

Figure 6.13 Geometry for measurement of the cross-spectral density on the input aperture of a lens by analysis of the power spectral density in the focal volume.

density in the focal volume is

$$
\begin{aligned}
S(x, y, z, \nu) = \frac{\nu^2}{c^2 z^2} \iiiint & W(x_1, y_1, x_2, y_2, \nu) P^*(x_1, y_1) P(x_2, y_2) \\
& \times \exp\left(i\pi\nu \frac{x_1^2 + y_1^2}{cF}\right) \exp\left(-i\pi\nu \frac{x_2^2 + y_2^2}{cF}\right) \\
& \times \exp\left(-i\pi\nu \frac{(x - x_1)^2 + (y - y_1)^2}{cz}\right) \\
& \times \exp\left(i\pi\nu \frac{(x - x_2)^2 + (y - y_2)^2}{cz}\right) dx_1\, dx_2\, dy_1\, dy_2
\end{aligned}
\tag{6.51}
$$

In the by now standard Δx, $\bar{x}$ parameterization, this transformation reduces to

$$
\begin{aligned}
S(x, y, z, \nu) = \frac{4\nu^2}{c^2 z^2} \iiiint & W(\Delta x, \Delta y, \bar{x}, \bar{y}, \nu) \\
& \times \exp\left[2i\pi\nu(\Delta x \bar{x} + \Delta y \bar{y}) \left(\frac{1}{cF} - \frac{1}{cz}\right)\right] \\
& \times P^*\left(\bar{x} + \frac{\Delta x}{2}, \bar{y} + \frac{\Delta y}{2}\right) P\left(\bar{x} - \frac{\Delta x}{2}, \bar{y} - \frac{\Delta y}{2}\right) \\
& \times \exp\left(2i\pi\nu \frac{x\Delta x + y\Delta y}{cz}\right) d\Delta x\, d\Delta y\, d\bar{x}\, d\bar{y}
\end{aligned}
\tag{6.52}
$$

Despite the fact that it projects a 5D distribution onto 4D, Eqn. (6.52) forms the basis for estimation of the cross-spectral density from focal measurements. Strategies for handling the mismatched dimensionality include

1. Taking advantage of the fact that W reduces to a 3D or 4D function in many optical systems

2. Using temporal variation of the pupil function or parallel nondegenerate optical systems to increase the dimensionality or sampling range of the focal volume
3. Applying generalized sampling and estimation strategies to infer W from discrete measurements on S.

These strategies are not exclusive and are often applied in combination. All three strategies are improved by design and coding of the aperture function to facilitate particular applications. Given that coding, sampling, and inversion strategies for Eqn. (6.52) are the focus of much of the remainder of this text, we cannot hope to fully analyze the possibilities in this section. We do, however, briefly overview examples of the first two basic strategies.

The first strategy focuses on reconstruction of the cross-spectral density arising from remote incoherent objects, as described by Eqn. (6.21). For such incoherent objects the cross-spectral density reduces to a 4D function over $(\Delta x, \Delta y, q, \nu)$ and reduces Eqn. (6.52) to

$$S(x, y, z, \nu) = \frac{4}{\lambda^2 z^2} \iiint W(\Delta x, \Delta y, q, \nu) B(\Delta x, \Delta y, q) \times \exp\left(-2i\pi\nu \frac{x\Delta x + y\Delta y + (1 - z/F)q}{cz}\right) d\Delta x\, d\Delta y\, dq \tag{6.53}$$

where the volume transfer function $B(\Delta x, \Delta y, q)$is defined as

$$B(\Delta x, \Delta y, q) = \int P^*\left[\frac{1}{2}\left(\frac{q}{\Delta x} - \tilde{q}\Delta y + \Delta x\right), \frac{1}{2}\left(\frac{q}{\Delta y} + \tilde{q}\Delta x + \Delta y\right)\right] \times P\left[\frac{1}{2}\left(\frac{q}{\Delta x} - \tilde{q}\Delta y - \Delta x\right), \frac{1}{2}\left(\frac{q}{\Delta y} + \tilde{q}\Delta x - \Delta y\right)\right] d\tilde{q} \tag{6.54}$$

and $\tilde{q} = -\bar{x}/\Delta y + \bar{y}/\Delta x$.

For the circular aperture described by $P(x, y) = \text{circ}\left(\sqrt{x^2 + y^2}/A\right)$, $B(\Delta x, \Delta y, q)$ is described in closed form as [80,106]

$$B(\Delta x, \Delta y, q) = \frac{2}{\Delta x^2 + \Delta y^2} \Re\left[\sqrt{(\Delta x^2 + \Delta y^2)A^2 - (\Delta x^2 + \Delta y^2 + 2|q|)^2}\right] \tag{6.55}$$

where $\Re[\ ]$ denotes the real part. $B(\Delta x, \Delta y, q)$ is well behaved except for a singularity at $(\Delta x = 0, \Delta y = 0)$. The cross section of $B(\Delta x, \Delta y, q)$ through the $\Delta y = 0$ plane is shown in Fig. 6.14.

We refer to the support of B as the *band volume* because, just as the aperture determines the 2D bandpass in focal imaging, we see in Section 6.4.2 that B is an effective transfer function for 3D imaging. The $B(\Delta x, \Delta y, q) = 0$ boundary for a

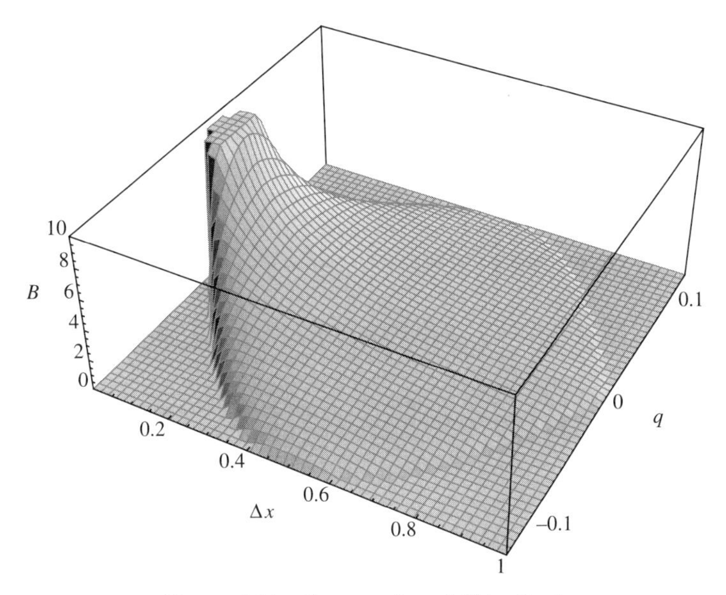

Figure 6.14 Cross section of $B(\Delta x, 0, q)$.

circular aperture is illustrated in Fig. 6.15. The figure is in units of A. The limited extent of the band volume restricts the range over which $W(\Delta x, \Delta y, q, \nu)$ is known by focal interferometery. The band volume fills a disk of radius A in the Δx, Δy plane. The bandpass along the q axis vanishes at the origin and at the edge of the Δx, Δy disk. The maximum q bandpass occurs at $\Delta x^2 + \Delta y^2 = A^2/2$, which yields $q_{\max} = A^2/8$.

Equation (6.53) may be inverted to estimate the bandlimited cross-spectral density on the lens aperture. This process is equivalent to imaging an incoherent object,

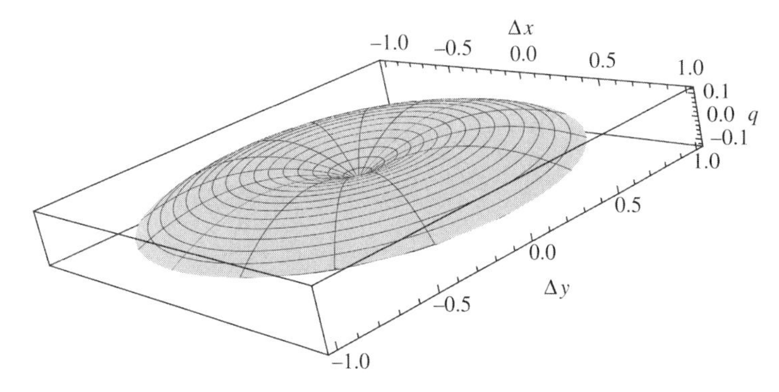

Figure 6.15 Band volume in Δx, Δy, q space for focal interferometry on a circular aperture lens. The Δx and Δy axes are in units of A aperture diameter. The q axis is in units of square amperes (A^2).

which is the focus of Section 6.4. Beyond simple inversion we discuss aperture coding strategies in Chapter 10 to reshape the transfer function B. Such strategies cannot increase the band volume, but they are effective in improving targeted image metrics. They may be used, for example, to reduce the need to sample $S(x, y, z, \nu)$ over the full focal volume or to improve mathematical conditioning of the sensor model for specific object classes.

We saw in Section 6.2 that $W(\Delta x, \Delta y, q, \nu)$ is often independent of q. If we limit our attention to the focal plane in this case, Eqn. (6.53) reduces to

$$S(x, y, \nu) = \frac{4}{\lambda^2 F^2} \iiint W(\Delta x, \Delta y, \nu)\tilde{B}(\Delta x, \Delta y) \times \exp\left(-2i\pi\nu \frac{x\Delta x + y\Delta y}{cz} \right) d\Delta x \, d\Delta y \tag{6.56}$$

where

$$\tilde{B}(\Delta x, \Delta y) = \int B(\Delta x, \Delta y, q)\, dq \tag{6.57}$$

Function $\tilde{B}(\Delta x, \Delta y)$, which is the primary focus of Section 6.4, is a smooth and well-behaved function. Its form for a diffraction limited circular aperture is given in Eqn. (6.67).

We next turn to focal interferometry strategy 2, temporal variation of the pupil function for 5D coherence sensing. 5D sensing is unnecessary for normal incoherent imaging, but 5D cross-spectral densities do arise for incoherent sources modulated by intervening scatters and for objects illuminated by partially coherent light.

Marks et al. [172] describe a mechanism for characterizing the 5D cross-spectral density using an *astigmatic coherence sensor* (ACS). The ACS uses a cylindrical lens assembly, schematically similar to the lens system of Fig. 6.16, to achieve fully 5D sampling of the cross-spectral density. The transmittance function of a cylindrical

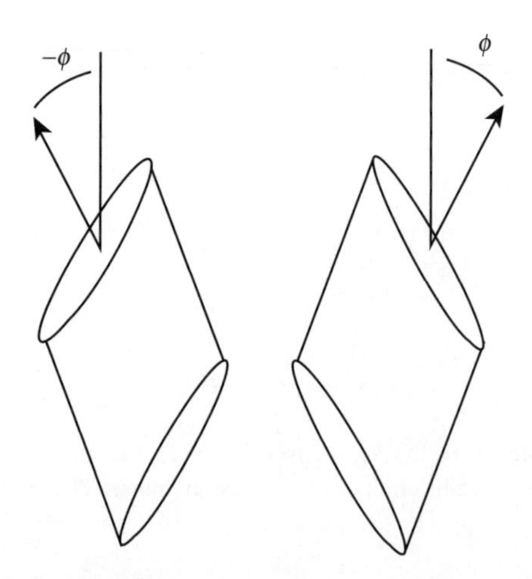

Figure 6.16 Cylindrical lens assembly for the astigmatic coherence sensor.

lens oriented with focal power along the x axis is $t(x, y) = \exp(i\pi x^2/F)$. If the transverse axis of the lens is rotated in the (x, y) plane by an angle ϕ, the transmittance becomes

$$t(x, y) = \exp\left(i\pi\frac{(x\cos\phi + y\sin\phi)^2}{\lambda F}\right) \tag{6.58}$$

The pair of lenses in Fig. 6.16 are rotated to positions ϕ and $-\phi$ such that the product of their transmittance functions is

$$t(x, y) = \exp\left(i2\pi\frac{(x^2\cos^2\phi + y^2\sin^2\phi)}{\lambda F}\right) \tag{6.59}$$

Substituting distinct x and y focal lengths in Eqn. (6.51) produces

$$\begin{aligned} S(\theta_x, \theta_y, \phi_x, \phi_y, \nu) = \kappa \int\!\!\int\!\!\int\!\!\int & W(x_1, y_1, x_2, y_2, \nu)P^*(x_1, y_1)P(x_2, y_2) \\ & \times e^{i\pi\nu(x_1^2 - x_2^2)\phi_x} e^{i\pi\nu(y_1^2 - y_2^2)\phi_y} \\ & \times e^{-i2\pi\nu[\theta_x(x_1 - x_2) + \theta_y(y_1 - y_2)]}\, dx_1\, dx_2\, dy_1\, dy_2 \end{aligned} \tag{6.60}$$

where

$$\phi_x = \left(\frac{2\cos^2\phi}{cF} - \frac{1}{cz}\right), \quad \phi_y = \left(\frac{2\sin^2\phi}{cF} - \frac{1}{cz}\right)$$

$\theta = x/cz$ and $\theta_y = y/cz$; ϕ_x is a defocus parameter that may be scaled by shifting the detector plane, adjusting focal length with a zoom lens mechanism, and/or adjusting the astigmatism.

The value of $W(x_1, y_1, x_2, y_2, \nu)$ may be recovered from Fourier analysis of $S(\theta_x, \theta_y, \phi_x, \phi_y, \nu)$. The value at x_1 and x_2, for example, is obtained from the spectral density at spatial frequencies $u_{\phi_x} = \nu(x_2^2 - x_1^2)/2 = \nu\Delta x\bar{x} = \nu q$ and $u_{\theta_x} = \nu(x_1 - x_2) = \nu\Delta x$. W can be reconstructed for (x_1, y_1, x_2, y_2) in the support of the pupil $P(x, y)$, which for a circular aperture is $\sqrt{x^2 + y^2} < A$. The resolution on the q manifold is determined by the sampling range for ϕ_x, $\Delta\phi = 2/cF - \Delta z/(cz_{\max}z_{\min})$, where $\Delta z = z_{\max} - z_{\min}$ is the range of z values over which one measures the power spectral density. If, for example, $z_{\max} = 2F$ and $z_{\min} = F/2$, then $\Delta\phi = 5/cF$ and the Fourier bandpass-limited resolution for q is approximately $\lambda F/5$. By a similar argument, the resolution with which one can estimate Δx is of order $\lambda f/\#$.

The significance of coherence measurement using Eqns. (6.52), (6.53), and (6.60) will become clearer in subsequent sections as we consider imaging transformations and modal decomposition of the cross-spectral density. For present purposes, it may be helpful to briefly consider likely characteristics of W on an aperture and the nature of the focal transformation. For example, we note from Eqn. (6.21) that a remote object consisting of a single-point radiator at $(x_0, y = 0, z_0)$ produces a cross-spectral density on the lens system aperture

$$W(\Delta x, \Delta y, \bar{x}, \bar{y}, \nu) = \frac{\kappa \nu^2}{z_0^2} e^{2\pi i(\Delta x x_0/\lambda z_0)} e^{-2\pi i[(\Delta x \bar{x} + \Delta y \bar{y})/\lambda z_0]} \tag{6.61}$$

A point object thus produces a harmonic cross-spectral density. The focal power spectral density, as the bandlimited Fourier transform of the cross-spectral density, localizes the image of the point object as tightly as possible given a finite aperture. As discussed in Section 6.4.2, the coherent impulse response for this localization is the Fourier transform of the band volume B.

The power spectral density in the focal volume for a point object is distributed as the magnitude squared of the coherent impulse response. As discussed in Section 6.6, the cross-spectral density forms a *nonnegative kernel* in Eqn. (6.60), ensuring that the power spectral density is everywhere nonnegative.

Analysis of the focal power spectral density as a coherence measurement is most useful in cases where the cross-spectral density is not well described by Eqns. (6.21) or (6.61). Examples include

1. Situations in which W is modulated by imaging system aberrations,
2. Situations in which the remote object is not an incoherent radiator, such as the case discussed in Section 6.2 of a secondary scatterer illuminated by partially coherent light
3. Situations in which W is transformed by propagation through inhomogeneous media

In each of these cases, the general form of the cross-spectral density due to a point generalizes from Eqn. (6.61) to $W(\mathbf{x}_1, \mathbf{x}_2, \nu) = \sum_n W_n \phi_n^*(\mathbf{x}_1, \nu)\phi(\mathbf{x}_2, \nu)$, where $\phi_n(\mathbf{x}, \nu)$ are the *coherent modes* of the field. Marks et al. [169] describe a method for imaging through a distortion by using an ACS to determine the coherent modes of the field. The 4D spatial sampling of the ACS is necessary to remove degeneracies in the power spectral density that could be created by different coherent-mode decompositions.

Cross-spectral density characterization using the ACS may be regarded within the general framework of applying coded aberrations and defocus to an imaging system to analyze unknown distortions called *phase diversity* [98]. Phase diversity is most commonly parameterized directly in the object density and the image distortion and analyzed using maximum likelihood methods [197].

6.4 FOURIER ANALYSIS OF COHERENCE IMAGING

Equation (6.18) is immediately useful in describing the impulse response and transfer function of imaging systems. While the result may be applied to imaging of objects in arbitrary coherence states, in most applications it is safe to assume that the source is spatially incoherent. This is certainly the case for self-luminous objects and diffusely illuminated objects. The present section accordingly focuses on incoherent objects.

Our immediate goal is to extend the Fourier analysis of Section 4.7 to the case of incoherent objects. We begin by considering 2D objects imaged from an object plane to a well-focused image plane satisfying the thin-lens imaging law [Eqn. (2.17)]. We describe the *point spread function* and the *optical transfer function* (OTF), which are the incoherent source analog of the coherent impulse response and transfer function. Incoherent imaging leads logically to discussions of multidimensional spatial and spectral imaging. We begin to consider these topics in this section by showing that the volume transfer function of Eqn. (6.54) is the 3D transfer function for incoherent imaging, and we relate volume transfer function to the OTF and to the *defocus transfer function* (which describes 2D imaging between misfocused planes).

6.4.1 Planar Objects

The coherent impulse response between the field on a image plane a distance z_1 in front of a lens of focal length F and the field on an object plane a distance z_2 behind a lens under the imaging condition that $1/z_1 + 1/z_2 = 1/F$ is presented in Eqn. (4.72). Referring to Eqn. (6.23), the incoherent impulse response [often referred to as the *point spread function* (PSF)] is simply the squared magnitude of the coherent impulse response. The effect of squaring on Eqn. (4.72) fortuitously eliminates shift-variant phase terms, producing the shift-invariant incoherent impulse response

$$h_{ic}(x, y) = |M|^2 |h_r(x, y)|^2 \tag{6.62}$$

where $h_r(x, y)$ is as given by Eqn. (4.73). The PSF is absolutely shift-invariant under the thin lens approximation, although as always we caution the student that this exact shift invariance is ultimately lost in nonparaxial optical systems.

We assume for simplicity that the field is quasimonochromatic such that $S(x, y, \nu) \approx f(x, y)\delta(\nu - \nu_0)$. In this case, the incoherent imaging transformation analogous to Eqn. (4.75) is

$$g(x', y') = \frac{l^2}{M^2} \int\int f\left(\frac{x}{M}, \frac{y}{M}\right) h_{ic}(x' - x, y' - y)\, dx\, dy \tag{6.63}$$

where we evaluate $h_{ic}(x, y)$ at a specific wavelength λ; $g(x', y')$ is the image irradiance produced for the object irradiance $f(x, y)$. We saw in Eqn. (4.73) that $h_r(x, y)$ is proportional to the Fourier transform of the pupil transmittance and, in Eqn. (4.76), that the coherent transfer function is proportional to the $P(x = -\lambda d_i u, y = -\lambda d_i v)$. Since the impulse response for the incoherent system is the square of the coherent impulse response, we know by the convolution theorem that the transfer function for incoherent imaging is the autocorrelation of the pupil transmittance:

$$\hat{h}_{ic}(u, v) = \iint P^*(-\lambda d_i u', -\lambda d_i v') P[-\lambda d_i(u' - u), -\lambda d_i(v' - v)] \, du' dv' \tag{6.64}$$

The maximum modulus of the autocorrelation occurs at $u = 0$, $v = 0$. Since one is usually most interested in relative values, the transfer function is most often considered in the normalized form

$$\mathcal{H}(u, v) = \frac{\hat{h}_{ic}(u, v)}{\hat{h}_{ic}(0, 0)} \tag{6.65}$$

$\mathcal{H}(u, v)$ is the *optical transfer function* (OTF). The OTF is a commonly used metric for image system analysis. The modulus of the OTF, the *modulation transfer function* (MTF), is an even more common metric.

For the canonical case of an incoherent imaging system with clear circular pupil of diameter A, we obtain the incoherent impulse response from the coherent PSF described by Eqn. (4.74). Estimating the spatial coherence length by λ, we find that this imaging system corresponds to the linear transformation

$$g(x', y') = \frac{A^4}{\lambda^2 M^2 d_i^4} \iint f\left(\frac{x}{M}, \frac{y}{M}\right) \times \operatorname{jinc}^2\left[\frac{A}{\lambda d_i}\left(\sqrt{(x' - x)^2 + (y' - y)^2}\right)\right] dx \, dy \tag{6.66}$$

The OTF for this system may be found in closed form by integrating Eqn. (6.64), which yields

$$\mathcal{H}(\mu) = \frac{2}{\pi} \Re \left[\arccos\left(\frac{\mu \lambda d_i}{A}\right) - \left(\frac{\mu \lambda d_i}{A}\right)\sqrt{1 - \left(\frac{\mu \lambda d_i}{A}\right)^2}\right] \tag{6.67}$$

where $\mu = \sqrt{u^2 + v^2}$. The OTF vanishes for $\mu > A/\lambda d_i$. The impulse response and MTF for a circular aperture are illustrated in Fig. 6.17. Compared with the corresponding plots for a coherent system a presented in Fig. 4.14, the support of the MTF is twice that of the coherent transfer function but the passband is no longer flat.

The MTF for an annular aperture, which was observed to produce a highpass coherent transfer function in Chapter 4, is shown in Fig. 6.18 for a lens with the center 0.9 radius component obscured. The MTF shows secondary peaks at high frequencies, but the maximum transfer function in the passband for incoherent systems is always at $u = 0$, $v = 0$.

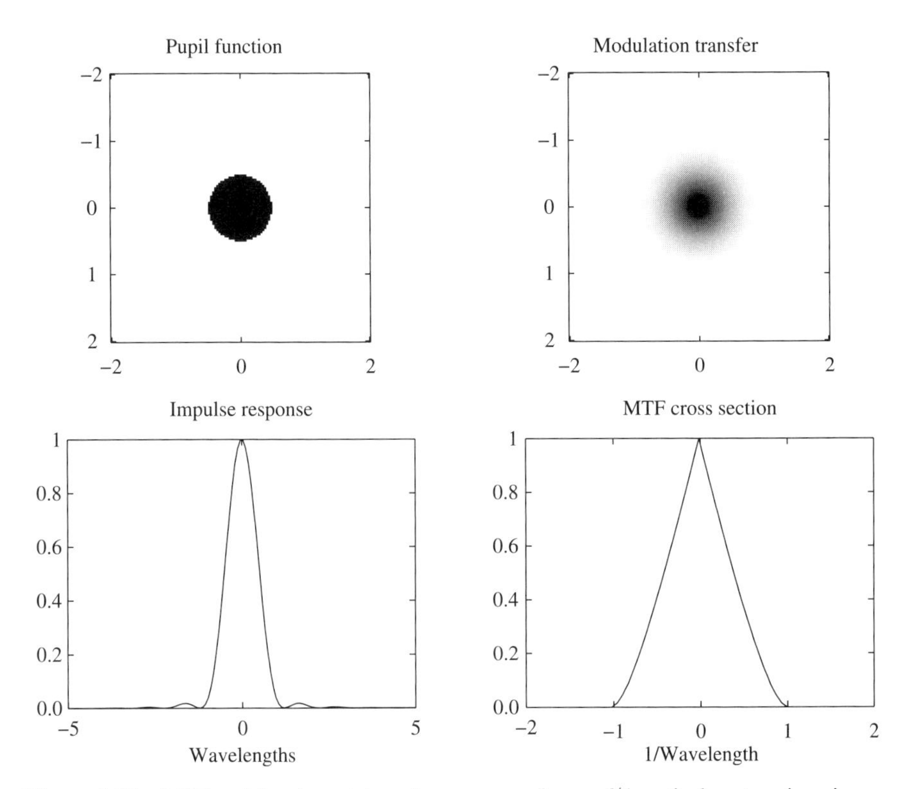

Figure 6.17 MTF and incoherent impulse response for an $f/1$ optical system imaging an object at infinity. As in Fig. 4.14, the distance between the first two zeros of the impulse response is 2.44 wavelenths. As a result of squaring, however, the full-width half-maximum is narrower and the passband is increased by a factor of 2. The passband is no longer flat, however.

Figure 6.19 shows the image obtained when the object of Fig. 4.16 is incoherently illuminated and imaged through a circular aperture. The lowpass image is obtained using a clear aperture, while the two highpass images correspond to the same annular apertures as considered in Fig. 4.16. Note that while the low-frequency component always dominates the incoherent image, it is possible to differentially increase the relative throughput of high-frequency components.

6.4.2 3D Objects

To this point we have considered transformations between fields distributed on planes. This section expands our attention to input–output relationships between object and image volumes. 3D analysis requires a careful distinction between coherence measures of the propagating optical field and coherence measures of the field generated locally by an object. We consider the mapping from an incoherent 3D

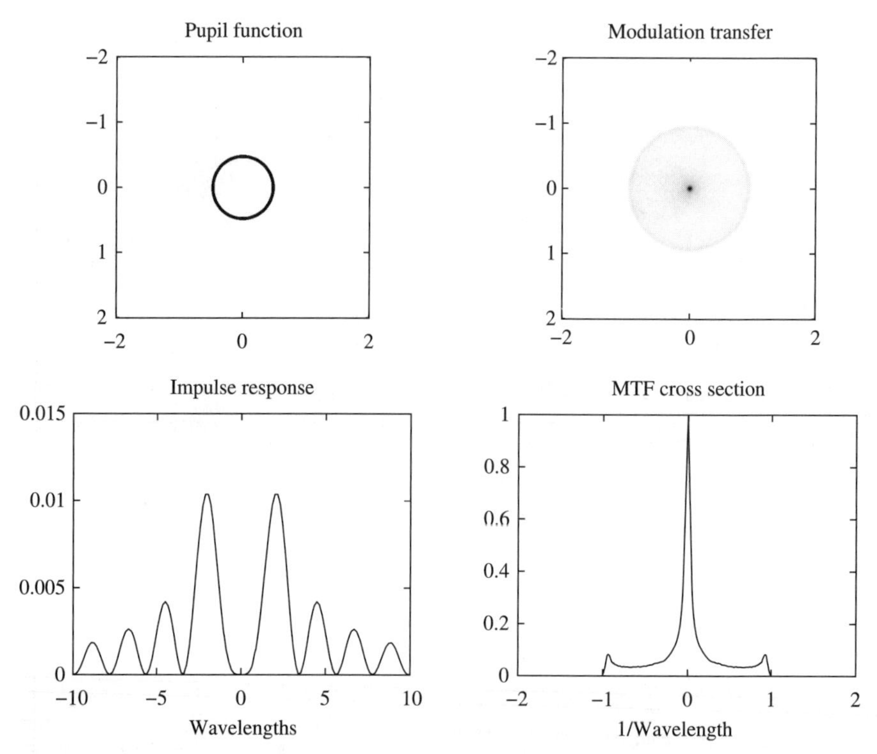

Figure 6.18 MTF and impulse response for an $F/1$ optical system imaging an object at infinity with an annular pupil. As in Fig. 4.15, the radius of the blocked center disk constitutes 90% of the radius of the full aperture. Note that for incoherent imaging, the imaging system no longer acts as a highpass filter.

object to the power spectral density detected by an imaging system. We assume that that the spectral density of the primary source is subject to the three-dimensional version of Eqn. (6.19):

$$W(x_1, y_1, z_1, x_2, y_2, z_2, \nu) = S(x_1, y_1, z_1, \nu)\delta(x_1 - x_2)\delta(y_1 - y_2)\delta(z_1 - z_2) \quad (6.68)$$

We do not consider such seven-dimensional versions of the cross-spectral density when considering measures of the optical field because, as we saw in Eqn. (6.17); knowledge of W on the five-dimensional $(x_1, x_2, y_1, y_2, \nu)$ manifold is sufficient to calculate W everywhere. This is not the case for a 3D primary source distribution, however, which is not subject to the Maxwell equations, and which is capable of independently radiating a signal at each point in 3D.

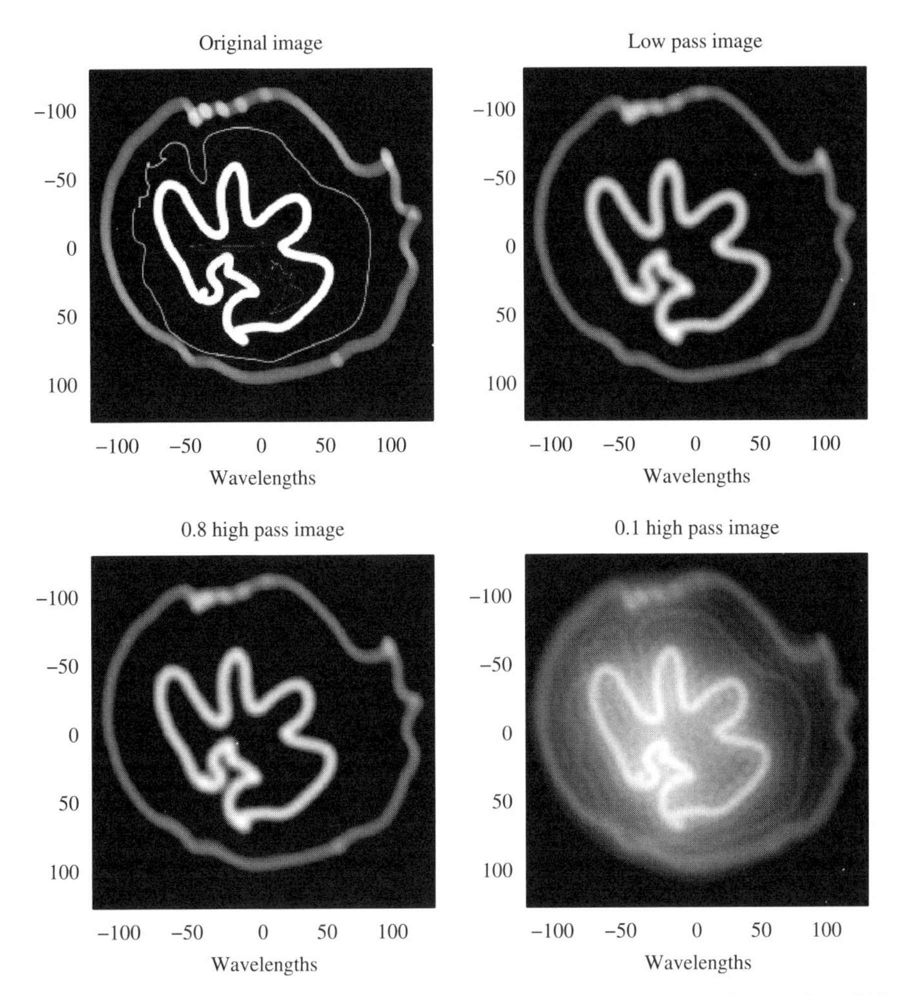

Figure 6.19 Effect of pupil filtering on in the imaging system corresponding to Figs. 6.17 and 6.18. Compare with Fig. 4.16.

Once emitted the object field and cross-spectral density become subject to the Maxwell equations and evolve according to

$$W_{\text{field}}(x'_1, y'_1, x'_2, y'_2, \nu) = \iiiint W_{\text{object}}(x_1, y_1, z_1, x_2, y_2, z_2, \nu) \times h_c(x_1, x'_1, y_1, y'_1, z_1, \nu) \times h_c(x_2, x'_2, y_2, y'_2, z_2, \nu)\, dx_1\, dy_1\, dx_2\, dy_2\, dz_1\, dz_2 \quad (6.69)$$

In particular, the cross-spectral density on the plane $z = 0$ due to a 3D incoherent primary source radiating the power spectral density $S(x, y, z, \nu)$ is

$$W(x_1, y_1, x_2, y_2, \nu) = \iiint S(x, y, z, \nu) h_c(x, x_1, y, y_1, z, \nu) \times h_c(x, x_2, y, y_2, z, \nu)\, dx\, dy\, dz \tag{6.70}$$

In the Fresnel approximation, Eqn. (6.70) yields

$$W(\Delta x, \Delta y, q, \nu) = \iiint \frac{S(x, y, z, \nu)}{\lambda^2 z^2} \times \exp\left[-i\frac{2\pi\nu}{cz}(x\Delta x + y\Delta y + q)\right] dx\, dy\, dz \tag{6.71}$$

where again $q = \bar{x}\Delta x + \bar{y}\Delta y$, $\bar{x} = (x_1 + x_2)/2$ and $\Delta x = x_1 - x_2$. Equation (6.71) is identical to Eqn. (6.21) with the addition of an integral over the longitudinal axis.

Equation (6.71) is an expression of the *van Cittert–Zernike theorem*, which states that the cross-spectral density radiated in the Fresnel or Fraunhofer regime of a spatially incoherent primary source is proportional to the spatial Fourier transform of the source distribution. The van Cittert–Zernike theorem is most frequently applied to radio wave imaging, particularly in the context of radio astronomy, but it has found use in optical imaging as well. For example, Marks et al. [173] used a rotational shear interferometeter to directly characterize $W(\Delta x, \Delta y, q, \nu)$. As discussed in Section 6.3.3, an RSI most easily measures $W(\Delta x, \Delta y, q = 0, \nu)$. The Fourier transform of $W(\Delta x, \Delta y, q = 0, \nu)$ with respect to Δx and Δy is

$$\begin{aligned} Q(\theta'_x, \theta'_y, \nu) &= \iint W(\Delta x, \Delta y, q = 0, \nu) e^{i(2\pi\nu/c)(\theta'_x \Delta x + \theta'_y \Delta y)}\, d\Delta x\, d\Delta y \\ &= \iiint S(\theta'_x, \theta'_y, \theta_z, \nu)\, d\theta_z \end{aligned} \tag{6.72}$$

Marks et al. used the ray integrals $Q(\theta'_x, \theta'_y, \nu)$ in the cone beam tomography algorithm described in Section 2.6 to reconstruct 3D objects [170]. Equation (6.72) is of interest again in Section 10.2 as an existence proof of an infinite depth of field imaging system.

Returning to focal systems, substituting Eqn. (6.71) into Eqn. (6.53) yields the transformation between the object power spectral density $S_o(x, y, z, \nu)$ to the left of a lens and the power spectral density $S_i(x, y, z, \nu)$ to the right

$$S_i(x', y', z', \nu) = \frac{4}{\lambda^2 z^2} \iiiiiint S_o(x, y, z, \nu) e^{-i(2\pi\nu/cz)(x\Delta x + y\Delta y + q)} B(\Delta x, \Delta y, q) \times \exp\left(-2i\pi\nu \frac{x'\Delta x + y'\Delta y + (1 - z'/F)q}{cz'}\right) d\Delta x\, d\Delta y\, dq\, dx\, dy\, dz$$

$$S_i(\theta_{x'}, \theta_{y'}, \theta_{z'}, \nu) = \iiint S_o(\theta_x, \theta_y, \theta_z, \nu) \times h(\theta_x + \theta_{x'}, \theta_y + \theta_{y'}, \theta_z + \theta_{z'}, \nu)\, d\theta_x\, d\theta_y\, d\theta_z \qquad (6.73)$$

where $B(\Delta x, \Delta y, q)$ is the volume transfer function of Eqn. (6.55) and $\theta_x = x/z$ and $\theta_y = y/z$. Similar definitions apply for the primed variables with the exception that $\theta_z = 1/z$ but $\theta_{z'} = (1/z' - 1/F)$. z and z' are both measured as positive distances from the plane of the lens.

The impulse response for mapping from the object volume to the image volume is

$$h(\theta_x, \theta_y, \theta_z, \nu) = \iiint e^{i2\pi(\theta_x u + \theta_y v + \theta_z w)} B(-\lambda u, -\lambda v, -\lambda w)\, du\, dv\, dw \qquad (6.74)$$

The transformation from the 3D object space of $(\theta_x, \theta_y, \theta_z)$ to the 3D image space $(\theta_{x'}, \theta_{y'}, \theta_{z'})$ is a shift-invariant linear transformation with impulse response $h(\theta_x, \theta_y, \theta_z, \nu)$ and transfer function

$$\hat{h}_{3D}(u, v, w, \lambda) = B(-\lambda u, -\lambda v, -\lambda w) \qquad (6.75)$$

Since θ_x and θ_y are dimensionless, the angular frequencies u and v are also dimensionless (θ_z is in units of inverse meters, w is in units of meters; λ is, of course, wavelength).

It is possible to measure $S_i(\theta_x, \theta_y, \theta_z, \nu)$ by scanning an imaging spectrometer through the focal volume of an imaging system. The general problem of estimating the $S_o(\theta_x, \theta_y, \theta_z, \nu)$ from such measurements is an *inverse problem* typical of tomographic analysis. Note that such an estimation process would be quite different from simply scanning through the focal volume and measuring the image field. Tomographic analysis reconstructs the object density rather than the field distribution the object produces.

In the present case, the *forward mapping* from the object spectral density to the image is a convolution. In principle, one could estimate $S_o(\theta_x, \theta_y, \theta_z, \nu)$ by deconvolution techniques as discussed in Section 8.5 Such techniques are not likely to be effective, however, because the 3D impulse response of Eqn. (6.74) does not have finite support along the longitudinal axis. In view of this challenge, a variety of techniques have been developed to image in three dimensions without directly inverting Eqn. (6.73), including the projection tomography and structured illumination (optical coherence tomography) strategies discussed in this chapter as well as wavefront coding and radiometric strategies discussed in Chapter 10. The most common conventional strategy is plane-by-plane analysis based on the 2D defocus transfer function discussed in Section 6.4.3.

Despite the challenges associated with the singularity in $B(\Delta x, \Delta y, q)$, it is possible to estimate the 3D resolution that one might expect to obtain by direct inversion of Eqn. (6.73). As depicted in Fig. 6.15, for a circular aperture of diameter A the domain of the band volume extends over $\sqrt{\Delta x^2 + \Delta y^2} \leq A$ and $|q| \leq A^2/8$. Assuming that our estimate of $S_o(\theta_x, \theta_y, \theta_z, \nu)$ is limited to this passband, the angular resolution of the imaging system is approximately $\delta\theta_x \approx \lambda/A$ and the minimum range resolution for an object at range z_o, assuming high-frequency transverse features, is $\delta z_\sigma \approx 8\lambda z_o^2/A^2$. Objects that are approximately 1 focal length from the aperture yield a range resolution of $\delta z \approx 8\lambda(f/\#)^2$, which is useful for 3D microscopy. As the object range extends beyond one focal length, however, the range resolution deteriorates as the square of the range. Substantially better range resolution is obtained using spatial tomography (Section 2.6) or spectral tomography (Sections 6.5 and 10.3.1).

6.4.3 The Defocus Transfer Function

Plane-to-plane imaging is based on the assumption that the object distribution is confined to a single plane, for example

$$S_o(\theta_x, \theta_y, \theta_z, \nu) = S_o(\theta_x, \theta_y, \nu)\delta(\theta_z - \theta_{z0}) \tag{6.76}$$

Under this assumption Eqn. (6.73) becomes

$$S_i(\theta_{x'}, \theta_{y'}, \theta_{z'}, \nu) = \iint S_o(\theta_x, \theta_y, \nu) h(\theta_x + \theta_{x'}, \theta_y + \theta_{y'}, \theta_z + \theta_{z0}, \nu)\, d\theta_x\, d\theta_y \tag{6.77}$$

To simplify our notation in the following discussion, let $f(x, y)$ represent $S_o(-\theta_x, -\theta_y, \nu_0)$ and $g_{\theta_z}(x', y')$ represent $S_i(\theta_{x'}, \theta_{y'}, \theta_{z'}, \nu_0)$. In this notation, the imaging transformation is

$$g_{\theta_z}(x', y') = \iint f(x, y) h_{\theta_z + \theta_{z0}}(x' - x, y' - y)\, dx\, dy \tag{6.78}$$

where

$$h_{\theta_z}(x, y) = \iiint e^{i2\pi\left(\theta_x u + \theta_y v + \theta_z w\right)} B(-\lambda u, -\lambda v, -\lambda w)\, du\, dv\, dw \tag{6.79}$$

If the imaging condition $\theta_z + \theta_{z0} = 0$ is satisfied, the plane-to-plane impulse response is

$$h_0(\theta_x, \theta_y, \lambda) = \iint e^{i2\pi(\theta_x u + \theta_y v)} B(-\lambda u, -\lambda v, -\lambda w)\, du\, dv\, dw \tag{6.80}$$

The OTF for the planar system is the 2D Fourier transform of the impulse response

$$\hat{h}(u, v, \lambda) = \int B(-\lambda d_i u, -\lambda d_i v, -\lambda w)\, dw \tag{6.81}$$

where, for consistency with Eqn. (6.64), we return to u, v in units of inverse spatial distance. It turns out, of course, that Eqn. (6.81) is equivalent to Eqn. (6.64).

More generally, we define the *defocus transfer function* [77] as the x, y Fourier transform of $h_{\theta_z}(x, y)$:

$$\hat{h}_{\theta_z}(u, v, \lambda) = \int e^{i2\pi\theta_z w} B(-\lambda d_i u, \ -\lambda d_i v, \ -\lambda w)\, dw \tag{6.82}$$

Substituting for B from Eqn. (6.54) and transforming from the q, $\tilde{q}$ plane back to the $\bar{x}$, $\bar{y}$ plane yields

$$\hat{h}_{\theta_z}(u, v, \lambda) = \iint e^{i2\pi\theta_z d_i(\bar{x}u+\bar{y}v)} P^*\left(\bar{x} - \frac{\lambda d_i u}{2}, \bar{y} - \frac{\lambda d_i v}{2}\right) \times P\left(\bar{x} + \frac{\lambda d_i u}{2}, \bar{y} + \frac{\lambda d_i v}{2}\right) d\bar{x}\, d\bar{y} \tag{6.83}$$

The *ambiguity function* of a two-dimensional function $P(x, y)$ is defined as [193]

$$A(x, y, \alpha, \beta) = \iint e^{i2\pi(\alpha x' + \beta y')} P^*\left(x' - \frac{x}{2}, y' - \frac{y}{2}\right) \times P\left(x' + \frac{x}{2}, y' + \frac{y}{2}\right) dx'\, dy' \tag{6.84}$$

Comparing Eqns. (6.83) and (6.84), we find $\hat{h}_{\theta_z}(u, v, \lambda) = A(\lambda d_i u, \lambda d_i v, \theta_z d_i u, \theta_z d_i v)$. One could equivalently derive the defocus transfer function by simply including the defocus distortion in the pupil function

$$P_{\theta_z}(x, y) = e^{\pi i(\theta_z/\lambda)(x^2+y^2)} P(x, y) \tag{6.85}$$

Substitution of this pupil into Eqn. (6.64) leads immediately to Eqn. (6.83).

Hopkins considers the defocus transfer function as a function of the wavefront curvature error w_{20} [120]. w_{20} is the distance between the actual focusing wavefront at the edge of the aperture and the wavefront that would focus without distortion at the image plane. The relationship between $\theta_z = (1/z_o + 1/z_i - 1/F)$ and w_{20} is $\theta_z = 2w_{20}/A^2$. Hopkins' notation allows for similar consideration of diverse wavefront aberrations, such as spherical aberration, coma, and astigmatism. The defocus transfer function is displayed as a function of u and w_{20} for a circular aperture in Fig. 6.20. The OTF is cylindrically symmetric in the u, v plane for a circular aperture; Fig. 6.20 shows the cross section along a single axis.

Hopkins [121] suggests that "tolerable" defocus at a given frequency u satisfies

$$\frac{|\hat{h}_{w_{20}}(u)|}{|\hat{h}_0(u)|} \geq 0.8 \tag{6.86}$$

Figure 6.21 illustrates the maximum defocus as a function of spatial frequency according to Hopkins' criterion.

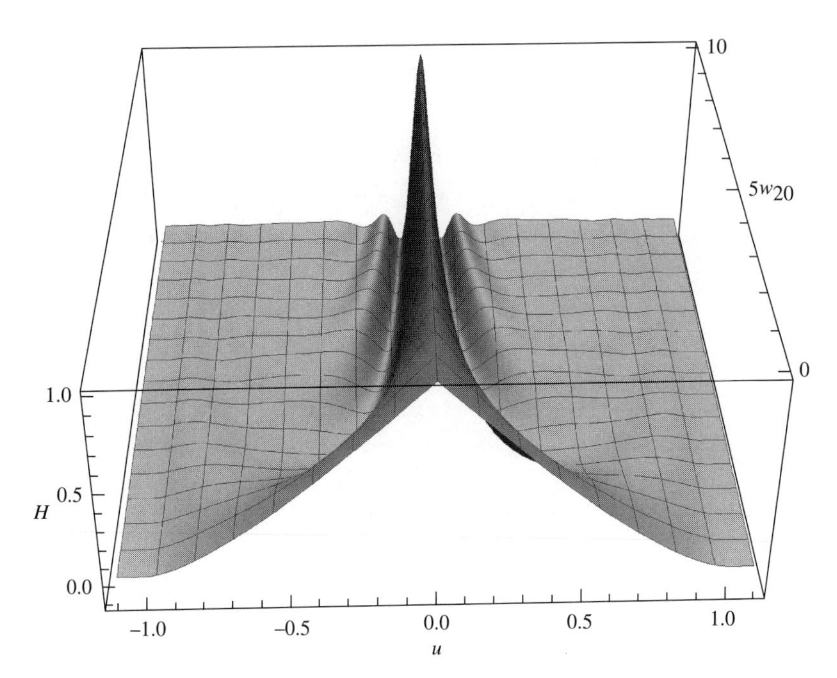

Figure 6.20 Defocus OTF for a circular pupil of diameter A as a function of the defocus parameter w_{20} (u is in units of $A/\lambda d_i$; w_{20} is in units of λ). The focal OTF of Fig. 6.17 is observed for $w_{20} = 0$.

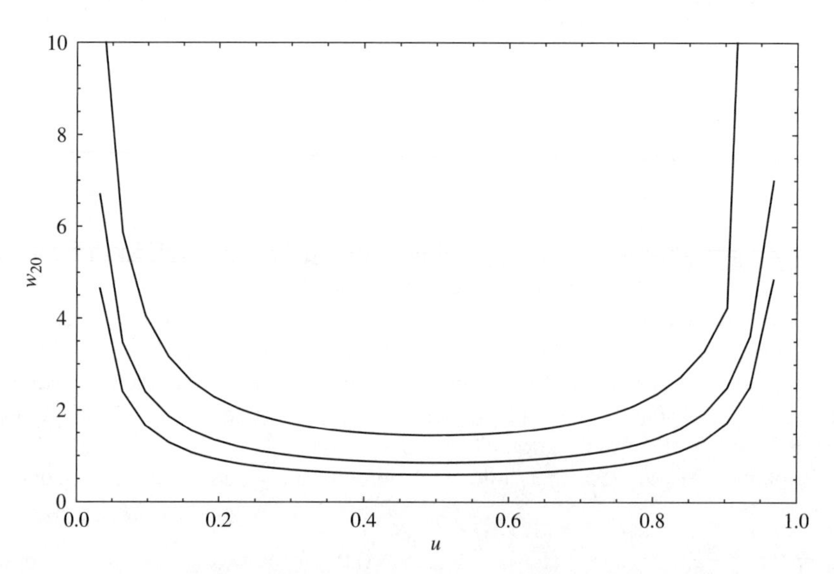

Figure 6.21 Defocus w_{20} such that $H_{w_{20}}(u, 0, \lambda) = \alpha H_0(u, 0, \lambda)$ as a function of spatial frequency (w_{20} is in units of λ; u is in units of $A/\lambda d_i$); $\alpha = 0.5$ in the topmost curve, 0.8 in the middle curve, and 0.9 in the bottom curve.

With z_i and F fixed, the change object position for a given defocus is

$$\Delta z_o = \frac{2w_{20}z_o^2}{A^2} \tag{6.87}$$

For $w_{20} \approx \lambda$, this suggests that the range of the object field is approximately $4\lambda z_o^2/A^2$. The object range is called the *depth of field* of the imaging system. The depth of field and strategies for extending it are the focus of Section 10.2.

6.5 OPTICAL COHERENCE TOMOGRAPHY

Optical coherence tomography (OCT) uses the encoding of spatial information in the scattered spectra and coherence of secondary sources [122]. Classic approaches to OCT use spatially coherent illumination with short temporal coherence length. As with projection tomography, OCT emphasizes subsurface volume imaging. Alternative strategies for optical 3D imaging include confocal scanning and optical projection tomography, but OCT images much deeper into scattering media because its axial PSF is sharp enough to overcome the exponential depth dependence of scattered light intensity [124].

The basic design of an OCT system is sketched in Fig. 6.22. A light source radiating spectral density $S(\nu)$ is coupled into a single-mode optical fiber. The light source is typically a point source such as light emitting diode or a single spatial mode source

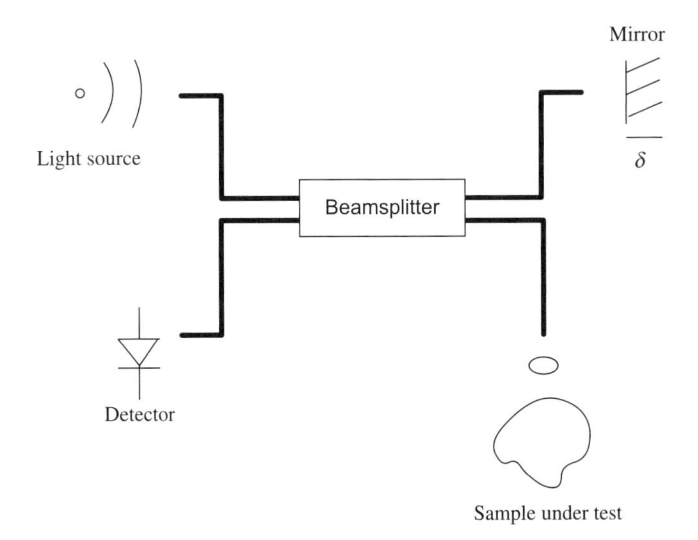

Figure 6.22 Single-spatial-mode optical coherence tomography system layout.

generated from a mode-locked laser. For reasons discussed in Section 6.6.3, the light source must be spatially coherent to efficiently couple into the fiber.

The source signal is separated by a fiber splitter into reference and probe signals. The reference signal is removed from the fiber and reflected by a mirror back toward the source. The mirror is on a translation stage such that a variable time delay $2\delta/c$ may be added to the reflected field, where δ is the spatial translation of the mirror. The probe signal is focused on a object under test. The signal retroreflected from the object and the retroreflected reference recombine in the fiber splitter, which directs half of the recombined light to a photodetector.

As with focal systems, the easiest approach to characterizing the response of a coherence system is to begin by characterizing the coherent response. In the OCT system, for example, assume that the source electromagnetic field is $E(t)$ (we neglect spatial degrees of freedom in single-mode systems). The signal reflected by the reference mirror is $\alpha E(t - 2\delta/c)$. The relationship between the returned probe field and the object under test is more complex; for simplicity we assume that the field reflected in the neighborhood of the probe focus is described by a scattering density $\sigma(z)$, where z is the longitudinal position relative to the probe focal point. The field returned from the object is then $\int \sigma(z)E(t - 2z/c)\, dz$.

The coherent response field at the output detector is thus

$$E_o(t) = \alpha E\left(t - \frac{2\delta}{c}\right) + \int \sigma(z)E\left(t - \frac{2z}{c}\right)dz \tag{6.88}$$

The output power spectral density corresponding to the coherent response is

$$\begin{aligned} S_o(\nu) = S(\nu)\Bigg[|\alpha|^2 &+ \int\int \sigma^*(z_1)\sigma(z_2)e^{4\pi i\nu(z_1-z_2/c)}dz_1dz_2 \\ &+ \alpha^* e^{4\pi i\nu(\delta/c)}\int \sigma(z)e^{-4\pi i\nu(z/c)}dz \\ &+ \alpha e^{-4\pi i\nu(\delta/c)}\int \sigma^*(z)e^{4\pi i\nu(z/c)}dz\Bigg] \end{aligned} \tag{6.89}$$

where $S(\nu)$ is the power spectral density of the illumination source. The second term in Eqn. (6.89) is the self-modulation of the spectrum scattered from the object; the third and fourth terms are intereferometric cross-modulation of the spectrum from the reference and probe arms.

The goal of the OCT system is to reconstruct the scattering density $\sigma(z)$ from measured data. Direct measurement of $S_o(\nu)$ is called *Fourier domain* OCT. $\sigma(z)$ is imaged from $S_o(\nu)$ by filtering and image postproscessing. A basic image appears in the Fourier transform of $S_m(\nu)$

$$\Gamma_m(\tau) = |\alpha|^2\Gamma(\tau) + \iint \Sigma(\Delta z)\Gamma\left(\tau - \frac{2\Delta z}{c}\right) d\Delta z$$
$$+ \alpha^* \int \sigma(z)\Gamma\left(\tau - \frac{2(\delta - z)}{c}\right) dz$$
$$+ \alpha \int \sigma^*(z)\Gamma^*\left(\tau - \frac{2(\delta - z)}{c}\right) dz \tag{6.90}$$

where $\Sigma(\Delta z) = \int \sigma^*(\bar{z})\sigma(\bar{z} - \Delta z)\, d\bar{z}$. In most cases, we assume that $\sigma(z)$ is uncorrelated and that $\Sigma(\Delta z) \approx \delta(\Delta z)$, in which case both of the first two terms of Eqn. (6.90) are localized around $\tau = 0$. A signal proportional to $\sigma(z)$ appears in $\Gamma_m(\tau)$ in the vicinity of $\tau = 2(\delta - z)/c$. The position of the signal may be shifted relative to the $\tau = 0$ interferents by setting δ appropriately.

Since $\Gamma(\tau)$ is the impulse response for the OCT imaging system, the logitudinal spatial resolution is approximately $c\tau_c/2$, where τ_c is the coherence length. Signal estimation consists of deconvolution of $\int \sigma(z)\Gamma\{\tau - [2(\delta - z)/c]\}\, dz$ using analytic methods as described in Sections 8.3 and 8.5. One may, of course, choose to shape $\Gamma(\tau)$ to assist in this deconvolution.

The first OCT studies relied on measurements of the output intensity as a function of δ; this approach is called *time-domain* OCT. Measurements in time-domain systems take the form

$$I(\delta) = \int S_m(\nu, \delta)\, d\nu$$
$$= \Gamma(0)|\alpha|^2 + \int \Sigma(\Delta z)\Gamma\left(\frac{2\Delta z}{c}\right) d\Delta z$$
$$+ \alpha^* \int \sigma(z)\Gamma\left(\frac{2(z - \delta)}{c}\right) dz + \alpha \int \sigma^*(z)\Gamma^*\left(\frac{2(z - \delta)}{c}\right) dz \tag{6.91}$$

Measurement of the mutual coherence and the power spectral density is, of course, equivalent. The decision as to which approach to take is based on physical design constraints and signal fidelity and SNR outcomes. As discussed in Chapter 9, measurement of a spectrum by multichannel dispersive spectroscopy generally produces better SNR than does measurement by Fourier transform spectroscopy. For this reason, spectral domain OCT is increasingly popular [48,54,148]. Of course, practical OCT involves numerous application-specific sampling, filtering, and processing strategies. As an example, Fig. 6.23 is a 3D optical coherence tomography image of a *Drosophila melanogaster* (fruit fly) sample commonly used for genetic studies. With current-generation Fourier domain OCT systems, such 3D datasets are acquired in $<$10 s, thus making such systems suitable for real-time clinical diagnostics (e.g., for imaging the cornea or retina in human patients) or for high-throughput phenotyping of small animals as illustrated here.

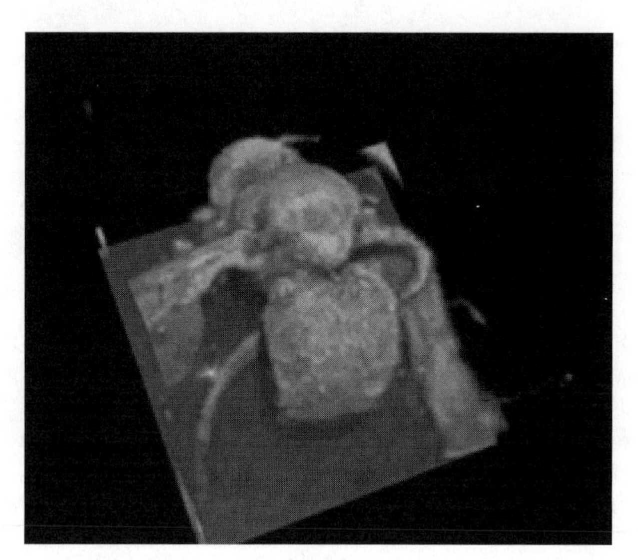

Figure 6.23 *Drosophila melanogaster* imaged by Fourier domain OCT. (Image courtesy of Bioptigen, Inc.)

Emerging OCT designs mix spatial and spectral tomography by considering the 3D shape of the illuminating beam and by combining data gathered by spatially scanning the illumination/return sensor head [209]. Many illumination and optical collection geometries may be considered in this case; the most obvious approach is a focal system in which light scattered from a spectrally broadband plane illumination wave is collected by an imaging optic. As always, the first step in analysing such a system is to consider the coherent system response. Modeling the spectrum of the illumination signal as $E(\nu)$ and the 3D scattering density as $\sigma(\mathbf{r})$, the coherent scattered field in the plane $z = 0$ under the Fresnel scattering model is

$$E(x, y, \nu) = \iiint e^{i(4\pi z\nu/c)}\sigma(x', y', z')h(x - x', y - y', z', \nu)\, dx'\, dy'\, dz' \qquad (6.92)$$

where $h(x, y, z, \nu)$ is the coherent impulse response for imaging from point (x, y, z) onto the output focal plane array. Creating interference of this field with the time-delayed reference signal, as in Eqn. (6.89), produces the spectral density

$$S_m(x, y, \nu) = S(\nu)\Big[I_r + I_o + \iiint e^{i\,[4\pi(\delta - z)\nu/c]}\sigma(x', y', z) \times h(x - x', y - y', z', \nu)\, dx'\, dy'\, dz + \text{c.c.}\Big] \qquad (6.93)$$

where I_r and I_o are the remitted reference and object intensities and c.c. denotes complex conjugation of the modulation term. As before, the cross-terms may be isolated by filtering on the spectral modulation at frequency $2\delta/c$. The returned signal in this case is the volume Fresnel transform of the object density, which may be processed with diffraction-limited transverse and coherence-limited longitudinal resolution to produce a true 3D image. Alternatively, one may choose to use a lens or coded aperture to shape the 3D system response for a more convenient 3D transformation.

6.6 MODAL ANALYSIS

Imagine that one seeks to create an optical sensor to distinguish two objects, "Alice" and "Bob." Ideally one might like to make a sensor such that all light scattered from Alice is collected on one detector while all light from Bob is collected on another detector. This is, in fact, more or less what happens in a focal system mapping light from a 2D object to a 2D image. Essentially all of the light radiated by a given pixel on the object is collected at the corresponding pixel on the image. In general, however, such isomorphic mappings are not possible. For example, it is not possible to create an optical sensor such that all light radiated from certain voxel in a volume source can be physically separated from light radiated by all other voxels. The present section explains why this is so.

Our first step is to relate the sampling theories of Chapter 3 to discrete representation of electromagnetic fields on boundaries and to *modes* and modal transformations. We then develop discrete representations of coherence functions and show that there is a *coherent-mode decomposition* corresponding to each particular realization of the cross-spectral density. We use the coherent-mode decomposition to comment on the incompressibility of the modal phase space. The "modal phase space" is the distribution of mode amplitudes on a given basis. The *incompressibility* of the phase space refers to physical constraints on the phase space transformations. In particular, in lossless linear optical systems the number of coherent modes is conserved and the greatest mode amplitude cannot be increased. The impossibility of increasing the largest mode amplitude means, for example, that where the light from Alice and light from Bob are mixed in overlapping modes, no optical system can abstract all light from Alice in a single mode while excluding all light from Bob.

6.6.1 Modes and Fields

We have analyzed many instances of field propagation through optical systems, including free-space diffraction as illustrated in Figs. 4.2 and 6.1 and propagation through lenses (Fig. 4.13) and transmittance masks (Fig. 4.8). Here we consider the general process of transformation of the electromagnetic field from an input plane parameterized by (x, y) to an output plane parameterized by (x', y') by such systems. We begin by considering the electromagnetic field in the input plane. We assume that the scalar amplitude of the field is $f(x, y, \nu) \in L^2(\mathbb{R}^3)$. We may represent $f(x, y, \nu)$ discretely either by confining it to a linear subspace (such as the space V_B of bandlimited functions), or we may use a multiscale basis to represent it on $L^2(\mathbb{R}^3)$.

In either case, we end up with a discrete representation

$$f(x, y, \nu) = \sum_{\mathbf{n}} f_{\mathbf{n}} \phi_{\mathbf{n}}(x, y, \nu) \tag{6.94}$$

where $\mathbf{n}$ is a vector index over position, order, scale, and/or dimension.

Discrete analysis has a long history in optical systems. In the optical tradition the basis components $\phi_{\mathbf{n}}$ are termed "modes" and $f_{\mathbf{n}}$ is regarded as the amplitude of the field in the $\mathbf{n}$th mode. A mode is an electromagnetic field distribution that satisfies boundary conditions and the Maxwell equations. Modal analysis is generally the basis for understanding waveguides and resonators, where boundary conditions severely constrain the spatial structure of modes.

As with all optical system analysis, the general goal of modal analysis is to apply knowledge of the field on an input boundary to calculate the state of the field on an output boundary. The field distribution of Eqn. (6.94) is transformed by diffraction, refraction, reflection, and absorption on propagation through the system. The output field $g(x', y', \nu)$ is

$$g(x', y', \nu) = \sum_{\mathbf{n}} f_{\mathbf{n}} \tilde{\phi}_{\mathbf{n}}(x', y', \nu) \tag{6.95}$$

where $\tilde{\phi}_{\mathbf{n}}(x', y', \nu)$ is the input component $\phi_{\mathbf{n}}(x, y, \nu)$ as transformed by the optical system. For free-space diffraction, $\tilde{\phi}_{\mathbf{n}}(x, y, \nu)$ is the Fresnel transform of $\phi_{\mathbf{n}}(x, y, \nu)$.

While the input field may be described using any basis, we are most interested in the basis vectors that propagate through the optical system. A mode, $\phi_{\mathbf{n}}(x, y, \nu) \in V_B$, propagating without attenuation in free space must be bandlimited such that $\hat{\phi}_{\mathbf{n}}(u, v, \nu)$ vanishes for all

$$\sqrt{u^2 + v^2} > \frac{\nu}{c} \tag{6.96}$$

Field components not in V_B are rejected on propagation through the optical system, and the rank of the field distribution is correspondingly reduced.

Typically, one chooses an orthonormal basis $\phi_{\mathbf{n}}(x, y, \nu)$ such that

$$\int \phi_{\mathbf{n}}^*(x, y, \nu) \phi_{\mathbf{m}}(x, y, \nu)\, dx\, dy = \delta_{\mathbf{nm}} \tag{6.97}$$

For $\phi(x, y, \nu) \in V_B$, we recall from Plancherel's theorem [Eqn. (3.20)] that

$$\begin{aligned} \int \tilde{\phi}_{\mathbf{n}}^*(x, y, \nu) \tilde{\phi}_{\mathbf{m}}(x, y, \nu)\, dx\, dy\, d\nu &= \int \hat{\tilde{\phi}}_{\mathbf{n}}^*(u, v, \tau) \hat{\tilde{\phi}}_{\mathbf{m}}(u, v, \tau)\, du\, dv\, d\tau \\ &= \int \hat{\phi}_{\mathbf{n}}^*(u, v, \tau) \hat{\phi}_{\mathbf{m}}(u, v, \tau) \end{aligned}$$

$$\times \exp\left(2\pi i\sqrt{\frac{\nu^2}{c^2} - u^2 - v^2 d}\right)$$

$$\times \exp\left(-2\pi i\sqrt{\frac{\nu^2}{c^2} - u^2 - v^2 d}\right) du\, dv\, d\tau$$

$$= \int \phi^*_{\mathbf{n}}(x, y, \nu)\phi_{\mathbf{m}}(x, y, \nu)\, dx\, dy\, d\nu$$

$$= \delta_{\mathbf{nm}} \tag{6.98}$$

In other words, free-space propagation of bandlimited distributions is a unitary transformation and preserves the orthogonality of a basis. In this case, the rank of the field is preserved and the coefficients $f_{\mathbf{n}}$ are good descriptors of $g \in V(\tilde{\phi})$. Orthogonality is also preserved for bandlimited fields in coherent imaging systems, although in this case the bandlimit is due to the aperture stop rather than the dispersion relationship.

In selecting basis functions for practical field analysis, one attempts to balance the utility of $\phi(x, y, \nu)$ in representing f and g against numerical and computational properties under Fresnel transformation. As discussed in Chapters 3 and 4, quasibandlimited eigenbases of the Fresnel transform, such as the Hermite–Gaussian and Laguerre–Gauaussian modes, are often used for coherent field analysis. At another extreme, Section 10.3.1 discusses modal decomposition in terms of prolate spheroidal functions, which are eigenfunctions of the bandlimited Fourier transform.

Compact bases on discrete lattices, like wavelets or B-spines, are more attractive for image plane analysis because one naturally represents an image as a discrete array of modal coefficients such that the array itself resembles the object. The projection of an image onto special functions like the Hermite–Gaussians or the prolate spheroidal functions looks nothing like the natural image. Unfortunately, choices of discrete lattice bases that remain attractive on optical propagation are slim. Several studies have attempted to address this deficiency. For example, Liebling et al. introduce *fresnelets*, which are Fresnel-transformed B-spines [154,155]. The fresnelets are naturally attractive for object expansion, but have no simple propagation features. One may alternatively consider lattices of Gaussian beams, a practice pioneered by Gabor [82]. Gabor functions do not unfortunately form wavelet bases, but more recent studies of Gabor analysis and frames, such as those by Shlivinski et al. [223] and Bastiaans [13], illustrate the power of this approach. While these strategies are useful in digital holography and device modeling, much simpler methods suffice for the present purposes. Our goals are simply to show that a basis of propagating modes remains a basis on propagation through a bandlimited system and to consider discrete transformation representations.

The concept of *time reversal* is essential in understanding the modal transformations on optical propagation. The assumption that a 3D field distribution is completely determined by 2D boundary conditions has been central to our discussion of field propagation. This assumption also applies for fields propagating from the output plane to the input plane of an optical system. Time reversal holds that $f(x, y, \nu)$ must result from propagating $g(x', y', \nu)$ backward through the system.

Time reversal symmetry holds generally in classical mechanics, electromagnetics, and quantum mechanics as long as coherent state descriptors apply. Time reversal symmetry is lost in systems with gain or loss (which one may regard as incompletely described quantum systems), in systems where boundary conditions are not fully specified (i.e., when some reflected or scattered field components are not included in the output field model) and in systems that are not static in time.

In cases where time reversal applies, all plane-to-plane mode transformations are unitary and modes remain orthogonal on propagation. The unitary nature of a particularly important mode transformation is discussed in Problem 6.13. Where time reversal does not apply, various constraints may still be specified on the resulting mode structure. These constraints are easier to describe in the context of coherence function transformations, which we address below (Sections 6.6.2–6.6.4).

6.6.2 Modes and Coherence Functions

We found in Eqn. (6.17) that the coherent impulse response can be used to form the transformation kernel for the cross-spectral density. Similarly, modes of the coherent field may be used to construct a discrete representation of the cross-spectral density. We model the electromagnetic field as a random process on a discrete spatiospectral basis described by Eqn. (3.2) with $f_{\mathbf{n}}$ as a random variable. Using the definition Eqn. (6.8) the cross-spectral density corresponding to the field of Eqn. (3.2) is

$$W(x_1, y_1, x_2, y_2, \nu) = \sum_{\mathbf{n}} \sum_{\mathbf{m}} W_{\mathbf{nm}}(\nu)\phi^*_{\mathbf{n}}(x_1, y_1, \nu)\phi_{\mathbf{m}}(x_2, y_2, \nu) \tag{6.99}$$

where $W_{\mathbf{nm}} = \langle f^*_{\mathbf{n}} f_{\mathbf{m}} \rangle$. Note that the matrix defined by the coefficients $W_{\mathbf{nm}}$ is Hermitian (e.g., $W_{\mathbf{nm}} = W^*_{\mathbf{mn}}$).

While the number of modes in Eqn. (6.99) may in general be infinite, it is not difficult to show that the effective number of propagating modes is finite in any given optical system. Our present goal is to show that the modal expansion of W may be diagonalized. As a first step, we seek modes $\psi(x_2, y_2, \nu)$ such that

$$\iint W(x_1, y_1, x_2, y_2, \nu)\psi(x_1, y_1, \nu)\, dx_1\, dy_1 = \Lambda\psi(x_2, y_2, \nu) \tag{6.100}$$

Representing these eigenmodes in the initial basis as

$$\psi(x, y, \nu) = \sum_{\mathbf{n}} \psi_{\mathbf{n}}\phi_{\mathbf{n}}(x, y, \nu) \tag{6.101}$$

reduces Eqn. (6.100) to the matrix form

$$\mathbf{W}^{\dagger}\psi = \Lambda\psi \tag{6.102}$$

As an Hermitian matrix $\mathbf{W}$ has real eigenvalues and may be expressed as

$$\mathbf{W} = \mathbf{U}\Lambda\mathbf{U}^{-1} \tag{6.103}$$

where $\mathbf{U}$ is a unitary matrix with rows equal to the mutually orthogonal eigenvectors of $\mathbf{W}$ and Λ is a diagonal matrix with Λ_{ii} equal to the eigenvalues of $\mathbf{W}$.

Substituting Eqn. (6.103) in Eqn. (6.99) yields

$$W(x_1, y_1, x_2, y_2, \nu) = \sum_{\mathbf{n}} \Lambda_{\mathbf{n}}(\nu)\psi_{\mathbf{n}}^{C^*}(x_1, y_1, \nu)\psi_{\mathbf{n}}^{C}(x_2, y_2, \nu) \tag{6.104}$$

where the *coherent modes* ψ^C are linear combinations of the original modes according to $\psi^C = \mathbf{U}\phi$. Equation (6.104) is called a *coherent-mode decomposition* of the cross-spectral density. By this decomposition we find that the cross-spectral density can always be reduced to a superposition of densities associated with a discrete set of perfectly coherent but mutually orthogonal modes. A more rigorous discussion of the coherent-mode decomposition without the assumption of a finite basis is presented by Wolf [249].

As discussed in Section 6.6.4, measurements of the power spectral density or irradiance may consist of projections of the cross-spectral density on particular modes. In particular, the projection on the coherent modes may be measured. Since such measurements must be nonnegative and since there is no significance to coherent modes with null amplitude, $\mathbf{W}$ is a positive definite matrix such that $\Lambda_{\mathbf{n}} > 0$ for all $\mathbf{n}$.

As an example, consider an object consisting of two mutually incoherent Gaussian sources in a plane. The sources are separated by a distance a and are of waist size $2a$. One of the sources is twice as intense as the other. The cross-spectral density for this source is

$$\begin{aligned} W(x_1, y_1, x_2, y_2, \nu) &= S(\nu)\phi_o\left(\frac{y_1}{2a}\right)\phi_o\left(\frac{y_2}{2a}\right)\left[2\phi_o\left(\frac{x_1 - 0.5a}{2\hat{a}}\right)\phi_o\left(\frac{x_2 - 0.5a}{2\hat{a}}\right)\right. \\ &\quad \left. + \phi_o\left(\frac{x_1 + 0.5a}{2\hat{a}}\right)\phi_o\left(\frac{x_2 + 0.5a}{2\hat{a}}\right)\right] \\ &= S(\nu)[2|\phi_-\rangle\langle\phi_-| + |\phi_+\rangle\langle\phi_+|] \end{aligned} \tag{6.105}$$

where $\phi_o(x)$ is the fundamental Gaussian mode of Eqn. (3.55), $|\phi_-\rangle = \phi_o(y/2a)\phi_o[(x - 0.5a)/2a)]$ and $|\phi_+\rangle = \phi_o(y/2a)\phi_o[(x + 0.5a)/2a]$. An orthonormal basis may be assembled from these modes using

$$\begin{aligned} |\phi_a\rangle &= \frac{|\phi_-\rangle + |\phi_+\rangle}{2^{3/4}\sqrt{1 + e^{-(\pi/8)}}} \\ |\phi_b\rangle &= \frac{|\phi_-\rangle - |\phi_+\rangle}{2^{3/4}\sqrt{1 - e^{-(\pi/8)}}} \end{aligned} \tag{6.106}$$

On this basis, the cross-spectral density matrix $\mathbf{W}$ is

$$\mathbf{W} \approx \begin{pmatrix} 3.55 & 0.52 \\ 0.52 & 0.69 \end{pmatrix} \tag{6.107}$$

We assemble coherent modes from the eigenfunctions of this matrix as described above; the eigenvalues are approximately 3.6 and 0.6. The coherent modes are

$$\begin{aligned} |\phi_1\rangle &\approx -0.98|\phi_a\rangle - 0.17|\phi_b\rangle \\ |\phi_2\rangle &= -0.17|\phi_a\rangle + 0.98|\phi_b\rangle \end{aligned} \tag{6.108}$$

and the cross-spectral density is

$$W(x_1, y_1, x_2, y_2, \nu) \approx S(\nu)[3.6|\phi_1\rangle\langle\phi_1| + 0.6|\phi_2\rangle\langle\phi_2|] \tag{6.109}$$

The irradiance, cross-spectral density, and coherent-mode distributions for this field are illustrated in Fig. 6.24.

The coherent-mode decomposition uniquely yields orthonormal decomposition modes, but this does not necessarily imply that it is more descriptive of the "true" source. In the specification of the current example, the overlapping Gaussian distributions might come from distinct lasers. The coherent-mode decomposition takes both sources into account to produce a decomposition into orthogonal modes. It is interesting to note that the amplitude of the stronger coherent mode is greater than the amplitude of either of the individual sources used to compose the joint source. We find, therefore, that it is possible to integrate power in a single coherent mode from multiple mutually incoherent sources. This is a subtle point; we will return to consider its significance after showing in Section 6.6.3 that it is impossible to efficiently combine power from mutually orthogonal modes.

6.6.3 Modal Transformations

We turn next to the effect of propagation through an optical system and transformations of coherent modes. As discussed relative to Eqn. (6.97), an orthogonal basis remains orthogonal on diffraction, meaning that the coherent-mode spectrum $\Lambda_{\mathbf{n}}$ remains invariant even as the spatial structure of the coherent modes evolves. Such invariance is particularly relevant in resonator and waveguide analysis, where one regards the modes as 3D spatial distributions. In sensor systems, however, one is more likely to seek to transform the modal structure of the field to enable effective information extraction. It is important to understand that the coherent-mode structure is a property of a single realization of the field, and that different objects may produce completely different coherent modes. Complete characterization of a field is equivalent to discovery of both the coherent-mode spectrum and the coherent modes. In many cases one has prior knowledge of the coherent modes; for example, in imaging an incoherent 2D object one may reasonably assume that the coherent modes correspond to the bandlimited Shannon basis functions. In cases where the coherent modes are not known a priori, however, one has no choice but to make measurements on a nonideal basis. We discuss measurement strategies in Section 6.6.4 after considering limits on modal transformations.

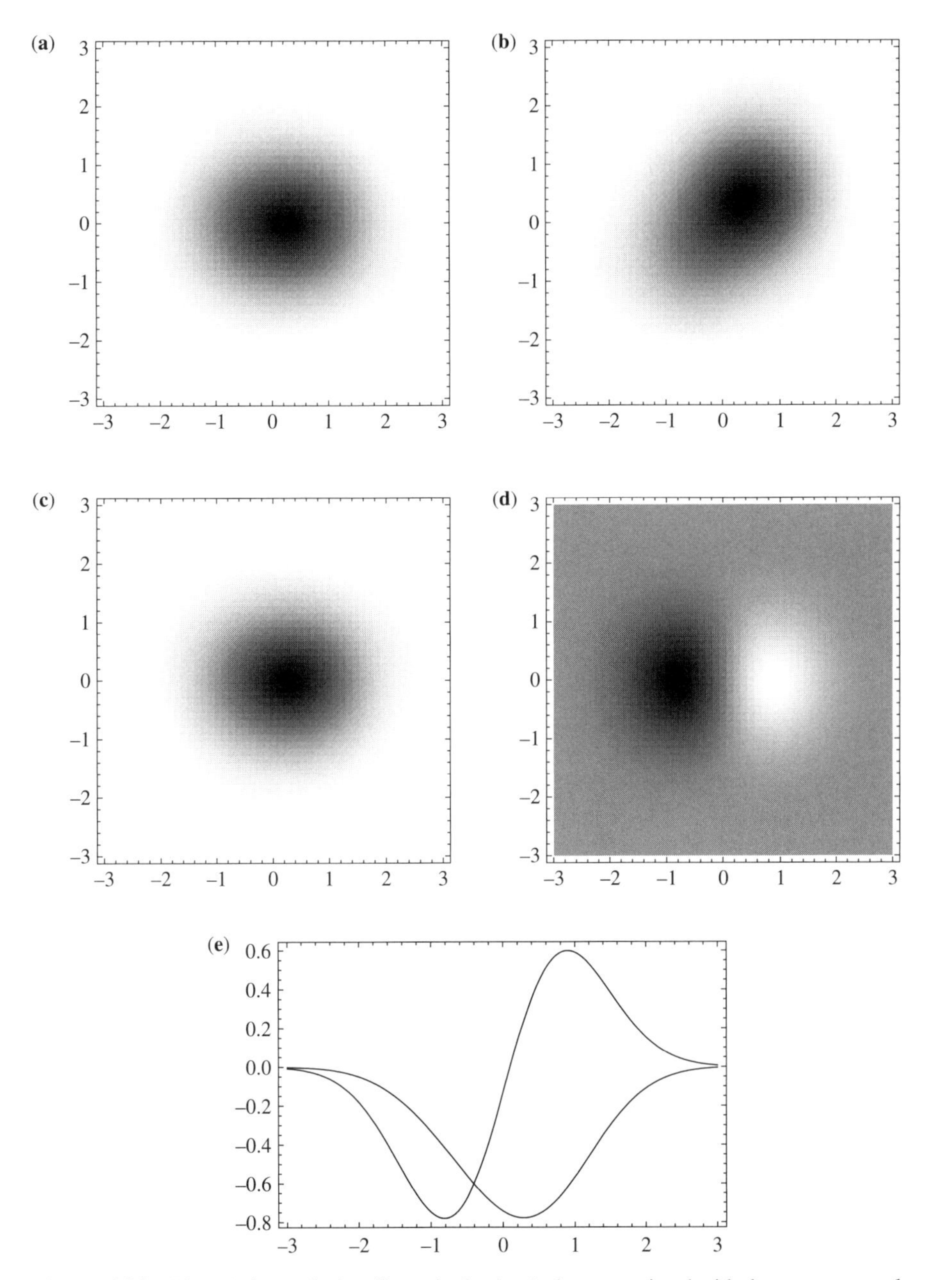

Figure 6.24 Plot (a) shows the irradiance in the (x, y) plane associated with the cross-spectral density of Eqn. (6.105), (b) is a cross section of the 5D function $W(x_1, y_1, x_2, y_2, \nu)$ at fixed ν in the plane $y_1 = 0$, $y_2 = 0$, (c) and (d) are the orthonormal coherent-mode distributions, and (e) plots the cross section of the coherent modes along the x axis.

In general, one may assume that an optical element exists to transform any individual mode into any other. We are most aware of the use of lenses to transform plane waves into focusing beams, but we can easily imagine the use of holograms to couple any pair of modes. Volume holograms, as discussed in Section 4.8, can approach 100% efficiency in such associations.

While mode-to-mode transformations are extremely useful, a more powerful mapping would map more than one input mode into a single output mode. For example, one might wonder whether it would be possible to make a device to focus two distinct input plane waves onto a single output focal spot. The answer to this question is "yes," but the efficiency with which light can be combined from two distinct modes depends on the relative coherence of the modes. Two mutually coherent plane waves of equal amplitude can be focused on the same spot with 100% efficiency. The maximum efficiency for focusing two mutually incoherent plane waves on the same spot, in contrast, is 50%.

The most basic geometry for two-mode integration is illustrated in Fig. 6.25, which shows two incident plane waves combined by a beamsplitter. The input modes $|\phi_1\rangle$ and $|\phi_2\rangle$ interfere in output modes $|\phi_a\rangle$ and $|\phi_b\rangle$. The mutual intensity of the input modes is described by three numbers: $J_{11} = I_1$, $J_{22} = I_2$, and $J_{12} = J_{21}^*$. In the bracket notation of Section 6.6.2, the mutual intensity of the input field is

$$J = (|\phi_1\rangle\,|\phi_2\rangle)\begin{pmatrix} I_1 & J_{12} \\ J_{21} & I_2 \end{pmatrix}\begin{pmatrix} \langle\phi_1| \\ \langle\phi_2| \end{pmatrix} \tag{6.110}$$

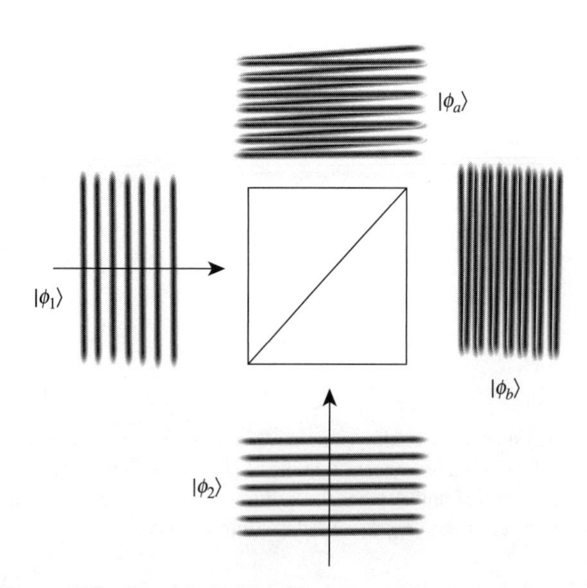

Figure 6.25 Use of a beamsplitter to combine two plane waves.

The beamsplitter transforms the input modes into output modes $|\phi_a\rangle$ and $|\phi_b\rangle$ such that

$$\begin{pmatrix} |\phi_a\rangle \\ |\phi_b\rangle \end{pmatrix} = \mathbf{T} \begin{pmatrix} |\phi_1\rangle \\ |\phi_2\rangle \end{pmatrix} \tag{6.111}$$

where, as discussed in Section 6.6.1 and in Problem 6.13, $\mathbf{T}$ is a unitary matrix.

The mutual intensity of the output field is described by the matrix

$$\mathbf{J} = \mathbf{T}^{\dagger} \begin{pmatrix} I_1 & J_{12} \\ J_{21} & I_2 \end{pmatrix} \mathbf{T} \tag{6.112}$$

In the particularly common case of a beamsplitter that sends half of the incident power from $|\phi_1\rangle$ onto output mode $|\phi_a\rangle$ and half onto output mode $|\phi_b\rangle$, a suitable form for $\mathbf{T}$ is

$$\mathbf{T} = \frac{1}{\sqrt{2}} \begin{pmatrix} 1 & 1 \\ 1 & -1 \end{pmatrix} \tag{6.113}$$

such that

$$\mathbf{J} = \frac{1}{2} \begin{pmatrix} I_1 + I_2 + 2\mathrm{Re}(J_{12}) & I_1 - I_2 - 2i\mathrm{Im}(J_{12}) \\ I_1 - I_2 + 2i\mathrm{Im}(J_{12}) & I_1 + I_2 - 2\mathrm{Re}(J_{12}) \end{pmatrix} \tag{6.114}$$

We note several important points:

1. $\mathbf{J}$ is Hermitian.
2. Power is conserved: $\mathrm{Tr}(\mathbf{J}) = I_1 + I_2$.
3. If the input modes are absolutely coherent and in phase, $J_{12} = \sqrt{I_1 I_2}$. If $I_1 = I_2 = I$ in this case then the output mutual intensity is described by

$$\mathbf{J} = \begin{pmatrix} 2I & 0 \\ 0 & 0 \end{pmatrix} \tag{6.115}$$

   The student may wish as an exercise to confirm that the input field is described by a single coherent mode in this case.
4. If the input modes are relatively incoherent, then $J_{12} = 0$. If $I_1 = I_2 = I$ in this case, then the output mutual intensity is described by

$$\mathbf{J} = \begin{pmatrix} I & 0 \\ 0 & I \end{pmatrix} \tag{6.116}$$

If $I_1 \neq I_2$, then

$$\mathbf{J} = \frac{1}{2}\begin{pmatrix} I_1 + I_2 & I_1 - I_2 \\ I_1 - I_2 & I_1 + I_2 \end{pmatrix} \tag{6.117}$$

meaning that output modes are partially coherent. The coherent modes for this system would correspond to the individual outputs for each input mode.

Transformations of the form $\mathbf{J}' = \mathbf{T}^{\dagger}\mathbf{J}\mathbf{T}$ with $\mathbf{T}$ unitary do not change the eigenvectors or eigenvalues of $\mathbf{J}$ and therefore maintain the coherent-mode structure unchanged. There are various mechanisms, however, by which an optical system can change the coherent modes and the modal eigenvalues. These include

- *Modal Filtering.* Various devices decrease the number of propagating modes in an optical system. A pinhole is the most common example. We have used pinholes several times to select a single spherical wave from a broader field distribution, most recently in our discussion of the two-pinhole interferometer. Such devices are called *spatial filters*. As discussed in Chapter 9, spatial filters are commonly used in spectroscopy to break degeneracies between spectral and spatial degrees of freedom. Complex spatial filters may be implemented using diffractive devices or coded apertures. In the example of Fig. 6.25, spatial filtering may consist simply of discarding the light on one of the output arms. In that case, we have combined two input modes into one output mode.
- *Absorption.* A device that differentially absorbs some of the light in an optical system breaks time reversal symmetry and restructures the coherent modes.
- *Modal Decorrelation.* Modes propagate on different paths through the optical system. If the structure of the optical system is time varying on these paths, the relative coherence of the modes may be destroyed. The most common example is propagation through turbulence. In the example of Fig. 6.25, the two output beams may experience different time-varying phase modulation. Such modulation can render the output beams relatively incoherent, which diagonalizes $\mathbf{J}$. In the case of the $50/50$ beamsplitter with incoherent input beams, this produces relatively incoherent output beams with equal irradiances of $(I_1 + I_2)/2$.

Other mechanisms for changing the coherent-mode structure include laser gain and nonlinear wavemixing.

For imaging and spectroscopy we focus on systems where optical components implement joint transformations on 10^6–10^{12} modes simultaneously. We begin consideration of such systems with the coherent mode decomposition of the cross-spectral density at the input

$$W(x_1, y_1, x_2, y_2, \nu) = \sum_n \lambda_n(\nu)\psi_n^*(x_1, y_1, \nu)\psi_n(x_2, y_2, \nu) \tag{6.118}$$

As discussed in Section 6.6.2, the mode amplitudes $\lambda_n(\nu)$ are real and positive and the modes are orthonormal. The coherent modes are transformed on propagation through an optical system such that the cross-spectral density at the output is

$$W(x'_1, y'_1, x'_2, y'_2, \nu) = \sum_n \lambda_n(\nu)\tilde{\psi}^*_n(x'_1, y'_1, \nu)\tilde{\psi}_n(x'_2, y'_2, \nu) \tag{6.119}$$

In view of the impact of optical loss and spatial filtering, the functions $\tilde{\psi}_n(x, y, \nu)$ are not necessarily orthogonal [251]. The cross-spectral density across the output aperture is described by the new coherent-mode decomposition

$$W(x'_1, y'_1, x'_2, y'_2, \nu) = \sum_n \Lambda_n(\nu)\Psi^*_n(x'_1, y'_1, \nu)\Psi_n(x'_2, y'_2, \nu) \tag{6.120}$$

where $\Psi_n(x, y, \nu)$ form a new set of orthonormal coherent modes and the functions Λ_n are new eigenvalues. Using the nodal orthonormality, projection of Eqn. (6.120) against the coherent modes yields

$$\iiiint W(x'_1, y'_1, x'_2, y'_2, \nu)\Psi^*_m(x'_1, y'_1, \nu) \times \Psi_m(x'_2, y'_2, \nu)\, dx'_1\, dy'_1\, dx'_2\, dy'_2 = \Lambda_m(\nu) \tag{6.121}$$

Substitution of the cross-spectral density of Eqn. (6.119) in the integral of Eqn. (6.121) yields

$$\sum_n |c_{nm}|^2 \lambda_n = \Lambda_m \tag{6.122}$$

where c_{nm} is the projection of the transformed input coherent modes on the output plane coherent modes:

$$c_{nm} = \iint \tilde{\psi}^*_n(x, y, \nu)\Psi_m(x, y, \nu)\, dx\, dy \tag{6.123}$$

We note again that the input coherent modes are orthonormal. Barring laser gain, nonlinear wavemixing, or sources within the optical system, the spectral density of a mode cannot increase in magnitude. This means that

$$\iint |\tilde{\psi}^*_n(x, y, \nu)|^2 dx\, dy \leq 1 \tag{6.124}$$

Since the output coherent modes are orthonormal and complete over the output space, we also find that

$$\iint |\tilde{\psi}^*_n(x, y, \nu)|^2 dx\, dy = \sum_m |c_{nm}|^2 \tag{6.125}$$

and

$$\sum_m |c_{nm}|^2 \leq 1 \tag{6.126}$$

Summing Eqn. (6.122) with respect m, this implies that

$$\sum_n \lambda_n \geq \sum_m \Lambda_m \tag{6.127}$$

meaning that the input power is greater than or equal to the output power.

Alternatively, we may express the output coherent modes in terms of the input coherent modes. Of course, in a system satisfying time reversal, the output and input coherent modes are matched one to one and $c_{nm} = \delta_{nm}$. If we consider the more general reversible transformation between any set of orthogonal input modes and orthogonal output modes, conservation of power on time reversal produces the complimentary relationship to Eqn. (6.126):

$$\sum_n |c_{nm}|^2 \leq 1 \tag{6.128}$$

While the structure of the coupling coefficients may change as a result of loss, filtering, and decoherence, all three of these mechanisms decrease, rather than increase, the effective value of the coupling coefficient between an input mode and an output mode. We therefore find that Eqn. (6.128) holds in any gain or source free optical system.

Equation (6.122) relates the power in the mth output mode to the power in input modes. $|c_{nm}|^2$ is the fraction of the nth input-mode power that is coupled into the mth output coherent mode. According to Eqn. (6.128), the coefficients $|c_{nm}|^2$ are a weighted distribution such that no output mode can be brighter than the brightest input mode, for example

$$\Lambda_{\max} \leq \lambda_{\max} \tag{6.129}$$

The maximal brightness is obtained when $c_{nm} = 1$ for n corresponding to the brightest input mode.

Equation (6.129) is an outcome of the second law of thermodynamics, which states that in isolated physical systems the total entropy can only remain constant or increase. The entropy of a system is a measure of the total number of physical states that it might occupy. If the exact state of the system is known, as in a coherent optical field, then the entropy is low. If the energy of the system is distributed over many different coherent modes, the entropy is higher. In terms of normalized coherent-mode amplitudes $\tilde{\lambda}_m = \lambda_m / \sum_n \lambda_n$, the entropy of an optical field may be expressed [190]

$$H = \sum_n \tilde{\lambda}_n \log \tilde{\lambda}_n \tag{6.130}$$

If, for example, $\tilde{\lambda}_1 = 1$ and $\lambda_{n>1} = 0$, then $H = 0$ and the entropy is a minimum. Alternatively, maximal entropy occurs for $\lambda_n = 1/N$ for all n, in which case $H = \log N$.

The more general statement of the second law in optical systems is

1. Any time reversable optical system leaves the coherent-mode amplitudes unchanged.
2. Nonreversible effects such as modal filtering, absorption, and modal decorrelation transform the coherent-mode amplitudes so as to increase the system entropy.

Returning to the combination of power from multiple modes into a single mode or a smaller number of modes, one may say as a corollary to Eqn. (6.129) that if energy is conserved, then $\Lambda_{\min} \geq \lambda_{\min}$. This means that the number of modes in the field is conserved when energy is conserved. In the example discussed at the beginning of this section, two mutually coherent plane waves at the input can be combined into a single output mode with 100% efficiency because both input signals are part of the same coherent mode of the cross-spectral density. If the input cross-spectral density consists of two modes, no single mode of the output cross-spectral density can exceed the stronger of the two input modes.

Conservation of modes may be related to the space–bandwidth product and the etendue of an optical system. The space–bandwidth product is the product of the spatial support over which an optical signal is observed and the spatial frequency bandwidth of the field. We saw in Section 3.6 that the number of discrete modes necessary to represent a signal is proportional to the space–bandwidth product. The *etendue*, $L \propto A^2\Omega$, of an optical system is the product of the solid angle Ω spanned by the wavevectors of radiation collected at the instrument aperture and the area A^2 of the entrace pupil. Of course, we know from many previous discussions that the angle of the wavevector is proportional to spatial frequency. Ω is thus a measure of the spatial bandwidth, and the etendue is proportional to the space–bandwidth product, which is proportional to the number of modes that propagate through the system. An energy conserving system must therefore conserve etendue.

6.6.4 Modes and Measurement

We have mentioned several times that measurements of the field take the form of irradiance or spectral density projections. Back in Section 2.1 we introduced the concept of a visibility function to describe these projections. We updated the visibility in terms of the coherent impulse response in Chapter 4 and the incoherent or partially coherent response in this chapter. With our understanding of coherence transformations, we are now ready to express the most general form of a first-order optical measurement as

$$m_i = \int\int h_i^*(x_1, y_1, \nu)W(x_1, y_1, x_2, y_2, \nu)h_i(x_2, y_2, \nu)\, dx_1\, dx_2\, dy_1\, dy_2\, d\nu \quad (6.131)$$

where $h_i(x, y, \nu)$ is a coherent impulse response or visibility for the ith measurement.

Substituting the coherent-mode decomposition of $\mathbf{W}$ into Eqn. (6.104) and representing the set of measurements as a vector $\mathbf{m}$ allows us to recast the measurement process as

$$\mathbf{m} = \mathbf{H}\Lambda \quad (6.132)$$

where

$$h_{ij} = \left| \int h_i(x, y, \nu)\psi_j^C(x, y, \nu)\, dx\, dy\, d\nu \right|^2 \quad (6.133)$$

and Λ is a vector of coherent-mode amplitudes.

Optical system design consists of selecting the measurement basis h_i within physical and economic limits to enable estimation of the object under observation. Ideally, design also consists of specifying the measurement matrix $\mathbf{H}$ in Eqn. (6.132), but this assumes that we know the coherent modes in advance. Common measurement scenarios in optical systems include the following cases:

1. *Focal systems*, where the coherent modes of the system are known and one has the physical capacity to create a mode-specific filter to separate the coherent modes in the measurement system. This scenario describes 2D focal imaging and conventional slit spectroscopy. In the imaging case, for example, the coherent modes are bandlimited focused spots spanning the object plane. A lens system maps the input spots to the output image, where each pixel is measured independently. In the ideal case, $\mathbf{H}$ is the identity matrix for this system. Criteria by which one might design $\mathbf{H}$ are discussed in Chapter 8, but for present purposes it is safe to say that if information is uniformly and independently distributed in the object pixels, then one can do no better than to design $\mathbf{H}$ to be an identity. If, on the other hand, object information is more subtly distributed, one may wish to consider generalized sampling strategies as dicussed in Section 7.5.
2. *Multidimensional systems*, where the coherent modes are known but one lacks the physical capacity to independently measure each coherent mode. Multispectral imaging, as discussed in Section 10.6, is the model system for this case. In principle, one could create a filter assembly to separate each spectral channel and then implement focal imaging on the coherent

modes, but physical implementation of such a system is impossible or inefficient, and one chooses to apply temporal multiplexing or generalized sampling instead. One can directly design **H** in this case.

3. *Interferometric systems*, where the coherent modes are unknown and must be discovered as part of the measurement process. Examples for this case include imaging through turbulent or inhomogeneous media and imaging under partially coherent illumination. A typical strategy for this case involves selection of an a priori modal decomposition. In some cases, the a priori basis may be "close" to a coherent mode decomposition. In other cases, the a priori basis may simply have convenient mathematical properties. The focal interferometry strategies discussed in Section 6.3.4 are an example of interferometric imaging on nearly orthogonal bases. Imaging with a rotational shear interferometer, on the other hand, applies a Fourier basis to an unknown cross-spectral density.

6.7 RADIOMETRY

The discipline of *radiometry* focuses on measuring the energy content of optical radiation and on mapping the flow of energy through optical systems. Radiometry describes the optical field using phenomenological functions related to the irradiance such as the *spectral radiance*, $B_\nu(x, s, \nu)$, which is the power radiated through point $x \in \mathbb{R}^3$ in direction $s \in \mathbb{S}^2$ per unit solid angle per unit wavelength. The SI unit of spectral radiance is watt per steradian per square meter per hertz. The spectral radiance may be integrated along the wavelength axis to produce the radiance, which is the radiant energy propagating through x in the s direction per unit solid angle. This section discusses the relationship between the spectral radiance and the cross-spectral density and describes the *constant radiance theorem*, which limits the ability of optical systems to focus partially coherent light. Unless otherwise indicated, the term *radiance* refers to the spectral radiance in subsequent discussion.

6.7.1 Generalized Radiance

As always in optical system analysis, propagation of the radiance from one boundary to the next is our first challenge. Propagation of the radiance is most easily derived from the wave equations satisfied by the cross-spectral density

$$\begin{aligned} \left[\nabla_1^2 + \left(\frac{2\pi\nu}{c}\right)^2\right] W(\mathbf{x}_1, \mathbf{x}_2, \nu) &= 0 \\ \left[\nabla_2^2 + \left(\frac{2\pi\nu}{c}\right)^2\right] W(\mathbf{x}_1, \mathbf{x}_2, \nu) &= 0 \end{aligned} \tag{6.134}$$

where ∇_1^2 is the 3D Laplacian with respect to the $\mathbf{x}_1$. These equations are derived by substituting the definition of the cross-spectral density into the wave equation [(Eqn. (4.19)].

The radiance is most often used to analyze diffuse objects with relatively low coherence, in which case one might assume that the cross-spectral density varies slowly with respect to $\bar{x}$. An object such that

$$\frac{\partial W(\Delta x, \bar{x}, \nu)}{\partial \bar{x}} \ll \frac{\partial W(\Delta x, \bar{x}, \nu)}{\partial \Delta x} \tag{6.135}$$

is said to be "quasihomogeneous" [165]. (*Homogeneous* in this context is synonymous with *spatially stationary.*) A quasihomogeneous object is spatially uniform on the scale of the coherence cross section. Given that the coherence cross section of an image field is approximately equal to the transverse resolution, a quasihomogeneous image must be slowly varying in comparison to the optical resolution. This is not generally a good assumption in imaging, but radiance is a sufficiently useful concept that we neglect for the moment difficulties at the limits of resolution.

Reparameterizing $W(x_1, x_2, \nu)$ in terms of Δx and $\bar{x}$ transforms Eqns. (6.134) into [198]

$$\nabla_{\Delta x} \cdot \nabla \bar{x} W = 0 \tag{6.136}$$

$$\left[\nabla^2_{\Delta x} + \frac{1}{4} \nabla^2_{\bar{x}} + \left(\frac{2\pi \nu}{c} \right)^2 \right] W(\Delta x, \bar{x}, \nu) = 0 \tag{6.137}$$

Neglecting the Laplacian with respect to $\bar{x}$ under the quasihomogeneous approximation reduces Eqn. (6.137) to

$$\left[\nabla^2_{\Delta x} + \left(\frac{2\pi \nu}{c} \right)^2 \right] W(\Delta x, \bar{x}, \nu) = 0 \tag{6.138}$$

which is solved by

$$W(\Delta x, x, \nu) = \int_{\Omega} \mathcal{B}(\mathbf{x}, s, \nu) e^{-(2\pi i \nu / c) s \cdot \Delta x} \, ds \tag{6.139}$$

where according to Eqn. (6.136)

$$s \cdot \nabla_x \mathcal{B}_\nu(x, s, \nu) = 0 \tag{6.140}$$

The "generalized radiance," as proposed by Walther [244], is defined by the inverse transform corresponding to Eqn. (6.139)

$$\mathcal{B}(\mathbf{x}, s, \nu) = \iint W(\Delta x, x, \nu) e^{(2\pi i \nu / c) s \cdot \Delta x} \, d\Delta x \tag{6.141}$$

where $\Delta x \in \mathbb{R}^2$. Walther's aim was to associate the phenomenological radiance with physical optics. Ideally, a physically based radiance satisfies four criteria:

1. $\mathcal{B}_\nu(\mathbf{x}, s, \nu)$ is a linear transform of $W(\Delta x, x, \nu)$.
2. $\mathcal{B}_\nu \geq 0$, assuring that the radiance from an object is positive.
3. $\mathcal{B}_\nu = 0$ outside the support of a radiating object.
4. $\cos\theta \int\int \mathcal{B}_\nu(x, s, \nu) = \mathcal{J}_\nu(s, \nu)$, where the *radiant intensity* $\mathcal{J}_\nu(s, \nu)$ is the physical optical energy flux along s.

Unfortunately, the generalized radiance does not satisfy these four criteria [166]. In fact, Friberg proved that no radiance function could satisfy these criteria [79]. The generalized radiance remains useful, however, for paraxial analysis of quasihomogeneous fields. Under the quasihomogenous approximation, the generalized radiance can be shown to be consistent with physical optics [165].

The attraction of the radiance in system analysis is encapsulated in Eqn. (6.140), which states that $\mathcal{B}(\bar{x}, s, \nu)$ is invariant along a ray in the s direction. In particular, Friberg demonstrated under the paraxial approximation that the generalized radiance may be propagated by the ABCD ray transfer matrice M according to

$$\mathcal{B}_\nu\left(\begin{bmatrix}\mathbf{x}'\\ s'\end{bmatrix}\right) = \mathcal{B}_\nu\left(M\begin{bmatrix}\mathbf{x}\\ s\end{bmatrix}\right) \tag{6.142}$$

As briefly discussed in Section 10.4.3, paraxial ray tracing of the radiance fields is widely applied in image rendering algoritms for computer graphics [152,101,199]. Propagation of the radiance by ray tracing is also useful for analysis astronomical sources and nonimaging optics [248].

6.7.2 The Constant Radiance Theorem

The radiance is sometimes termed the "brightness" of the field. The goal of this section is to show that no linear optical system can increase brightness, a result called *the constant radiance theorem.* The constant radiance theorem expresses the primary difference between a diffuse source, such as a fluorescent lightbulb or a flame, and a laser. While the fluorescent source may emit several watts and the laser just a few milliwatts, the laser can be focused on a much brighter spot. In contrast, no linear optical system can make a fluorescent source brighter than the irradiance on the irradiating surface.

The generalized radiance is the *Wigner distribution function* of the coherent field

$$\mathcal{B}(\mathbf{x}, s, \nu) = \int\int \left\langle E^*\left(x - \frac{\Delta x}{2}, \nu\right) E\left(x + \frac{\Delta x}{2}, \nu\right)\right\rangle e^{(2\pi i\nu/c)s\cdot\Delta x}\, d\Delta x \tag{6.143}$$

The Wigner function developed as a "phase space" representation of quantum mechanical states [259]. As a space–frequency distribution, the Wigner function is closely related to windowed Fourier transformations and wavelet analysis.

The relationship of the Wigner distribution function and the generalized radiance was developed by Bastiaans [11]. In particular, Bastianns [12] considers modal transformations of the generalized radiance. Substituting Eqn. (6.118) into Eqn. (6.143), we find that the generalized radiance corresponding to the coherent mode decomposition of the cross-spectral density is

$$\mathcal{B}_\nu(x, s, \nu) = \sum_n \lambda_n \mathcal{B}_{n,\nu}(x, s, \nu) \tag{6.144}$$

where

$$\mathcal{B}_{n,\nu}(x, s, \nu) = \iint \psi_n^*\left(x - \frac{\Delta x}{2}, \nu\right) \psi_n\left(x + \frac{\Delta x}{2}, \nu\right) e^{(2\pi i\nu/c)s\cdot\Delta x}\, d\Delta x \tag{6.145}$$

The Wigner distributions of the coherent modes satisfy the orthogonality relationship [11]

$$\begin{aligned}\iiiint \mathcal{B}_{n,\nu}(x, s, \nu)\mathcal{B}_{m,\nu}(x, s, \nu)\, dx\, ds &= \left|\iint \psi_n^*(x, \nu)\psi_m(x, \nu)\, dx\right|^2 \\ &= \begin{cases} 1 & m = n \\ 0 & m \neq m \end{cases}\end{aligned} \tag{6.146}$$

Equations (6.129), (6.144), and (6.146) broadly constrain radiance transformations in optical systems. According to Eqn. (6.144), the radiance is the weighted sum of the radiances of the coherent modes. According to Eqn. (6.146), the brightness at any point in the output field will be associated primarily with a single coherent mode. According to Eqn. (6.129), the brightness of the brightest such mode cannot be increased in a linear optical system.

PROBLEMS

6.1 *Coherence Time and Cross Section.* Consider a light source emitting the blackbody spectrum

$$S(\nu) = \frac{2h\nu^3}{c^2}\frac{1}{e^{(h\nu/k_bT)} - 1} \tag{6.147}$$

where h is Planck's constant, k_b is Boltzmann's constant, and T is temperature.

(a) Plot $S(\nu)$ for $T = 5000$ K. What is the wavelength corresponding to the maximum of $S(\nu)$?

(b) Estimate the coherence time for this source.

(c) Suppose that the source is a 1-mm-diameter disk. Derive an expression for the cross-spectral density at a range of 100 m. Plot the cross-spectral density as a function of Δx and λ for $\Delta y = 0$.

(d) Estimate the coherence cross section [the range of Δx over which $W(\Delta x, \Delta y, \nu)$ is nonnegligible] at 100 m.

6.2 *Temporal Coherence and Bandwidth*

(a) A Michelson interferometer scans the path delay δ over range D with a sampling period Δ. Estimate the wavelength range and resolution that this instrument achieves when used as a spectrometer.

(b) A two-point interferometer with pinhole separation Δx, as illustrated in Fig. 6.6, is sampled on a focal plane at a range of d from the pinholes. The pinholes are illuminated by spatially coherent light. The spatial sampling period is Δ, and the sampling range is D. Estimate the wavelength range and resolution that this instrument achieves when used to estimate the power spectral density of the illumination.

6.3 *Spectral Encoding in Scattered Partial Coherence.* Two unresolved point scatteres separated by $2a$ in the plane transverse to the optical axis generate the spectrum $(\lambda) = S_o(\lambda)[1 + \text{jinc}(a\alpha/\lambda)]$.

(a) Use Fourier arguments to quantify the accuracy with which one might estimate a from this spectrum. Comment on the significance of this result for remote target superresolution.

(b) What is the scattered cross-spectral density if the scatters are displaced by a along the optical axis rather than in the transverse plane? Generate a plot similar to Fig. 6.3 showing the spectral density observed along the optical axis as a function of a for this case.

6.4 *3D and 2D Transfer Functions.* Show that the incoherent transfer function, $H_{ic}(u, v)$ as described by Eqn. (6.64), satisfies

$$H_{ic}(u, v, \lambda) = \frac{1}{2(\lambda d_i)^2} \int B(-\lambda d_i u, \ -\lambda d_i v, \ q) \, dq \tag{6.148}$$

where $B(x, y, q)$ is defined according to Eqn. (6.54).

6.5 *RSI Fringe Frequency.* The point source of Fig. 6.10 was illuminated at $\lambda =$ 532 nm and $\lambda = 633$ nm.

(a) What is the fringe period in Fig. 6.10(a) and (b)?

(b) Generate a plot of simulated data showing the irradiance at some particular detection pixel as a function of longitudinal delay δ for this source.

(c) Assuming 20-μm detector pixels, what is the maximum source position θ_x could one observe with this instrument at $\theta = 2^\circ$ for each of the two illumination wavelengths?

6.6 *Coherent and Incoherent Imaging*

(a) Given that the impulse response for incoherent imaging is the square of the impulse response for coherent imaging, does an incoherent imaging system implement a linear transformation?

(b) Does the higher bandpass of an incoherent imaging system yield higher spatial resolution than the comparable coherent imaging system?

(c) Is it possible for the image of a coherent object to be an exact replica? (For instance, can the image be equal, modulo phase constants, to the input object with no blurring?)

(d) Is it possible for the image of an incoherent object to be an exact replica?

6.7 *MTF and Resolution.* An $f/2$ imaging system with circular aperture of diameter $A = 1000\lambda$ is distorted by an aberration over the pupil such that

$$P(x, y) = e^{i\pi\alpha x^2 y} \tag{6.149}$$

where $\alpha = 10^{-4}/A\lambda F$. Plot the MTF for this system. Estimate its angular resolution.

6.8 *Coherent and Incoherent Imaging* (suggested by J. Fienup). The interference of two coherent plane waves produces the field

$$\psi(x, z) = \tfrac{1}{2}[1 + \cos(2\pi u_0 x)]e^{i2\pi w_0 z} \tag{6.150}$$

The interfering field is imaged as illustrated in Fig. 6.26. As we have seen, the coherent imaging system is bandlimited with maximum frequency $u_{\max} = A/\lambda d_i$. Experimentally, one observes that the image of the field is uniform (e.g., there is no harmonic modulation) for $A < u_o \lambda d_i$. If, however, one places a "diffuser" at the object plane, then one observes harmonic modulation in the image irradiance as long as $A > u_o \lambda d_i/2$. The diffuser is described by the transmittance $t(x) = e^{i\phi(x)}$, where $\phi(x)$ is a random process. Diffusers are typically formed from unpolished glass.

(a) Explain the role of the diffuser in increasing the imaging bandpass for this system.

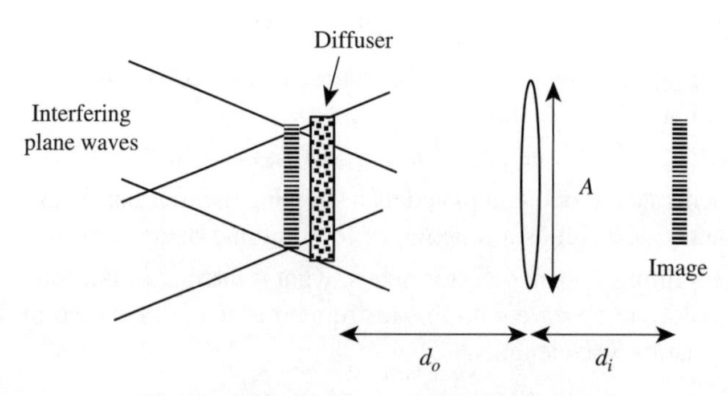

Figure 6.26 Geometry for Problem 6.8.

(b) Without the diffuser, the image of the interference pattern is insensitive to defocus. Once the diffuser is inserted, the image requires careful focus. Explain this effect.

(c) Model this system in one dimension in Matlab by considering the coherent field

$$f(x) = e^{-(x^2/\sigma^2)}[1 + \cos(2\pi u_o x)] \tag{6.151}$$

for $\sigma = 20/u_o$. Plot $|f|^2$. Lowpass-filter f to the bandpass $u_{\max} = 0.9u_o$ and plot the resulting function.

(d) Modulate $f(x)$ with a random phasor $t(x) = e^{i\phi(x)}$ (simulating a diffuser). Plot the Fourier transforms of $f(x)$ and of $f(x)t(x)$. Lowpass-filter $f(x)t(x)$ with $u_{\max} = 0.9u_o$ and plot the resulting irradiance image.

(e) Phase modulation on scattering from diffuse sources produces "speckle" in coherent images. Discuss the resolution of speckle images.

6.9 *OCT.* A Fourier domain OCT system is illuminated by a source covering the spectral range 800–850 nm. One uses this system to image objects of thickness up to 200 μm. Estimate the longitudinal spatial resolution one might achieve with this system and the spectral resolution necessary to operate it.

6.10 *Resolution and 3D Imaging.* Section 6.4.2 uses $\delta\theta_x \approx \lambda/A$ and $\delta z \approx 8\lambda z^2/A^2$ to approximate the angular and range resolution for 3D imaging using Eqn. (6.73). Derive and explain these limits. Compare the resolution of this approach with projection tomography and optical coherence tomography.

6.11 *Bandwidth.* Section 6.6.1 argues that the spatial bandwidth of the field does not change under free-space propagation. Nevertheless, one observes that coherent, incoherent, and partially coherent images blur under defocus. The blur suggests, of course, that bandwidth is reduced on propagation. Explain this paradox.

6.12 *Coherent-Mode Decomposition.* Consider a source consisting of three mutually incoherent Gaussian beams at focus in the plane $z = 0$. The source is described by the cross-spectral density

$$\begin{aligned} W(x_1, y_1, x_2, y_2, \nu) = S(\nu)\Bigg[& \phi_0\left(\frac{x_1}{\Delta}\right)\phi_0\left(\frac{y_1 - 0.5\Delta}{\Delta}\right)\phi_0\left(\frac{x_2}{\Delta}\right)\phi_0\left(\frac{y_2 - 0.5\Delta}{\Delta}\right) \\ & + \phi_0\left(\frac{x_1 - 0.5\Delta}{\Delta}\right)\phi_0\left(\frac{y_1 + 0.5\Delta}{\Delta}\right)\phi_0\left(\frac{x_2 - 0.5\Delta}{\Delta}\right) \\ & \times \phi_0\left(\frac{y_2 + 0.5\Delta}{\Delta}\right) + \phi_0\left(\frac{x_1 + 0.5\Delta}{\Delta}\right)\phi_0\left(\frac{y_1 + 0.5\Delta}{\Delta}\right) \\ & \times \phi_0\left(\frac{x_2 + 0.5\Delta}{\Delta}\right)\phi_0\left(\frac{y_2 + 0.5\Delta}{\Delta}\right)\Bigg] \end{aligned} \tag{6.152}$$

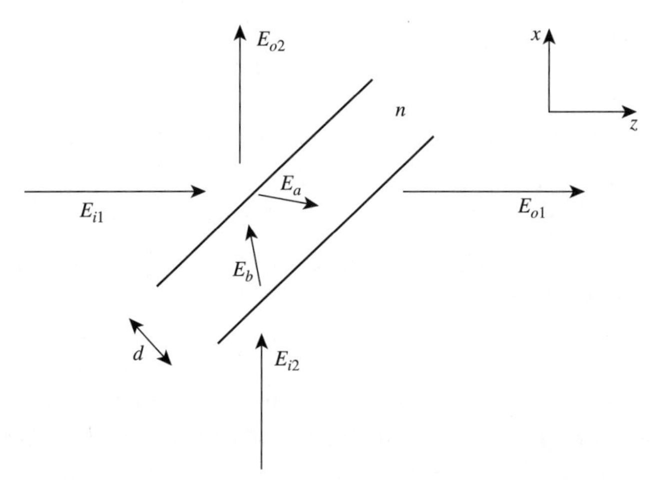

Figure 6.27 Pellicle beamsplitter geometry.

(a) In analogy with Fig. 6.24, plot the irradiance, cross sections of the cross-spectral density, and the coherent modes in the plane $z = 0$.

(b) Assuming that $\Delta = 100\lambda$, plot the coherent modes and the spectral density in the plane $z = 1000\lambda$.

6.13 *Time Reversal and Beamsplitters.* Consider the pelicle beamsplitter as illustrated in Fig. 6.27. The beamsplitter may be illuminated by input beams from the left or from below. Assume that the illuminating beams are monochromatic TE polarized plane waves (e.g., **E** is parallel to the y axis). The amplitude of incident wave on ports 1 and 2 are E_{i1} and E_{i2}. The beamsplitter consists of a dielectric plate of index $n = 1.5$. Write a computer program to calculate the matrix transformation from the input mode amplitudes to the output plane wave amplitudes E_{o1} and E_{o2}. Show numerically for specific plate thicknesses ranging from 0.1λ to 0.7λ that this transformation is unitary.

6.14 *Wigner Functions.* Plot the 2D Wigner distributions corresponding to the 1D Hermite–Gaussian modes described by Eqn. (3.55) for $n = 0,2,7$. Confirm that the Wigner distributions are real and numerically confirm the orthogonality relationship given in Eqn. (6.146).

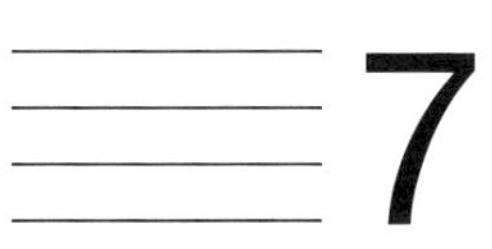

7

SAMPLING

> If a function $f(t)$ contains no frequencies higher than W cps, it is completely determined by giving its ordinates at a series of points spaced $1/2$ W seconds apart. This is a fact which is common knowledge in the communication art.
>
> —C. Shannon [219]

7.1 SAMPLES AND PIXELS

"Sampling" refers to both the process of drawing discrete measurements from a signal and the representation of a signal using discrete numbers. It is helpful in computational sensor design and analysis to articulate distinctions between the various roles of sampling:

- *Measurement sampling* refers to the generation of discrete digital values from physical systems. A measurement sample may consist, for example, of the current or voltage returned from a CCD pixel.
- *Analysis sampling* refers to the generation of an array of digital values describing a signal. An analysis sample may consist, for example, of an estimated wavelet coefficient for the object signal.
- *Display sampling* refers to the generation of discrete pixel values for display of the estimated image or spectrum.

Hard separations between sampling categories are difficult and unnecessary. For example, there is no magic transition point between measurement and analysis samples in the train of signal filtering, digitization, and readout. Similarly, analysis samples themselves are often presented in raw form as display samples. In the context of system analysis, however, one may easily distinguish measurement, analysis, and display.

Optical Imaging and Spectroscopy. By David J. Brady

A mathematical and/or physical process is associated with each type of sample. The *measurement* process implements a mapping from the continuous object signal $\mathbf{f}$ to discrete measurement data $\mathbf{g}$. In optical imaging systems this mapping is linear:

$$g_i = \int \mathbf{f}(\mathbf{x}) h_i(\mathbf{x})\, d\mathbf{x} \tag{7.1}$$

$\mathbf{g}$ consists of measurement samples. The *analysis* process transforms the measurements into data representative of the object signal. Specifically, the analysis problem is

Given $\mathbf{g}$, derive the set of values $\mathbf{f_a}$ that best represent $\mathbf{f}$.

The postdetection analysis samples $\mathbf{f_a}$ may correspond, for example, to estimates of the basis coefficients f_n from Eqn. (7.2), to estimates of coefficients on a different basis, or to parameters in a nonlinear representation algorithm. We may assume, for example, that $\mathbf{f}$ is well represented on a basis $\psi_n(\mathbf{x})$ such that

$$f(\mathbf{x}) = \sum_n f_n \psi_n(\mathbf{x}) \tag{7.2}$$

and Eqn. (7.1) reduces to

$$\mathbf{g} = \mathbf{Hf} \tag{7.3}$$

where $h_{ij} = \int h_i(\mathbf{x}) \psi_j(\mathbf{x})\, dx$. The vector $\mathbf{f}$ with elements f_n consists of analysis samples, and the analysis process consists of inversion of Eqn. (7.3). Algorithms for implementing this inversion are discussed in Chapter 8.

Display samples are associated with the processes of *signal interpolation* or *feature detection*. A display vector consists of a set of discrete values $\mathbf{f_d}$ that are assigned to each modulator in liquid crystal or digital micromirror array or that are assigned to points on a sheet of paper by a digital printer. Display samples may also consist of principal component weights or other feature signatures used in object recognition or biometric algorithms. While one may estimate the display values directly from $\mathbf{g}$, the direct approach is unattractive in systems where the display is not integrated with the sensor system. Typically, one uses a minimal set of analysis samples to represent a signal for data storage and transmission. One then uses interpolation algorithms to adapt and expand the analysis samples for diverse display and feature estimation systems.

Pinhole and coded aperture imaging systems provide a simple example of distinctions between measurement, analysis, and display. We saw in Eqn. (2.38) that $\mathbf{f_a}$ naturally corresponds to projection of $f(x, y)$ onto a first-order B-spline basis. Problem 3.12 considers estimation of display values of $f(x, y)$ from these projections using biorthogonal scaling functions. For the coded aperture system, measurement samples are described by Eqn. (2.39), analysis samples by Eqn. (2.38), and display samples by Eqn. (3.125). Of course, Eqn. (3.125) is simply an algorithm for estimating $f(x)$; actual plots or displays consist of a finite set of signal estimates. In subsequent discussion, we refer to both display samples and to Haar or impulse function elements as picture elements or *pixels* (voxels for 3D or image data cube elements).

We developed a general model of measurement, analysis, and display for coded aperture imaging in Chapter 2, but our subsequent discussions of wave and coherence imaging have not included complete models of continuous to discrete to display processes. This chapter begins to rectify this deficiency. In particular, we focus on the process of measurement sample acquisition. Chapter 8 focuses on analysis sample generation, especially with regard to codesign strategies for measurement and analysis. In view of our focus on image acquisition and measurement layer design, this text does not consider display sample generation or image exploitation.

Section 7.2 considers sampling in focal imaging systems. Focal imaging is particularly straightforward in that measurements are isomorphic to the image signal. Section 7.3 generalizes our sampling model to include focal spectral imaging. Section 7.4 describes interesting sampling features encountered in practical focal systems. The basic assumption underlying Sections 7.2–7.4, that local isomorphic sampling of object features is possible, is not necessarily valid in optical sensing. In view of this fact, Section 7.5 considers "generalized sampling." Generalized sampling forsakes even the attempt maintain measurement/signal isomorphism and uses deliberately anisomorphic sensing as a mechanism for improving imaging system performance metrics.

7.2 IMAGE PLANE SAMPLING ON ELECTRONIC DETECTOR ARRAYS

Image plane sampling is illustrated in Fig. 7.1. The object field $f(x, y)$ is blurred by an imaging system with shift-invariant PSF $h(x, y)$. The image is sampled on a 2D detector array. The detector pitch is Δ, and the extent of the focal plane in x and y is X and Y.

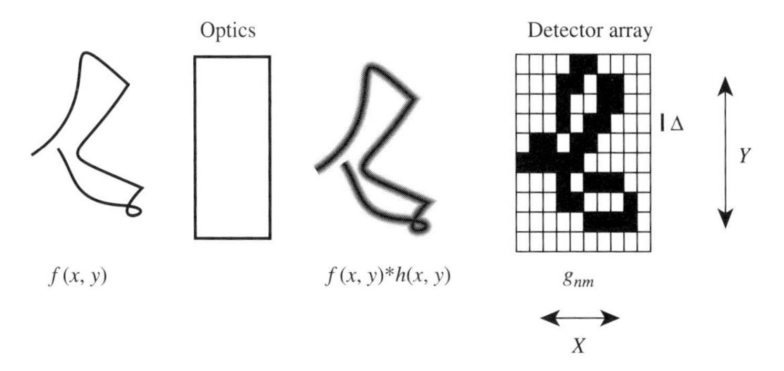

Figure 7.1 Image plane sampling.

The full transformation from the continuous image to a discrete two-dimensional dataset $\mathbf{g}$ is modeled as

$$g_{nm} = \int_{-\infty}^{\infty}\int_{-\infty}^{\infty}\int_{-X/2}^{X/2}\int_{-Y/2}^{Y/2} f(x, y)h(x' - x, y' - y) \times p(x' - n\Delta, y' - m\Delta)\, dx'\, dy'\, dx\, dy \quad (7.4)$$

where $p(x, y)$ is the pixel sampling function. For rectangular full fill factor pixels, for example, $p(x, y) = \mathrm{rect}(x/\Delta)\mathrm{rect}(y/\Delta)$.

Several assumptions are implicit in the sampling model of Eqn. (7.4). The object distribution, the optical PSF, and the pixel sampling function are in general all dependent on the optical wavelength λ. For simplicity, we assume in most of this section that the field is quasimonochromatic such that we can neglect the wavelength dependence of these functions. We also focus on irradiance imaging, meaning that $f(x, y)$ and $h(x, y)$ are nonnegative. We neglect, for the moment, the possibility of 3D object distributions. We also neglect complexity in the pixel sampling function, such as crosstalk, shading, and nonuniform response.

The function $\mathbf{g}$ consists of discrete samples of the continuous function

$$g(x'', y'') = \int_{-\infty}^{\infty}\int_{-\infty}^{\infty}\int_{-(X/2)}^{X/2}\int_{-(Y/2)}^{Y/2} f(x, y)h(x' - x, y' - y) \times p(x' - x'', y' - x'')\, dx'\, dy' dx\, dy \quad (7.5)$$

$g(x, y)$ is, in fact, a bandlimited function and can, according to the Whittaker–Shannon sampling theorem [Eqn. (3.92)], be reconstructed in continuous form from the discrete samples g_{nm}. To show that $g(x, y)$ is bandlimited, we note from the convolution theorem that the Fourier transform of $g(x, y)$ is

$$\hat{g}(u, v) = \hat{f}(u, v)\hat{h}(u, v)\hat{p}(u, v) \quad (7.6)$$

$\hat{h}(u, v)$ is the optical transfer function (OTF) of Section 6.4.1, and its magnitude is the optical modulation transfer function. In analogy with the OTF, we refer to $\hat{p}(u, v)$ as the *pixel transfer function* (PTF) and to the product $\hat{h}(u, v)\hat{p}(u, v)$ as the *system transfer function* (STF). Figure 7.2 illustrates the magnitude of the OTF [e.g., the *modulation transfer function*, (MTF)] for an object at infinity imaged through an aberration-free circular aperture and the PTF for a square pixel of size Δ.

One assumes in most cases that the object distribution $f(x, y)$ is not bandlimited. Since the pixel sampling function $p(x, y)$ is spatially compact, it also is not bandlimited. As discussed in Section 6.4.1, however, the optical transfer function $\hat{h}(u, v)$ for a well-focused quasimonochromatic planar incoherent imaging system is limited to a bandpass of radius $1/\lambda f/\#$. Lowpass filtering by the optical system means that

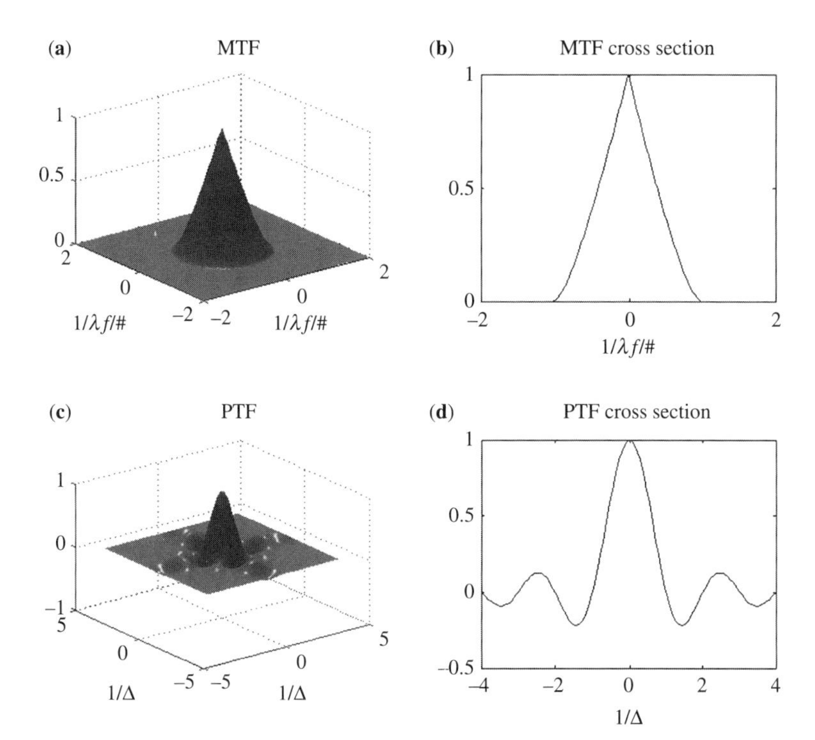

Figure 7.2 MTF for a monochromatic aberration-free circular aperture and PTF for a square pixel of size Δ: (a) MTF; (b) MTF cross section; (c) PTF; (d) PTF cross section.

aliasing is avoided for

$$\Delta \leq \frac{\lambda f/\#}{2} \tag{7.7}$$

The factor of 2 in Eqn. (7.7) arises because the width of the OTF is equal to the autocorrelation of the pupil function. We saw in Section 4.7 that a coherent imaging system could be characterized without aliasing with a sampling period equal to $\lambda f/\#$; the distinction between the two cases arises from the relative widths of the coherent and incoherent transfer functions as illustrated in Figs. 4.14 and 6.17.

The goals of the present section are to develop familiarity with discrete analysis of imaging systems, to consider the impact of the pixel pitch Δ and the pixel sampling function $p(x, y)$ on system performance, and to extend the utility of Fourier analysis tools to discrete systems. We begin by revisiting the sampling theorem. Using the Fourier convolution theorem, the double convolution in Eqn. (7.4) may be represented as

$$g_{nm} = \int_{-\infty}^{\infty}\int_{-\infty}^{\infty} e^{2\pi iun\Delta} e^{2\pi ivm\Delta} \hat{f}(u, v)\hat{h}(u, v)\hat{p}(u, v)\, du\, dv \tag{7.8}$$

The discrete Fourier transform of $\mathbf{g}$ is

$$\hat{g}_{n'm'} = \frac{1}{N^2} \sum_{-N/2+1}^{N/2} \sum_{-N/2+1}^{N/2} e^{i2\pi nn'/N} e^{i2\pi mm'/N} g_{nm}$$

$$= \int_{-\infty}^{\infty} \int_{-\infty}^{\infty} e^{-\pi i[(n'/N)-u\Delta]} e^{-\pi i[(m'/N)-v\Delta]}$$

$$\times \frac{\sin[\pi(n' - uX)]}{\sin\{\pi[(n'/N) - u\Delta]\}} \frac{\sin[\pi(m' - vY)]}{\sin\{\pi[(m'/N) - v\Delta]\}}$$

$$\times \hat{f}(u, v)\hat{h}(u, v)\hat{p}(u, v)\, du\, dv \tag{7.9}$$

Under the approximation that

$$\frac{\sin[\pi(n' - uX)]}{\sin\{\pi[(n'/N) - u\Delta]\}} \approx \sum_n (-1)^n N \operatorname{sinc}(uX - n' - nN) \tag{7.10}$$

We find that $\hat{g}_{n'm'}$ is a projection of $\hat{g}(u, v)$ onto the Shannon scaling function basis described in Section 3.8. Specifically

$$\hat{g}_{n'm'} \approx \sum_{n=-\infty}^{\infty} \sum_{m=-\infty}^{\infty} \hat{g}\left(\frac{n' + nN}{X}, \frac{m' + mN}{Y}\right) \tag{7.11}$$

Since $\hat{g}_{n'm'}$ is periodic in n' and m', $\hat{\mathbf{g}}$ tiles the discrete Fourier space with discretely sampled copies of $\hat{g}(u, v)$. We previously encountered this tiling in the context of the sampling theorem, as illustrated in Fig. 3.4.

Figure 7.3 is a revised copy of Fig. 3.4 to allow for the possibility of aliasing. $\hat{g}[(n' + nN)/X, (m' + mN)/Y]$ is a discretely sampled copy of $\hat{g}(u, v)$ centered

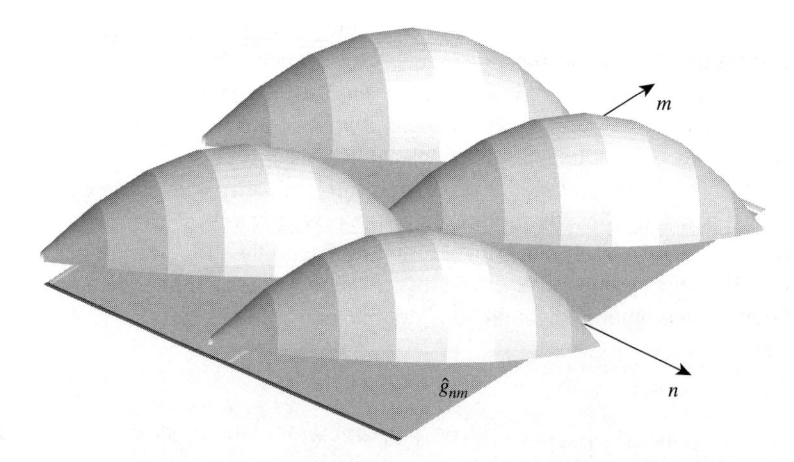

Figure 7.3 Periodic Fourier space of a sampled image.

on $n' = -nN$. The Fourier space separation between samples is $\delta u = 1/X$, and the separation between copies is $N/X = 1/\Delta$. The value of $\hat{g}_{n'm'}$ within a certain range is equal to the sum of the nonvanishing values of $\hat{g}[(n' + nN)/X, (m' + mN/Y)]$ with that range. If more than one copy of $\hat{g}$ is nonzero within any range of $n'm'$, then the measurements are said to be *aliased* and an undistorted estimation of $f(x, y)$ is generally impossible. Since the displacement between copies of $\hat{g}(u, v)$ is determined by the sampling period Δ, it is possible to avoid aliasing for bandlimited $\hat{g}$ by selecting sufficiently small Δ. Specifically, there is no aliasing if $|\hat{g}(u, v)|$ vanishes for $|u| > 1/2\Delta$. If aliasing is avoided, we find that

$$\hat{g}_{nm} = \hat{f}\left(\frac{n}{X}, \frac{m}{Y}\right)\hat{h}\left(\frac{n}{X}, \frac{m}{Y}\right)\hat{p}\left(\frac{n}{X}, \frac{m}{Y}\right) \tag{7.12}$$

Figures 7.4 and 7.5 illustrate the STF for $\Delta = 2\lambda f/\#$, $\Delta = 0.5\lambda f/\#$ and $\Delta = \lambda f/\#$. Figure 7.4 plots the STF in spatial frequency units $1/\Delta$. The plot is a cross section of the STF as a function of u. For rotationally symmetric systems, STF and MTF plots are typically plotted only the positive frequency axis. Relative to the critical limit for aliasing, the STF of Fig. 7.4(a) is undersampled by a factor of 4, and Fig. 7.4(b) is undersampled by a factor of 2. Figure 7.4(c) is sampled at the Nyquist frequency. The aliasing limit is $1/2\Delta$ in all of Fig. 7.4(a)–(c). The STF above this limit in (a) and (b) transfers aliased object features into the sample

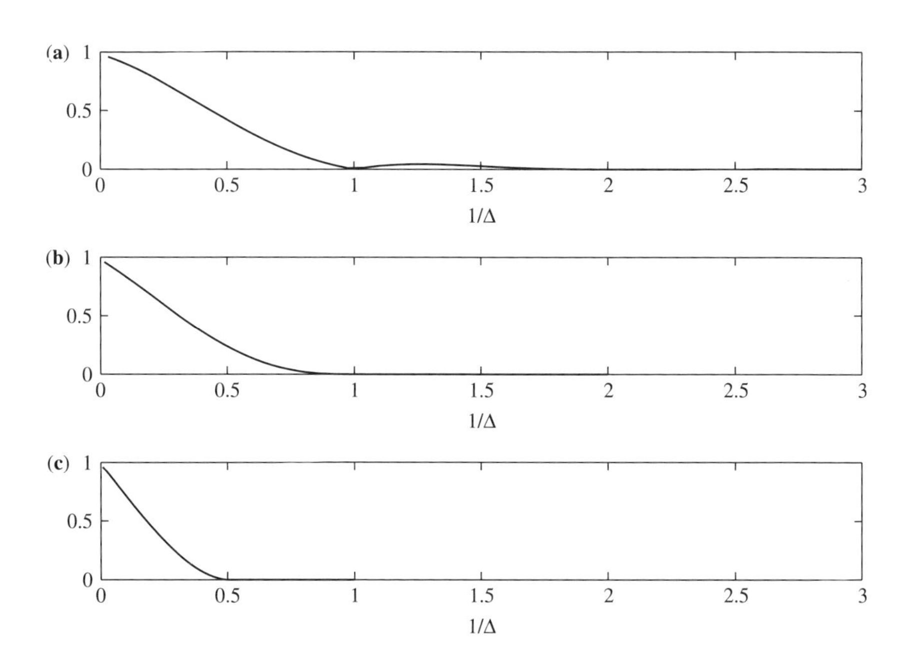

Figure 7.4 Imaging system STF for various values of Δ: (a) $\Delta = 2\lambda f/\#$; (b) $\Delta = \lambda f/\#$; (c) $\Delta = 0.5\lambda f/\#$.

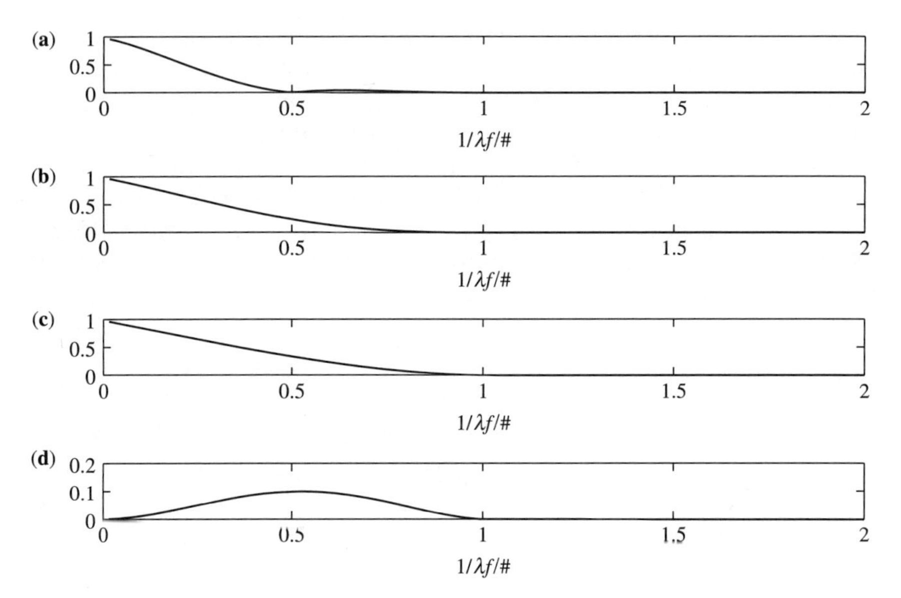

Figure 7.5 Imaging system STF for various values of Δ: (a) $\Delta = 2\lambda f/\#$; (b) $\Delta = \lambda f/\#$; (c) $\Delta = 0.5\lambda f/\#$; (d) = (c) – (b).

data. Because of the shape of the pixel and optical transfer functions, the aliased data are strongly attenuated relative to the low frequency passband.

Figure 7.5 plots the same STFs as in Fig. 7.4 in spatial frequency units of $1/\lambda f/\#$. In these units, the band edge for all three STFs is at $u = \pm 1/\lambda f/\#$. The strongly undersampled system of Fig. 7.5(a) aliases frequencies in the range $1/\lambda f/\# > |u| > 1/4\lambda f/\#$. The STF of Fig. 7.5(b) is aliased over the range $1/\lambda f/\# > |u| > 1/2\lambda f/\#$. The structure of the STF is similar for the critically sampled and the $2\times$ undersampled systems; Fig. 7.5(d) illustrates the difference between the two. The relatively modest increase in STF and the attenuated magnitude of the MTF may lead one to question whether the $4\times$ increase in focal plane pixels is worth the improvement in the system passband.

Various definitions of the system modulation transfer function have been applied to sampled imaging systems. Several studies favor an attenuated system MTF proposed by Parks et al. [196]. We describe Parks' MTF presently, but suggest that a simpler definition is more appropriate. We define the system MTF to be the magnitude of the STF defined above. We prefer this definition because it clearly separates the influence of measurement sampling from the influence of analysis or display sampling on the MTF. A simple example is helpful in clarifying these distinctions.

Consider the signal $f(x, y) = \frac{1}{2}\{1 + \cos[2\pi(u_o x + v_o y) + \phi]\}$. According to Eqn. (7.4), the measured samples for this signal are

$$g_{nm} = \hat{h}(0,0)\hat{p}(0,0) + |\hat{h}(u_o, v_o)||\hat{p}(u_o, v_o)| \times \cos(2\pi(u_o n\Delta + v_o m\Delta) + \phi + \phi_h + \phi_s) \tag{7.13}$$

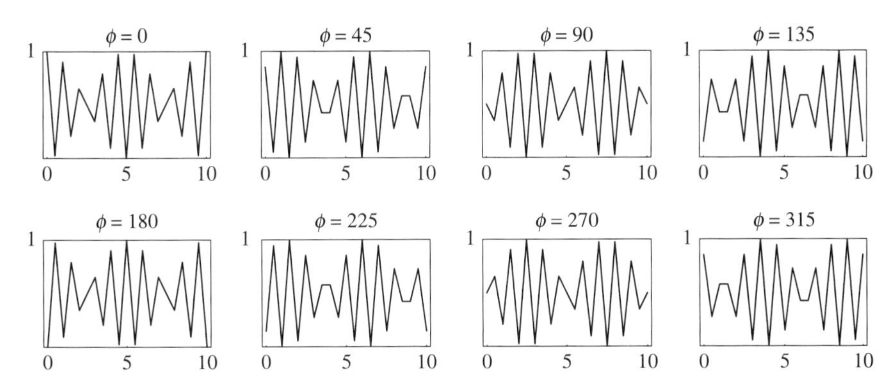

Figure 7.6 Sample data for a harmonic image at $u_0 = 0.9$ Nyquist (e.g., $u_0\Delta = 0.45$) for various relative phases between the image and the sampling grid. The horizontal axis is in units of the Nyquist period, 2Δ.

where ϕ_h and ϕ_s are the phases of the optical and pixel transfer functions at (u_0, v_0). Figure 7.6 illustrates this sample data for various relative phases between the input signal and the sampling lattice. As noted by Parks [196] and in many other studies, the fringe visibility (e.g., the ratio of the peak-to-valley difference to the average signal value) varies as a function of ϕ. The sample data are not periodic in n unless u_0 is rational. Under the assumption that the sample data is the display image, Parks accounts for the variation in the peak-to-valley signal by noting that the mean peak-to-valley range when ϕ is accounted as a uniformly distributed random variable is reduced by a factor $\text{sinc}(u_o\Delta)$ relative to the peak-to-valley range in the input. Parks accordingly suggests that a factor $\text{sinc}(u_o\Delta)$ should be added to the STF.

The addition of a $\text{sinc}(u_o\Delta)$ factor to the system MTF is erroneous because it neglects the interpolation step implicit in the sampling theorem. As noted in Eqn. (7.5), the sample data g_{nm} describe discrete samples of the bandlimited function $g(x, y)$ in direct accordance with the Shannon sampling theorem. Interpolation of $g(x, y)$ from the sample data using the sampling theorem [Eqn. (3.92)] produces the plots shown in Fig. 7.7. As expected, the reconstructed plots accurately reflect the input frequency and phase. Of course, there is some difficulty in producing these plots because the sampling theorem relies on an infinite sum of interpolation functions at each reconstruction point. Figure 7.7 approximates the sampling theorem by summing over the 100 sample points nearest the origin to generate an estimated signal spanning 20 sample cells. Lessons learned from this exercise include the following:

1. The sampling period Δ determines the aliasing limits for the signal but does not directly affect the MTF.
2. Separation and careful utilization of measurement, analysis, and display samples is important even for simple samples drawn from a Cartesian grid. As our measurements become more sophisticated, this separation becomes increasingly essential.

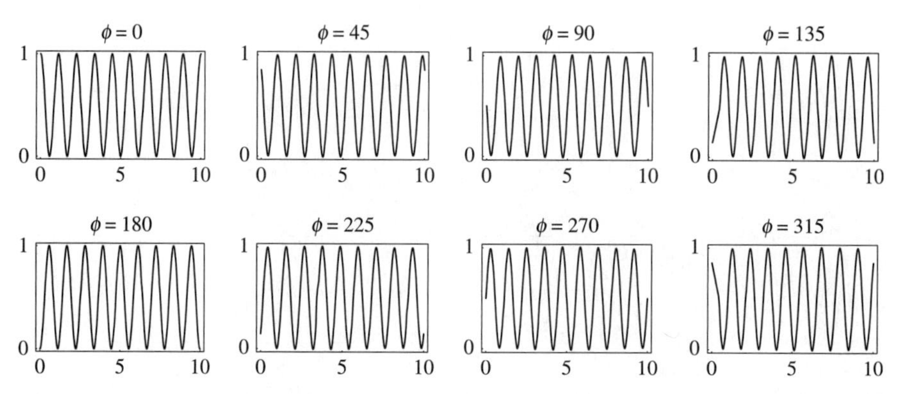

Figure 7.7 Signals reconstructed using Shannon interpolation from the sample data of Fig. 7.6. The frequency for all signals remains $u_o = 0.9$ Nyquist, and the horizontal axis is again in units of the Nyquist period.

3. Better display image samples are obtained by processing measurement samples. Optimal estimation of the continuous signal $f(x)$ at x depends on multiple measurement sample values in a neighborhood of x. Of course, this may mean that the resolution of the image display device should exceed the resolution of the sample data. Higher resolution on display is not uncommon, however, especially if one chooses to print the image.
4. Accurate signal interpolation from measurement samples is enabled by prior knowledge regarding the signal and the measurement system. In this case, we know that the signal is bandlimited by the optical MTF and thus expect the sampling theorem to apply. If the sampling theorem applies, Fig. 7.7 does more than smooth Fig. 7.6; it actually recovers the signal from sampling distortion.
5. It may be undesirable in terms of data and communication efficiency to transmit a "best" image estimate from the image capture device. It may be more useful to record compressed sample data along with known system priors for later image estimation.
6. The process of sampled image recording and estimation is deeper and more subtle than one might first suppose. Knowledge of the sampling period is sufficient to estimate $g(x, y)$, but the sampling theorem itself is computationally unattractive. One expects that a balance between computational complexity and signal fidelity could be achieved using wavelets with compact support rather than the Shannon scaling function. One may consider how sampling systems can then be designed to match interpolation functions.
7. All of these points rely on the existence of careful and accurate models for the physical measurement process.

The particular interpolation method used above, based on a finite support version of the Shannon generating function, is windowed sinc interpolation. The art of interpolation is itself a fascinating and important topic, most of which is unfortunately beyond the scope of this text. Interpolation in the presence of noise and balanced analysis of computational complexity and image fidelity are particularly important considerations. See Refs. 234 and 230 for relatively recent overviews. While the primary focus of this text is on image capture system design, over the next several chapters we regularly return to joint design of optical coding, sampling, and signal interpolation. For the present purposes, it is most important to understand that post-measurement interpolation is necessary, natural, and powerful.

While optical imaging systems are always bandlimited, undersampling relative to the optical Nyquist limit is common in digital imaging systems. One may well ask whether our analysis of Shannon interpolation applies in undersampled systems. The short answer is "Yes." Undersampling aliases signals into the reconstructed bandwidth but does not reduce the MTF. This is illustrated in Fig. 7.8, which shows sample data and the reconstructed signal for various phase shifts for a signal at $u_o = 1.5$ Nyquist. As expected, the signal reconstructs faithfully the aliased signal corresponding to the frequency -0.5 Nyquist.

While mindful of the fact that the ideal display image is interpolated from measurement data, we find direct visualization of the measured data sufficient in

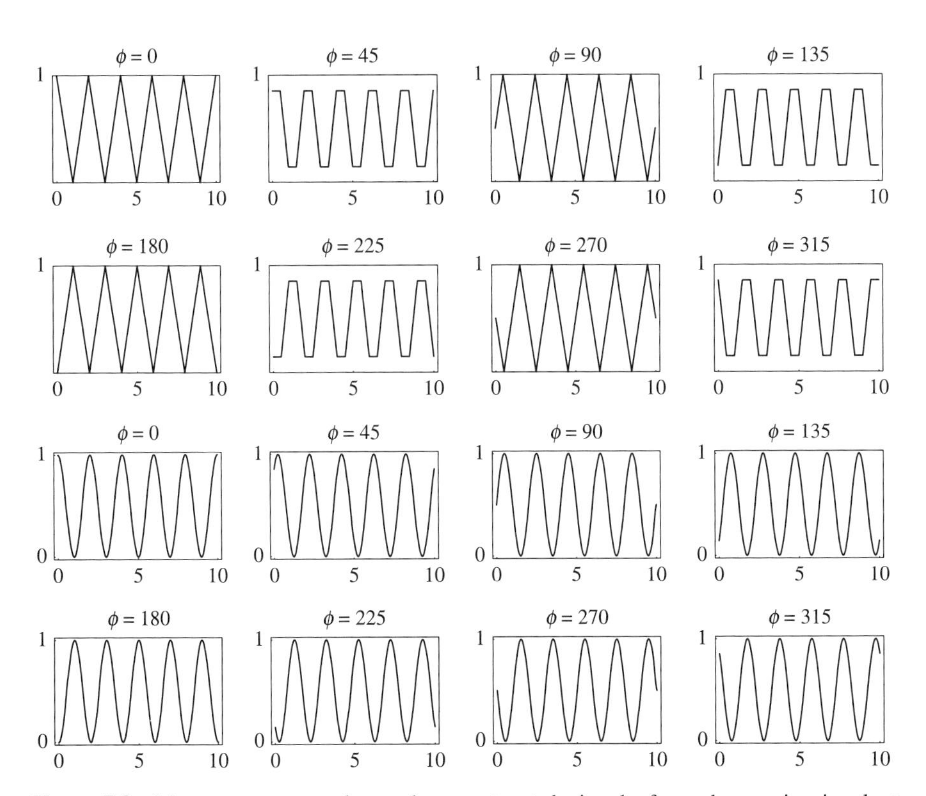

Figure 7.8 Measurement samples and reconstructed signals for a harmonic signal at $u_0 = 1.5$ Nyquist.

the remainder of this section. Figures 7.9 and 7.10 present simulated measurement data for various sampling rates. The object field is illustrated at top. The second row illustrates the image as lowpass filtered by a diffraction limited incoherent imaging system. The third row is a discrete representation sampled at $\Delta = (\lambda f/\#)/2$. The fourth row is "rect" sampled with 100% fill factor at

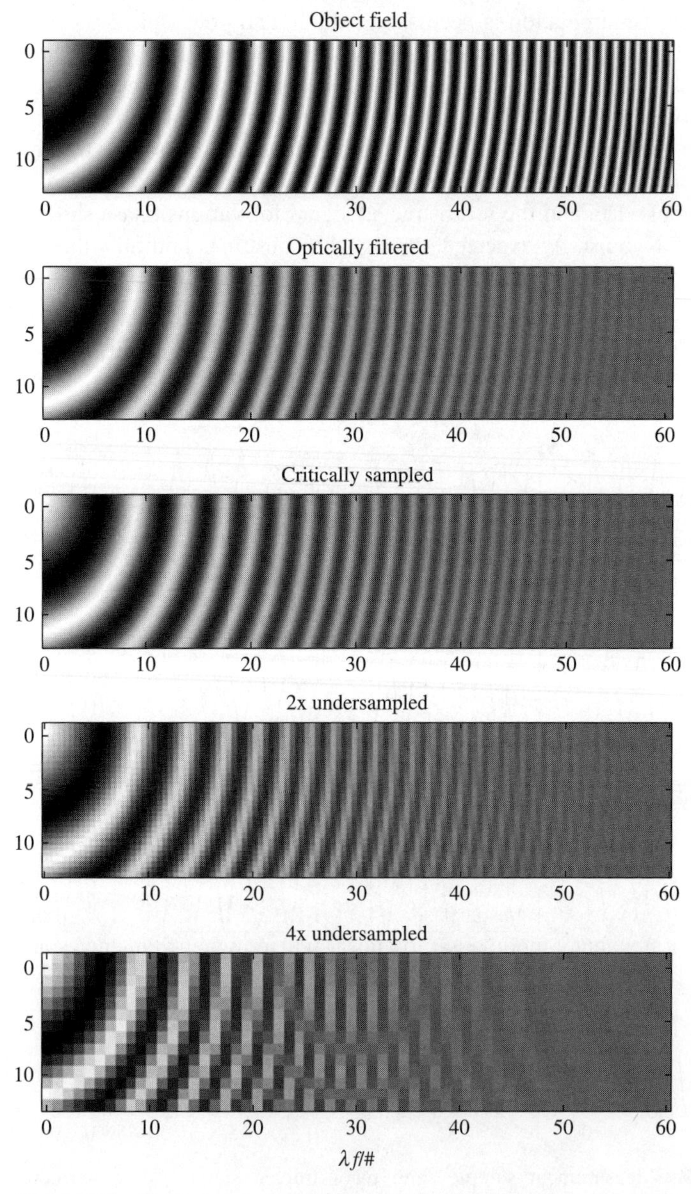

Figure 7.9 Optical and pixel filtering in incoherent imaging systems.

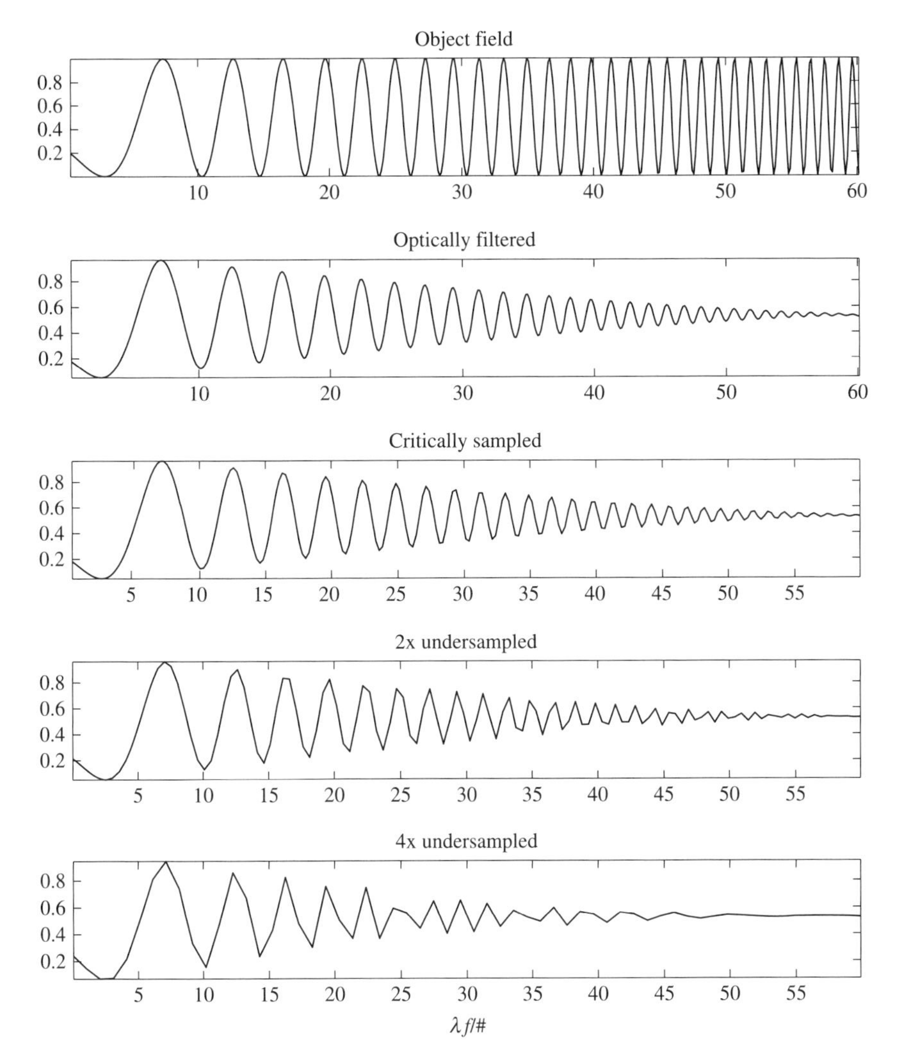

Figure 7.10 Cross sections of the images shown in Fig. 7.9.

$\Delta = \lambda f/\#$. The fifth row is $4\times$ undersampled. Aliasing in the fourth and fifth rows causes low frequency modulation to appear where higher frequencies are present in the object. While these figures clearly illustrate aliasing and loss of contrast as the sampling rate falls with fixed field of view and magnification, selection of a sampling rate for practical systems is not as simple as setting $\Delta = (\lambda f/\#)/2$. In fact, most current digital imaging systems deliberately select $\Delta > (\lambda f/\#)/2$.

Ambiguity and *noise* are the primary factors limiting the fidelity with which one can estimate the true object from measurement samples. Ambiguity takes many forms and becomes increasingly subtle as we move away from isomorphic imaging measurements, but for the present purposes aliasing is the most obvious form of ambiguity. Barring some prior knowledge regarding the spectral structure of an image, one cannot disambiguate whether low frequencies in the fifth row of Fig. 7.9 come from actual object features or from aliased high-frequency features.

However, aliasing is by no means the only or even the primary barrier to high-fidelity object estimation. One may often choose to tolerate some aliasing in favor of other system parameters, particularly given that the optical transfer function is itself highly attenuated in the aliasing range.

We begin exploring the impact of noise by estimating the strength of the signal as a function of sampling period. Figure 7.11 shows an imaging system mapping an object at range R onto an image. An object patch of size $d \times d$ is mapped onto a focal plane pixel, which we assume to be of symmetric size $\Delta \times \Delta$. This means that the magnification of the system is $\Delta/d = F/R$. In remote sensing systems d is called the *ground sample distance* [75]. Letting P represent the mean object irradiance, the mean signal power from the ground sample patch passing through the entrance aperture A is Pd^2A^2/R^2. The mean signal measured from a single focal plane pixel is thus

$$\langle g_{nm} \rangle = P\tau \frac{d^2A^2}{R^2} = P\tau \frac{\Delta^2}{(f/\#)^2} \tag{7.14}$$

where τ is the pixel integration time. This simple result indicates a tension between high signal values obtained via a large pixel pitch and high resolution obtained via a small pixel pitch. No such tension is evident for the $f/\#$, however, where small values are universally desirable. Small $f/\#$ is associated with aberration and spatial variance in the optical MTF, however, so optical design will ultimately need to balance optical processing (to achieve small $f/\#$), computational processing, and signal values.

For a photon noise–dominated system the signal of Eqn. (7.11) corresponds to a mean pixel SNR of

$$\text{SNR} = \frac{\langle g \rangle}{\sigma_g} = \sqrt{\frac{\eta P \tau}{h\nu}} \frac{\Delta}{f/\#} \tag{7.15}$$

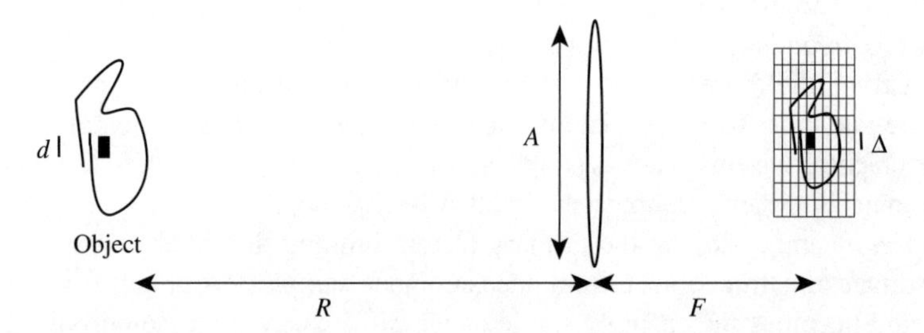

Figure 7.11 Imaging geometry showing ground sampling distance.

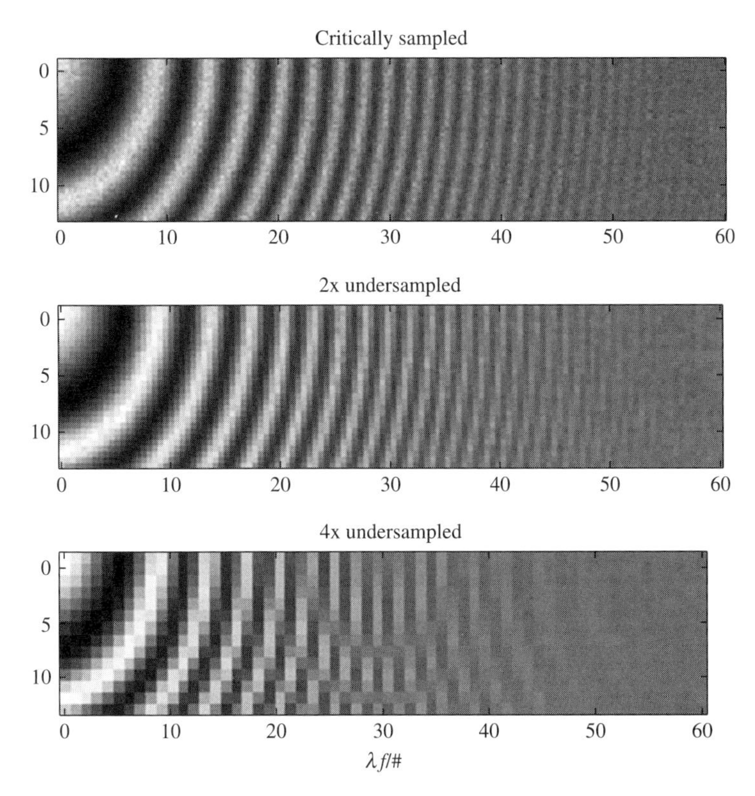

Figure 7.12 The sampled images of Fig. 7.9 with Poisson noise added under the assumption that at critical sampling $\langle g_{nm} \rangle = 200$ electrons.

In some cases τ may be a function of Δ; for example, if the total focal plane readout bandwidth remains constant then τ is proportional to Δ^2. Most often, τ is determined by the frame rate needed to avoid motion blur in the scene. We assume that each pixel integrates for essentially the full frame period. This is not possible for all types of focal planes, and again the effective integration time may depend on the pixel pitch and density.

A thermal noise dominated system yields

$$\mathrm{SNR} = D^* P \sqrt{\tau} \frac{\Delta}{(f/\#)^2} \tag{7.16}$$

The effect of Poisson noise on the sampled data is illustrated in Fig. 7.12, which shows the sampled images of Fig. 7.9 with simulated photon noise added. The contrast and uniformity of the subsampled images are better than for the critically sampled data, indicating that for low photon fluxes better image metrics may be obtained with larger pixel pitch. Of course, for Poisson noise one could achieve the advantage of a larger pitch adaptively by binning adjacent pixels or using more advanced forms of computational filtering. A larger pixel pitch offers diverse benefits to the focal plane array designer, however.

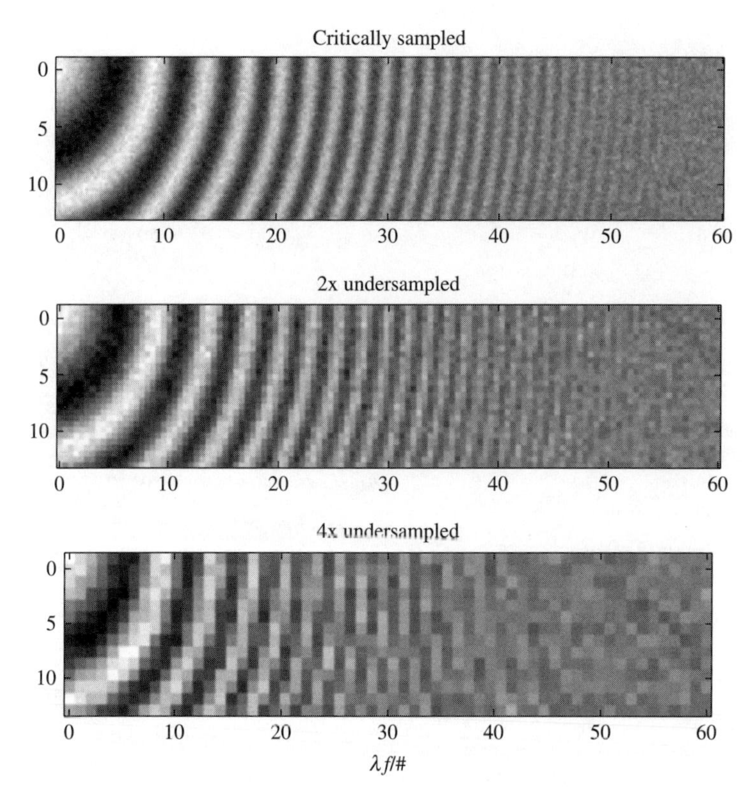

Figure 7.13 The sampled images of Fig. 7.9 with Johnson noise added with a variance of 5% of the mean signal.

The potential advantage of a larger pixel pitch is less clear for Johnson noise, as illustrated in Fig. 7.13. While the SNR scaling as a function of Δ is the same for the two cases, Johnson noise is uncorrelated with the image structure and tends to wash out high-contrast high-frequency features. While Poisson noise is related to the signal value and Johnson noise is related to the detector area, multiple noise sources may be considered that are not correlated to the signal value or to Δ. For noise sources independent of Δ, such as read or $1/f$ noise, there may be little cost in going to finer pixel pitches.

7.3 COLOR IMAGING

As discussed in Chapter 6, complete description of the optical field requires consideration of multidimensional coherence and spectral density functions. The power spectral density distributed across a 2D image plane, $S(x, y, \lambda)$, is the simplest example of a multidimensional optical distribution. A 3D image of the power spectral density is commonly termed an *optical data cube*. The optical data cube consists of a stack of 2D images with each image corresponding to an object distribution at a particular wavelength.

This section generalizes the sampling model of Eqn. (7.4) to spectral images. Our initial goal is to understand how to generate measurement samples. The problem we confront is that, while the spectral data cube is distributed over 3D, available detectors sample distributions over 2D detector arrays. The most common strategies for overcoming this problem are as follows:

1. Interlacing diverse spectral projections of the data cube across a 2D array. This approach produces the measurement model

$$g_{nm} = \iiiint\!\!\int f(x, y, \lambda) h(x' - x, y' - y, \lambda) t_{nm}(\lambda) \times p(x' - n\Delta, y' - m\Delta)\, dx'\, dy' dx\, dy\, d\lambda \tag{7.17}$$

   The spectral sampling function $t_{nm}(\lambda)$ is encoded at different pixels by placing microfilters over each physical pixel or by application of coding masks.

2. Temporally modulating the spectral structure of the pixel response. This approach produces the measurement model

$$g_{nmk} = \iiiint\!\!\int f(x, y, \lambda) h(x' - x, y' - y, \lambda) \times p(x' - n\Delta, y' - m\Delta, \lambda) t_k(\lambda)\, dx'\, dy' dx\, dy\, d\lambda \tag{7.18}$$

   where $t_k(\lambda)$ is a spectral filter and k indexes measurement time. Temporal modulation may be implemented interferometrically or by using electrically tunable spectral filters.

3. Temporally modulating the field of view. For example, the "pushbroom" strategy isolates a single spatial column of the object data cube by imaging on a slit and then disperses the slit spectrally as described in Section 9.2. The measurement model for this approach is

$$g_{nmk} = \iiiint\!\!\int f(x - k\Delta, y, x + \alpha\lambda) h(x' - x, y' - y, \lambda) \times p(x' - n\Delta, y' - k\Delta, \lambda) t(x - k\Delta)\, dx'\, dy' dx\, dy\, d\lambda \tag{7.19}$$

   Temporal modulation and pushbroom sampling each measure a plane the 3D data cube in timesteps indexed by k. A narrow pass temporally varying filter measures the color plane $f(x, y, \lambda = k\Delta\lambda)$ in the kth timestep. The pushbroom strategy measures the spatio-spectral plane $f(x = k\Delta, y, \lambda)$ in the kth timestep.

4. Multichannel spatial sampling separates the optical signal into multiple detector arrays and measures multiple images in parallel. For example, a prism assembly can separate red, green, and blue channels for parallel image capture.

There are advantages and disadvantages to each spectral image sampling strategy. The interlaced sampling approach is relatively simple and produces compact imaging systems but gives up nominal spatial resolution. The temporal modulation strategy gives up temporal resolution on dynamic scenes. Spatially parallel sampling requires multiple focal planes, relatively complex optics, and advanced algorithms for multiplane registration and data cube integration.

The most common approach to spectral imaging is an interlaced imaging strategy using three different microfilters over a 2D detector array to produce the familiar RGB (red–green–blue) spectral images of "color cameras." This strategy is based on the Bayer filter, consisting of a 2D array of absorption filters. Exactly one such filter is placed over each sensor pixel. Denoting the red filter as R, the green filter as G, and the blue filter as B, the Bayer filter assigns microfilters to pixels in the pattern [16]

$$\begin{array}{cccccccc} \ddots & \vdots & \vdots & \vdots & \vdots & \vdots & & \\ \cdots & \mathrm{R} & \mathrm{G} & \mathrm{R} & \mathrm{G} & \mathrm{R} & \mathrm{G} & \cdots \\ \cdots & \mathrm{G} & \mathrm{B} & \mathrm{G} & \mathrm{B} & \mathrm{G} & \mathrm{B} & \cdots \\ \cdots & \mathrm{R} & \mathrm{G} & \mathrm{R} & \mathrm{G} & \mathrm{R} & \mathrm{G} & \cdots \\ \cdots & \mathrm{G} & \mathrm{B} & \mathrm{G} & \mathrm{B} & \mathrm{G} & \mathrm{B} & \cdots \\ \cdots & \mathrm{R} & \mathrm{G} & \mathrm{R} & \mathrm{G} & \mathrm{R} & \mathrm{G} & \cdots \\ \cdots & \mathrm{G} & \mathrm{B} & \mathrm{G} & \mathrm{B} & \mathrm{G} & \mathrm{B} & \cdots \\ & \vdots & \vdots & \vdots & \vdots & \vdots & \vdots & \ddots \end{array} \tag{7.20}$$

Motivated by the peak in the human visual response to green, the Bayer filter collects twice as many green samples as red and blue. Many details related to human vision may be considered in the measurement and display of color images.

The Bayer system collects the three color planes:

$$g_{nm}^{r} = \iiiint\!\!\int f(x, y, \lambda) h(x' - x, y' - y, \lambda) \times t_r(\lambda) s(x' - 2n\Delta, y' - 2m\Delta)\, dx'\, dy' dx\, dy\, d\lambda \tag{7.21}$$

$$g_{nm}^{g} = \iiiint\!\!\int f(x, y, \lambda) h(x' - x, y' - y, \lambda) \times t_g(\lambda) s(x' - n\Delta - m\Delta, y' - n\Delta + m\Delta)\, dx'\, dy'\, dx\, dy\, d\lambda \tag{7.22}$$

$$g_{nm}^{b} = \iiiint\!\!\int f(x, y, \lambda) h(x' - x, y' - y, \lambda) \times t_b(\lambda) s(x' - 2n\Delta, y' - 2m\Delta)\, dx'\, dy'\, dx\, dy\, d\lambda \tag{7.23}$$

$t_r(\lambda)$ is the filter transmittance as a function of wavelength for the red spectral channel; similar filters are assumed for the green and blue channels.

The effect of interlaced sampling can be visualized using the lattices of Fig. 7.14. The sampling lattices are the spatial maps of the measurement points. The lattice vectors define unit steps from one sampling point to the next. For a monochrome

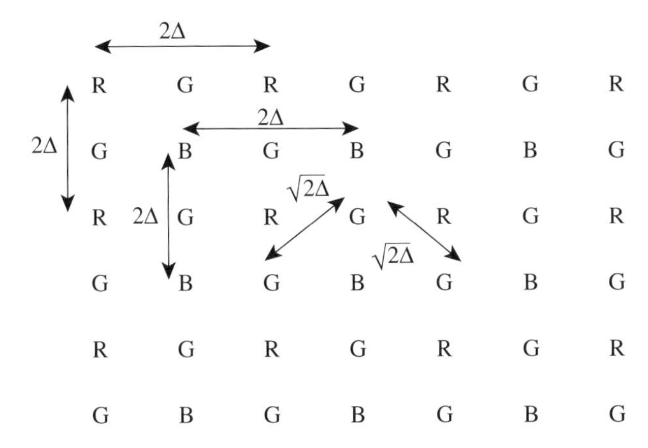

Figure 7.14 Sampling lattices for interlaced color imaging.

imager, the lattice vectors are $\Delta\mathbf{i}_x$ and $\Delta\mathbf{i}_y$. For the Bayer filter, the red and blue sampling lattice vectors are $2\Delta\mathbf{i}_x$ and $2\Delta\mathbf{i}_y$. The sampling lattice vectors for the green image are $\Delta\mathbf{i}_x + \Delta\mathbf{i}_y$ and $\Delta\mathbf{i}_x - \Delta\mathbf{i}_y$.

The STF for the red, green, and blue images is $\hat{h}(u, v)\hat{s}(u, v)$, the same as for a monochromatic imager, and the discrete Fourier transforms of Eqns. (7.21), (7.22), and (7.23) produce periodically replicated copies of the image spectra, as in Eqn. (7.11). The Fourier space unit cells for the red, green, and blue images differ as a result of the different sampling periods.

As illustrated in Fig. 7.15, reduced sampling rates for the color planes shrink the aliasing (Nyquist) frequency limit. The limit for the red and blue channels is reduced by a factor of 2. The aliasing window for the green data plane is reduced by a factor of

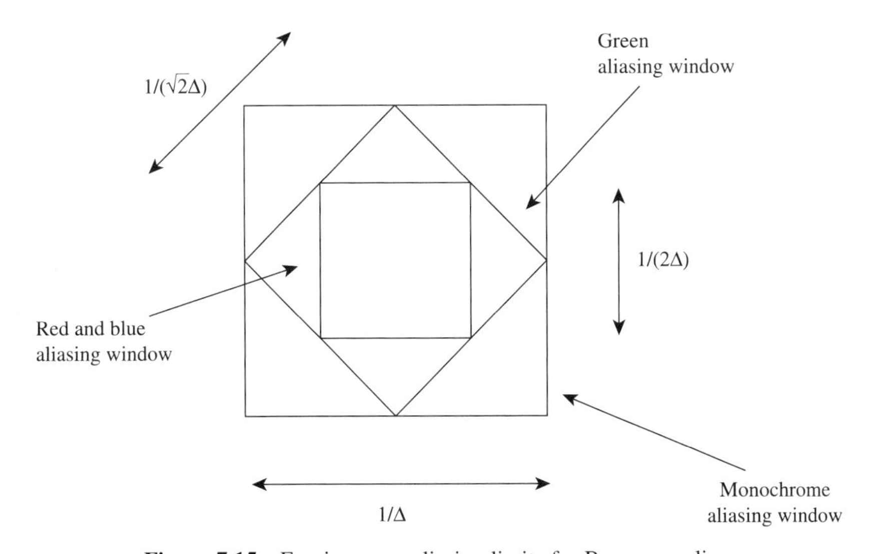

Figure 7.15 Fourier space aliasing limits for Bayer sampling.

$\sqrt{2}$ and rotated by 45°. Along the u and v axes the aliasing limit is the same for the green and monochrome images, while along the $u = v$ diagonal the aliasing boundary for the green is half the monochrome limit.

At the simplest level, image interpolation for the interlaced color capture system proceeds in the same way as for the monochrome system. The uniformly sampled data in each color plane are used to interpolate a finer display sample grid for that color plane, and then the RGB planes are stacked to produce a color image. Ideally, the interpolated signal resolution and structure are chosen to match the display device.

Since the aliasing limit on the reciprocal lattice is reduced for interlaced color imaging, the effective spatial resolution also decreases. One wonders, however, if this reduction is justified. In independently reconstructing each color plane, one implicitly assumes that each color plane is independent. We know empirically, however, that this is not true. If the planes were independent, then the image in the red channel would not tend to resemble the image in the blue channel. In practice, red, green, and blue planes in an RGB image tend to look extremely similar. If we can utilize the interdependence of the color planes, perhaps we can obtain spectral images without loss of spatial resolution. To explore this possibility, however, we need to generalize our sampling models (Section 7.5) and develop nonlinear signal estimation methods (Section 8.5).

While we had to work through a simple challenge to sample the 3D data cube on a 2D plane, we have been able with interlaced measurement to maintain isomorphic sampling in the sense that each measurement g_{nm} corresponds to a sample point in the 3D spatio-spectral data cube. Such isomorphisms are not always possible, however, and even when possible, they are not always a good idea. For example, if we choose the filters $t_i(\lambda)$ not as simple RGB filters but with some more complex multichannel response, we can increase the light sensitivity and perhaps the spatial resolution of the imager.

7.4 PRACTICAL SAMPLING MODELS

In practical cameras, transduction of an optical signal consisting of the irradiance or power spectral density incident on a focal plane [$f(\mathbf{x})$ in Eqn. (7.1)] to a discrete array of analytical samples occurs through a sequence of embedded sampling and processing steps. The details of these steps differ based on the nature of the focal plane and the application. As an example, typical consumer visible light imaging systems apply the following steps in readout circuitry:

1. *Sensor Data Readout.*
2. *Black-Level Correction and Digital Gain.* Black-level correction subtracts the mean dark current from the readout prior to amplification.
3. *Bad Pixel Correction.* A small percentage of pixels, especially in microbolometer and active pixel cameras, are defective because of manufacturing errors. These pixels are identified and data from them are dropped for the life of the focal plane. Rather than return a null for the bad pixel data, the readout circuit interpolates a value from neighboring pixels.

4. *Green–Red/Green–Blue Compensation.* The 3D structure of color filters and microlenses induces crosstalk between pixel values. A difference in the sensitivity of green pixels horizontally adjacent to red and green pixels horizontally adjacent to blue is one aspect of this crosstalk. Statistical properties of the crosstalk are calibrated and used to estimate true signal values.
5. *White Balance.* The filter response and detector quantum efficiency differ for red, green, and blue channels. The known response is used to adjust estimated signal values.
6. *Nonuniformity Correction.* Pixel responses are nonuniform across arrays because of slight variations in manufacturing parameters. Since fixed pattern noise may vary as a function of temperature and other environmental conditions, and since accurate characterization is challenging, it is difficult to completely eliminate this effect. However, it can be roughly characterized and is corrected on readout.
7. *Antishading.* As discussed below, spatially varying gain is applied to correct radial variations in the image brightness due to the microlens array and aberrations.
8. *Denoising.* Processing to this point reflects corrections due to known system characteristics. Denoising uses strategies discussed in Section 8.5 to remove image features due to random fluctuations or uncharacterized system variations.
9. *RGB Interpolation.* Simple algorithms interpolate the RGB image from the Bayer sampled measurements. More sophisticated approaches may apply nonlinear inference strategies discussed in Section 10.6.
10. *Image Enhancement.* Digital contrast and brightness correction, high- or lowpass filtering, deblurring and other high-level enhancements may be applied to image data prior to final readout.

As one might expect from this complex array of processing steps, the final digital signals may not be quite linear in the irradiance or "radiometrically calibrated." As discussed in Section 5.7, pixel-level analog–digital signal transduction remains under active development. We will generally neglect the full complexity of signal transduction in the remainder of this text, but given our focus on optical system design and analysis, it is worth commenting here on the impact of optical prefilters on the sampled signal.

Microlens arrays are used to increase the effective fill factor of focal plane arrays. Figure 5.12 shows cylindrical lenslets consistent with interline CCD readout; in general, microlens arrays will be patterned in two dimensions to match photodiode distributions on active pixel arrays. The geometry of a microlens array is illustrated in Fig. 7.16. The microlenses are affixed to a focal plane array with photodiode cross section δ. The microlens pitch matches the pixel pitch Δ. The microlens focal length is F. The goal of the microlenses is to collect as much light as possible onto the photodiodes. Of course, we know from the constant radiance theorem generally and from Eqn. (6.53) specifically that there are limits on how much a lens can

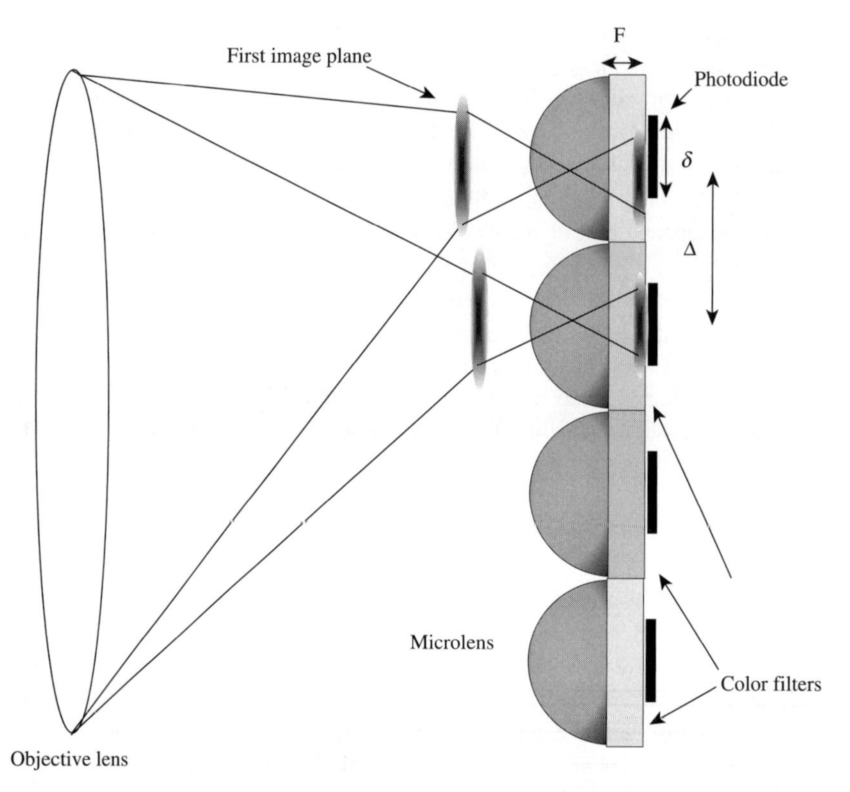

Figure 7.16 Shading due to microlenses. Field curvature at the focal plane causes microlens focal spots to shift in scale and position across the field, leading to variation in quantum efficiency.

do to localize partially coherent light. If the image striking the microlens array is incoherent, then the microlenses cannot increase the radiance on the photodiodes.

We saw in Eqn. (6.30) that the coherence cross section of an incoherent image is approximately $\lambda f/\#$. According to Eqn. (6.53), the spectral density on a photodiode at the focal plane of the microlens is the Fourier transform of the image cross-spectral density. With reference to Eqn. (6.30), we model the cross spectral density on the microlens aperture as

$$W(\Delta x, \Delta y, \nu) = S(\nu)\mathrm{jinc}\left(\frac{\sqrt{\Delta x^2 + \Delta y^2}}{\lambda f/\#}\right) \tag{7.24}$$

The Fourier transform produces a focal spot with cross section $F/f/\#$. This cross section is less than photodiode cross section if the

$$f/\# \geq \frac{F}{\delta} \tag{7.25}$$

If we set the microlens focal length to the pixel pitch, efficient light collection limits the system $f/\#$ to Δ/δ, for example, 1 divided by the square root of the focal plane fill factor without the microlenses. A fill factor of 50%, for example, would require at least $f/1.4$ optics to take full advantage of the microlenses. It is not uncommon for this effect to limit practical cameras to $f/2$ optics.

On the other hand, one can imagine sub-$f/1$ microlenses achieving modest improvements in effective fill factor even for low-$f/\#$ objective optics. An additional issue with the microlenses, however, leads to more trouble if the lenses are effective in concentrating the irradiance. Microlenses, along with the 3D structure of the color filters and the detector array, also affect the image measurement model through shading. Typical imaging systems produce field curvature across the objective lens focal plane, as illustrated on an absurd scale in Fig. 7.16. Light at the edge of the field may undergo phase modulation that makes it less coherent (increasing the size of the microlens focal spot) and that shifts the position of its focal spot on the photodiodes. We illustrate this effect in Fig. 7.16 by showing the light distribution from the objective focal spot shifted relative to the photodiode when relayed by the microlens at the edge of the field. Shading reduces the effective fill factor and corresponding quantum efficiency toward the edges of the field. The net effect may be modeled by a shift variant pixel mask t_{nm} such that

$$g_{nm} = t_{nm} \int_{-\infty}^{\infty} \int_{-\infty}^{\infty} \int_{-(X/2)}^{X/2} \int_{-(Y/2)}^{Y/2} f(x, y) h(x' - x, y' - y) \times s(x' - n\Delta, y' - m\Delta)\, dx'\, dy'\, dx\, dy \tag{7.26}$$

Figure 7.17 Effect of shading on measured data. The image in (a) is "true," and in (b) the image shows shading simulated by a Gaussian pixel mask.

Figure 7.17 shows shading in simulated data with a pixel mask $t_{nm} = \exp(-n^2/\sigma^2)\exp(-m^2/\sigma^2)$. Assuming that the affect is well characterized, shading may be ameliorated in pixel nonuniformity correction. One may also consider objective lens design and microlens or focal plane layouts to reduce this effect.

The 3D structure of the focal plane optics and optoelectronics introduce additional effects, known broadly as *pixel crosstalk.* Scattering at the microlens and color filter junctions and on the focal plane, as well as charge diffusion between pixels, causes crosstalk. These effects may be modeled in the pixel sampling function $s(x, y)$, which may be space-variant across the array and may overlap between color channels.

7.5 GENERALIZED SAMPLING

This section considers in more detail the spaces and generalized sampling strategies introduced in Section 3.9. Our goal is to understand the philosophy of modern measurement design as applied to optical systems and to explore challenges and opportunities in generalized sampling. The term "generalized sampling" dates to Papoulis' work showing that one can reconstruct a bandlimited function from sub-Nyquist rate samples over multiple nonredundant narrowband channels [194]. As discussed in Section 10.4, Papoulis' result is the basis for digital superresolution in multiple aperture imaging systems.

A discussion of modern generalized sampling begins with concepts of dimensionality. An optical field is a distribution $f(\mathbf{x})$ over D dimensional space, where $\mathbf{x} \in \mathbb{R}^D$ and f assumes a scalar value in $\mathbb{R}$ or $\mathbb{C}$ at each value of $\mathbf{x}$. Example values of D are 1 for a spectrum, 2 for a monochrome image, 3 for a color image, and 5 for the cross-spectral density on an aperture. We refer to $f(\mathbf{x})$ as the object field (or simply the object) in this text and to $\mathbb{R}^D$ as the object embedding space. A digital image of the object is a discrete representation of $f(\mathbf{x})$, $\mathbf{f} \in \mathbb{R}^N$ or $\mathbf{f} \in \mathbb{C}^N$. N is the number of data values in the digital image.

Mechanisms by which the continuous distribution $f(\mathbf{x})$ is mapped onto a discrete list were introduced in Chapter 3. Typically, $\mathbf{f}$ is an *ordered list* with coefficients $\mathbf{f}_\mathbf{n}$ where $\mathbf{n} \in \mathbb{N}^D$ assumes N distinct values. In the case of Shannon or wavelet sampling, the order parameter $\mathbf{n}$ corresponds to a spatial position in $\mathbb{R}^D$ and, potentially, wavelet order. For analytic samples, $\mathbf{f}$ may alternatively describe the representation of $f(\mathbf{x})$ in special functions, such as the Hermite–Gaussians, with an index parameter corresponding to order alone.

In the remainder of this section, we focus on ordered representations of $\mathbf{f}$. Although we recognize that one may choose to interpolate this data for final display, we refer to the components of $\mathbf{f}$ in this representation as "pixels." For an ordinary set of data values the order may not matter, but in an image one assumes that adjacent pixels are somehow related. For a 1D signal, the coefficients of $\mathbf{f}$, f_n, are indexed by a scalar parameter n with the assumption that f_n and $f_{n\pm m}$ are statistically dependent for m less than some coherence length.

7.5.1 Sampling Strategies and Spaces

Repeating Eqn. (7.3), optical sampling strategies satisfy the measurement model

$$\mathbf{g} = \mathbf{H}\mathbf{f} \tag{7.27}$$

where $\mathbf{g} \in \mathbb{R}^M$ and $\mathbf{H}$ is an $M \times N$ matrix. The measurement data $\mathbf{g}$ may also be an ordered list, but the dimension of the measurement order parameter may be different from that of the display order parameter. The measurements do not necessarily have a natural display space, but in the following discussion we assume that the coefficients of $\mathbf{g}$ are $g_{\mathbf{n_g}}$ with $\mathbf{n_g} \in \mathbb{N}^{D_g}$ such that $\mathbf{n_g}$ assumes assumes M distinct values. D_g is the "embedding dimension" for the measurements.

Sampling strategies may be categorized in terms of the relationship between the object and measurement embedding spaces as follows:

- *Conventional Measurement*: $D_g = D$. This case covers the traditional understanding of Shannon sampling (e.g., the actual signal value of f is measured at fixed points in $\mathbb{R}^D$). It also covers Shannon sampling on linear transformations of $f(\mathbf{x})$, such as the Fourier or Radon transform. The most common conventional sampling model is

$$g_{\mathbf{n}} = \int f(\mathbf{x}) h(\mathbf{x} - \Delta_{\mathbf{n}})\, d\mathbf{x} \tag{7.28}$$

- *Dimension Increasing Measurement*: $D_g > D$. This case corresponds to Papoulis' multiband sampling. As an example, imagine a system sampling wavelet coefficients of $\mathbf{f}$. Measurements in such a system are indexed by position in the object embedding space as well as wavelet level. The dimension of the measurement embedding space is twice the dimension of the object embedding space. Multiple aperture imaging, as discussed in Section 10.4, is a dimension increasing system. A typical sampling model is

$$g_{\mathbf{nm}} = \int f(\mathbf{x}) h_{\mathbf{m}}(\mathbf{x} - \Delta_{\mathbf{n}})\, d\mathbf{x} \tag{7.29}$$

- *Dimension Reducing Measurement*: $D_g < D$. The Bayer sampling strategy of Section 7.3 is an example of this case; the 3D optical data cube is projected onto an interlaced 2D array of measurements. A typical sampling model is

$$g_{\mathbf{n}} = \int f(\mathbf{x}) h(\mathbf{P}\mathbf{x} - \Delta_n, (\mathbf{I} - \mathbf{P})\mathbf{x})\, d\mathbf{x} \tag{7.30}$$

  where $\mathbf{n} \in \mathbb{N}^{D_g}$. $\mathbf{P}$ is a projection operator selecting a subspace of $\mathbb{R}^D$. The sampling kernel is shift-invariant on this subspace, but shift-variant in the complementary subspace corresponding to $\mathbf{I}-\mathbf{P}$. For example, Bayer sampling uses regularly spaced samples in each color plane, but integrates shift-variant projections along the color axis. We discuss more advanced spectral projection and image inference strategies for dimension reducing measurements in Section 10.6.

- *Dimensionless Measurement*: $D_g = M$. In this case $\mathbf{g}$ is not ordered. Measurements consist of global projections of $\mathbf{f}$, and the effective dimension of the measurement embedding space is **M.** The measurement model is

$$g_n = \int f(\mathbf{x})h_n(\mathbf{x})\, d\mathbf{x} \tag{7.31}$$

Each of these sampling strategies is represented in the physical examples presented in this text. While each may include diverse and sophisticated kernels, typical kernels, and design goals differ in systems implemented to date. In the typical conventional system the sampling function integrates the signal using a compact kernel. In dimension increasing systems physical constraints such as difficulty in obtaining sufficiently small pixel pitch or temporal bandwidth lead one to use multiple non-degenerate sampling kernels, again spanning compact support. Dimension reducing systems may apply a sampling kernel with unbounded support over one dimension on a lattice with finite support over other dimensions. For example, the sampling kernel may have compact spatial support but be unbounded in time or color. The typical dimensionless kernel has unbounded support in all embedding dimensions.

Following a relatively quiet period between Shannon's work in the 1940s, research in sampling theory has been vibrant since the mid-1970s. Powerful new results have been obtained very recently in the development of *compressive sampling* based on dimensionless measurements. In view of the continuing rapid development of this field, our discussion here will necessarily be incomplete. We hope, however, to provide the reader with insight into how sampling structure shapes optical sensors.

A sampling system is compressive if the number of signal values inferred from the data, N, is greater than the number of measurements, M. As illustrated in Figs. 7.6 and 7.7, interpolation to create more display signal values than measurement values is always a good idea. In this sense, compressive sampling has a very long history in sampling system design. More generally, one may design the conventional sampling kernel to work in combination with image constraints to enhance decompressive inference [3]. As an example, Fig. 7.18 illustrates the use of a multilobe PSF to enable decompressive image restoration. This particular example is not physically accurate in that the model PSF consists of random discrete points, but it illustrates the general concept well. The point cloud at the upper left of Fig. 7.18 is convolved with this PSF, which we assume models an optical system. The convolved image is detected by an array under the assumption that each 4×4 block of image pixels maps onto a single detector image. We refer to this process as *downsampling*, which is identical to fourth-order Haar averaging. The simulation illustrated in Fig. 7.18 then uses the Matlab function `deconvlucy` (see Section 8.5) with $4\times$ upsampling. Surprisingly, the upsampled image corresponds to the original input.

Compressive sampling may occur with each type of generalized sampling. Sensor system design must consider both the structure of measurement sampling and the

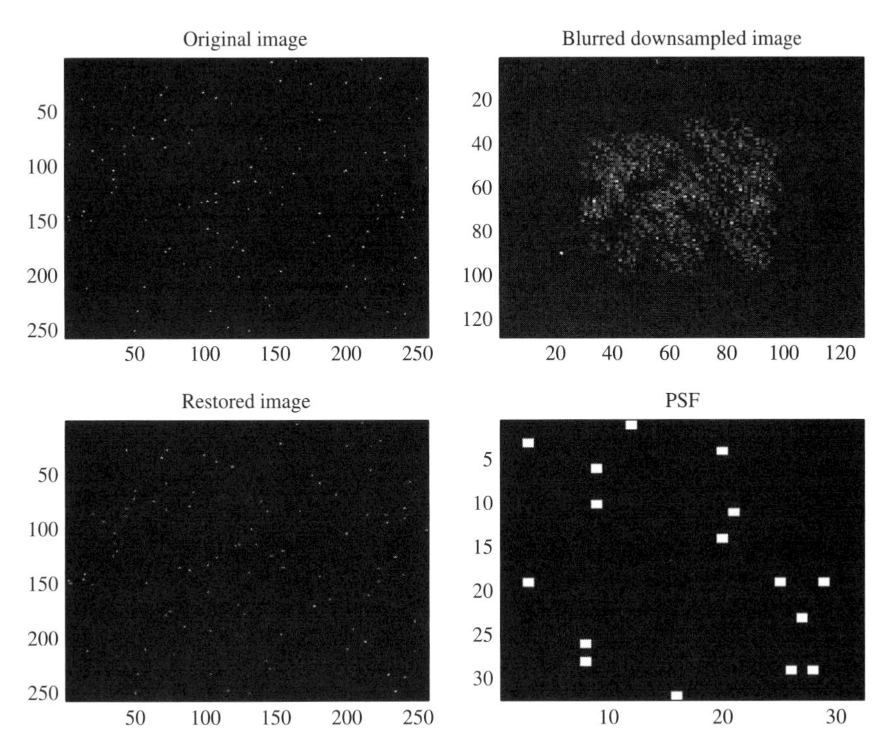

Figure 7.18 Image restoration from downsampled data using a blurred PSF. The point cloud in the original image is convolved with the PSF at lower right prior to downsampling by a factor of 4 in each dimension. Upsampled Richardson–Lucy deconvolution is used to estimate the restored image at lower left. All images were zeropadded for processing. The zero padding is retained in the downsampled image but has been cropped for the original and restored images.

process of signal inference from sample data. Most of our discussion of signal inference is reserved for Chapter 8, but general comments on the nature of inference algorithms are helpful in comparing sampling strategies. The easiest case is, of course, conventional Shannon sampling. With regularly sampled Shannon data, the display signal consists of either the samples themselves, as in Fig. 7.6, or, more accurately, interpolated sample data as in Fig. 7.7. As originally envisioned, dimension increasing measurement is designed for well-conditioned linear transformations of Shannon data. Such transformations may be inverted using methods described in Section 8.2.1 to obtain signal representation samples. More recent studies, however, have demonstrated utility in simple compressive sampling and inference strategies for multichannel sampling systems [202,203].

Dimension reducing and dimensionless data are not generally well conditioned for signal estimation on the object embedding space. These systems rely on the nonlinear inference algorithms described briefly in Section 8.5. Before we consider *how* these data are inverted, however, it is useful to consider *why* we believe that they may be invertible. Before doing this, we must consider more carefully the meaning of *well-conditioned.*

Selection of the sampling kernels $h(\mathbf{x}, \mathbf{n})$ and of the signal representation basis determine the nature of the signals that can be characterized by a sensor system. As discussed in Section 3.6, a set of functions $\{h(\mathbf{x}, \mathbf{n})\}$ generates a vector space V_H consisting of all linear superpositions of the sampling kernels. Each sampling function is a point in the vector space. Measurement sampling consists of projecting the signal $f(x)$ onto V_H.

In generalized sampling systems, one chooses a basis $\{\phi_i(\mathbf{x})\}$ for signal display. The display signal $f \in L^2(\mathbb{R}^D)$ is represented as

$$f(x) = \sum_i f_i \phi_i(\mathbf{x}) \tag{7.32}$$

We refer to the vector space generated by the display basis as the "canonical" space V_c. The image or spectrum represented on V_c is the signal that an end user accepts as "the image."

Ideally one expects that $V_H \subset V_c$, in which case all measurement data may be accounted for in estimating the display signal. In this case, we may assume that the analytical sample vector $\mathbf{f}$ in Eqn. (7.27) has components corresponding to the display coefficients f_i from Eqn. (7.32) (although one may sometimes choose to store analytic samples on a different basis). The measurement matrix $\mathbf{H}$ in this case has coefficients

$$h_{\mathbf{n}i} = \int h(\mathbf{x}, \mathbf{n}) \phi_i(\mathbf{x}) \, d\mathbf{x} \tag{7.33}$$

It is also helpful to define the vector space $V_\perp$, corresponding to the null space of $\mathbf{H}$ on V_c and the set V_f, corresponding to points in V_c that are possible objects. Properties of $V_\perp$ include

$$\begin{aligned} V_H \cap V_\perp &= 0 \\ V_H \oplus V_\perp &= V_c \end{aligned} \tag{7.34}$$

$V_f \subset V_c$ is not a vector space because it violates linear superposition; the sum of two images is not necessarily an image. As is the case with art, it is difficult to explicitly define V_f. One may imagine a "I know it when I see it" hyperfunction capable of classifying elements of V_c as "images" or "not images." Such a hyperfunction is not usually available, but in most cases one is confident in claiming that elements of V_f are sparse on V_c and almost all elements of V_c are not in V_f.

The origin of aliasing is elegantly described in terms of the spaces V_H, V_c, $V_\perp$, and V_f. For conventional sampling systems reconstructing the optical and pixel bandlimited image of Eqn. (7.5) the sampling kernel is $\delta(x - n\Delta)$ and the display basis is sinc $(x/\Delta - i)$. The measurement matrix $\mathbf{H}$ is the identity, and V_c is the space of functions with bandlimit $1/2\Delta$. One may alternatively assume a sampling kernel rect $(x/\Delta - n)$, which reflects the measurement model of Eqn. (7.4). With this choice, the measurement matrix is no longer an identity and the analytic samples are interpolated values of $g(x)$ rather than $f(x)$. Note that the sampling function is not bandlimited in either case, $\delta(x) \notin V_B$ and $\mathrm{rect}(x) \notin V_B$. This means, of course, $V_H \not\subset V_c$.

In general, we may separate V_H into $V_{HC} = V_H \cap V_c$ and an orthogonal component $V_{H\bar{C}}$. A particular signal $f(x)$ may be similarly decomposed as

$$f(x) = f_c(\mathbf{x}) + f_{\bar{c}}(\mathbf{x}) \tag{7.35}$$

where $f_c(\mathbf{x}) \in V_c$ and $f_{\bar{c}}(\mathbf{x})$ is in the orthogonal complement of V_c in $L^2(\mathbb{R}^D)$. A measurement takes the form

$$\begin{aligned} \mathbf{g} &= \int \mathbf{h}(\mathbf{x}) f(\mathbf{x})\, d\mathbf{x} \\ &= \mathbf{H}_c \mathbf{f}_c + \mathbf{H}_{\bar{c}} \mathbf{f}_{\bar{c}} \end{aligned} \tag{7.36}$$

where $h_{\bar{c},ij} = \int h_i(\mathbf{x}) \phi_{\bar{c},j}(\mathbf{x})\, d\mathbf{x}$ and $\phi_{\bar{c},j}(\mathbf{x})$ is a basis vector for $V_{H\bar{C}}$. $\mathbf{g}$ is an M-dimensional vector and $\mathbf{H}_c$ under conventional sampling is an $M \times M$ matrix. $\mathbf{H}_{\bar{C}}$ is an $M \times N$ matrix with N dependent on the number of expansion coefficients used to represent $f_{\bar{c}}$. While somewhat more advanced methods, as discussed in Section 8.2.1, are used in practice, formally one inverts Eqn. (7.36) as

$$\mathbf{H}_c^{-1} \mathbf{g} = \mathbf{f}_c + \mathbf{H}_c^{-1} \mathbf{H}_{\bar{c}} \mathbf{f}_{\bar{c}} \tag{7.37}$$

The rightmost term in Eqn. (7.37) is the aliased signal; by choosing a display basis that does not fully capture the signal or the sampling functions, one naturally induces aliasing artifacts. Where we have previously encountered aliasing only in the context of Shannon sampling, Eqn. (7.37) generalizes aliasing analysis to any set of sampling and display functions. The Whittaker–Shannon sampling theorem assumes that $\mathbf{f}_{\bar{c}} = \mathbf{0}$, thus eliminating the aliased display component. Aliasing is avoided in general if $f(x) \in V_c$ or $\mathbf{H}_{\bar{c}} = \{0\}$. Aliasing is unavoidable if neither condition is satisfied.

Aliasing arises when V_H contains elements not in V_c. The complementary problem that V_c contains elements not in V_H is far more common and troubling in optical sensor systems. In this case, $f \in V_c$ may be decomposed as

$$f = f_H + f_\perp \tag{7.38}$$

where $f_H \in V_H$ and $f_\perp \in V_\perp$. The measurement process projects f onto V_H according to

$$g = \boldsymbol{H} f_H \tag{7.39}$$

The measurement samples g may be processed to estimate analytical samples f_H, but no information is available about $f_\perp$ unless prior knowledge can be applied to infer data in the null space from data in the range. Display signals may consist of f_H or of $f \notin V_c$ calculated using prior constraints.

Projection onto V_H allows us to separate the processes of measurement, analysis, and signal estimation. The measurement process is likely to characterize f well if $f \notin V_\perp$. If f falls completely in the null space, then no information is obtained from the measurement. The relationship between the projection of f on V_H and the utility of the measurement is captured in the *restricted isometry property* (RIP) of Candes and Tao [42]. **H** satisfies RIP if the following relationship applies for nearly all **f**:

$$(1 - \delta_k)||\mathbf{f}||^2 \leq ||\mathbf{Hf}||^2 \leq (1 + \delta_k)||\mathbf{f}||^2 \tag{7.40}$$

One assumes that the sampling functions $h_n(x)$ are orthonormal in Eqn. (7.40), in which case the rows of $\boldsymbol{H}$ will also be orthonormal. The basic idea of the **RIP** is that if the Euclidean length of the signals are preserved, then the signals cannot lie in the null space of **H** and that measurement cannot reduce the phase space of the signal. If $\delta_k \ll 1$, then Eqn. (7.40) simply implies that $f \in V_H$. Candes and Tao define RIP for f sparse on some basis and allow nonnegligible values of δ_k. To understand the idea of an incompressible phase space, imagine that f_1 and f_2 are distinct prospective signals. For the measurement system to distinguish between f_1 and f_2, we require that $||\boldsymbol{H}(f_1 - f_2)||^2 \sim ||f_1 - f_2||^2$. The RIP ensures that this condition is satisfied.

The RIP is a measure of the conditioning of the measurement system. If this property is satisfied for $\delta_k \ll 1$, then $\boldsymbol{H}$ is well conditioned for estimation of $f \in V_f$. More detailed analysis of conditioning and V_H using singular value decomposition is presented in Section 8.4. For the present purposes, it is sufficient to note that $f \notin V_\perp$ suggests that inversion is possible. Inversion strategy depends on the value of δ_k. For $\delta_k \ll 1$, linear inversion of Eqn. (7.27) provides a good estimate of **f**. For finite δ_k, nonlinear estimators are applied to find $f_\perp$ given f_H. In each case, the system design challenge is to codesign $\boldsymbol{H}$ and the image estimation algorithm to achieve high-fidelity imaging. We consider linear and nonlinear approaches in turn.

7.5.2 Linear Inference

Linear projection attempts to maximize the overlap between V_H and V_f. The estimated signal in this case is the projection of $f(x)$ onto V_H. Karhunen–Loeve decomposition, also known as *principal component analysis* (PCA), is a basic linear technique for linear projection. PCA calculates a set of basis vectors that can be used to approximate **f**. The PCA estimate achieves the least mean square error in estimating **f** while maximizing the variance of V_H for a fixed number of measurements [129]. The number of measurements may be much smaller than N, so encoding the data as linear combinations of the basis vectors transforms them to a lower-dimensional space. This dimension reduction can be used to minimize sensor resources and bandwidth and to improve visualization and feature extraction.

Principal components are data-dependent vectors. Given a representative set of signal vectors $\mathbf{F} = \{\mathbf{f}_1, \mathbf{f}_2, \ldots \mathbf{f}_Q\}$, the principal components (or **K–L** vectors) are the eigenvectors of the covariance $\sum = (\mathbf{F} - \bar{\mathbf{f}})(\mathbf{F} - \bar{\mathbf{f}})'$, where $\bar{\mathbf{f}}$ is the mean signal. The principal components are ordered over the magnitude of their eigenvalues, meaning that the first principal component corresponds to the largest eigenvalue. Projection of a given signal $\mathbf{f}_i$ onto the PCA basis takes the form

$$f_{iH} = \mathbf{H}'\mathbf{H}(f_i - \bar{f}) + \bar{f} \tag{7.41}$$

where $\mathbf{H}$ in this case is an $M \times N$ matrix with each row corresponding to a principal component. $\mathbf{H}$ is formed from the principal components corresponding to the $\mathbf{M}$ largest eigenvalues.

Imaging systems based on compressive linear projections are considered by Neifeld and Shankar [186]. Figure 7.19 shows an example of an image training set. The face images were divided into 8×8 pixel blocks; each block was used as an example of a 64-element signal vector. The 16 strongest features arising from KL decomposition of the covariance are shown in Fig. 7.19(b). One may obtain image fidelity improvements in addition to compressive sampling advantages due to feature-specific sampling. As an example, Fig. 7.20 illustrates reconstruction of a face from conventional pixel sampling and from KL feature-specific sampling in the presence of additive white Gaussian noise.

Figure 7.19 (a) Faces used for training KL vectors in a feature specific imager; (b) first 16 principal components for $N = 64$ block representation space. (From Neifeld and Shankar [185] © 2003 Optical Society of America. Reprinted with permission.)

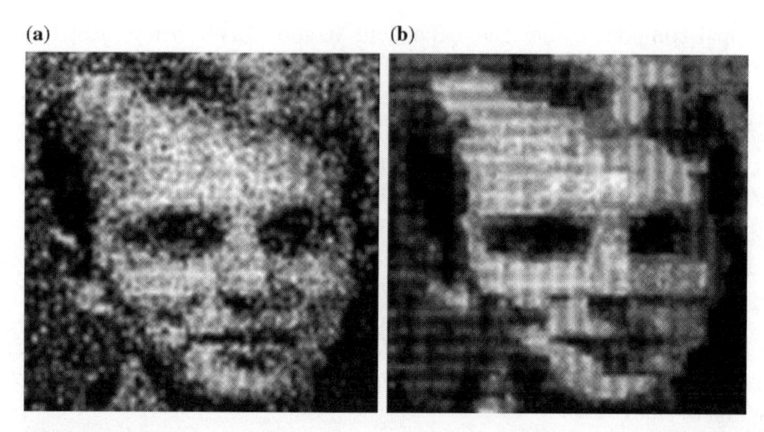

Figure 7.20 Reconstructed images using (a) conventional imaging and (b) KL feature-specific imaging. White Gaussian noise is added to both images, at this noise level the feature-specific imager produces lower mean-square error due to filtering on significant principal components. (From Neifeld and Shankar [185] © 2003 Optical Society of America. Reprinted with permission.)

The primary challenges of PCA-based imaging are that V_f may not be sufficiently characterized to enable principal component estimation, that the principal components (which may contain negative components) may not be easily measured with optical hardware, and that V_f may not be compact on a linear space. Confronted with these challenges, the system designer seeks to to minimize the cost of measurement within physical constraints while maximizing system performance. Typical physical constraints focus on the range and distribution of coefficients h_{ij}. With irradiance measurements, for example, one often requires $0 \geq h_{ij} \leq 1$. On the other hand, designers may effectively achieve negative weights by double sampling and electronic processing as discussed in Section 9.3.

Neifeld and colleagues describe and compare several linear projection strategies in addition to KL decomposition, including measurement matrices based on Hadamard and wavelet decompositions. They also discuss sampling based on independent component analysis (ICA). ICA uses linear sampling but applies nonlinear techniques in identifying data-dependent features. The sampling strategy outlined in Ref. 185 is a block-based dimension reducing approach, but the analysis in that paper and subsequent work (especially Ref. 185) applies as well to dimensionless design. Various alternative feature design strategies may be considered to match physical constraints. For example, Hamza and Brady [109] describe compressive spectroscopy based on "nonnegative matrix factorization," which maintains nonnegativity in both measurement vectors and estimated signals.

7.5.3 Nonlinear Inference and Group Testing

Nonlinear signal inference enables us to estimate $f(x) \in V_c$ even when V_H does not span V_c. More specifically, we estimate $f_H(x) \in V_H$ and $f_\perp(x) \in V_\perp$ such that

$f_H(x)$ satisfies $\mathbf{g} = \mathbf{Hf}$. The null space component $f_\perp(x)$ is inferred from $f_H(x)$ using a "prior" constraint on the signal. A prior expresses some knowledge regarding the nature of the signal known even before the measurement is made. Common priors include

- Knowledge that the signal is nonnegative
- Knowledge that nonzero signal values are sparse
- Knowledge that the signal is smooth
- Knowledge of the principal components of the signal class
- Knowledge of a probability density function on signal values or signal features

Given a prior, one infers the signal from joint linear estimation and prior enforcement. This approach is particularly attractive in view of the fact that since V_f is not a linear space, no linear estimation strategy can be efficient. In the ideal system only one possible signal will jointly satisfy $\mathbf{g} = \mathbf{Hf}$ and the prior constraint.

The two most common priors are that $\mathbf{f}(x)$ must be "image-like" or that $\mathbf{f}(x)$ must be sparse on some basis. "Image-like" means that $f(x)$ has some ad hoc properties associated with images, such as smoothness, local connectivity, or fractal structure. We descibe minimum variance and deconvolution algorithms for enforcing the image-like prior in Section 8.5. The remainder of this section focuses on sampling and nonlinear inference based on sparsity priors. Sparsity and image-like priors represent two endpoints of a spectrum of constraints. Sparsity emphasizes the independence of each data value on an appropriate basis, image-like priors emphasize the statistical dependence of data embedded in low-dimensional spaces. One imagines that more sophisticated priors combining sparsity and connectivity constraints will emerge from continuing research.

Nonlinear sparse signal estimation is closely related to "group testing" [62]. A group testing system seeks to determine the state of a set of objects using a set of measurements. In the current context, the objects are the elements of $\mathbf{f}$ and the measurements are the elements of $\mathbf{g}$. A single test, for instance, a measurement, consists of a weighted sum of the elements of $\mathbf{f}$, $g_i = \sum h_{ij} f_j$. The test is a group test because $\mathbf{H}$ is not an identity. Group testing is "nonadaptive" if the outcome of one test cannot be used to inform future tests. Adaptive group testing is somewhat more measurement-efficient but is more difficult to implement in optical systems. Group testing is separated into "combinatorial" systems, under which the combination of priors and code design ensures that one is logically certain to correctly characterize $\mathbf{f}$ from $\mathbf{g}$, and "probalistic" systems, under which one infers $\mathbf{f}$ from $\mathbf{g}$ according to likelihood functions with some possibility of error.

The classic group testing system seeks to characterize a point object occupying one of N positions. In imaging the object is a spatial impulse; in spectroscopy the object is an isolated spectral feature. Naively, one might characterize such a source by measuring all N pixels or spectral channels. Using combinatorial group testing, however, one can characterize such a source using $\log_d(N)$ measurements, where d is the dynamic range of h_{ij}. A design for $\mathbf{H}$ achieving this measurement efficiency

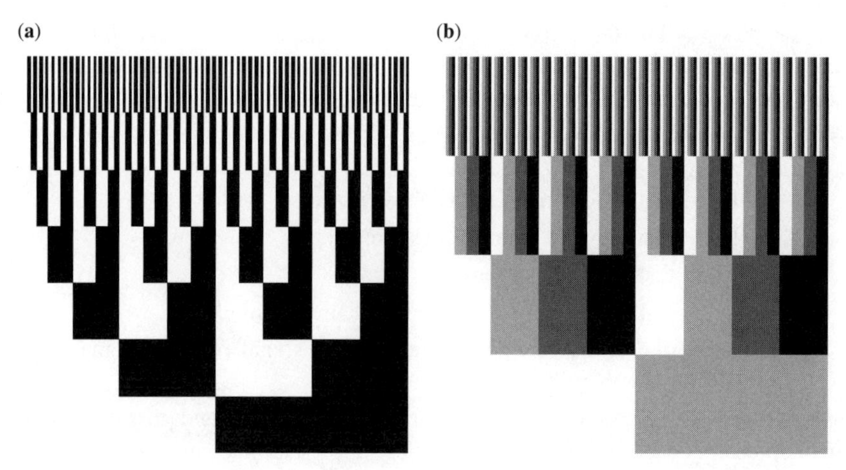

Figure 7.21 Sampling matrices for (a) a binary-coded single-object group test and (b) a four-level dynamic range single-object testing ($N = 128$ in both cases). Seven measurements are sufficient to uniquely identify all object positions in the binary case, in (b) three four-level rows and one binary row enable object localization and power estimation.

takes the form

$$h_{ij} = \text{mod}\left[\text{floor}\left(\frac{j-1}{d^{i-1}}\right), d\right] \tag{7.42}$$

where floor(x) is the greatest integer less than or equal to x and $i \in \{1, \log_d N\}$, $j \in \{1, N\}$ Plots of $\mathbf{H}$ for $d = 2$ and $d = 4$ for $N = 128$ are shown in Fig. 7.21. The goal of the sampling code design is to ensure that the signature associated with each object position is unique. In Fig. 7.21(a) each column corresponds to an object position. Taking black as 0 and white as 1, the code for the first object position is 1111111 measured along the column. The second object position from the left corresponds to the code 0111111. The third column produces 1011111. The last column on the right corresponds to the code 0000000 (producing a null measurement). In general, an object located in position i radiating power S produces a signature $S\mathbf{h}_i$, where $\mathbf{h}_i$ is the ith column vector. 2^M different signatures are produced for $h_{ij} \in [0, 1]$, of which $2^M - 1$ produce a measurable signature. Figure 7.21(b) may be analyzed similarly for a four-level code. In this case, an object located in position i radiating power S produces a signature also produces signature $S\mathbf{h}_i$, but h_{ij} is now drawn from [0, 1, 2, 3]. The first three rows of the code sketched in Fig. 7.21(b) produce $4^3 = 64$ unique measurement signatures. A binary code in the fourth row drawn from $h \in [3, 4]$ disambiguates the 64 codes into 128 states and allows unambiguous estimation of S.

Group testing codes are particularly easy to implement in optical spectroscopy. Figure 7.22 illustrates a spectrograph design used to implement the sampling matrices of Fig. 7.21 by Potuluri et al. [206], who used a "measurement efficient optical

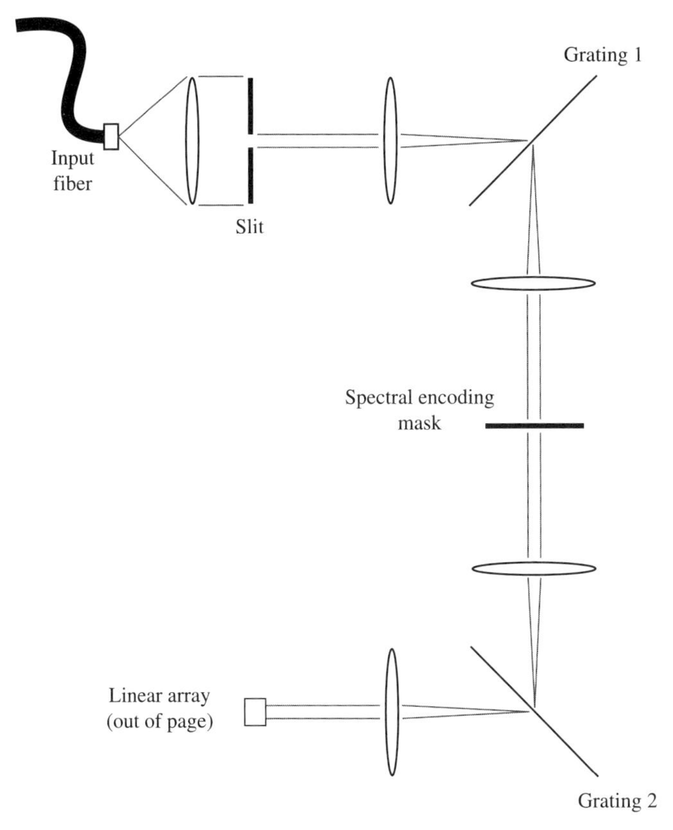

Figure 7.22 Optical spectrograph for implementing group testing/generalized sampling strategies.

wavemeter" to characterize a source in 256 wavelength bands with seven or fewer measurements. The source in this case is a slit or pinhole; the goal is to identify the wavelength of the source in $\log_d N$ measurements. The collimator and frontend grating map the source onto the spectral encoding mask such that each color channel occupies a different transverse column; for instance, if the source power spectrum is $S(\lambda)$, then the power spectrum illuminating the mask is $f(x, y) = S(x - \alpha\lambda)$. Section 9.2 describes how a dispersive spectrograph implements this transformation. For the present purposes it is sufficient to note that if the mask transmission $t(x, y)$ is described by one of the codes of Fig. 7.21, then the spectral density on the plane immediately after the mask can be represented discretely as $t_{ij} f_j$, where $f_j \approx S(\lambda = j\Delta/\alpha)$ is the signal in the jth spectral channel. Optical components following the mask decollimate the light to focus on an output slit, which in this case is occupied by a detector array distributed along the y axis. The effect of the backend optics is to sum the spectral channels transmitted by the mask along the rows such that the measured irradiance on the ith detector is $g_i = \sum_j t_{ij} f_j$.

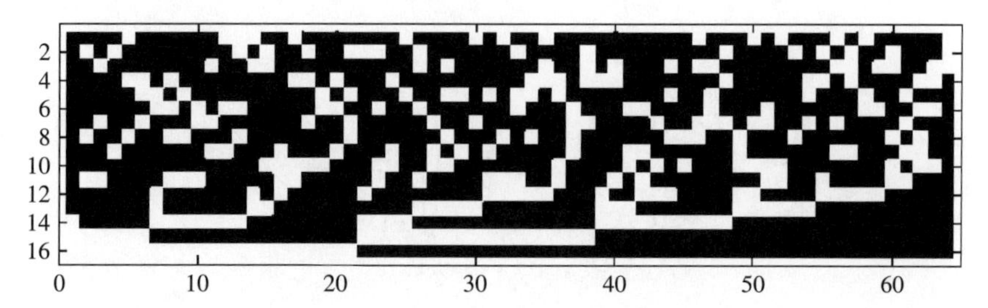

Figure 7.23 UD_2 code for $N = 64$ with 16 codewords. White corresponds to pixel value 1, and black corresponds to 0.

The spectrograph of Fig. 7.22 can implement any linear transformation on the input spectrum under the constraint that $0 \leq t_{ij} \leq 1$. Potuluri demonstrated the binary codes of Fig. 7.21 as a means of minimizing the number of detectors needed to characterize the spectrum. Minimizing the number of detectors is particularly important in wavelength ranges like the near and midwave infrared, where detector cost scales linearly with the number of elements in the array. Dispersive systems similar to Fig. 7.22 have been combined with dynamic spatial light modulators to implement adaptive group testing [68].

The single target group testing problem is generalized in binary "superimposed codes" introduced by Kautz and Singleton [135]. Rather than implementing $\boldsymbol{g} = \boldsymbol{Hf}$ as a linear product, sensors using superimposed codes implement a logical OR operation on $g_i = \cup h_{ij} f_i$, where we assume that $f_j \in \{0, 1\}$. Superimposed codes uniquely discriminate up to k nonzero values of $\mathbf{f}$ among N pixels, effectively allowing $k \ll N$ possible signal components rather than just 1. To achieve this objective, the inclusive logical OR of all subsets of up to k columns of $\mathbf{H}$ must be mutually distinct. A set of codewords (or equivalently $\mathbf{H}$ with the codewords as columns) with this property is called a *uniquely decipherable code* of order k (UD_k). A UD_2 code for $N = 64$ consisting of 16 codewords is illustrated in Fig. 7.23. Up to two objects in 64 elements correspond to 2049 object states. The measurement states corresponding to these object states are illustrated in Fig. 7.24. As with the single-object codes, the design goal is to ensure that each object state produces a unique measurement state. Superimposed codes have been applied in diverse signal analysis and communication applications. Zheng et al. [262] embedded the codes of Fig. 7.23 in a fiber sensor network to track up to two individuals walking across a floor.

7.5.4 Compressed Sensing

Unfortunately, design of UD_k codes is not possible for significant values of k, and the superimposed code approach does not enable robust reconstruction of analog signal values. It turns out, however, that probabilistic group testing enables k-sparse signal estimation without the necessity of careful code or inference algorithm design. Compressive sampling is a probalistic group testing method describing both the nature of sampling codes necessary for sparse signal reconstruction and algorithms for signal inference.

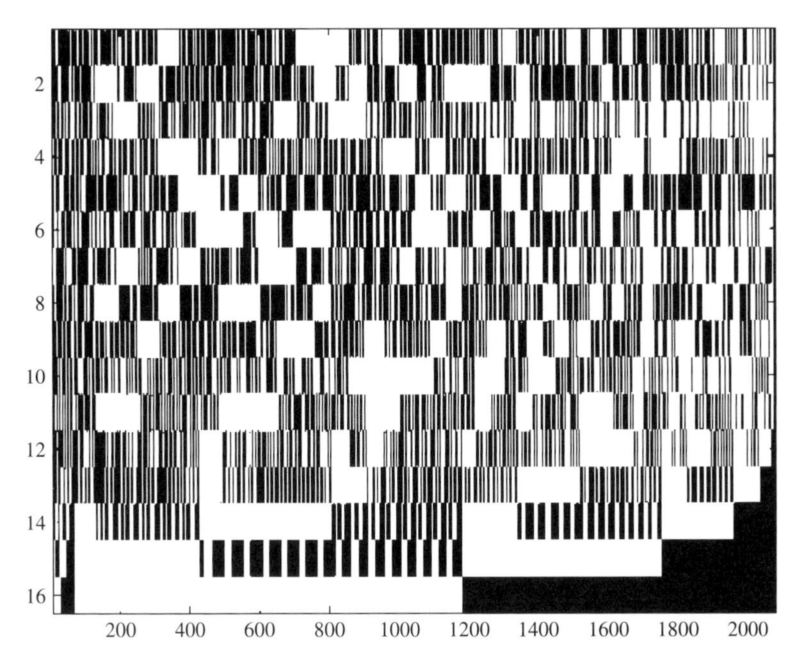

Figure 7.24 Measurement states correspond to all possible object states for the code of Fig. 7.23.

Potuluri et al. used multiscale wavelet codes in the spectrograph of Fig. 7.22 to demonstrate probabilistic group testing for compressive spectroscopy [205,29], but the estimated signal fidelity suffered from incomplete development of both the sampling scheme and the inference algorithms. These issues are resolved in the compressed sensing (CS) theory of Donoho [59], Candes and Tao [40], and Candes et al. [41]. CS explores the estimation of sparse signals using dimensionless measurements on an "incoherent" sampling basis. We begin our brief exploration of CS by defining terms:

A signal $f(x) \in V_c$ is k-sparse in the basis Ψ of V_c if the expansion

$$f(x) = \sum_i f_i \psi_i(x) \tag{7.43}$$

includes at most k nonvanishing coefficients f_i.

The coherence between two different bases of V_c, Φ, and Ψ is

$$\mu(\Phi, \Psi) = \sqrt{\boldsymbol{N}} \max_{k,j} |\langle \phi_k | \psi_j \rangle|^2 \tag{7.44}$$

The coherence quantifies the largest correlation between any two elements of Φ and ψ. If the two bases happen to share a basis vector, the coherence is $\sqrt{n}$. Since the two bases span the same vector space, the coherence cannot be 0. The range of coherence values is $\mu(\Phi, \Psi) \in [1, \sqrt{n}]$. In the case that $\mu(\Phi, \Psi) = 1$, the bases Φ and ψ are said to be incoherent. As an example, in the finite support approximation of Section 3.7, bases of V_B include the Shannon basis $\psi_j = \sqrt{2B}\,\mathrm{sinc}(2Bx - j)$ and the Fourier basis $\phi_k = \sqrt{1/X}\exp(2\pi ikx/X)$. These two bases are incoherent on V_B.

Compressed sensing uses measurements on a subspace spanned by elements of Φ to characterize $\mathbf{f}$ k-sparse on Ψ. As a simple example, suppose that the measurement matrix $\mathbf{H}$ is constructed by randomly selecting M vectors ϕ_k from Φ such that $\mathbf{H} = \sum_m^M |\phi_m\rangle\langle\phi_m|$. CS theory combines k-sparsity and coherence in the following theorem [38]:

If the representation of $\mathbf{f}$ is k-sparse in ψ and if

$$M \geq kC\mu(\mathbf{\Phi}, \mathbf{\Psi}) \log N \tag{7.45}$$

for some positive constant C, then the estimated signal

$$\mathbf{f}_e = \arg\min_{\mathbf{f}} ||\mathbf{f}||_1$$

such that

$$\mathbf{H}\mathbf{f}_\mathbf{e} = \mathbf{g} \tag{7.46}$$

satisfies $\mathbf{f}_\mathbf{e} = \mathbf{f}$ with overwhelming probability.

$||\mathbf{f}||_1 = \sum_j |f_j|$ is the l_1 norm of the representation $\mathbf{f}$ of $f(x)$ on Ψ.

As discussed in Section 8.5, the basic strategy expressed by Eqn. (7.46) is characteristic of nonlinear signal estimation from linear measurements. One augments the requirement that the estimated signal must be self-consistent with the measured data with a constraint based on prior knowledge. In this case, the prior knowledge is that the signal is sparse. The brilliant insight of CS theory is a proof that k-sparsity, low coherence measurement, and l_1 minimization are linked with high probability.

(a)

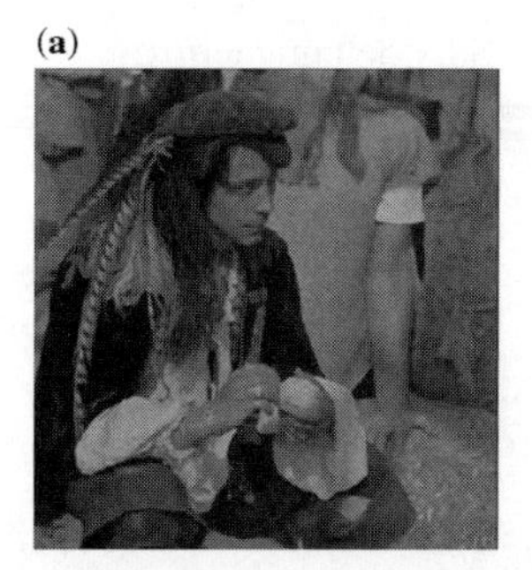

(b)

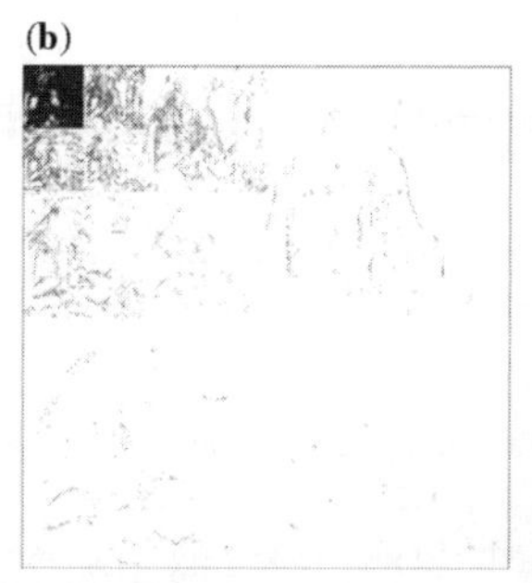

(c)

Figure 7.25 Image recovery from compressively sampled data. Image (a) is formed from $k =$ 25,000 nonzero Haar wavelet coefficients on an $N = 1024^2$ pixel image. Image (b) shows the averages and details of Haar decomposition. Image (c) is an exact copy of (a) reconstructed using l_1 minimization from $m = 70{,}000$ samples corresponding to projections onto randomly selected noiselet basis vectors. (From Candes and Romberg [38] © 2007 IOP Publishing, Ltd. Reprinted with permission.)

Figure 7.25 is an example of image sampling and restoration from compressive samples. The representation basis, on which the image is sparse by design, is the Haar wavelet. The noiselets described by Coifman et al. [50] are applied as the sampling functions. The coherence between the noiselets and the Haar wavelets is $\sqrt{2}$.

Compressed sensing theory has profound implications for measurement theory generally and for optical imaging and spectroscopy in particular. Many interesting questions remain unanswered and unexplored, however. The theory as outlined thus far assumes that one has prior knowledge of the basis Ψ on which $\mathbf{f}$ is k-sparse. Examples in this section and later in the text will often assume that $\mathbf{f}$ is sparse on a particular wavelet basis. The selection of this basis is often arbitrary, although "best basis" algorithms may be applied to optimize $\mathbf{f}$ to some image-related prior. Clearly, where one has prior knowledge of Ψ CS may be applied to great effect. Examples in which $\mathbf{f}$ is sparse on the display basis or in which the Fourier transform of $\mathbf{f}$ is sparse are more common at radio—rather than optical—frequencies, but in such cases the "fast Fourier sampling" subbranch of CS is of great value [91].

In the absence of precise knowledge of the sparse basis, one may wonder how to design the measurement operator $\mathbf{H}$. Guidance in this regard is provided by the restricted isometry property of Eqn. (7.40). RIP and CS are closely related to "dimensionality reduction." The idea of dimensionality reduction is that if signals are projected onto lower-dimensional subspaces using incoherent operators, then the basic topology of V_f, meaning Euclidean length and distance, will tend to be preserved. Since in most imaging applications the display basis for $\mathbf{f}$ is not sparse (e.g., images are rarely like the point clouds of Fig. 7.18), the restricted isometry principle provides a rough guide to evaluating the performance of measurement operators without knowledge of Ψ.

Compressed sensing theory has been directly applied to optical imaging in the Rice University single-pixel camera [228]. As illustrated in Fig. 7.26, the Rice

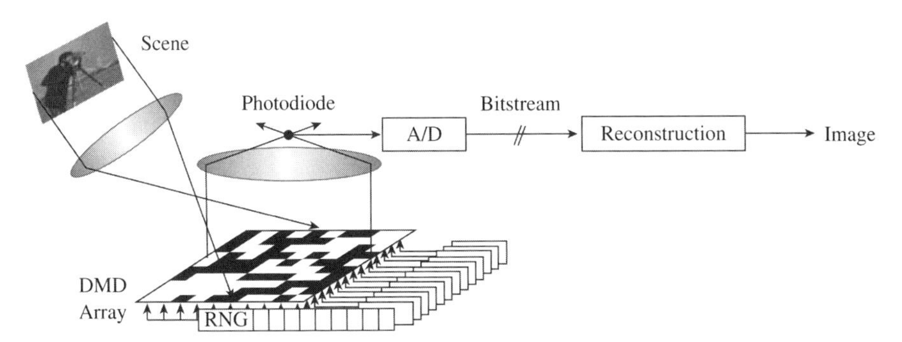

Figure 7.26 Rice single-detector compressive sampling camera. A scene imaged onto a digital micromirror array is relayed onto a single photodiode after weighting by the mirror reflectivities. (From Takhuar et al., *Computational Imaging IV*, SPIE, Vol. 6065 [228]. © 2006 SPIE. Reprinted with permission.)

camera places a digital mirror device (DMD) in an image plane. The image plane is relayed by the mirrors onto a single photodetector, which sums the irradiance from all relayed pixels. A basic model for the camera assumes that the *j*th mirror element can be switched between a state that transmits the *j*th pixel to the photodetector and a state that which prevents the *j*th pixel from hitting the photodetector. The irradiance striking the photodetector in the *i*th timestep is $g_i = \sum_j h_{ij} f_i$, where $h_{ij} \in \{0, 1\}$. The state of the measurement projector h_{ij} can be changed arbitrarily from one timestep to the next to implement a full measurement $\mathbf{g} = \mathbf{Hf}$.

Figure 7.27 shows an image reconstructed using CS processing on the Rice camera. l_1 minimization was implemented using the "basis pursuit" algorithm [45]. The figure assumes *k*-sparse representation on the Haar basis and uses random projections with approximately half of the pixels transmitting in the design of $\mathbf{H}$; *k*-sparsity on the Haar basis is confirmed in Figs. 7.27(b) and (c).

In comparing dynamic range and image fidelity measures under photon counting noise for the single-pixel camera with conventional pixel arrays, raster-scanned imagers, and full-rank multiplex imaging systems, Duarte et al. find that the per measurement dynamic range required by the single-pixel system is $N/2$ times greater than for conventional pixel array [63]. The mean-square error (MSE) is

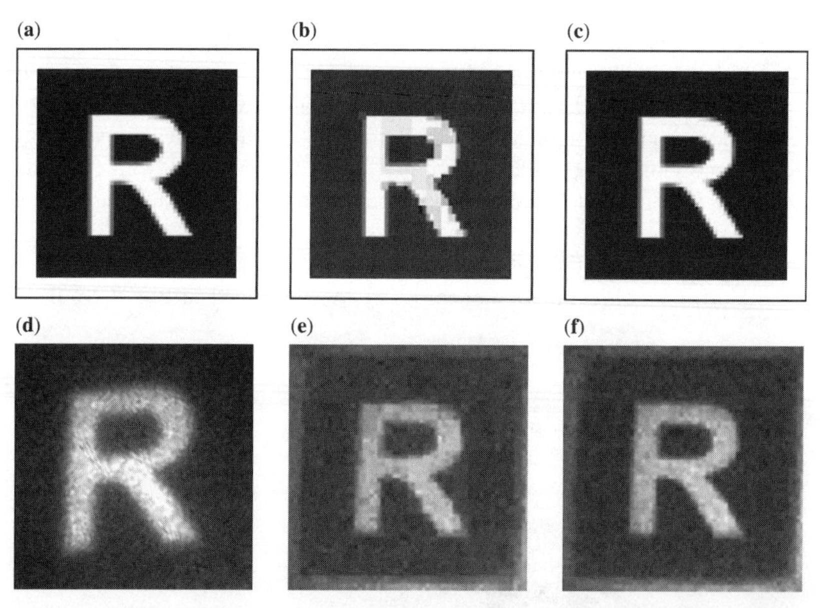

Figure 7.27 A 64 × 64 image reconstructed using the Rice single-pixel compressive camera: (a) ideal image; (b) 400 largest wavelets; (c) 675 largest wavelets; (d) image on DMD; (e) 1600 measurements; (f) 2700 measurements. parts (b) and (c) show the digital image as projected onto limited numbers of Haar wavelets; (d) is a 320 × 240 camera image of the irradiance incident on the DMD/image plane; (e) and (f) show l_1 minimization constrained reconstructions for 1600 and 2700 measurements of random pixel maps. (From Takhuar et al., *Computational Imaging IV*, SPIE, Vol. 6065 [228] © 2006 SPIE. Reprinted with permission.)

estimated at $3C_N^2M$ times greater, where C_N is a constant. The precise error of compressive image estimation is somewhat difficult to estimate; however, most algorithms outperform linear estimation limits. Section 9.3 describes a simple example comparing error reduction strategies for multiplex and isomorphic imaging systems. In view of the increased MSE, one is unlikely to prefer a single-pixel camera for applications where pixel arrays are readily available. On the other hand, in extreme ultraviolet, x-ray, infrared, and terahertz systems where pixel arrays are not available, the CS approach may be attractive. Of course, one may also imagine compressive designs even in the visible range that are somewhere in between single-pixel and full-array sampling. Consideration of such arrays raises interesting questions involving the relationship between sampling and pixel size.

Duarte's analysis implicitly assumes that the light collection efficiency of the Rice camera is equal to the collection efficiency of the pixel array. The validity of this assumption depends on the etendue of the camera. As discussed in Section 6.6.3, etendue must be conserved in power conserving optical systems. Since the solid angle accepted by a camera is inversely proportional to the square of the $f/\#$, the etendue is proportional to

$$L = \frac{A^2}{(f/\#)^2} \tag{7.47}$$

where A is the entrance aperture diameter. The quantum efficiency of the Rice camera for light collected onto the DMD is thus

$$\eta = \frac{A_{\mathrm{pd}}^2 (f/\#)_{\mathrm{DMD}}^2}{A_{\mathrm{DMD}}^2 (f/\#)_{\mathrm{pd}}^2} \tag{7.48}$$

where A_{pd} is the aperture diameter for the photodiode, A_{DMD} is the area of the DMD, and the $f/\#$ terms describe the collection optics for each system. In order to maintain optical throughput, the etendue of the photodiode must equal that of the DMD, meaning, for example, that the area of the photodiode equals that of the DMD array or that the $f/\#$ for the photodiode is much less than that for the DMD collection optics. In practice, the latter strategy is adopted because the input $f/\#$ accepted by the DMD modulator is naturally modest because of the limited scan range of the mirrors. In either case, however, the etendue of the photodiode must greatly exceed the etendue of a conventional pixel.

The digital mirror array of Fig. 7.26 is an example of a *spatial light modulator* (SLM), such as an electrically programmable transmission mask. One may imagine building single-pixel cameras with an SLM that accepts a smaller $f/\#$ than the DMD array, such as a liquid crystal display or a micromechanical shutter array. In this case, η may be dominated by the ratio of the photodiode and modulator areas. Operating in a regime where the noise equivalent power is

$$\mathrm{NEP} = \frac{\sqrt{A^2 \Delta f}}{D^*} \tag{7.49}$$

The SNR for a single measurement of the Rice camera is

$$\begin{aligned} \mathrm{SNR} &= \frac{D^*}{A_{\mathrm{pd}}\sqrt{\Delta f}} \frac{A_{\mathrm{pd}}^2}{A_{\mathrm{SLM}}^2} P \\ &= \frac{D^*}{\sqrt{\Delta f}} \frac{A_{\mathrm{pd}}}{A_{\mathrm{SLM}}^2} P \end{aligned} \tag{7.50}$$

where P is the incident power per pixel. In addition to the factors identified by Duarte, one finds that the SNR for the CS camera may be reduced by both the potential increase in the readout frequency necessary for single-pixel imaging and the increase in sensor area needed for efficient light collection.

The generalized sampling story told in this section has rapidly launched our discussion into coding and signal estimation strategies and physical constraints. Prior to continuing this discussion, the reader may benefit from the background on coding and signal estimation provided in Chapter 8 and on physical system design provided in Chapter 9. We return to generalized and compressive sampling system design in Chapter 10.

PROBLEMS

7.1 *Pixel Transfer Function.* Consider an active pixel sensor with 25% fill factor. Suppose that the pixel pitch is 8 μm x and in y. The circuit designer may choose between a compact 4×4-μm photodiode or a photodiode integrating the photocurrent from 16 distinct 1×1-μm patches placed randomly within the pixel cell. Plot and compare the pixel transfer functions for these two approaches.

7.2 *Aliasing and Noise.* Consider a one-dimensional system imaging the object $f(x) = 1 + \cos(\alpha x^2)$ over the range $x = [-20, 20]$. Let $\alpha = 0.15$. $f(x)$ is imaged through a rectangular aperture with pupil function $P(x) = \mathrm{rect}(x/A)$ at wavelength $\lambda = 1$. The resulting image is sampled on an electronic detector array with rectangular pixels on a pixel pitch $\Delta = 0.5$.

(a) Plot $f(x)$ over its range.

(b) Plot the OTF, PTF, and STF under the following circumstances:

- $f/1$ optics, full fill factor
- $f/1$ optics, 50% fill factor
- $f/1$ optics, 10% fill factor
- $f/2$ optics, full fill factor

(c) Plot the sample data for $f/1$ optics and full fill factor with no noise.

(d) Plot the sample data for $f/1$ optics and full fill factor with additive Gaussian noise added to each data point with a standard deviation of 0.1.

(e) Plot the sample data for $f/2$ optics and full fill factor with additive Gaussian noise added to each data point with a standard deviation of 0.1.

7.3 *Interpolation.* Consider the function $f(x) = 1 + \cos(\alpha x^2 + \beta x)$ over the range $x = [-40, 40]$. Let $\alpha = 0.05$ and $\beta = \pi$.

(a) Plot $f(x)$ over its range.

(b) Suppose that $f(x)$ is sampled using rectangular pixels at full fill factor on a pitch of $\Delta = 0.5$. Plot the measured data g_n.

(c) Use truncated sinc interpolation to estimate the continuous function $g(x)$ corresponding to the sampled data. Plot $g(x)$.

(d) Discuss the statement "$g(x)$ is lowpass-filtered with respect to $f(x)$." What is the filter function?

(e) Do you observe any aliasing artifacts in $g(x)$?

7.4 *Aliasing.* Choose a natural image of size $N \times N$ pixels for $N \geq 512$. Use the Haar and cubic spline scaling functions to produce averages reducing N by a factor of 2, a factor of 4, and a factor of 8. Can you find aliasing artifacts in the resulting images? Illustrate them in your report. Compare aliasing in Haar and cubic spline downsampling.

7.5 *RGB Interpolation and Aliasing.* Consider a color object in red, green, and blue color channels with

$$f_r(x, y) = f_g(x, y) = f_b(x, y) = 1 + \cos\left[\alpha(x^2 + y^2)\right] \tag{7.51}$$

for x, y in the range $(-255, 256)$ and $\alpha = 0.002$. The object is sampled on focal plane with pixel pitch $\Delta = 1$ covered by the Bayer pattern.

(a) Plot the measured data for the red, green, and blue channels.

(b) Interpolate the sampled data to 512×512 RGB pixels.

(c) Combine the images to show an RGB image of the object. Discuss any aliasing and registration issues that you encounter. How might you resolve registration issues?

7.6 *Microlenses.* Consider a focal plane with a pixel pitch of 10 μm and a 25% fill factor. Microlenses with focal length equal to the pixel pitch are used to create the effect of a full fill factor.

(a) What is the minimum useful objective *f*/# for imaging with this focal plane?

(b) Suppose that the focal plane is used to image a ground sample distance of 10 cm of earth illuminated by sunlight. The pixel integration time is 10 ms. Estimate the photon noise–limited SNR. Estimate the thermal noise–limited SNR under the assumption that $D^* = 10^{11}$ jones.

7.7 *Deconvolution and Decompression.* The image in Fig. 7.18 is a random sparse array of points (approximately 100 points on a 256×256 array). The PSF is also a random set of sparse points (approximately 25 on a 32×32 array). The blurred downsampled image was generated using the function `imfilter` with the PSF followed by Haar decomposition. Reconstruction was implemented using the `deconvlucy` function.

(a) Write Matlab code to replicate Fig.7.18 using your own randomly generated data.

(b) Simulate the reconstruction process in the beginning with normally distributed additive noise with a noise level of 10^{-6} of the signal maximum. At what noise level does the reconstruction degrade? Do results differ with Poisson noise?

(c) What happens if you use this method on a natural image rather than a sparse point cloud?

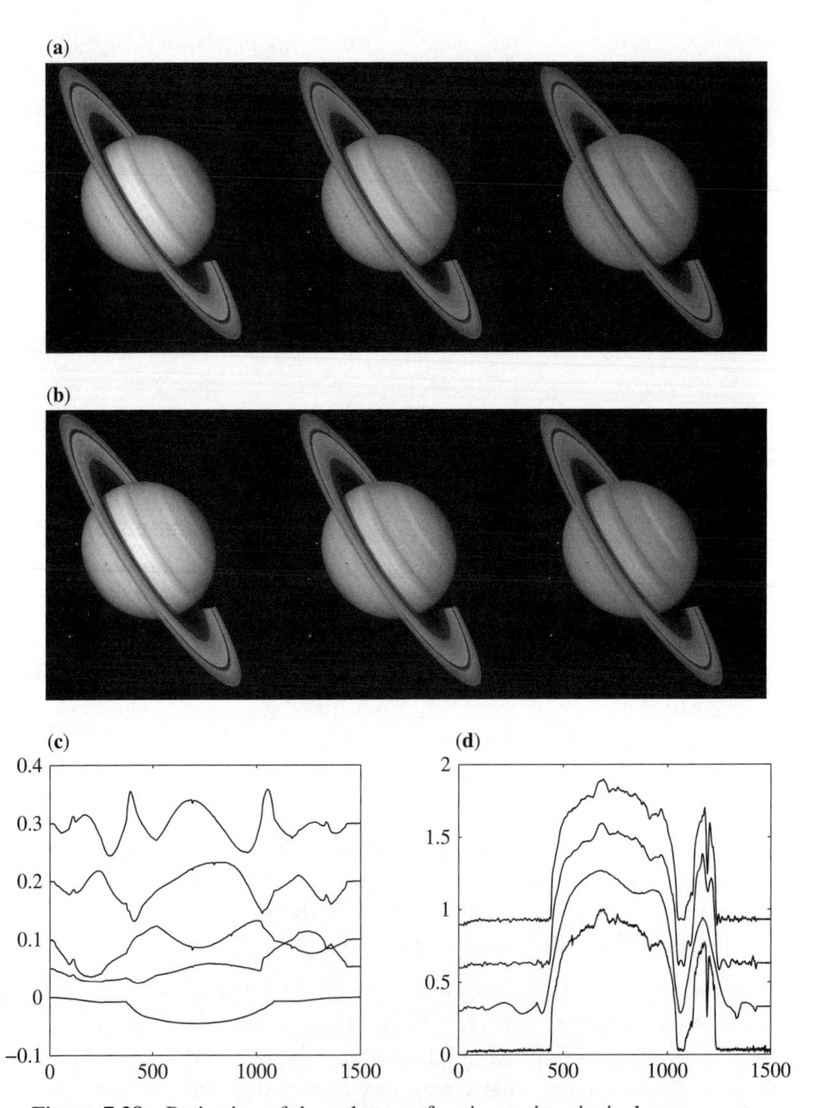

Figure 7.28 Projection of the columns of an image in principal components.

7.8 *PCA Compression.* As a simple example of PCA compression, one may use the rows or columns of an image as a representative set of 1D signals. As an example, Fig. 7.28 shows the Matlab image `saturn.png`. The 1500×3600 matrix consisting of the red, green, and blue color planes, illustrated in Fig. 7.28(a), was decomposed using the `svd` function. Figure 7.28(c) shows the column eigenvectors corresponding to the five largest eigenvalues, ordered from the first eigenvector on the bottom to the fifth at the top. One expects these eigenvectors to reflect the large-scale features of the columns. Figure 7.28(d) shows the 700th column vector at the bottom, and the 700th column vector projected on the first 10, 50, and 100 eigenvectors. Figure 7.28(b) shows all of the columns projected onto the first 100 eigenvectors. In this sense, the data in Fig. 7.28(b) are compressed by $15\times$ relative to the baseline image in (a).

(a) Generate the analog of Fig. 7.28 for the 300th and 500th columns of (a).

(b) What happens to the reconstructed signal quality if you add Poisson noise to the signal prior to projection on the principal components?

(c) Generate the analog of Fig. 7.28(b) using only the first 10 principal components of the columns.

7.9 *Multiplex Sampling and Etendue*

(a) Consider a single-pixel compressive imaging camera using a spatial light modulator with pixel pitch Δ. The camera operates at a center wavelength λ_0. Use conservation of etendue to estimate the minimum pixel area necessary to efficiently collect the light transmitted by the modulator.

(b) The compressive spectrometer design of Fig. 7.22 also uses a spatial pattern to encode a measurement matrix. In this case, pixels of approximate size $\lambda f/\#$ may efficiently collect the weighted signal. Explain the difference between the two systems.

8

CODING AND INVERSE PROBLEMS

> In retrospect, I am guessing that my own innocent half-page letter in the June 1949 IRE Proceedings may have had a beneficial effect.
>
> —M. Golay [95]

8.1 CODING TAXONOMY

Coding is the process of structuring data for transmission between a source and a receiver. This chapter considers transmission of information between an object and an optical sensor system as a coding problem. Coding is most advanced in the context of "algebraic coding theory," which is the mathematical basis of modern communication systems. The system structure for a communications system is illustrated in Fig. 8.1. The goal of the system is to transmit data from a source to a receiver. In algebraic coding systems the source data is a set of algebraic symbols, such as a text file in ASCII symbols. In many cases, the source may encode its data for compression or encryption. The source-encoded data are transmitted over a channel, such as a telephone line or radio connection, to a receiver. Prior to transmission, a channel encoder writes the source data in a code specific to properties of the channel. The goal of channel encoding is to maximize the probability that the data will be received without error, loss, or interception. The encoding process is reversed by a channel decoder and source decoder at the receiving end of the channel.

Figure 8.2 is an analogous diagram for optical sensing. The goal of the optical system is to relay data describing a physical object from the object to a sensor. The input to the system is the object itself; the output is an "image" of the object. We place image in quotes to respect the many forms that an image may take in a computational sensor system, including a description of the 2D or 3D distribution of an object's density or composition, an object's emission, fluorescence, or reflectance

Optical Imaging and Spectroscopy. By David J. Brady

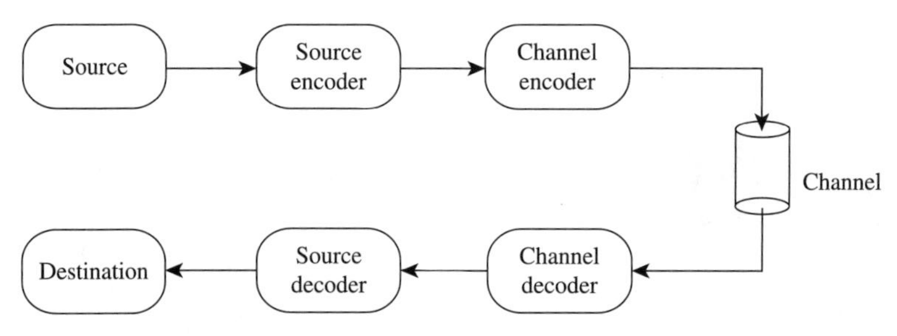

Figure 8.1 Coding in communication systems.

spectrum, or even, in the example of a bar code reader or a character recognition system, a digital identifier of the object.

The object encoder in Fig. 8.2 excites a response from the the object state. The system designer may control this encoder, for example, through selection of temporal, spectral, or spatial illumination patterns. The channel encoder translates the object state into patterns in the radiating optical field. For remote sensing, the object and channel encoders may consist simply of sunlight scattering from the object. In microscopy, the channel encoder may consist of optical elements, such as reference structures, nanoparticles, or spatial or spectral filters, in the near field of the object. For optical sensor systems, the channel is the medium through which the field propagates. In this text, the channel is most often unrestricted space, but one may imagine systems in which the channel consists of a waveguide or fiber bundle.

The channel decoder consists of the optical and optoelectronic components that transform the field transmitted over the optical channel into digital data. Most commonly the channel decoder is the lens system that forms a conventional 2D image of the object. In computational systems, however, digital data from the channel decoder are seldom the final image. In these systems a source decoder in the form of digital signal processing produces the final image.

The primary difference between communication and sensor systems is that the former produces an exact replica of the input source at the destination while the

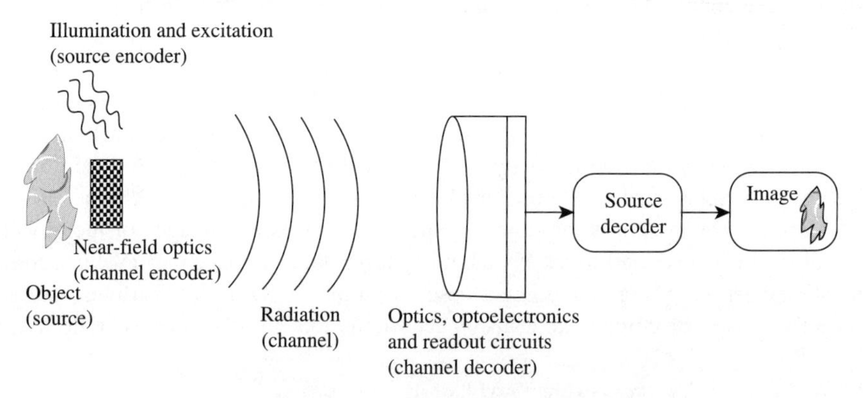

Figure 8.2 Coding in optical sensors.

latter produces a digital description of a physical object. Performance metrics in algebraic coding theory are relatively straightforward. The source data are a set of symbols. The goals are for the receiver to replicate these symbols without error and without some unfriendly receiver intercepting them. These goals are challenging because noise and distortions in the transmission channel corrupt the transmitted signal. Algebraic coding structures are used to avoid error and interception even in the presence of noise and eavesdropping. Performance metrics for algebraic coding include the rate of data transmission, the probability of error, the probability of interception, and the computational complexity of encoding and decoding. These metrics may be balanced against each other, codes with high transmission rates tend to produce greater error than do codes with lower rates and more redundancy.

Performance metrics and the goals of code design are more difficult to define for optical sensing. The source data are a unique physical object. The overall system goal is to transmit and estimate a set of symbols describing this object. The sensor does not exactly replicate the object. Common measures of system performance include the mean-square error between a set of object properties and the corresponding image properties. Alternatively, a weighted sum of errors based on the information value of each image property or a utility measure based on the decision value of the image as a whole might be considered, or the information transfer capacity between the object state and the image might be evaluated.

As increasingly sophisticated computational sensor systems emerge, performance metrics have become more diverse and application-specific. A chemometric imaging system might be evaluted on the basis of its chemical classification and concentration accuracy, a biometric system might be evaluated in terms of human identification metrics, or an imaging system might be evaluated according to the visual appearence of specific types of objects. For example, infrared imaging systems are evaluated according to the range at which a human observer can find a human target in the image. These complex performance-specific measures are often difficult to accurately model in hierachical sensor systems in which the channel coding and decoding (e.g., the optoelectronic subsystem) is largely distinct from the object decoding (i.e., digital signal processing).

Sensor system coding generally predates algebraic coding; indeed, Golay's work on coded aperture spectroscopy [94] is a primary precursor to algebraic coding theory. Today sensor system coding is less advanced than algebraic coding, but substantial progress has occurred recently, and great promise is clear in continuing research. Unfortunately, we are unable to present a definitive analysis of optimal sensor codes and decoding (e.g., image estimation) algorithms in this text. We are able to describe the structure of physical-layer coding in optical systems and the motivation for nontrivial coding and physical components for code implementation.

With reference to the optical sensor model expressed in Eqns. (7.1) and (7.3), this chapter focuses on *coding* as the process of designing the continuous or discrete measurement operators $\mathbf{h}(\mathbf{x}, \mathbf{x}')$ or $\mathbf{H}$ to optimize system performance. A model of $\mathbf{H}$ and of the object and data sampling structure is called a "forward model." Inversion of the forward model to estimate the object is the "inverse problem." The basic idea of deliberate coding of the forward problem to match advanced inverse algorithms is "integrated sensing and processing," as developed by Dennis Healy in a series of DARPA programs.

In traditional focal imaging one seeks a physical system such that $\mathbf{h}(\mathbf{x}, \mathbf{x}') \approx \delta(\mathbf{x} - \mathbf{x}')$ or $\mathbf{H} \approx \mathbf{I}$. We refer to the point-to-point mappings described by such identity responses as *isomorphisms*. In many cases, isomorphic mapping may in fact be the optimal imaging system code. In other cases, *multiplex* sensing, under which each measurement depends jointly on multiple object points, may be preferred. Multiplex sensing is of particular interest in situations in which

- Isomorphic mapping is not possible. Imaging over an extended depth of field, as discussed in Section 10.2, as well as tomographic and multidimensional imaging, as discussed in Sections 2.6, 6.4.2, 6.5, and 10.6, fall in this category.
- Isomorphic mappings are imperfect. Chapters 4 and 6 describe band limits on focal mappings. In addition to these Fourier limits, imperfections in optical fabrications and the failure of the paraxial approximation produce blurring and aberration in imaging systems. Code design balances the impact of image blurring and distortion by joint optimization of optical design and image processing algorithms.
- Optical components for isomorphic mappings are impractical or unavailable. Coded aperture imaging, as discussed in Section 2.5, is an example of such a situation.
- The cost per measurement is high. In some spectral ranges the cost of detectors or physical implementation limits may restrict the number of measurements. In such cases deliberate blurring of the PSF enables compressive measurement.
- The combination of prior object knowledge and nonlinear reconstruction algorithms produce better results than do the linear estimators. As an example, Ashok and Neifeld [5] consider deliberate blurring to enable subpixel point target identification and tracking. Alternative examples may be found in diverse feature and target identification systems.

Code design for optical systems balances physical constraints on achievable mappings against mathematically desireable code properties. Coding strategies implemented in optical sensor systems may be organized according to the physical significance of code parameters and the number of parameters per reconstructed pixel element. By a "code parameter," we mean a physical variable set by the system designer. A code parameter may correspond to direct modulation of the optical signal in a spatial light modulator or to diverse lens design parameters. In some systems, each value of $\boldsymbol{H}$ is independently set by the code designer, but in most optical systems $\boldsymbol{H}$ is implicitly determined by a relatively small number of design parameters. The range of coding strategies for optical systems is captured in the following taxonomy:

- *Pixel coding* refers to systems that directly and locally modulate data in an image plane. Pixel code parameters correspond directly to elements of $\mathbf{H}$. The physical significance of a code value may correspond, for example, to the transmission of a mask pixel. If each value of $\mathbf{H}$ can be independently selected, the number of code values greatly exceeds the number of signal pixels reconstructed. Pixel coding is commonly used in spectroscopy and spectral imaging. Structured spatial and temporal modulation of object illumination is also an example of pixel coding. In imaging systems, focal plane foveation and some forms of embedded readout circuit processing may also be considered as pixel coding. The impulse response of a pixel coded system is shift-variant. Physical constraints typically limit the maximum value or total energy of the elements of $\mathbf{H}$.

- *Convolutional coding* refers to systems with shift-invariant impulse reponse $\mathbf{h}(\mathbf{x}-\mathbf{x}')$. As we have seen in imaging system analysis, convolutional coding is exceedingly common in optical systems, with conventional focal imaging as the canonical example. Further examples arise in dispersive spectroscopy. We further divide convolutional coding into *projective coding*, under which code parameters directly modulate the spatial structure of the impulse response, and *Fourier coding*, under which code parameters modulate the spatial structure of the transfer function. Coded aperture imaging and computed tomography are examples of projective coding systems. Section 10.2 describes the use of pupil plane modulation to implement Fourier coding for extended depth of field. The number of code elements in a convolutional code corresponds to the number of resolution elements in the impulse response. Since the support of the impulse response is usually much less than the support of the image, the number of code elements per image pixel is much less than one.
- *Implicit coding* refers to systems where code parameters do not directly modulate $\mathbf{H}$. Rather, the physical structure of optical elements and the sampling geometry are selected to create an invertible measurement code. Reference structure tomography, van Cittert–Zernike-based imaging, and Fourier transform spectroscopy are examples of implicit coding. Spectral filtering using thin-film filters is another example of implicit coding. More sophisticated spatiospectral coding using photonic crystal, plasmonic, and thin-film filters are under exploration. The number of coding parameters per signal pixel in current implicit coding systems is much less than one, but as the science of complex optical design and fabrication develops, one may imagine more sophisticated implicit coding systems.

The goal of this chapter is to provide the reader with a context for discussing spectrometer and imager design in Chapters 9 and 10. We do not discuss physical implementations of pixel, convolutional, or implicit codes in this chapter. Each coding strategy arises in diverse situations; practical sensor codes often combine aspects of all three. In considering sensor designs, the primary goal is always to compare system performance metrics against design choices. Accurate sampling and signal estimation models are central to such comparisons. We learned how to model sampling in Chapter 7, the present chapter discusses basic stragies for signal estimation and how these strategies impact code design for each type of code.

The reader may find the pace of discussion a bit unusual in this chapter. Apt comparison may be made with Chapter 3, which progresses from traditional Fourier sampling theory through modern multiscale sampling. Similarly, the present chapter describes results that are 50–200 years old in discussing linear estimation strategies for pixel and convolutional coding in Sections 8.2 and 8.3. As with wavelets in Chapter 3, Sections 8.4 and 8.5 describe relatively recent perspectives, focusing in this case on regularization, generalized sampling, and nonlinear signal inference. A sharp distinction exists in the impact of modern methods, however. In the transition from Fourier to multiband sampling, new theories augment and extend Shannon's basic approach. Nonlinear estimators, on the other hand, substantially replace and revolutionize traditional linear estimators and completely undermine traditional approaches to sampling code design. As indicated by the hierarchy of data readout and processing steps described in Section 7.4, nonlinear processing has become ubiquitous even in the simplest and most isomorphic sensor systems. A system designer refusing to apply multiscale methods can do reasonable, if unfortunately constrained, work, but competitive design cannot refuse the benefits of nonlinear inference.

While the narrative of this chapter through coding strategies also outlines the basic landscape of coding and inverse problems, our discussion just scratches the surface of digital image estimation and analysis. We cannot hope to provide even a representative bibliography, but we note that more recent accessible discussions of inverse problems in imaging are presented by Blahut [21], Bertero and Boccacci [19], and Barrett and Myers [8]. The point estimation problem and regularization methods are well covered by Hansen [111], Vogel [241], and Aster et al. [6]. A modern text covering image processing, generalized sampling, and convex optimization has yet to be published, but the text and extensive websites of Boyd and Vandenberghe [24] provide an excellent overview of the broad problem.

8.2 PIXEL CODING

Let $\mathbf{f}$ be a discrete representation of an optical signal, and let $\mathbf{g}$ represent a measurement. We assume that both $\mathbf{f}$ and $\mathbf{g}$ represent optical power densities, meaning that f_i and g_i are real with $f_i,\ g_i \geq 0$. The transformation from $\mathbf{f}$ to $\mathbf{g}$ is

$$\mathbf{g} = \mathbf{Hf} + \mathbf{n} \tag{8.1}$$

where $\mathbf{n}$ represents measurement noise. Pixel coding consists of codesign of the elements of $\mathbf{H}$ and a signal estimation algorithm.

The range of the code elements h_{ij} is constrained in physical systems. Typically, h_{ij} is nonnegative. Common additional constraints include $0 \leq h_{ij} \leq 1$ or $\sum_i h_{ij} \leq 1$. Design of $\mathbf{H}$ subject to constraints is a *weighing design* problem. A classic example of the weighing design problem is illustrated in Fig. 8.3. The problem is to determine the masses of N objects using a balance. One may place objects singly or in groups on the left or right side. One places a calibrated mass on the

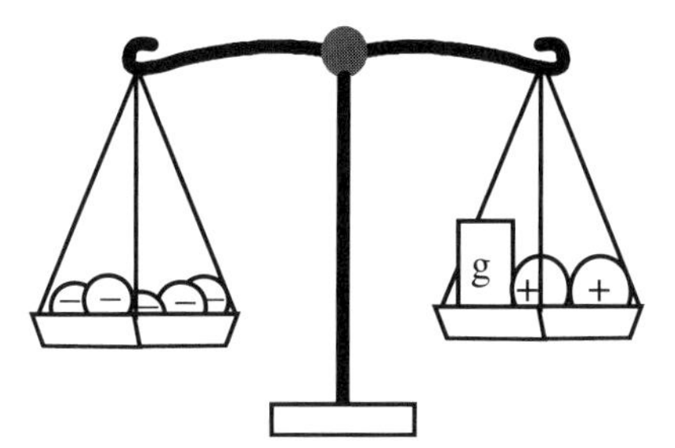

Figure 8.3 Weighing objects on a balance.

right side to balance the scale. The ith measurement takes the form

$$g_i + \sum_j h_{ij} m_j = 0 \tag{8.2}$$

where m_j is the mass of the jth object. h_{ij} is $+1$ for objects on the right, -1 for objects on the left and 0 for objects left out of the ith measurement. While one might naively choose to weigh each object on the scale in series (e.g., select $h_{ij} = -\delta_{ij}$), this strategy is just one of many possible weighing designs and is not necessarily the one that produces the best estimate of the object weights. The "best" strategy is the one that enables the most accurate estimation of the weights in the context of a noise and error model for measurement. If, for example, the error in each measurement is independent of the masses weighed, then one can show that the mean-square error in weighing the set of objects is reduced by group testing using the Hadamard testing strategy discussed below.

8.2.1 Linear Estimators

In statistics, the problem of estimating **f** from **g** in Eqn. (8.1) is called *point estimation*. The most common solution relies on a regression model with a goal of minimizing the difference between the measurement vector $\mathbf{Hf_e}$ produced by an estimate of **f** and the observed measurements **g**. The mean-square regression error is

$$\varepsilon(\mathbf{f_e}) = \langle (\mathbf{g} - \mathbf{Hf}_e)'(\mathbf{g} - \mathbf{Hf}_e) \rangle \tag{8.3}$$

The minimum of ε with respect to $\mathbf{f_e}$ occurs at $\partial\varepsilon/\partial\mathbf{f_e} = 0$, which is equivalent to

$$-\mathbf{H}'\mathbf{g} + \mathbf{H}'\mathbf{H}\mathbf{f_e} = 0 \tag{8.4}$$

This produces the ordinary least-squares (OLS) estimator for **f**:

$$\mathbf{f_e} = (\mathbf{H}'\mathbf{H})^{-1}\mathbf{H}'\mathbf{g} \tag{8.5}$$

So far, we have made no assumptions about the noise vector $\mathbf{n}$. We have only assumed that our goal is to find a signal estimate that minimizes the mean-square error when placed in the forward model for the measurement. If the expected value of the noise vector $\langle \mathbf{n} \rangle$ is nonzero, then the linear estimate $\mathbf{f_e}$ will in general be biased. If, on the other hand

$$\langle \mathbf{n} \rangle = 0 \tag{8.6}$$

and

$$\langle \mathbf{nn}' \rangle = \sigma^2 \mathbf{I} \tag{8.7}$$

then the OLS estimator is unbiased and the covariance of the estimate is

$$\Sigma_{f_e} = \sigma^2 (\mathbf{H}'\mathbf{H})^{-1} \tag{8.8}$$

The *Gauss–Markov theorem* [147] states that the OLS estimator is the *best linear unbiased estimator* where "best" in this context means that the covariance is minimal. Specifically, if $\Sigma_{\tilde{f}_e}$ is the covariance for another linear estimator $\tilde{\mathbf{f}}_\mathbf{e}$, then $\Sigma_{\tilde{f}_e} - \Sigma_{f_e}$ is a positive semidefinite matrix.

In practical sensor systems, many situations arise in which the axioms of the Gauss–Markov theorem are not valid and in which nonlinear estimators are preferred. The OLS estimator, however, is a good starting point for the fundamental challenge of sensor system coding, which is to codesign $\mathbf{H}$ and signal inference algorithms so as to optimize system performance metrics. Suppose, specifically, that the system metric is the mean-square estimation error

$$\sigma_e^2 = \frac{1}{N} \operatorname{trace}\left(\Sigma_{f_e}\right) \tag{8.9}$$

where $\mathbf{H}'\mathbf{H}$ is an $N \times N$ matrix. If we choose the OLS estimator as our signal inference algorithm, then the system metric is optimized by choosing $\mathbf{H}$ to minimize $\operatorname{trace}[(\mathbf{H}'\mathbf{H})^{-1}]$.

The selection of $\mathbf{H}$ for a given measurement system balances the goal of minimizing estimation error against physical implementation constraints. In the case that $\sum_j h_{ij} \leq 1$, for example, the best choice is the identity $h_{ij} = \delta_{ij}$. This is the most common case for imaging, where the amount of energy one can extract from each pixel is finite.

8.2.2 Hadamard Codes

Considering the weighing design constraint $|h_{ij}| \leq 1$, Hotelling proved in 1944 that for $h_{ij} \in [-1, 1]$

$$\sigma_e^2 \geq \frac{\sigma^2}{N} \tag{8.10}$$

under the assumptions of Eqn. (8.6). The measurement matrix $\mathbf{H}$ that achieves Hotelling's minimum estimation variance had been explored a half century earlier by Hadamard. A Hadamard matrix $\mathbf{H_n}$ of order n is an $n \times n$ matrix with elements $h_{ij} \in \{-1, +1\}$ such that

$$\mathbf{H_n}\mathbf{H_n'} = n\mathbf{I} \tag{8.11}$$

where $\mathbf{I}$ is the $n \times n$ identity matrix. As an example, we have

$$\mathbf{H}_2 = \begin{bmatrix} + & + \\ + & - \end{bmatrix} \tag{8.12}$$

If $\mathbf{H_a}$ and $\mathbf{H_b}$ are Hadamard matrices, then the Krönecker product $\mathbf{H_{ab}} = \mathbf{H_a} \otimes \mathbf{H_b}$ is a Hadamard matrix of order ab. Applying this rule to $\mathbf{H}_2$, we find

$$\mathbf{H}_4 = \begin{bmatrix} + & + & + & + \\ + & - & + & - \\ + & + & - & - \\ + & - & - & + \end{bmatrix} \tag{8.13}$$

Recursive application of the Krönecker product yields Hadamard matrices for $n = 2^m$. In addition to $n = 1$ and $n = 2$, it is conjectured that Hadamard matrices exist for all $n = 4m$, where m is an integer. Currently (2008) $n = 668$ ($m = 167$) is the smallest number for which this conjecture is unproven.

Assuming that the measurement matrix $\mathbf{H}$ is a Hadamard matrix $\mathbf{H'H} = N\mathbf{I}$, we obtain

$$\Sigma_{f_e} = \frac{\sigma^2}{N}\mathbf{I} \tag{8.14}$$

and

$$\sigma_e^2 = \frac{\sigma^2}{N} \tag{8.15}$$

If there is no Hadamard matrix of order N, the minimum variance is somewhat worse.

Hotelling also considered measurements $h_{ij} \in 0, 1$, which arises for weighing with a spring scale rather than a balance. The nonnegative measurement constraint $0 < h_{ij} < 1$ is common in imaging and spectroscopy. As discussed by Harwit and Sloane [114], minimum variance least-squares estimation under this constraint is achieved using the Hadamard S matrix:

$$S_n = \tfrac{1}{2}(1 - \mathbf{H}_n) \tag{8.16}$$

Under this definition, the first row and column of $\mathbf{S}_n$ vanish, meaning that $\mathbf{S}_n$ is an $(n-1) \times (n-1)$ measurement matrix. The effect of using the S matrix of order n rather than the bipolar Hadamard matrix is an approximately four-fold increase in the least-squares variance.

Spectroscopic systems often simulate Hadamard measurement by subtracting S-matrix measurements from measurements based on the complement $\tilde{S}_n = (\boldsymbol{H}_n + 1)/2$. This difference isolates $g = \boldsymbol{H}_n \boldsymbol{f}$. The net effect of this subtraction is to increase the variance of each effective measurement by a factor of 2, meaning that least squares processing produces a factor of 2 greater signal estimation variance. This result is better than for the S matrix alone because the number of measurements has been doubled.

8.3 CONVOLUTIONAL CODING

As illustrated in Eqns. (2.30), (4.75), and (6.63), convolutional transformations of the form

$$g(x, y) = \iint f(x', y')h(x - x', y - y')dx'dy' + n(x, y) \tag{8.17}$$

where $n(x, y)$ represents noise, are common in optical systems. We first encountered the coding problem, namely, design of $h(x, y)$ to enable high fidelity estimation of $f(x, y)$, in the context of coded aperture imaging. The present section briefly reviews both code design and linear algorithms for estimation of $f(x, y)$ from coded data.

The naive approach to inversion of Eqn. (8.17) divides the Fourier spectrum of the measured data by the system transfer function according to the convolution theorem [Eqn. (3.18)] to obtain an estimate of the object spectrum

$$\hat{f}_{\text{est}}(u, v) = \frac{\hat{g}(u, v)}{\hat{h}(u, v)} \tag{8.18}$$

As we saw in Problem 2.10, this approach tends to amplify noise in spectral ranges where $|h(u, v)|$ is small.

In 1942, Wiener proposed the alternative deconvolution strategy based on minimizing the mean-square error

$$\begin{aligned} e^2 &= \iint \left\langle (f - f_{\text{est}})^2 \right\rangle dx\, dy \\ &= \left\langle \iint (\hat{f} - \hat{f}_{\text{est}})^2\, du\, dv \right\rangle \end{aligned} \tag{8.19}$$

Noting that $\varepsilon(u, v) = \langle (\hat{f} - \hat{f}_{\text{est}})^2 \rangle$ is nonnegative everywhere, one minimizes e^2 by minimizing $\varepsilon(u, v)$ at all (u, v). Supposing that $\hat{f}_{\text{est}} = \hat{w}(u, v)\hat{g}(u, v)$, we find

$$\begin{aligned} \varepsilon(u, v) &= \left\langle |\hat{f}(u, v)|^2 \right\rangle - \left\langle \hat{f}\hat{w}^*(\hat{h}^*\hat{f}^* + \hat{n}^*) \right\rangle \\ &\quad - \left\langle \hat{f}^*\hat{w}(\hat{h}\hat{f} + \hat{n}) \right\rangle + \left\langle |\hat{w}|^2 |(\hat{h}\hat{f} + \hat{n})|^2 \right\rangle \\ &= |[1 - \hat{w}(u, v)\hat{h}(u, v)]|^2 S_f(u, v) + |\hat{w}|^2 S_n(u, v) \end{aligned} \tag{8.20}$$

where we assume that the signal and noise spectra are uncorrelated such that $\langle \hat{f}(u, v)\hat{n}^*(u, v)\rangle = 0$. $S_n(u, v)$ and $S_f(u, v)$ are the statistical expectation values of the power spectral density of noise and of the signal, $S_f(u, v) = \langle|\hat{f}(u, v)|^2\rangle$ and $S_n(u, v) = \langle|\hat{n}(u, v)|^2\rangle$. Setting the derivative of $\varepsilon(u, v)$ with respect to $\hat{w}$ equal to zero yields the extremum $-\hat{h}[1 - \hat{w}(u, v)\hat{h}(u, v)]^* S_f(u, v) + \hat{w}^* S_n(u, v) = 0$. The minimum mean-square error estimation filter is thus the *Wiener filter*

$$\hat{w}(u, v) = \frac{\hat{h}^*(u, v)S_f(u, v)}{|\hat{h}(u, v)|^2 S_f(u, v) + S_n(u, v)} \tag{8.21}$$

The Wiener filter reduces to the direct inversion filter of Eqn. (8.18) if the signal-to-noise ratio $S_f/S_n \gg 1$. At spatial frequencies for which the noise power spectrum becomes comparable to $|\hat{h}(u, v)|^2 S_f(u, v)$, the noise spectrum term in the denominator prevents the weak transfer function from amplifying noise in the detected data.

Substituting in Eqn. (8.20), the mean-square error at spatial frequency (u, v) for the Wiener filter is

$$\varepsilon(u, v) = \frac{S_f(u, v)}{1 + |\hat{h}(u, v)|^2 [S_f(u, v)/S_n(u, v)]} \tag{8.22}$$

Convolutional code design consists of selection of $\hat{h}(u, v)$ to optimize some metric. While minimization of the mean-square error is not the only appropriate design metric, it is an attractive goal. Since the Wiener error decreases monotonically with $|\hat{h}(u, v)|^2$, error minimization is achieved by maximizing $|\hat{h}(u, v)|^2$ across the target spatial spectrum.

Code design is trivial for focal imaging, where Eqn. (8.22) indicates clear advantages for forming as tight a point spread function as possible. Ideally, one selects $h(x, y) = \delta(x, y)$, such that $\hat{h}(u, v)$ is constant. As discussed in Section 8.1, however, in certain situations design to the goal $h(x, y) = \delta(x, y)$ is not the best choice. Of course, as discussed in Sections 8.4 and 8.5, one is unlikely to invert using the Wiener filter in such situations.

Figure 8.4 illustrates the potential advantage of coding for coded aperture systems by plotting the error of Eqn. (8.22) under the assumption that the signal and noise power spectra are constant. The error decreases as the order of the coded aperture increases, although the improvement is sublinear in the throughput of the mask. The student will, of course, wish to compare the estimation noise of the Wiener filter with the earlier SNR analysis of Eqns. (2.47) and (2.48).

The nonuniformity of the SNR across the spectral band illustrated in Fig. 8.4 is typical of linear deconvolution strategies. Estimation error tends to be particularly high in near nulls or minima in the MTF. Nonlinear methods, in contrast, may utilize relationships between spectral components to estimate information even from bands where the system transfer function vanishes. Nonlinear strategies are also more effective in enforcing structural prior knowledge, such as the nonnegativity of optical signals.

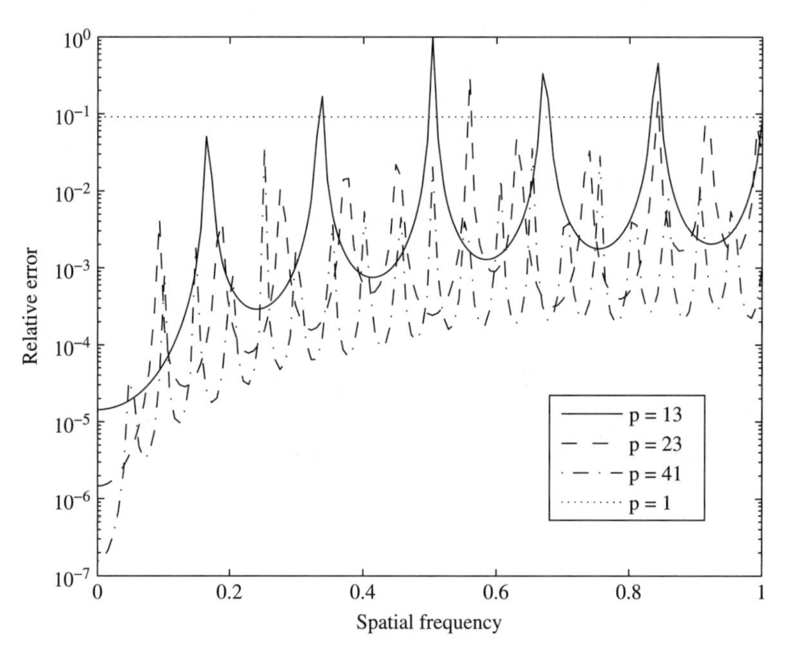

Figure 8.4 Relative mean-square error as a function of spatial frequency for MURA coded apertures of various orders. The MURA code is described by Eqn. (2.45). We assume that $S_f(u, v)$ is a constant and that $S_f(u, v)/S_n(u, v) = 10$.

The Wiener filter is an example of *regularization*. Regularization constrains inverse problems to keep noise from weakly sensed signal components from swamping data from more strongly sensed components. The Wiener filter specifically damps noise from null regions of the system transfer function. In discrete form, Eqn. (8.17) is implemented by Toeplitz matricies. Hansen presents a recent review of deconvolution and regularization with Toeplitz matrices [112]. We consider regularization in more detail in the next section.

8.4 IMPLICIT CODING

A coding strategy is "explicit" if the system designer directly sets each element h_{ij} of the system response $\boldsymbol{H}$ and "implicit" if $\boldsymbol{H}$ is determined indirectly from design parameters. Coded aperture spectroscopy (Section 9.3) and wavefront coding (Section 10.2.2) are examples of explicit code designs. Most optical systems, however, rely on implicit coding strategies where a relatively small number of lens or filter parameters determine the large-scale system response. Even in explicitly coded systems, the actual system response always differs somewhat from the design response.

Reference structure tomography (RST; Section 2.7) provides a simple example of the relationship between physical system parameters and sensor response. Physical parameters consist of the size and location of reference structures. Placing one reference structure in the embedding space potentially modulates the visibility for all sensors. While the RST forward model is linear, optimization of the reference structure against coding and object estimation metrics is nonlinear. This problem is mostly academic in the RST context, but the nonlinear relationship between optical system parameters and the forward model is a ubiquitous issue in design.

The present section considers coding and signal estimation when $\boldsymbol{H}$ cannot be explicitly encoded. Of course an implicitly encoded system response is unlikely to assume an ideal Hadamard or identity matrix form. On the other hand, we may find that the Hadamard form is less ideal than we have previously supposed. Our goals are to consider (1) signal estimation strategies when $\boldsymbol{H}$ is ill-conditioned and (2) design goals for implicit ill-conditioned $\boldsymbol{H}$.

The $m \times n$ measurement matrix $\mathbf{H}$ has a singular value decomposition (SVD)

$$\boldsymbol{H} = \boldsymbol{U}\Lambda\boldsymbol{V}' \tag{8.23}$$

where $\boldsymbol{U}$ is an $m \times m$ unitary matrix. The columns of $\boldsymbol{U}$ consist of orthonormal vectors $\boldsymbol{u}_i$ such that $\boldsymbol{u}_i \cdot \boldsymbol{u}_j = \delta_{ij}$. $\{\boldsymbol{u}_i\}$ form a basis of $\mathbb{R}^m$ spanning the *data space*. $\mathbf{V}$ is similarly an $n \times n$ unitary matrix with columns v_i spanning the *object space* $\mathbb{R}^n$. Λ is an $m \times n$ diagonal matrix with diagonal elements λ_i corresponding to the singular values of $\boldsymbol{H}$ [97]. The singular values are nonnegative and ordered such that

$$\lambda_1 \geq \lambda_2 \geq \cdots \geq \lambda_n \geq 0 \tag{8.24}$$

The number of nonzero singular values r is the *rank* of $\boldsymbol{H}$ and the ratio greatest singular value to the least nonzero singular value λ_1/λ_r is the *condition number* of $\boldsymbol{H} \cdot \boldsymbol{H}$ is said to be ill-conditioned if the condition number is much greater than 1.

Inversion of $\boldsymbol{g} = \boldsymbol{Hf} + \boldsymbol{n}$ using the SVD is straightforward. The data and object null spaces are spanned by the $m-r$ and $n-r$ vectors in $\boldsymbol{U}$ and $\boldsymbol{V}$ corresponding to null singular values. The data range is spanned by the columns of $\boldsymbol{U}_r = (\boldsymbol{u}_1, \boldsymbol{u}_2, \ldots, \boldsymbol{u}_r)$. The object range is spanned by the columns of $\boldsymbol{V}_r = (\boldsymbol{v}_1, \boldsymbol{v}_2, \ldots, \boldsymbol{v}_r)$. The generalized or *Moore–Penrose pseudoinverse* of $\boldsymbol{H}$ is

$$\boldsymbol{H}^{\dagger} = \boldsymbol{V}_r\Lambda_r^{-1}\boldsymbol{U}_r^T \tag{8.25}$$

One obtains a naive object estimate using the pseudoinverse as

$$\begin{aligned} \boldsymbol{f}_{\text{naive}} &= \boldsymbol{H}^{\dagger}\boldsymbol{g} \\ &= P_{V_H}\boldsymbol{f} + \sum_{i=1}^{r} \frac{\boldsymbol{u}_i \cdot \boldsymbol{n}}{\lambda_i}\boldsymbol{v}_i \end{aligned} \tag{8.26}$$

where $P_{V_H}\boldsymbol{f}$ is the projection of the object onto V_H. The problem with naive inversion is immediately obvious from Eqn. (8.26). If noise is uniformly distributed over the data space, then the noise components corresponding to small singular values are amplified by the factor $1/\lambda_i$.

Regularization of the pseudoinverse consists of removing or damping the effect of singular components corresponding to small singular values. The most direct regularization strategy consists of simply forming a psuedoinverse from a subset of the singular values with λ_i greater than some threshold, thereby improving the effective condition number. This approach is called *truncated SVD reconstruction*.

Consider, for example, the shift-coded downsampling matrix. A simple downsampling matrix takes Haar averages at a certain level. For example, 4× downsampling is effectively a projection up two levels on the Haar basis. A 4× downsampling matrix takes the form

$$\boldsymbol{H} = \begin{matrix} \ddots & \vdots & \vdots & \vdots & \vdots & \vdots & \vdots & \vdots & \vdots & \vdots & \vdots & \vdots & \vdots & \\ \cdots & \frac{1}{4} & \frac{1}{4} & \frac{1}{4} & \frac{1}{4} & 0 & 0 & 0 & 0 & 0 & 0 & 0 & 0 & \cdots \\ \cdots & 0 & 0 & 0 & 0 & \frac{1}{4} & \frac{1}{4} & \frac{1}{4} & \frac{1}{4} & 0 & 0 & 0 & 0 & \cdots \\ \cdots & 0 & 0 & 0 & 0 & 0 & 0 & 0 & 0 & \frac{1}{4} & \frac{1}{4} & \frac{1}{4} & \frac{1}{4} & \cdots \\ & \vdots & \vdots & \vdots & \vdots & \vdots & \vdots & \vdots & \vdots & \vdots & \vdots & \vdots & \vdots & \vdots & \ddots \end{matrix} \tag{8.27}$$

In general, downsampling by the factor d projects $\boldsymbol{f}$ from $\mathbb{R}^n$ to $\mathbb{R}^{n/d}$.

Digital superresolution over multiple apertures or multiple exposures combines downsampled images with diverse sampling phases to restore $\boldsymbol{f} \in \mathbb{R}^n$ from d different projections in $\mathbb{R}^{n/d}$. We discuss digital superresolution in Section 10.4.2. For the present purposes, the shift-coded downsampling operator is useful to illustrate regularization. By "shift coding" we mean the matrix that includes all single pixel shifts of the downsampling vector. For 4× downsampling the shift coded operator is

$$\boldsymbol{H} = \begin{matrix} \ddots & & \vdots & \vdots & \vdots & \vdots & \vdots & \vdots & \vdots & \vdots & \\ \cdots & & \frac{1}{4} & \frac{1}{4} & \frac{1}{4} & \frac{1}{4} & 0 & 0 & 0 & 0 & 0 & \cdots \\ \cdots & 0 & 0 & \frac{1}{4} & \frac{1}{4} & \frac{1}{4} & \frac{1}{4} & 0 & 0 & 0 & \cdots \\ \cdots & 0 & 0 & 0 & \frac{1}{4} & \frac{1}{4} & \frac{1}{4} & \frac{1}{4} & 0 & 0 & \cdots \\ \cdots & 0 & 0 & 0 & 0 & \frac{1}{4} & \frac{1}{4} & \frac{1}{4} & \frac{1}{4} & 0 & \cdots \\ \cdots & 0 & 0 & 0 & 0 & 0 & \frac{1}{4} & \frac{1}{4} & \frac{1}{4} & \frac{1}{4} & \cdots \\ & & \vdots & \vdots & \vdots & \vdots & \vdots & \vdots & \vdots & \vdots & \vdots & \ddots \end{matrix} \tag{8.28}$$

The singular value spectrum of a 256 × 256 shift-coded 4× downsample operator is illustrated in Fig. 8.5. Only one set of singular vectors is shown because the data and object space vectors are identical for Toeplitz matrices (e.g., matrices representing shift-invariant transformations) [112]. This singular value spectrum is typical of many measurement systems. Large singular values correspond to relatively low-frequency features in singular vectors. Small singular values correspond to singular vectors containing high-frequency components. By truncating the basis, one effectively lowpass-filters the reconstruction.

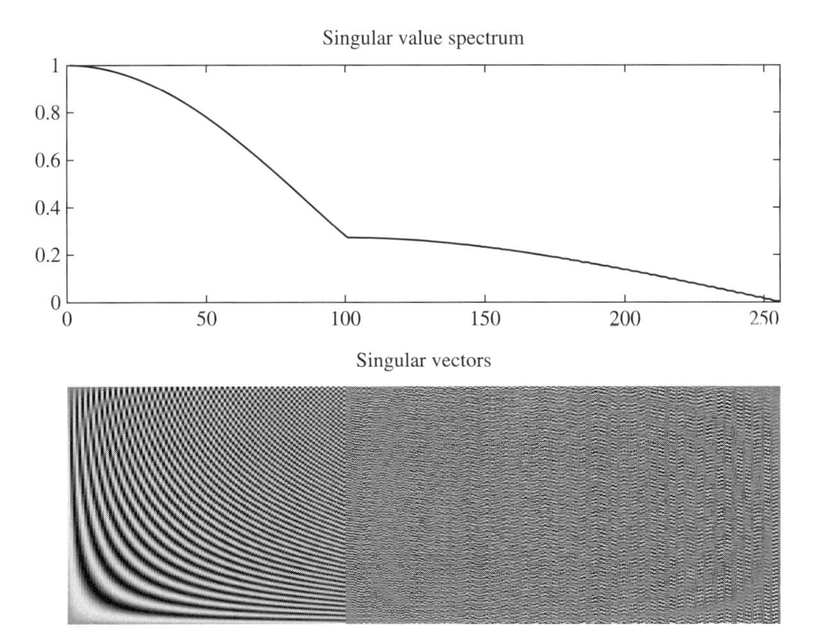

Figure 8.5 Singular values of a 256 × 256 shift-coded 4× downsample operator.

Transformations of images are greatly simplified if the system operator is separable in Cartesian coordinates. A separable downsampling operator may operate on an image $\boldsymbol{f}$ with a left operator $\boldsymbol{H}_l$ for vertical downsampling and a right operator $\boldsymbol{H}_r$ for horizontal downsampling. As an example, Fig. 8.6(a) shows a particular image consisting of a 256 × 256-pixel array. We model measured data from this image as

$$\boldsymbol{g} = \boldsymbol{H}_l \boldsymbol{f} \boldsymbol{H}_r' + n \tag{8.29}$$

The least mean-square estimate of the image for shift-coded 4× downsampling with $\sigma^2 = 10^{-4}$ normally distributed additive noise is illustrated in Fig. 8.6(b). As expected, the mean-square error is enormous because of the ill-conditioned measurement operators. Figure 8.6(c) is a truncated SVD reconstruction from the same data using the first 125 of 256 singular vectors. One observes both artifacts and blurring in the truncated SVD image; the loss of spatial resolution is illustrated in a detail from the center of the image in Fig. 8.7.

The mean-square error in the truncated SVD reconstruction (0.037) exceeds the measurement variance by more than two orders of magnitude. The MSE includes effects due to both noise and reconstruction bias, however. Since the truncated SVD reconstruction is not of full rank, image components in the null space of the reconstruction operator are dropped and lead to bias in the estimated image. One may consider that the goal of truncated SVD reconstruction is to measure the projection of $\boldsymbol{f}$ on the subspace spanned by the high singular value components. In this case,

Figure 8.6 A 256×256 image reconstructed using linear least-squares and truncated SVD: (a) original; (b) least-squares reconstruction MSE = 51.4; (c) truncated SVD MSE = $4.18e-003$.

one is more interested in the error between the estimated projection and the true projection, $||P_{V_H}\boldsymbol{f} - P_{V_H}\boldsymbol{f}_e||_2$. For the image of Fig. 8.6(c) the mean-square projection error is 3.3×10^{-4}, which is $3\times$ larger than the measurement variance. The vast majority of the difference between the reconstructed image and the original arises from bias due to the structure of the singular vectors. As discussed in Section 8.5, it might be possible to remove this bias using nonlinear inversion algorithms.

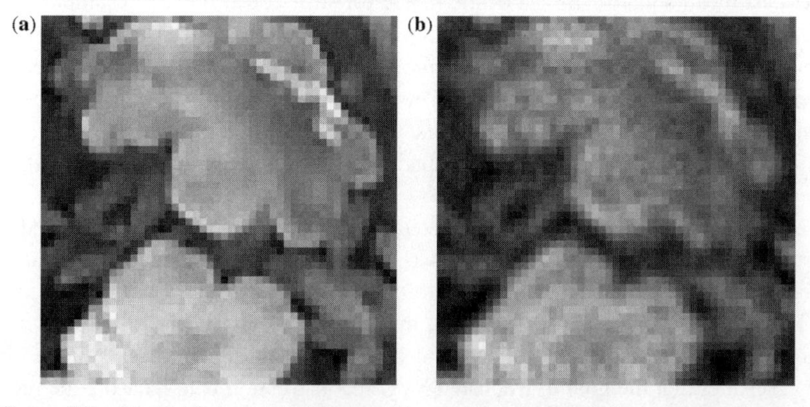

Figure 8.7 Detail of the original image (a) and the truncated SVD reconstruction (b).

Tikhonov regularization addresses the noise sensitivity of the pseudoinverse by constraining the norm of the estimated signal. The basic idea is that since noise causes large fluctuations, damping such fluctuations may reduce noise sensitivity. The goal is to find f_e satisfying

$$f_e = \arg\min_{f_e}\left\{ ||\mathbf{g} - \mathbf{H}f_e||_2^2 + \lambda_o^2 ||f_e||_2^2 \right\} \tag{8.30}$$

The solution to this constraint may be expressed in terms of the singular vectors as

$$f_e = \sum_{i=1}^{r} \frac{\lambda_i^2}{\lambda_i^2 + \lambda_o^2} \frac{\mathbf{u}_i \cdot \mathbf{g}}{\lambda_i} \mathbf{v}_i \tag{8.31}$$

Tikhonov regularization adjusts the pseudoinverse in a manner extremely similar to the adjustment that the Wiener filter makes to deconvolution. Singular components with $\lambda_i \gg \lambda_o$ are added to the estimated signal as with the normal pseudoinverse. Components with $\lambda_i \ll \lambda_o$ are damped in the reconstruction. In the limit that $\lambda_o \rightarrow 0$, the Tikhonov solution is the pseudoinverse solution (or least squares in

Figure 8.8 Reconstruction of the 4× downsampled shift coded system using Tikhonov regularization. Detail images at the bottom compare the same original and reconstructed regions as in Fig. 8.7.

the case of a rectangular system matrix). One may expect the Tikhonov solution to resemble the order-k truncated SVD solution in the range that $\lambda_k \approx \lambda_o$. Figure 8.8 is a Tikhonov reconstruction the data from Fig. 8.6 with $\lambda_o = 0.3$. There is no Tikhonov regularization parameter that obtains MSE comparable to the truncated SVD for this particular image, but one may expect images with more high-frequency content to achieve better Tikhonov restoration. Just as estimation of $S_n(u, v)$ is central to the Wiener filter, determination of λ_o is central to Tikhonov regularization. Tikhonov regularization is closely related to Wiener filtering, both are part of a large family of similar noise damping strategies. Since our primary focus here is on the design of $\boldsymbol{H}$, we refer the reader to the literature for further discussion [111].

The nominal design goal for implicit coding is basically the same as for pixel and convolutional coding: making the singular spectrum flat. Hadamard, Fourier transform, and identity matrices perform well under least-squares inversion because their singular values are all equal. Any measurement matrix formed of orthogonal row vectors similarly achieves uniform and independent estimation of the singular values (with the measurement row vectors forming the object space singular vectors). For the reasons listed in Section 8.3, however, there are many situations where unitary $\boldsymbol{H}$ is impossible or undesirable.

For implicit coding systems in particular, one seeks to optimize sensor system performance over a limited range of physical control parameters. Relatively subtle changes in sampling strategy may substantially impact signal estimation. As an example, consider again a $4\times$ downsampling system. Suppose that one can implement any 8 element shift invariant sampling code with four elements equal to $\frac{1}{4}$ and four elements equal to 0. The downsampling code 11110000/4 with SVD spectral illustrated in Fig. 8.6 is one such example, but there 70 different possible codes. Figure 8.9 plots the singular values for three such codes for a 128×128 measurement matrix. The 11110000 code produces the largest singular values for low-frequency singular vectors but lower singular values in the midrange of frequency response. The other example codes produce fewer low-frequency singular vectors and yield higher singular values in midrange. Figure 8.10 shows the $\lambda_o = 0.3$ Tikhonov reconstruction of the detail region shown in Figs. 8.7 and 8.8 for these codes with $\sigma^2 = 10^{-4}$. The MSE is higher for the noncompact PSFs, but one can argue that the Tihonov reconstruction using the 11100100 code captures features missed by the 11110000 code. Truncated SVD reconstruction using the disjoint codes produces artifacts due to the higher-frequency structure of the singular vectors. At this point, we argue only that code design matters, leaving our discussion for how it might matter to the next section.

More generally, we may decompose $\boldsymbol{f}$ in terms of the object space singular vectors as

$$\boldsymbol{f} = \sum_i f_i^{\mathrm{SV}} v_i \tag{8.32}$$

We may similarly decompose $\boldsymbol{g}$ in terms of the data space singular vectors. On these bases, the measurement take the form

$$\boldsymbol{g}^{\mathrm{SV}} = \Lambda \boldsymbol{f}^{\mathrm{SV}} + \boldsymbol{U}^T \boldsymbol{n} \tag{8.33}$$

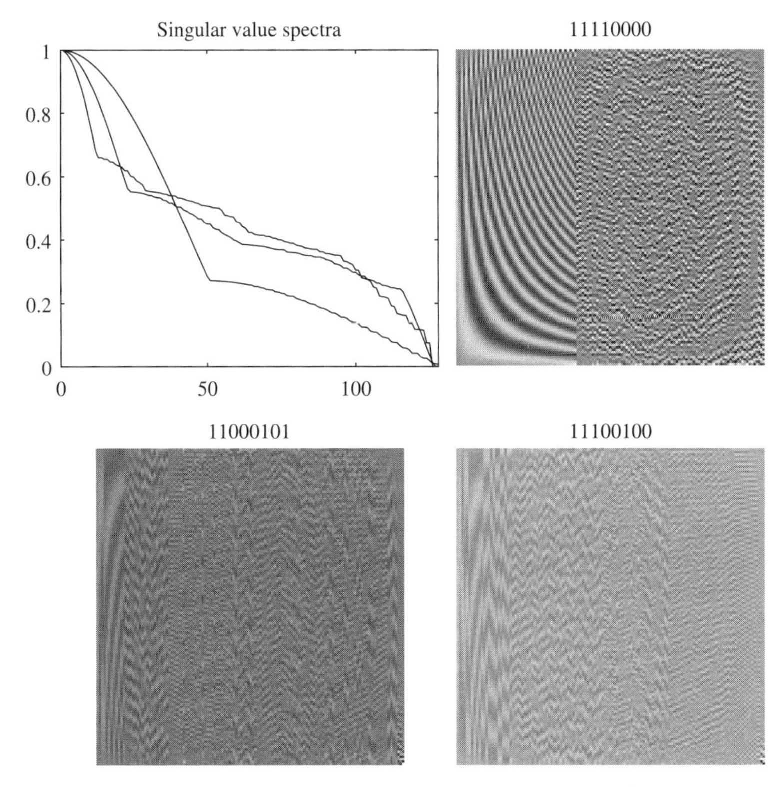

Figure 8.9 Singular value spectra for Toeplitz matrix sampling using eight-element convolutional codes. The code elements listed as 1 are implemented as $\frac{1}{4}$ so that the singular values are comparable to those in Fig. 8.5.

Since identically and independently distributed zero mean noise maintains these properties under unitary transformation, one obtains the covariance statistics of Eqn. (8.8) on least-squares inversion of Eqn. (8.33). In fact, since $\mathbf{\Lambda}$ is diagonal, each singular value component can be independently estimated with variance

$$\sigma^2_{f^{\mathrm{SV}}_{i,e}} = \frac{\sigma^2}{\lambda_i^2} \tag{8.34}$$

The significance of this variance in the estimated image depends on how singular value estimates are synthesized in the inversion process. One certainly expects to neglect components with $\sigma \gg \lambda$, but linear superposition of the remaining singular vectors is only one of many estimation algorithms.

One may confidently say that optical measurement effectively consists of measuring the singular value components f_i^{SV} for $\lambda_i > \sigma$. One has less confidence in asserting how one should design the structure of the singular vectors or how one should estimate $\boldsymbol{f}$ from the singular value components. Building on our discussion from

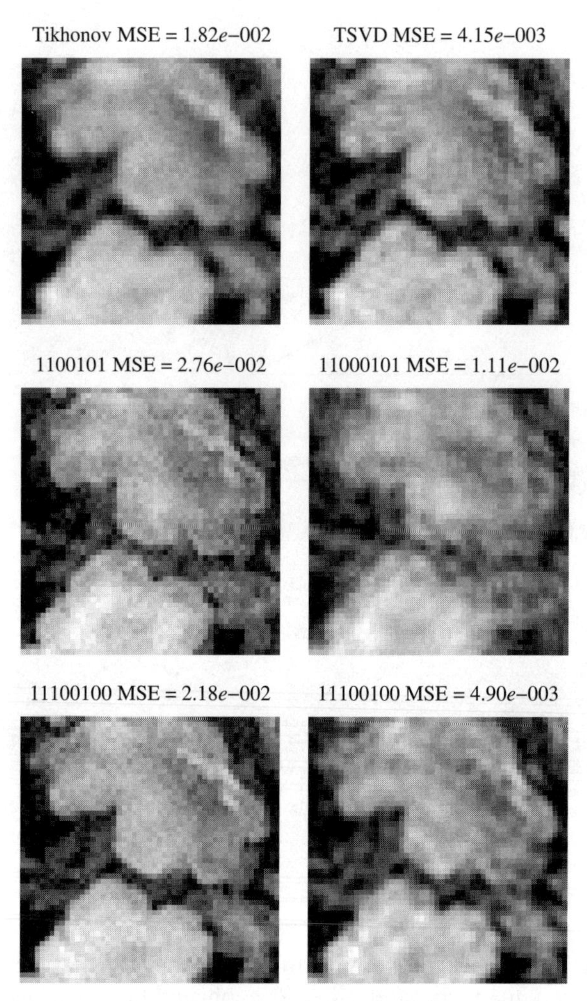

Figure 8.10 Tikhonov and truncated SVD reconstruction of the detail region of Fig. 8.7. Tikhonov reconstruction with $\lambda_0 = 0.3$ is illustrated on the left; the top image corresponds to the 11110000 code. The SVD on the right used the first 125 of 256 singular vectors from the left and right.

Section 7.5.4, one generally seeks to design $\boldsymbol{H}$ such that $\boldsymbol{f} \notin V_\perp$ and such that distinct images are mapped to distinct measurements. So long as these requirements are satisfied, one has some hope of reconstructing $\boldsymbol{f}$ accurately.

Truncated SVD data are anticompressive in the sense that one obtains fewer measurement data values than the number of raw measurements recorded. As we see with the reconstructions in this section, this does not imply that the number of estimated pixels is reduced. One may ask, however, why not measure the SVD projections directly? With this question we arrive at the heart of optical sensor design. One is unlikely to have the physical capacity to implement optimal object space projectors in a measurement system. Physical constraints on $\boldsymbol{H}$ determine the structure of the measurements. Optical sensor design consists of optimizing the singular values and singular vectors within physical constraints to optimize signal estimation. To understand the full extent of this problem, one must also consider the possibility of nonlinear image estimation, which is the focus of the next section.

8.5 INVERSE PROBLEMS

As discussed in Section 7.5, a generalized sampling system separates the processes of measurement, analysis, and display sampling. Generalized measurements consist of multiplex projections of the object state. With the exception of principal component analysis, the signal estimation algorithms mentioned in Section 7.5 bear little resemblence to the estimation algorithms considered thus far in the present chapter. As we have seen, however, linear least squares is only appropriate for well-conditioned measurement systems. Regularization methods, such as the Wiener filter and truncated SVD reconstruction, have wider applicability but produce biased reconstructions. The magnitude of the bias may be expected to grow as the effective rank (the number of useful singular values) drops.

Regularized SVD reconstruction differs sharply in this respect from compressed sensing. As discussed in Section 7.5.4, a compressively sampled sparse signal may be reconstructed without bias even though the measurement operator is of low rank. The present section considers similar methods for estimation of images sampled by ill-conditioned operators.

Prior to considering estimation strategies, it is useful to emphasize lessons learned in Section 8.4. Specifically, no matter what type of generalized sampling one follows in forward system design, the singular vectors of the as-implemented measurement model provide an excellent guide to the data that one actually measures. One may regard design of the singular vectors as the primary goal of implicit coding. Evaluation of the quality of the singular vectors depends on the image estimation algorithm.

Image estimation and analysis from a set of projections $f_i^{\mathrm{SV}} = \langle v_i, \boldsymbol{f} \rangle$ is an extraordinarily rich and complex subject. One can imagine, for example, that each singular vector could respond to a feature in a single image. One might in this case identify the image by probablistic analysis of the relative projections of the measurements. Once identified, the full image might be reconstructed on the basis of a single measurement value. One can imagine many variations on this theme targeting specific image features. As the primary focus of this text is the design of optical systems to estimate mostly unconstrained continuous images and spectra, however, we limit our attention to estimation more evolutionary revisions to least-squares methods.

As discussed at the end of Section 8.1, inverse problems have a long history and an extensive biography. The main objectives of the present section are to present a few examples to prepare the reader for design and analysis exercises in this and succeeding chapters. Inversion algorithms continue to evolve rapidly in the literature; the interested reader is well advised to explore beyond the simple presentation in this text.

We focus here on the two most popular strategies for image and spectrum estimation: *convex optimization* and *maximum likelihood* methods.

8.5.1 Convex Optimization

The *inverse problem* returns an estimated image f_e given the measurements $g = Hf + n$. Optimization-based estimation algorithms augment the measurements with an *objective function* $\gamma(f_e)$ describing the quality of the estimated image on the basis of prior knowledge. The objective function returns a scalar value. The optimization-based inverse problem may be summarized as follows

$$\mathbf{f}_e = \arg\min_{\mathbf{f}} \gamma(\mathbf{f})$$

such that

$$\mathbf{Hf} = \mathbf{g} \tag{8.35}$$

Image estimation using an objective function consists of finding the image estimate $\mathbf{f_e}$ consistent with the measurements that also minimizes the objective function.

The core issues in optimization-based image estimations are (1) selection of the objective function and (2) numerical optimization. The objective function may be derived from

- *Physical Constraints*. Unconstrained estimators may produce images that violate known physical properties of the object. The most common example in optical systems is nonnegativity. Optical power spectra and irradiance values cannot be negative, but algebraic and Wiener filter inversion commonly produces negative values from noisy data. Optimization of least-squares estimation with an objective function produces a better signal estimate than does truncation of nonphysical values.
- *Functional Constraints*. Natural objects do not consist of assortments of independent random pixels (commonly called "snow" in the age of analog television). Rather, pixel values are locally and globally correlated. Local correlation is often described as "smoothness," and pixels near a given pixel are likely to have similar values. Global correlation is described by sharpness, and edges tend to propagate long distances across an image. An objective function can enforce smoothness by limiting the spatial gradient of a reconstructed image and sharpness by constraining coefficients in wavelet or "curvelet" decompositions. Sparsity, as applied in compressive sampling, is also a functional constraint.
- *Feature Constraints*. At the highest level, image inference may be aware of the nature of the object. For example, knowledge that one is reconstructing an image of a dog may lead one to impose a "dog-like" constraint. Such higher-order analysis lies at the interface between computational imaging and machine vision and is not discussed here.

Constrained least-squares estimators provide the simplest optimization methods. Lawson and Hanson [146] present diverse algorithms for variations on the least-squares estimation problem, including the algorithm for nonnegative estimation implemented in Matlab as the function `lsqnonneg`. `lsqnonneg` is a recursive algorithm designed to move the ordinarly least-squares solution to the nearest nonnegative solution.

The least-gradient (LG) algorithm described by Pitsianis and Sun [31] provides a useful example of constrained least-squares methods. LG is closely related to well-known least squares with quadratic inequality (LSQI) minimization problems. The signal estimated by the LG agorithm is

$$\mathbf{f}_{\mathrm{LG}} = \arg\min_{\mathbf{f}} \gamma(\mathbf{f}) = \|\nabla \mathbf{f}\|_2$$

such that

$$\mathbf{H}\,\mathbf{f} = \mathbf{g} \tag{8.36}$$

where ∇ denotes the discrete gradient operation. When discretized over equispaced samples of a signal, the gradient may be the backward difference $\nabla_k \mathbf{f} = f_k - f_{k-1}$, or the forward difference, or the central difference. In matrix form, ∇ is an $(N-1)\times N$ bidiagonal matrix:

$$\nabla = \begin{bmatrix} -1 & 1 & 0 & \cdots & 0 \\ 0 & -1 & 1 & \cdots & 0 \\ \vdots & \vdots & \ddots & \ddots & \vdots \\ 0 & 0 & \cdots & -1 & 1 \end{bmatrix}$$

We obtain the LG solution in two steps. First, we find a particular least-squares solution $\mathbf{f}_p$ to the linear equation $\mathbf{H}\ \mathbf{f} = \mathbf{g}$. The general solution to the equation can then be described as $\mathbf{f} = \mathbf{f}_p + \mathbf{N}\ \mathbf{c}$, where $\mathbf{N}$ spans the null space of $\mathbf{H}$, and $\mathbf{c}$ is an arbitrary coefficient vector. The problem described by Eqn. (8.36) reduces to a linear least-squares problem without constraints:

$$\mathbf{f}_{\mathrm{LG}} = \arg\min_{\mathbf{c}} \|\nabla(\mathbf{N}\ \mathbf{c} - \mathbf{f}_p)\|_2^2$$

The solution is expressed

$$\mathbf{f}_{\mathrm{LG}} = \mathbf{f}_p - \mathbf{N}(\mathbf{N}^T \nabla^T \nabla \mathbf{N})^{-1} (\nabla \mathbf{N})^T\ \nabla \mathbf{f}_p \tag{8.37}$$

where we assume that the $\nabla\mathbf{N}$ is of full rank in columns. The general solution [Eqn. (8.37)] does not depend on the selection of a particular solution $\mathbf{f}_p$ to the measurement equation. More advanced strategies than ordinary least-squares inversion include QR factorization of the measurement matrix. Other approaches, like `lsqnonneg`, require iterative processing.

Figures 8.11 and 8.12 plot example LG reconstructions using the signal of Fig. 3.9. The measurement operator shown in Fig. 8.11 takes the level 0 Haar averages. (The function is modeled using 1024 points. The measurement operator consists of a 16×1024 matrix; 64 continuous values in each row are 1.) The measurement operator is a $64\times$ downsample matrix. Figure 8.11(a) shows the true function and the least-squares inversion from the downsampled data. Figure 8.11(b) is the LG reconstruction. For these measurements, LG estimation may be simply regarded as interpolation on sampled data.

Figure 8.12 considers the same data with the rect(x) sampling kernel replaced by sinc($8x$). The measurement operator is again 16×1024. As shown in Fig. 8.12(b), the least-squares inversion reflects the structure of the singular vectors of the measurement operator. The LG operator uses null space smoothing to remove the naive structure of the singular vectors. The efficacy of LG and other constrained least-squares methods depends on the structure of the sampled signal space. For example, the sinc($8x$) sampling function may achieve better results on sparse signals, as illustrated in Fig. 8.13, which compares Haar and sinc kernel measurement for a signal consisting of two Gaussian spikes.

The ability to implement computationally efficient spatially separable processing is a particular attraction of linear constrained reconstruction. For example, the shift-coded downsample operator of Section 8.4 may be inverted simply by operating

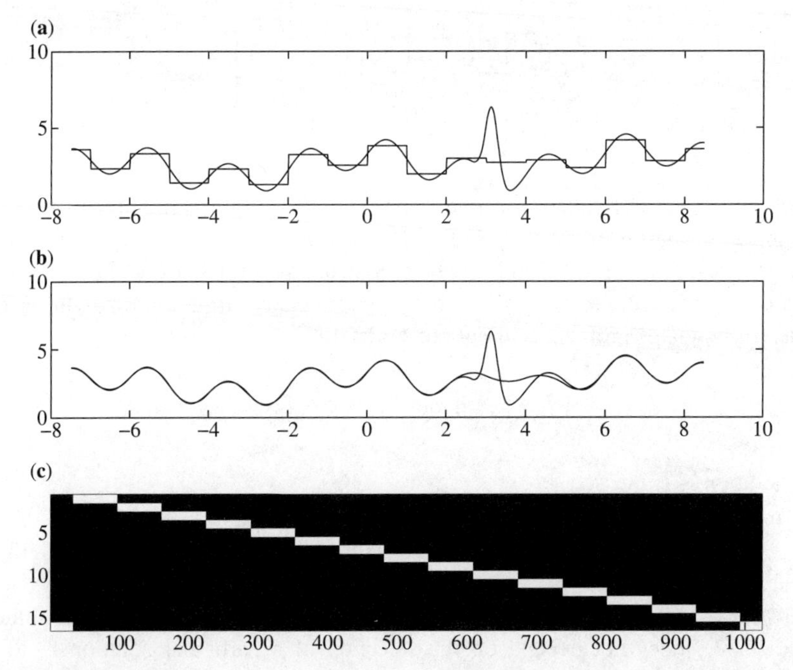

Figure 8.11 Reconstructions of the signal of Fig. 3.9 as sampled on the Haar basis of order 0: (a) the true function and the least-squares estimate; (b) the least gradient; (c) the measurement operator $\boldsymbol{H}$.

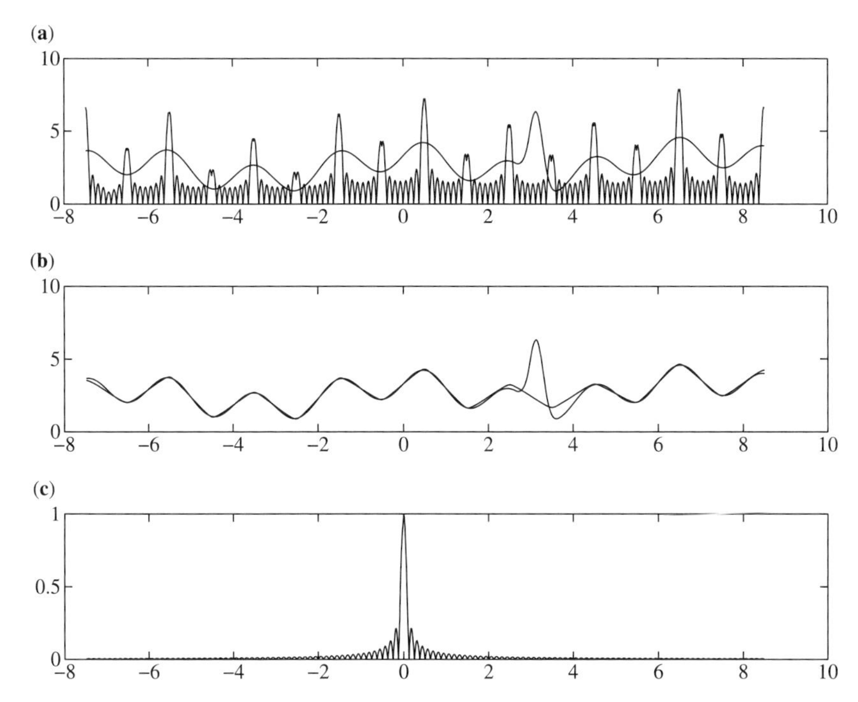

Figure 8.12 Reconstructions of the signal of Fig. 3.9 as captured by sampling function sinc(8x): (a) the true function and the least-squares estimate; (b) the true function and the least-gradient reconstruction; (c) shows the sampling function.

Eqn. (8.29) from the left and right by the using the LG operator of Eqn. (8.37). Figure 8.14 uses this approach to demonstrate a slight improvement in image fidelity under LG smoothing of the Tikhonov regularized image of Fig. 8.8. Of course, the shift-coded downsample operator does not have a null space, but Fig. 8.14 treats the 156 singular vectors corresponding to the smallest singular values as the null space for LG optimization.

Equation (8.35) is a *convex optimization* problem if $\gamma(\boldsymbol{f})$ is a convex function. A set of points V_f, such as the domain of input objects, is convex if for all $\boldsymbol{f}_1, \boldsymbol{f}_2 \in V_f$

$$\alpha \boldsymbol{f}_1 + (1 - \alpha)\boldsymbol{f}_2 \in V_f \tag{8.38}$$

for $0 \leq \alpha \leq 1$. The point $\alpha \boldsymbol{f}_1 + (1 - \alpha)\boldsymbol{f}_2$ is on the line segment between $\boldsymbol{f}_1$ and $\boldsymbol{f}_2$ at a distance $(1 - \alpha)\|\boldsymbol{f}_1 - \boldsymbol{f}_2\|$ from $\boldsymbol{f}_1$ and $\alpha\|\boldsymbol{f}_1 - \boldsymbol{f}_2\|$ from $\boldsymbol{f}_2$.

$\gamma(\boldsymbol{f})$ is a *convex function* if V_f is a convex set and

$$\gamma(\alpha \boldsymbol{f}_1 + (1 - \alpha)\boldsymbol{f}_2) \leq \alpha\gamma(\alpha \boldsymbol{f}_1) + (1 - \alpha)\gamma(\boldsymbol{f}_2) \tag{8.39}$$

for all $\boldsymbol{f}$ in the domain of $\gamma()$ with $0 \leq \alpha \leq 1$. $\|\boldsymbol{Hf} - \boldsymbol{g}\|_2^2$ and $\|\boldsymbol{f}\|_1$ are example convex functions.

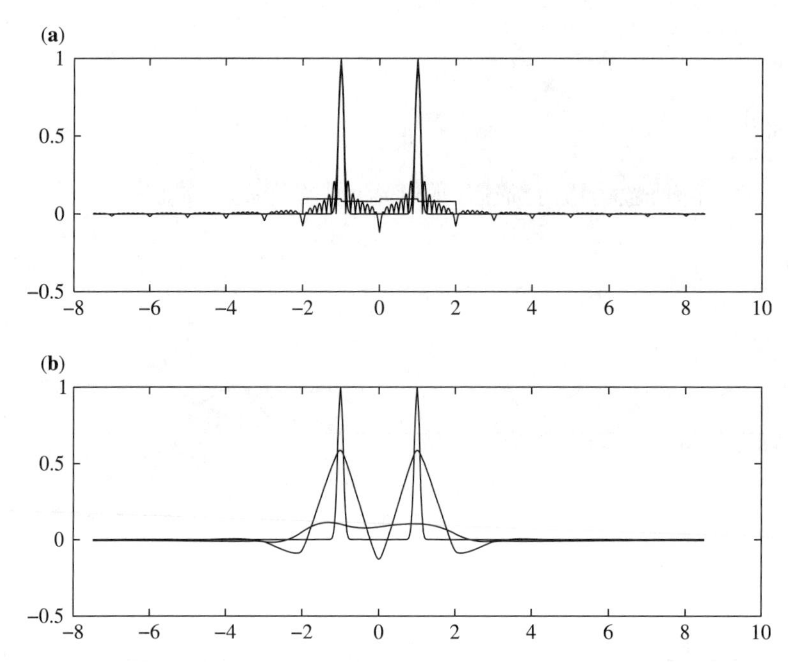

Figure 8.13 Reconstructions of the signal of a pair of isolated Gaussian signals as captured by zeroth-order Haar function and by sampling function sinc($8x$) (shown in Fig. 8.12): (a) the true function and the least-squares estimate for each sampling function; (b) the true function and the least-gradient reconstructions.

The basic idea of convex optimization is illustrated in Fig. 8.15. Figure 8.15(a) illustrates a convex function as a density map over a convex region in 2D. Figure 8.15(b) shows a nonconvex set in the 2D plane. Optimization is implemented by a search algorithm that moves from point to point in V_c. Typically, the algorithm analyzes the gradient of $\gamma(\boldsymbol{f})$ and moves interatively to reduce the current value of γ.

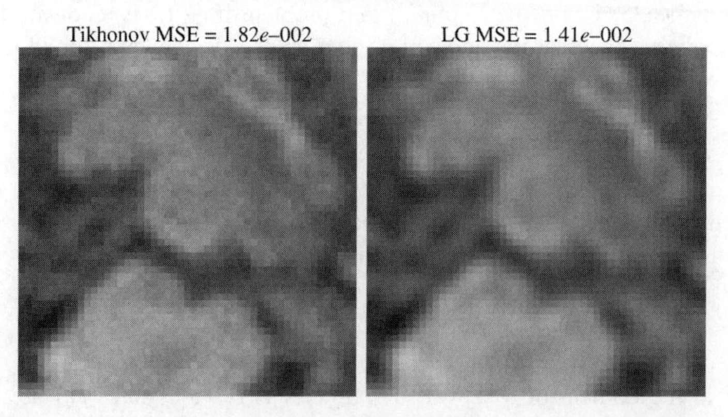

Figure 8.14 Least-gradient reconstruction of the Tikhonov regularized image of Fig. 8.8.

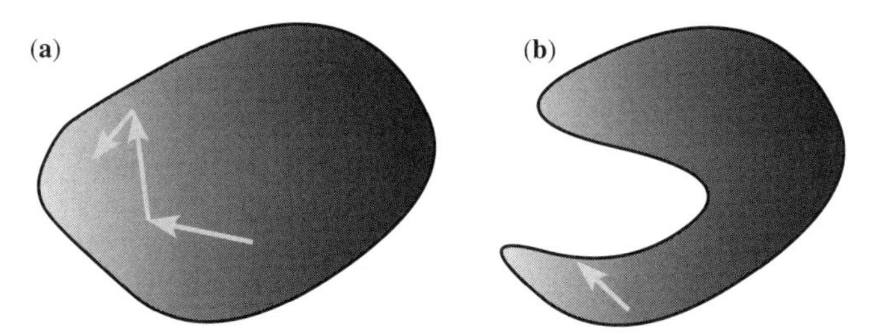

Figure 8.15 Boundary (a) outlines a convex set in 2D. Minimization of a convex function over this set finds the global minimum. Boundary (b) outlines a nonconvex set. Minimization of a convex function over this set may be trapped in a local minima.

If V_c and $\gamma(f)$ are convex, it turns out that any local minima of the objective function discovered in this process is also the global minimum over V_c [24]. If, as illustrated in Fig. 8.15(b), V_c is not convex, then the search may be trapped in a local minimum. Simple gradient search algorithms converge slowly, but numerous fast algorithms have been developed for convex optimization [24].

Equation (8.35) is a *constrained optimization problem*, with optimization of the objective function as the goal and the forward model as the constraint. A general approach to solving the constrained problems reduces Eqn. (8.35) to the *unconstrained optimization problem*

$$\mathbf{f}_e = \arg \min_{f_e} \{\|\boldsymbol{g} - \boldsymbol{H}\boldsymbol{f}_e\|_2^2 + \lambda_0^2 \gamma(\boldsymbol{f}_e)\} \tag{8.40}$$

This problem is a nonlinear regularization comparable to Tikhonov regularization [Eqn. (8.30)]. Compressive imaging may, in particular, be viewed as Tikhonov regularization using the l_1 norm. For the case $\lambda_0 = 0$, Eqn. (8.40) reduces to the psuedoinverse. In practice, one may attempt to jointly satisfy the forward model and the constraint by iteratively minimizing Eqn. (8.40) for decreasing values of λ_0. Algorithms under which this iteration rapidly converges have been developed [96], leaving rapid solution of the unconstrained minimization problem as the heart of convex optimization.

A linear constraint with a quadratic objective function provides the simplest form of convex optimization problem. As observed for Eqn. (8.36), this problem can be solved algebraically. One may find, of course, that the algebraic problem requires advanced methods for large matricies. At the next level of complexity, many convex optimization problems provide differentiable objectives. These problems are solved by gradient search algorithms, usually based on "Newton's method" for conditioning the descent. At a third level of complexity, diverse algorithms mapping optimization problems onto linear programming problems, interior point methods and interative shrinkage/thresholding algorithms may be considered.

Software for convex optimization and inverse problems is summarized on the Rice University compressive sensing Website (`www.dsp.ece.rice.edu/cs/`), on the Caltech l_1-magic site (`www.acm.caltech.edu/l1magic/`), on Boyd's webpage (`www.stanford.edu/boyd/cvx/`), and on Figueiredo's website (`www.lx.it.pt/mtf/`).

One may imagine many objective functions for image and spectrum estimation and would certainly expect that as this rapidly evolving field matures, objective functions of increasing sophistication will emerge. At present, however, the most commonly applied objective functions are the l_1 norm emerging from the compressive sampling theory [59,39,40] and the *total variation* (TV) objective function [212]

$$\gamma_{\mathrm{TV}}(\boldsymbol{f}) = \sum_{i,j=1}^{N-1} \sqrt{(f_{i+1,j} - f_{ij})^2 + (f_{i,j+1} - f_{ij})^2} \tag{8.41}$$

The l_1 objective is effective if the signal is sparse on the analysis basis and the TV objective is effective if the gradient of the signal is sparse. Since TV is often applied to image data, we index $\boldsymbol{f}$ in 2D in Eqn. (8.41). The first term under the root analyzes the discrete horizontal gradient and the second, the vertical gradient.

As illustrated in Figs. 7.25 and 7.27, the l_1 objective is often applied to signals that are not sparse in the display basis. One assumes, however, that there exists a useful basis on which the signal is sparse. Let $\boldsymbol{\theta} = \boldsymbol{W}\boldsymbol{f}$ be a vector describing the signal on the sparse basis. The optimization problem may then be described as

$$\boldsymbol{\theta} = \arg\min_{\boldsymbol{\theta}} \|\boldsymbol{\theta}\|_1$$

such that

$$\mathbf{HW}\,\boldsymbol{\theta} = \boldsymbol{g} \tag{8.42}$$

Determination of the sparse basis is, of course, a central issue under this approach. Current strategies often assume a wavelet basis or use hyperoptimization strategies to evaluate prospective bases.

We consider a simpler example here, focusing on the atomic discharge spectrum of xenon. Atomic discharge spectra consist of very sharp discrete features, meaning that they are typically sparse in the natural basis. Figure 8.16(a) shows the spectrum of a xenon discharge lamp measured to 0.1 nm resolution over the spectral range 860–930 nm. The spectrum was collected by the instrument described by Wagadarikar et al. [243]. Measured data extended slightly beyond the display limits; 765 data sample experimental values were used for the simulations shown in Fig. 8.16. Figure 8.16(b) is the spectral estimate reconstructed from 130 random projections of the spectrum. The reconstruction used the Caltech l_1-magic program `l1eq_example.m`. Typical results have reported that sparse signals consisting of K features require approximately $3K$ random projections for accurate reconstruction. While the xenon spectrum contains only four features over this range, each feature is approximately 0.5 nm wide in these data, suggesting that there are 20–30 features in the spectrum. The experimental spectrum, including background noise, was presented to the simulated measurement system.

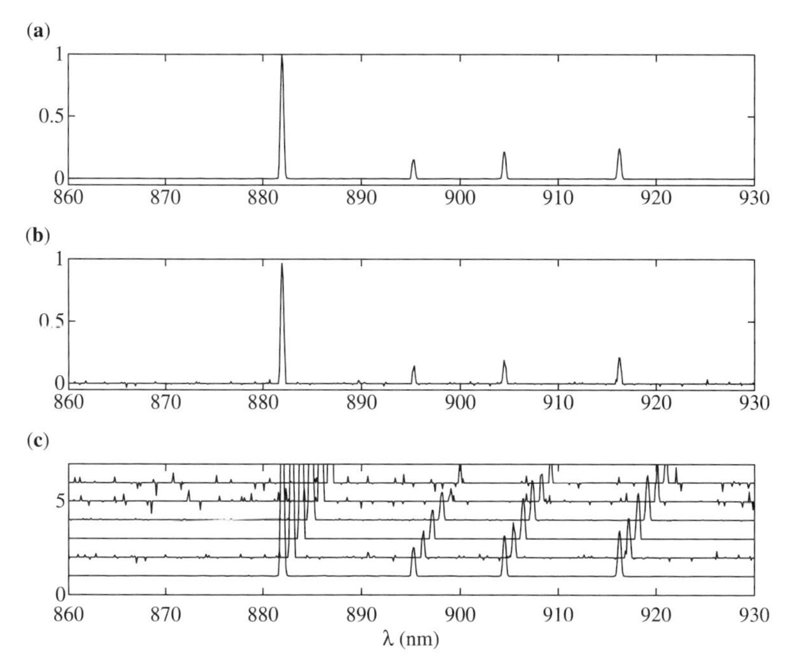

Figure 8.16 (a) Discharge spectrum of xenon measured measured by Wagadarikar et al. [243]; (b) reconstruction using l_1 minimization from 130 random projections; (c) reconstruction baseline detail for several strategies.

Figure 8.16(c) shows baseline details for diverse measurement and reconstruction data. The plot 1 baseline is the experimental data, which has slight noise features on the baseline. The plot on the 2 baseline is the reconstructed data from (b). The 3 baseline shows the reconstruction obtained from 130 projections if the baseline noise is thresholded off of the experimental data prior to simulated measurement. The 4 baseline shows the reconstructed data from the noisy experimental data if 200 projections are used. The 5 baseline shows the reconstruction from 100 projections, and the 6 baseline shows the reconstruction from 90 random projections. The random projections used the normal distribution measurement operator generated by the original l_1-magic program. As illustrated in the figure, estimated signal degregation is rapid if the sample density falls below a critical point. Note that each sucessive trace in Fig. 8.16 is shifted to the right by 1 nm to aid visualization.

A second example uses the TV objective function and the two-step iterative shrinkage/thresholding algorithm (TWIST) [20]. As discussed in [76], the original iterative shrinkage/thresholding algorithm combines maximum likelihood estimation with wavelet sparsity. We briefly review maximum likelihood methods in Section 8.5.2. For the present purposes, we simply treat TWIST as a blackbox optimizer of the TV objective.

We use TWIST to consider again the $4\times$ downsample shift code. Rather than force model consistency with the full measurement operator, however, we focus on the optimization problem

$$f_e = \arg\min_f \gamma_{TV}(f)$$

such that

$$f_{i,e}^{\mathrm{SV}} = g_i^{\mathrm{SV}} \quad \text{for all } i \leq r \tag{8.43}$$

where r is the rank of the truncated SVD and $f_{i,e}^{\mathrm{SV}}$ and g_i^{SV} are the projections onto the singular vectors discussed in Section 8.4. This optimization forces consistency with the high-singular-value vectors, treating those vectors as generalized measurement projectors.

Reconstruction under this algorithm is illustrated in Fig. 8.17, which analyzes the same image as in Fig. 8.6 using the sampling codes 11110000 and 11100100. The first 125 out of 256 singular vectors are used in each case. In comparison with Figs. 8.8 and 8.10, we observe that truncated SVD reconstruction augmented by TWIST optimization substantially improves the image in each case. While the disjoint code performs worse under truncated SVD and Tikhonov reconstruction, it

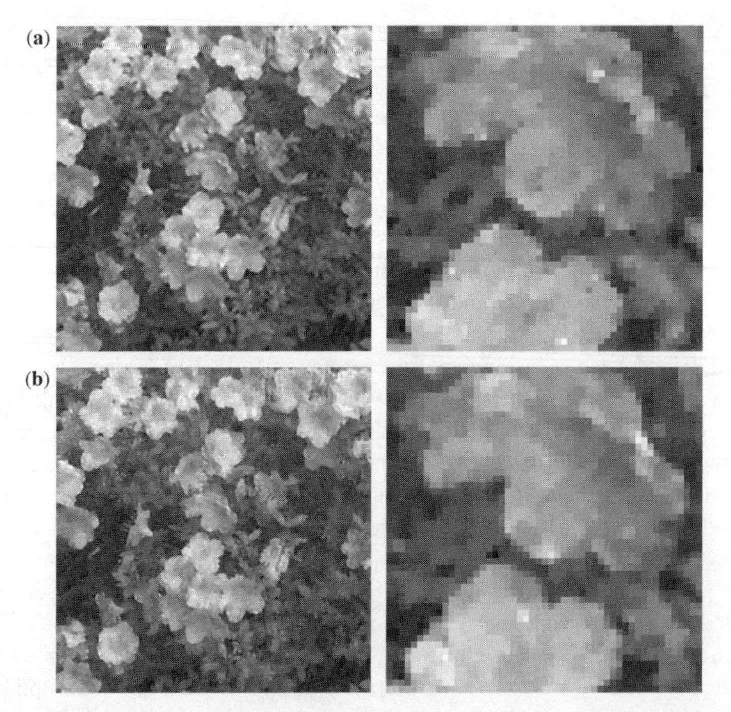

Figure 8.17 Reconstruction of the image of Fig. 8.6 using SVD/TWIST optimization satisfying Eqn. (8.43) for the first 125 singular vectors: (a) MSE $= 2.94e-003$ using the 11110000 shift code; (b) MSE $= 2.65e-003$ using the 11100100 code.

yields modestly better performance after TV optimization. The normally distributed measurement variance is 10^{-4} as in previous images. Reconstruction was implemented using Bioucas-Dias and Figueiredo's `demo_l2_TV.m` code distributed at `www.lx.it.pt/mtf/`.

Implementation of the constrained inverse problems using the truncated SVD is unnecessary in direction-coded (e.g., pixel-coded) systems, but may enable substantial noise reduction in implicitly coded systems.

8.5.2 Maximum Likelihood Methods

One strategy for selecting among the infintely many solutions to the ill-posed or ill-conditioned measurement

$$g(y) = \int f(x)h(x, y)\, dx \tag{8.44}$$

focuses on finding the mostly likely image $f(x)$ given measurements $g(y)$. To find the *maximum likelihood* image, one must evaluate the probability of $f(x)$ given $g(y)$. *Expectation–maximization* (EM) algorithms evaluate this probability by treating $g(y)$, $f(x)$, and $h(x, y)$ as probability distributions in Bayesian statistical analysis. The joint density of the signal on the interval $x \in (x, x + dx)$ and the output data on the interval $y \in (y, y + dy)$ is $f(x)h(x, y)\, dx\, dy$. Conversely, we may write this density in terms of the output data as $g(y)\tilde{h}(x, y)\, dx\, dy$, where $\tilde{h}(x, y)$ is the inverse impulse response. Equating these two expressions of the same probability and substituting the forward model for $g(y)$ yields

$$\tilde{h}(x, y) = \frac{f(x)h(x, y)}{\int f(x)h(x, y)\, dx} \tag{8.45}$$

Given that $\tilde{h}(x, y)$ is the inverse response, we then find

$$f(x) = \int \frac{g(y)f(x)h(x, y)}{\int f(x)h(x, y)\, dx}\, dy \tag{8.46}$$

Equation (8.46) may be viewed as a version of Bayes' theorem expressing a self-consistency requirement on the reconstructed object. Unfortunately, one would need to know the object distribution to verify that it satisfies this equation.

Equation (8.46) was independently introduced to optical image analysis by Richardson [210] and Lucy [158]. The Richardson–Lucy algorithm attempts to find the image by recursive application of Eqn. (8.46). Starting with an initial non-negative guess for $f(x)$, further solutions are attempted according to

$$f^{r+1}(x) = f^{r}(x) \int \frac{g(y)h(x, y)}{\int f^{r}(x)h(x, y)\, dx}\, dy \tag{8.47}$$

For Poisson noise–dominated signals, this recursion can be shown to coverge to the maximum likelihood estimate of $f(x)$. Whatever the nature of the noise, each sucessive estimate is nonnegative if one starts with nonnegative $f(x)$, $g(y)$ and $h(x, y)$.

The Richardson–Lucy algorithm generalizes to diverse EM algorithms based on similar statistical arguments. These algorithms may include hidden parameters, such as unknown PSFs, in the physical model. EM algorithms recursively apply "estimation" based on Eqn. (8.47) to find $f^r(x)$ and "maximization" based on Eqn. (8.45) to find the likelihood function [56]. The formal analysis of EM algorithms to image processing is developed in Ref. 226 and subsequent publications; Blahut [21] provides an excellent introduction to Richardson–Lucy and EM algorithms. While the preceding discussion focuses on 1D signals, the Richardson–Lucy algorithm extends trivially to higer dimensions.

Discrete coding of Richardson–Lucy algorithms based on Eqn. (8.47) is straightforward. For the special case of 2D shift-invariant kernels, Matlab includes the Richardson–Lucy algorithm in the function `deconvlucy`, with arguments consisting of the measurement data and kernel and the number of iterations for signal reconstruction. Determining the number of iterations is a particular challenge for EM algorithms; in many cases human observers tune the algorithm to meet qualitative image quality metrics.

Expectation–maximization algorithms are sometimes attractive in comparison to l_1-norm and total variation (TV) methods because they naturally enforce nonnegativity and process nonseparable 2D and 3D kernels relatively easily and efficiently. For system design, our interest focuses on the deliberate coding of the system PSF to improve system performance under deconvolution. As illustrated in Fig. 8.18, naive Richardson–Lucy deconvolution is not competitive on the simple example problem of the last two sections. As we saw with the IST algorithm, however, maximum likelihood and wavelet or TV sparsity algorithms are not orthogonal.

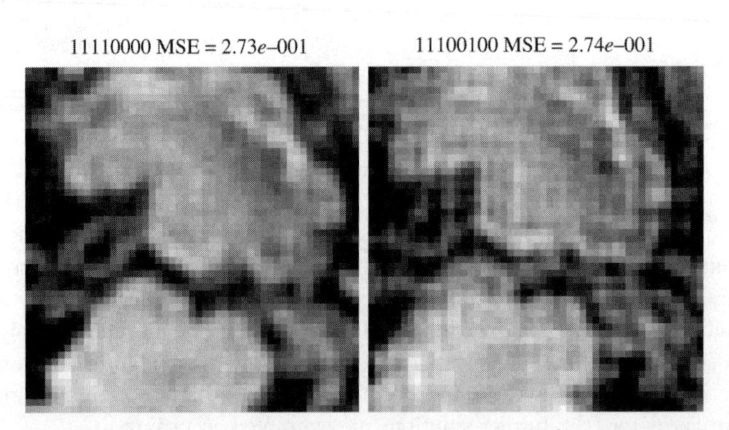

Figure 8.18 Richardson–Lucy deconvolution of an image with $\sigma^2 = 10^{-4}$ normally distributed noise for a 4×4-block code (left) and a code formed from the outer product of the 11100100 operator. Both PSFs are divided by 16 to normalize as in previous examples. The edge of the image was zero padded by 100 pixels on each side to avoid boundary effects.

Choi and Schulz incorporate TV regularization in a Richardson–Lucy algorithm to achieve results comparable to those achieved for the shift-coded IST inverse problem [47].

The reader thus leaves this chapter with demonstrations that the impact of coding and inverse problems is enormous, but with little clear guidance beyond a general understanding that singular values and singular vectors may be used to roughly understand the SNR and structure of data transfer in implicit sensors. The best design strategy in current design practice is to jointly tune algorithms and sampling codes in simulation prior to optical system construction. For examples, we proceed to Chapters 9 and 10.

PROBLEMS

8.1 *Deconvolution and Coded Apertures.* Consider again Problem 2.10.

(a) Decode your image using the Wiener filter corresponding to your coded aperture. Assume normally distributed independent noise in each measurement pixel. What is the noise spectral density? Assume uniform image spectral density. Plot Wiener filter reconstructed images for various SNR levels. How do your results compare with results using the decoding filter of Chapter 2?

(b) Decode your image using Richardson–Lucy deconvolution. Assume normally distributed independent noise in each measurement pixel. What is the noise spectral density? Assume uniform image spectral density. Plot reconstructed images for various SNR levels.

8.2 *Hadamard Coding.* Consider the measurement system $\boldsymbol{g} = \boldsymbol{H}\boldsymbol{f} + \boldsymbol{n}$ with $\boldsymbol{f}, \boldsymbol{g} \in \mathbb{R}^{127}$ and $\boldsymbol{H}$ a 127×127 matrix. The zero-mean noise $\boldsymbol{n}$ may be drawn from a normal distribution or may be Poisson distributed about the expected values g_{ij}. Assuming that in the normal case the measurement variance is $10^{-4}\langle f_{ij}^2 \rangle$ and in the Poisson case that $\langle f_{ij}^2 \rangle = 1000$, use Monte Carlo simulations to characterize the least-squares estimation variance for $\boldsymbol{H}$ corresponding to

(a) The order 128 Hadamard S matrix

(b) A random 127×127 matrix with values uniformly drawn from $[-1, 1]$

(c) The level 3 Haar wavelet transform of $\boldsymbol{f}$

8.3 *Regularizaton.* Operating from the left and right on a 127×127 natural image using the measurement matrices of Problem 8.2, find the mean-square error for least-squares image estimation, truncated SVD estimation, and Tikhonov regularization. Make sure to use the same random matrix in all experiments and document your regularization parameters.

8.4 *Deconvolution.* Using a 256×256 monochrome natural image sampled by a 3×3 shift-coded downsampling operator, compare the Wiener filter and Richardson–Lucy deconvolution assuming that $\sigma^2 = 10^{-3}\langle f^2 \rangle$ normally distributed independent pixel noise is added. Repeat the comparison for Poisson noise with $\langle f \rangle = 1000$.

8.5 *LG Interpolation.* Consider the Matlab image camera `trees.tif`.

(a) Downsample the image by factors of 2 and 4 using the Haar wavelet. Plot the resulting images. For each of the downsampled images, upsample the image using linear, cubic spline, and least-gradient interpolation (using the null space smoothing function posted on the course Website for LG interpolation).

(b) Suppose that the image is downsampled according to

$$g_{ij} = f_{(2i+1)(2j+1)} + f_{(2i+1)(2j+6)} + f_{(2i+6)(2j+1)} + f_{(2i+6)(2j+6)} \tag{8.48}$$

for $(i, j) \in ((0, 0),$ `size(trees)`$/2)$. Plot the measurement data. Use the LG algorithm to estimate $\boldsymbol{f}$ from $\boldsymbol{g}$. Plot the resulting image.

8.6 *Convex Functions.* Prove that $\|\boldsymbol{HF} - \boldsymbol{g}\|_2^2$ and $\|\boldsymbol{f}\|_1$ are convex functions for $\boldsymbol{f} \in \mathbb{R}^N$.

8.7 *Compressive Sampling.* NIST maintains a database of atomic spectra at `physics.nist.gov/PhysRefData/ASD/`. Use the lines and intensities of the noble gases over the spectral range 550–600 nm to create a spectral library. Design a random matrix compressive sampling system to characterize these spectra and simulate the system using one or more of the convex optimization codes described in Section 8.5.1. How many measurements are required to characterize these signals? Test your sensor on other atoms, such as sodium. What happens when you add noise to the measurements?

8.7 *Sparse Reconstruction.* Download the gradient projection for sparse reconstruction code from `www.lx.it.pt/mtf/GPSR/`.

(a) Run `demo_continuation.m` and `demo_image_deblur`. Try changing the number of spikes in the `demo_continution` code to 10, 50, 100, and 300.

(b) Run the `deblur` code for all three impulse responses in the code.

(c) Deblur the same images using the Richardson–Lucy algorithm.

8.8 *Shift Coding.* Generate copies of Figs. 8.6, 8.8, 8.14, 8.17, and 8.18 corresponding to three different images. Example images might include the Shepp–Logan phantom or cameraman images distributed with Matlab or images from your own library. How much variability in MSE and other reconstruction characteristics do you observe? Do you observe differences between discrete images like the phantom and continuous natural images?

SPECTROSCOPY

Parmi les appareils utilises en analyse spectrale, il est maintenant classique de distinguer, d'une part, les spectrometres, o le spectre est explore dans le temps, element par element et, d'autre part, les spectrographes qui permettent d'obtenir des informations simultanement sur tous les elements du spectre.

—A. Girard [93]

9.1 SPECTRAL MEASUREMENTS

A *spectrometer* analyzes the power spectral density $S(\nu)$ of an optical signal. Of course, optical signals are functions of space, time, and polarization as well as wavelength or frequency, but we limit our discussion in this chapter to instruments that measure the power spectral density in a single optical mode or the average spectral density over a range of modes, returning to the larger issue of *spectral imaging* [determination of $S(\mathbf{r}, \nu)$] in Section 10.6.

Spectral information is isolated from optical signals by three mechanisms:

- *Spatial Dispersion. Dispersive elements*, such as prisms and gratings, redirect refracting and diffracting waves as a function of wavelength. These elements are combined with spatial filters to produce spectrally informative spatial patterns. Sections 9.2 and 9.3 integrate the dispersive elements described in Chapter 4 in spectrometer designs. The most recent trend in dispersive spectrograph design focuses on the use of volume diffractive structures, such as photonic crystals, multiplex holograms, and multilayer diffractive optical elements, to produce 2D dispersion patterns. 2D dispersion radically reduces the volume required to achieve a given resolution in a dispersive spectrometer. 2D design is discussed in Section 9.8.
- *Interferometry*. The general distinction between dispersive and interferometric spectrometers is that dispersive spectrometers modulate light in image planes,

Optimal Imaging and Spectroscopy. By David J. Brady

whereas interferometric spectrometers modulate light in Fourier or modal space. For example, the Fourier transform (FT) spectrometer briefly described in Section 6.3.1 filters certain spectral modes based on a phase matching criterion. We expand our discussion of FT spectroscopy in Section 9.4 and discuss "multi-beam" or resonant interferometric spectroscopy in Sections 9.5–9.8.

- *Resonance.* Spectrally sensitive absorption, transmission, and reflection may arise in optical devices and materials as a result of microscopic optical and electronic properties rather than macroscopic optical design. Resonant effects are created by optical cavities and nanostructures or by quantum mechanical materials processes. These effects are used to create spectroscopic filters. Thin-film filters, organic dyes, metal nanoparticles, and semiconductor quantum dots are the most common such devices in current practice. Spectral analysis may be implemented by using electronic detectors with intrinsic spectral sensitivity, as in the Foveon X3 direct image sensor. Under the rubics *photonic crystals*, *metamaterials*, *quantum dots*, *plasmonics*, and volume holography, current research focuses on coding the spectral response of materials that are artificially structured in multiple dimensions. Structured materials for spectroscopy are discussed in Sections 9.6 and 9.8.

The careful reader will notice that classification boundaries between dispersive, interferometric, and resonant devices are not absolute. In fact, differences between the three mechanisms exist primarily in the geometries of optical axes and the spatial scale of interferometric effects. Filters utilize interferometry in micrometer- and nanometer-scale cavities, dispersive elements use micron scale diffractive components but implement coding in transverse planes rather than in fine structure. Interferometric spectrometers may be viewed as large scale tunable multiplex filters.

Dispersive and interferometric strategies for spectrometer design are illustrated in Fig. 9.1. While color channels are illustrated by red, green, and blue planes in Fig. 9.1(a), we are in general interested in substantially more than just three channels. As suggested by Girard in this chapter's opening quote, one may design a spectrometer to measure each spectral channel, or linear combinations of the spectral channels, serially in time or spatially in parallel. One may imagine an instrument that completely separates the spectral channels spatially, as illustrated in Fig. 9.1(b). In view of the constant radiance theorem, however, such an instrument must maintain the full radiance of each input color channel in the detector plane. This effectively means that the area of the exit aperture of the spectrograph must exceed the input aperture by a factor of N, where N is the number of spectral channels. This approach is commonly adopted in fiber spectrometers. More commonly, however, one cannot afford such a large exit aperture, and the approach illustrated in Fig. 9.1(c) is adopted. In this case, a slit is introduced as an entrance stop. Referring again to the constant radiance theorem, reducing the spatial extent of the object inevitably means discarding some of its energy. This tradeoff and mechanisms for alievating it using coding and spatially overlapping dispersed channels are discussed in Sections 9.2 and 9.3.

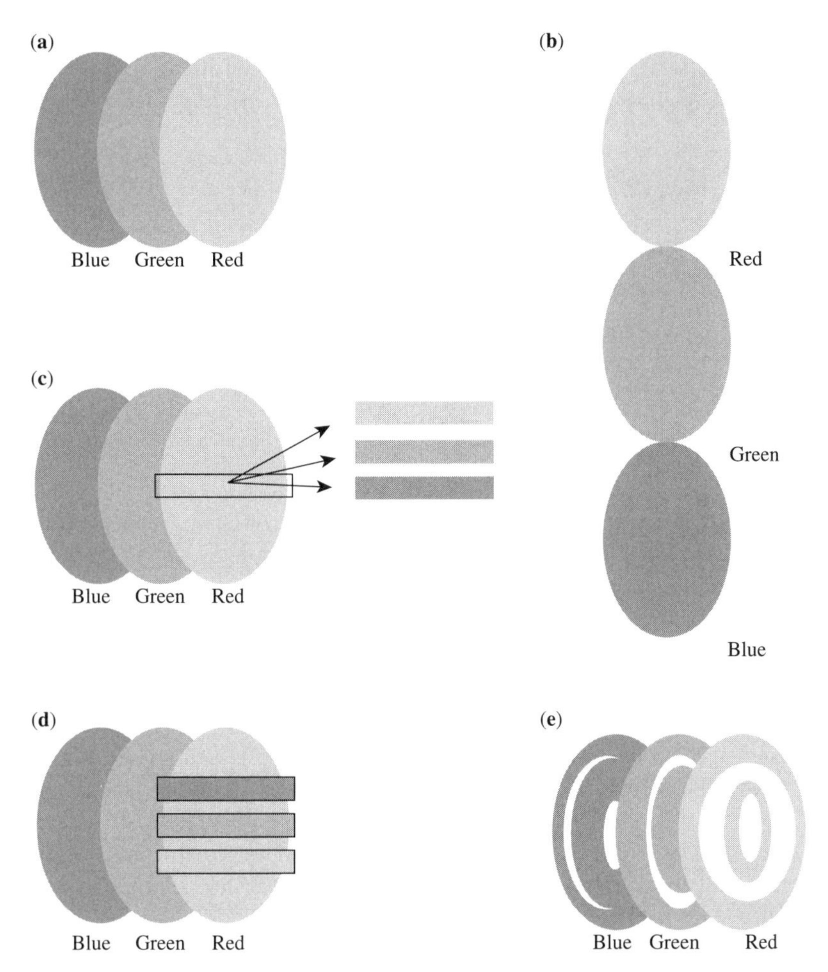

Figure 9.1 Encoding strategies for spectrometer design: (a) color planes in an optical signal; (b) spatially separated dispersion; (c) spatially filtered dispersion; (d) spectral filtering; (e) interferometric multiplexing.

Resonant filtering based on a spatially dispersed array of filters is illustrated in Fig. 9.1(d). One may alternatively utlize a single spectral filter tuned as a function of time, a strategy that we discuss in Section 9.7. Figure 9.1(e) shows ring patterns in each spectral channel as typically observed in interferometric instruments. These systems commonly measure the output at the central point as a function of time as the interferometer is scanned. As indicated by the ring patterns, however, it is also possible to discover spectral information from the spatial distribution of light in interferometers.

Other than differences in the physical encoding mechanism, dispersive, interferometric, and resonant spectroscopy are quite similar. To obtain spatial or temporal modulation without increasing the aperture, each system must filter out a fraction of the incident energy. All three approaches are capable of spatially parallel and/or temporally serial measurement. Design choices between various strategies are driven by four factors:

1. *Properties of Target Objects*. Determining factors include whether these objects are single- or multiple-mode, self-luminous or optically excited, or spatially homogeneous. Spectroscopy for physical science and telecommunications applications is often single-mode; biological objects tend to be diffuse.
2. *Properties of Available Optoelectronic Detectors*. Detector technology is often the determining factor for spectrometer design. Prior to the twentieth century, spectra were observed by eye or using photochemical recording. Electronic recording using discrete detectors blossomed over the period from 1950–1970. Electronic recording enabled Fourier transform spectroscopies, which require digital postprocessing. Since the mid-1980s, 2D detector arrays have been gradually introduced to spectroscopy. Current design strategies use 1D and 2D detector arrays for the visible range and single-channel detectors for regions of the spectrum where large-scale detector arrays are expensive or noisy. One may anticipate that detector arrays will eventually grow to dominance in the infrared as well as the visible.
3. *Properties of Available Optics and Modulators*. Materials and fabrication technologies for dispersion and imaging components vary across spectral ranges. The diversity of established optical materials and grating fabrication technologies is much more limited at spectral extremes than in the visible. For this reason, systems that utilize very simple components, such as interferometers, may be expedient. At the other end of complexity, continuing advances in spatial light modulator technologies enable adaptive coding spectroscopic strategies. Integration of micromirror arrays or liquid crystal devices may significantly impact design.
4. *System Performance Metrics*. Where multiple design strategies are feasible, design decisions are based on comparative system performance. The performance of a spectrometer is measured by
   - *Spectral resolution* $\delta\lambda$, which is often expressed in terms of the *resolving power*, which is the ratio of mean operating wavelength to the resolution, $R = \bar{\lambda}/\delta\lambda$.
   - *Spectral range* $\Delta\lambda$, the difference between the minimum and maximum resolvable wavelength. One typically desires a broad spectral range (e.g., 350–1000 nm) in absorption or fluorescence instruments. Raman instruments generally require wavenumbers ranging within only a few thousand [A *wavenumber* is an optical frequency measured in reciprocal centimeters (cm^{-1}). The frequency corresponding to one wavenumber is the frequency at which the wavelength of light is 1 cm (e.g., 30 GHz); 1000 wavenumbers is a spectral range of 30 THz, corresponding to a wavelength range of 100 nm for a 1 μm

center wavelength. Wavenumbers are commonly used to describe Raman spectra.]

- *Etendue L*, the product of the solid angle collected at the instrument aperture and the area of the entrance pupil. The étendue quantifies the light collection capacity of the instrument.
- *Instrument volume.*
- *Spectral throughput*, the fraction of the incident power spectral density within the instrument's etendue that is absorbed in each measurement. The spectral throughput for narrowpass filters is $\delta\lambda/\Delta\lambda$. The spectral throughput for dispersive and interferometric instruments is generally $\geq \frac{1}{2}$. The utility of higher spectral throughput depends on the noise characteristics of the detection system and estimation algorithm, which brings us to *signal fidelity or information metrics* such as SNR or transinformation.

As discussed in the following sections, design consists of tradeoffs between these metrics.

Before launching into our survey of basic spectrometer designs, one must emphasize that spectroscopy is an enormous, diverse, and well-developed discipline with many different design strategies. This chapter reviews the landscape of potential approaches, but many interesting designs must unfortunately be neglected. In particular, guided-wave, coded source absorption, diverse component-specific designs, and a vast array of ingenious implementation strategies are ignored. Even within the approaches discussed, our review is cursory. Thin-film filter design, photonic crystals, plasmonic filters, and liquid crystal devices may each be fruitfully explored in much greater detail. The analytic approach presented here is quite general, however, and will hopefully assist students and researchers in the consideration of more detailed designs.

9.2 SPATIALLY DISPERSIVE SPECTROSCOPY

A device using optical elements to physically isolate spectral channels is called a *spectrograph*. A spectrometer combines a light collection system, a spectrograph, and a optoelectronic detection and processing system to computationally estimate the power spectral density. As illustrated in Fig. 9.2, the spectrograph for a simple dispersive spectrometer consists of a spatial filter (a slit), a dispersive element, and an imaging system. The dispersive element is typically a diffraction grating, although prisms may also be used. The spectrograph images the spatial filter onto a detector array after transmission through the dispersive element. The dispersive element induces a wavelength-dependent shift in the position of the image.

The resolving power, etendue, and volume of the dispersive spectrometer depend on the slit width a, the grating period Λ, and the focal length F. The 4F spectrometer of Fig. 9.2 is identical to the Vanderlugt correlator of Problem 4.14 with the grating assuming the role of the filter. The impulse response for propagation from the input plane of the spectrometer to the output plane is the imaging PSF as

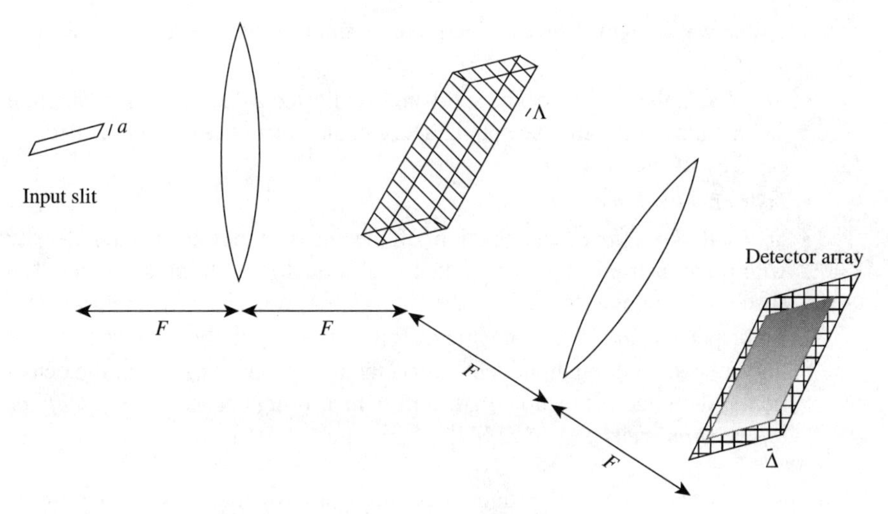

Figure 9.2 A dispersive spectrometer based on a 4F optical system with an entrance slit of width D and a transmission diffraction grating of period Λ.

filtered by the grating transmittance. Assuming, after Eqn. (4.55), that the grating transmittance is

$$t_g(x) = \sum_q \eta_q e^{2\pi i q(x/\Lambda)} \tag{9.1}$$

we apply the Fourier shift theorem to find the optical impulse response

$$h_g(x, y) = \sum_q \eta_q h\left(x - \frac{q\lambda F}{\Lambda}, y\right) \tag{9.2}$$

where $h(x, y)$ is the PSF of the imaging system without the grating.

Letting $t(x, y) = \text{rect}(x/a)\text{rect}(y/A)$, represent the transmission function of the slit, the detected signal may be modeled as

$$\begin{aligned} g_n &= \sum_q \eta_q \iiiint S(\lambda) t(x, y) h\left(x' - x - \frac{q\lambda F}{\Lambda}, y' - y\right) p(x' - n\Delta, y')\, dx\, dy\, dx' dy' d\lambda \\ &= \sum_q \eta_q \int S(\lambda) h_l(\lambda - n\Delta_l)\, d\lambda \end{aligned} \tag{9.3}$$

where Δ is the pixel pitch, $p(x)$ is the pixel sampling function, $\Delta_l = \Lambda\Delta/qF$, and

$$h_l(\lambda) = \iiiint t(x, y) h(x' - x, y' - y) p\left(x' + \frac{qF\lambda}{\Lambda}, y'\right) dx\, dy\, dx' dy' \tag{9.4}$$

We have assumed a 1D detector array; we expand our analysis to 2D detectors in Section 9.3. The system impulse response $h_1(\lambda)$ is called the *instrument function* in spectroscopy literature.

In practice, the bend induced in the optical path by the grating, shearing between the wavelength channels and conventional aberrations induce substantial shift variance and distortion in the instrument function. Sophisticated designs, often incorporating reflective gratings and optics, have been developed for spectrometer systems to overcome these challenges. Systems using volume transmission gratings, as illustrated in Fig. 9.2, have become popular since the mid-1990s; the primary advantages are that the Bragg effects discussed in Section 4.8 enable high-efficiency single-order diffraction and that volume phase holograms are easily manufactured.

While we must be mindful of the complexity of real system analysis, we return again to this text's focus on abstract system models. While Eqn. (9.3) may be invertible even with multiple diffraction orders, we assume for simplicity that either Bragg matching or an "order sorting" spatial filter is used to eliminate all but the $q = 1$ order, which reduces Eqn. (9.3) to

$$g_n = \int S(\lambda) h_l(\lambda - n\Delta_l)\, d\lambda \tag{9.5}$$

This dispersive spectrometer is the second of the two classes mentioned by Girard in the quote at the beginning of this chapter; it maps the 1D spectral signal onto a parallel 1D array of pixels. If $h_1(x)$ is suitably compact, then g_n is a spatial image of the spectral density.

As discussed Chapter 7, g_n are samples of the continuous function $g(x)$. Assuming that aliasing and interpolation issues are managed as discussed in Chapter 7, the spectrometer measures a image of $S(\lambda)$ with system transfer function

$$\hat{h}_l(u) = \hat{t}\left(\frac{\Lambda u}{F}\right)\hat{h}\left(\frac{\Lambda u}{F}\right)\hat{p}\left(\frac{\Lambda u}{F}\right) \tag{9.6}$$

As with imaging systems, the resolution of the estimated spectrum depends on the aliasing limit, $u_{\text{aliasing}} = 1/(2\Delta_l)$ and on the system bandpass. Factors in the system STF are illustrated in Fig. 9.3, assuming a rectangular slit of width $25\lambda f/\#$. For the slit width illustrated in the figure, $\hat{t}(\Lambda u/F)$ approaches zero much faster than the optical or pixel transfer functions, meaning that $\hat{h}_l(u) \approx \hat{t}(\Lambda u/F)$ (and that the impulse response $h_l(\lambda)$ is effectively an image of the slit). One may well ask, why not use a narrower slit to broaden the STF? The answer, of course, is that a narrower slit increases the spectral resolution at a cost of decreased light collection efficiency.

The slit transfer function is $\hat{t}(\Lambda u/F) = \text{sinc}(a\Lambda u/F)$. According to the Fourier uncertainty relationship [Eqn. (3.35)], this system bandpass implies a resolution of order

$$\delta\lambda \approx \frac{a\Lambda}{F} \tag{9.7}$$

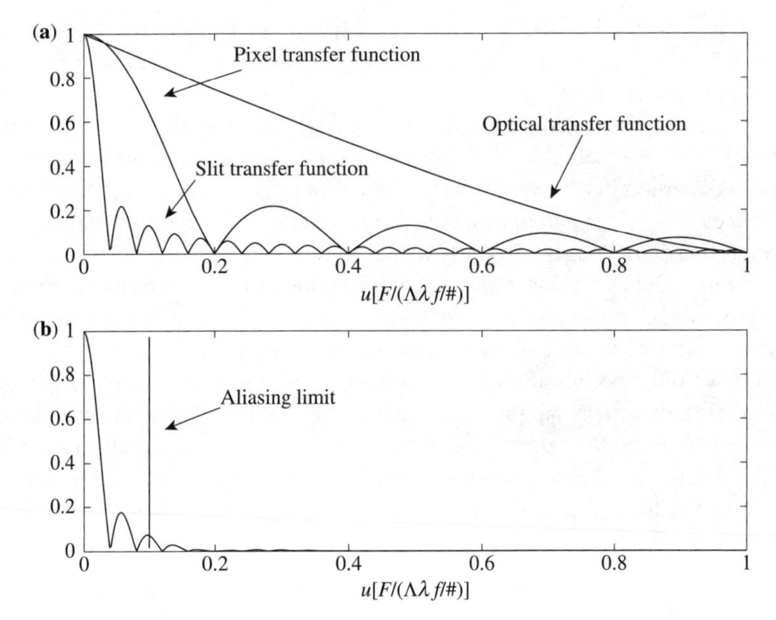

Figure 9.3 Transfer functions for slit-based dispersive spectroscopy. The slit width is $25\lambda f/\#$, and the pixel width is $10\lambda f/\#$. Plot (b) shows the STF, which is the product of the individual transfer functions.

and a resolving power of

$$R = \frac{\lambda F}{a\Lambda} \tag{9.8}$$

Of course, the minimum resolvable slit width is $\lambda f/\#$, meaning that if we make the slit as small as possible, $R = A/\Lambda = N_g$, where N_g is the number of grating periods within the aperture [126]. In practice, the resolving power is usually much less than the diffraction-limited value.

The etendue of the dispersive spectrometer is approximately

$$L \approx \frac{\pi^2 A a}{2(f/\#)^2} \tag{9.9}$$

The product of L and R is known as the *efficiency* of a spectrograph. The efficiency of a grating spectrometer is

$$E = \frac{\pi^2 \lambda F^2}{2\Lambda (f/\#)^3} \tag{9.10}$$

We note that the efficiency is independent of slit width, indicating a fundamental tradeoff between resolution and light collection in slit-based instruments. While there are possibilities for ingenious folding, the volume of a slit spectrometer may be approximated $V = \pi^2 F^3/4(f/\#)^2$, meaning that the efficiency is approximately

$$E \approx \frac{\lambda V^{2/3}}{\Lambda (f/\#)^{5/3}} \tag{9.11}$$

It is, therefore, possible to improve both the étendue and the resolving power by increasing the volume, which leads to well-known associations between spectrometer size and system performance. Designs in the next several sections challenge the "bigger is better" mantra.

Since one seldom uses a slit that challenges the optical resolution limit, there is little motivation to use small pixels in spectroscopic focal planes. Figure 9.3 assumes a slit width of $10\lambda f/\#$, which mostly avoids aliasing and does not substantially degrade the STF. The many sidelobes of the slit transfer function and the resulting bit of aliasing indicate that an apodized slit, such as $t(x) = e^{-x^2/a^2}$, might have some advantages.

9.3 CODED APERTURE SPECTROSCOPY

The basic geometry for a coded aperture dispersive spectrometer is illustrated in Fig. 9.4. The only visible change relative to Fig. 9.2 is that we have replaced the slit with a coded aperture. Just as replacing a pinhole with a coded aperture in Chapter 2 enabled us to increase throughput for projective imaging, a coded aperture dispersive spectrometer avoids the dependencies between resolving power, etendue, and volume derived in Section 9.2.

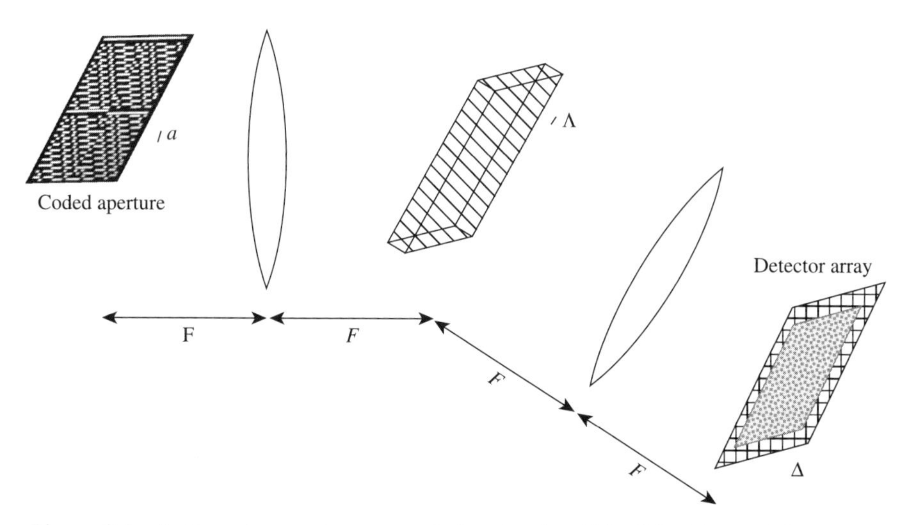

Figure 9.4 A dispersive spectrometer replacing the slit of Fig. 9.2 with a coded aperture.

In addition to the coded aperture, we now find it useful to assume that the readout detector array is two-dimensional. With a 2D detector and again assuming only one diffracted order, Eqn. (9.3) becomes

$$g_{nm} = \int S(\lambda) h_l(\lambda - n\Delta_l, m\Delta)\, d\lambda \tag{9.12}$$

where

$$h_l(\lambda, y'') = \iiiint t(x, y) h(x' - x, y' - y) p\left(x' + \frac{F\lambda}{\Lambda}, y' - y''\right) dx\, dy\, dx' dy' \tag{9.13}$$

As in Eqn. (2.35), we model the coded aperture transmittance as a discrete binary code modulating identical mask pixels, for example

$$t(x, y) = \sum_{ij} t_{ij} \tau(x - ia_x, y - ja_y) \tag{9.14}$$

In contrast with the slit spectrometer, a_x is the width of the coded aperture pixel rather than the full input aperture size. To simplify our analysis, we assume that $a_x = \alpha_x \Delta$ for $\alpha_x \in \mathbb{Z}$. We also assume that the impulse responses are separable in x and y, for instance, that $\tau(x, y) = \tau_x(x)\tau_y(y)$, $p(x, y) = p_x(x)p_y(y)$ and $h(x, y) = h_x(x)h_y(y)$. Defining a reduced instrument function

$$h_{lr}(\lambda) = \iint \tau_x(x) h_x(x' - x) p_x\left(x' + \frac{F\lambda}{\Lambda}\right) dx\, dx' \tag{9.15}$$

We find a discrete measurement model analogous to Eqn. (2.39)

$$g_{nm} = \sum_{ij} t_{ij} \kappa_{m - a_y j} S_{n + \alpha_x i} \tag{9.16}$$

where

$$S_n = \int S(\lambda) h_{lr}(\lambda - n\Delta_l)\, d\lambda \tag{9.17}$$

and

$$\kappa_m = \iint \tau_y(y) h_y(y' - y) p_y(y' - m\Delta)\, dy\, dy' \tag{9.18}$$

The optical system images the transmission mask onto the focal plane along the y axis. In general, one might desire a pixel sampling pitch smaller than the mask pitch to ensure Nyquist sampling of the y-axis image. Under this approach one can correct for misalignments between the input mask and the focal plane [243]. To simplify our analysis, we assume that either perfect alignment between the coded aperture and the focal plane or corrected alignment using digital interpolation enables us to assume $\alpha_y = 1$ and $\kappa_m = \delta_m$, in which case Eqn. (9.16) becomes

$$g_{nm} = \sum_i t_{im} S_{n-\alpha_x i} \tag{9.19}$$

While Eqn. (9.19) is similar in form to the coded aperture imaging measurement model, there are important differences: (1) since the coded modulation occurs in an image plane of the sensor, the impact of diffraction is much less for coded aperture spectroscopy; and (2) Eqn. (9.19) reflects a one-dimensional convolution rather than the 2D convolution of (2.39). One may choose to invert Eqn. (9.19) using convolutional coding along the dispersion direction, as described by Mende et al. [177], but it is also possible to implement better-conditioned pixel codes along the axis transverse to dispersion.

We focus specifically on *independent column coding* [89]. With this strategy, columns of the coding matrix t_{ij} are selected to be maximally orthogoal under the nonnegative weighting constraint. Letting g_n be the vector of measurements corresponding to the nth column of the measurement data and S_n be the N-dimensional vector with coefficients $S_{n-a_x i}$, where N is the dimension of t_{ij}, we may express Eqn. (9.16) as

$$g_n = TS_n \tag{9.20}$$

where T is a matrix with coefficients t_{ij}. This measurement may be inverted using the least-squares or least-gradient methods described in Chapter 8 to produce an estimate of S_n corresponding to spectrum samples spanning the range from S_{n-N} to S_{n-1}. Reconstructing the columns over the range from $n = 1$ to $n = N$ produces one or more estimates of the spectral density samples over the range from S_{1-N} to S_{N-1}.

As an example, Fig. 9.5 illustrates spectral reconstruction from experimental data in with a 48-element Hadamard S-matrix code. Motivations for selecting this code arise from the SNR issues discussed in Section 8.2.2 and in Refs. 114 and 243. The image in Fig. 9.5(a) shows the raw CCD data from the spectrometer. Curvature in the image arises from the nonlinearity of the grating equation [Eqn. (4.58)] with respect to angle. This curvature is corrected by an anamorphic transformation in Fig. 9.5(b). Least-gradient signal estimation using an upsampled calibrated measurement code is implemented on each column of the corrected image to produce the spectral estimates in Fig. 9.5(c). The spectral estimates in the ith row are shifted to the right by i to produce the aligned spectral data shown in Fig. 9.5(d). The columns of this matrix are averaged to produce the spectral density estimate shown in Fig. 9.5(e). These particular data correspond to the spectrum of a xenon discharge lamp.

After processing, the independent column code spectrometer returns an estimate of S_n over the range discussed above. S_n is a discrete sample of a continuous function filtered by the system transfer function

$$\hat{h}_{lr}(u) = \hat{\tau}_x\left(\frac{\Lambda u}{F}\right)\hat{h}_x\left(\frac{\Lambda u}{F}\right)\hat{p}_x\left(\frac{\Lambda u}{F}\right) \tag{9.21}$$

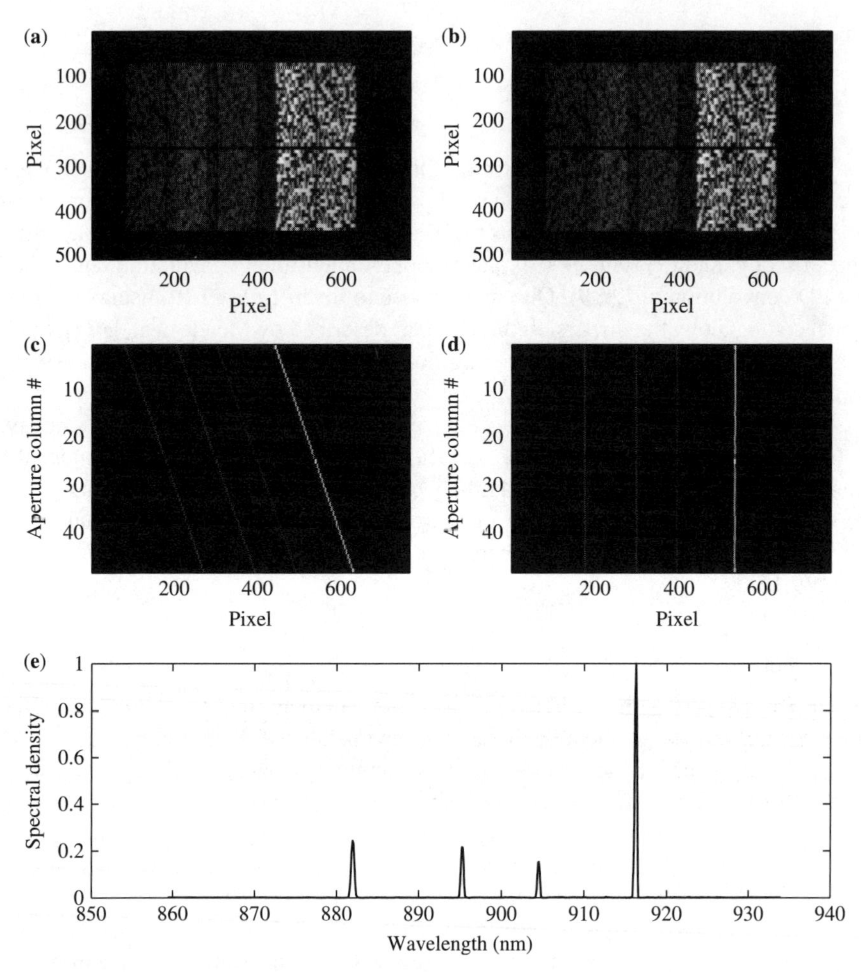

Figure 9.5 Spectral estimation using an independent column code spectrometer: (a) raw CCD image; (b) smile-corrected CCD image; (c) spectral estimates from each column of aperture code; (d) aligned spectral estimates; (e) spectral estimate.

The coding pixel function $\tau(x)$ replaces the slit transmittance in the STF of the coded aperture system. As discussed momentarily, coded aperture systems take advantage of aperture mask features approaching the diffraction limit. The difference between the transfer function for a typical coded aperture system and a slit is illustrated in Fig. 9.6. In the coded aperture system, one must be careful to select $\alpha_x \geq 2$ to avoid aliasing the reconstructed spectrum.

Using the same arguments as those used to derive Eqn. (9.7), the spectral resolution of the coded aperture system is $\delta\lambda = a\Lambda/F = \alpha_x\Delta\Lambda/F$. For the same mask feature size, the etendue is for the coded aperture system is a factor of $N/2$ higher

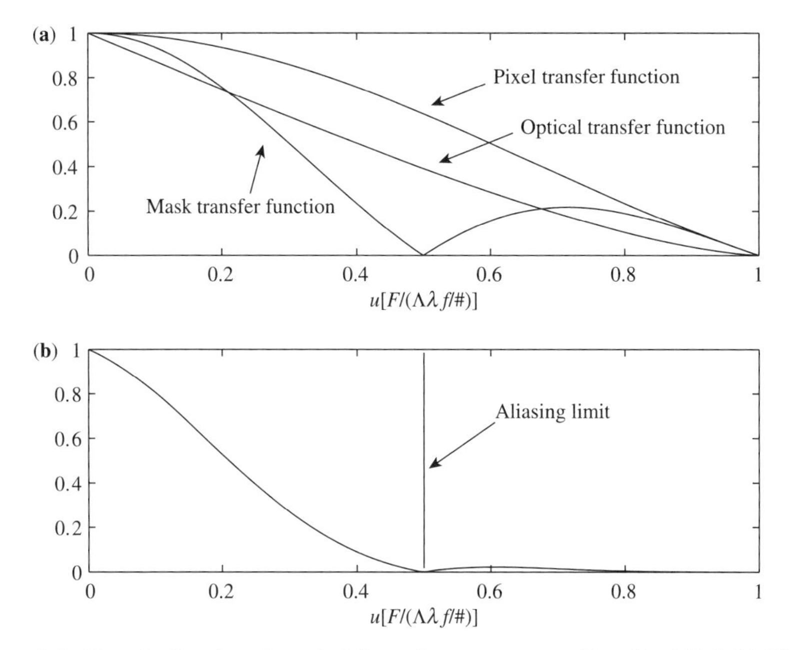

Figure 9.6 Transfer functions for coded dispersive spectroscopy. The slit width is $2\lambda f/\#$, and the pixel width is $\lambda f/\#$. Plot (b) shows the STF, which is the product of the individual transfer functions. The pixels undersample the optical resolution limit but sample near the Nyquist limit for the mask feature size.

than for a slit, specifically

$$L = \frac{\pi^2 NAa}{4(f/\#)^2} \tag{9.22}$$

The factor of $\frac{1}{2}$ is introduced based on an expectation that mean code transmittance is $\frac{1}{2}$. The efficiency of the coded aperture system is

$$E \approx \frac{\lambda NV^{2/3}}{2\Lambda(f/\#)^{5/3}} \tag{9.23}$$

Unlike the slit spectrometer, the étendue of a coded aperture spectrometer can be increased without reducing the spectral resolving power, and the spectral resolving power can be increased without reducing the étendue. In each case, these effects are acheived by increasing the order of the code N in proportion to any decrease in the code feature size a. In principle, one could reduce a to the diffraction limited value for a coded aperture system and attain the $R = N_g$ resolving power limit. In practice, one is more likely to select $a \geq 2\Delta$ to ensure Nyquist rate sampling of the spectral data. It is interesting to note that one can maintain spectral efficiency E in smaller volume spectrometers by increasing N in inverse proportion to $V^{2/3}$.

Etendue and spectral efficiency are not in themselves good metrics of the spectrum returned by an instrument. Ultimately, analysis should be based on the performance of an instrument in the context of the specific task for which it is designed. Such metrics will depend on the nature of the objects under analysis. For example, for reasons discussed momentarily, coded aperture spectrometers are particularly useful in the analysis of systems dominated by additive noise or signals in which the components that one wishes to measure are the strongest features. Conventional slit spectrometers, or codes intermediate between a full coded aperture and a slit, may be optimal in analyzing objects where shot noise from a background feature dominates relatively weak signatures from features of interest.

In the case of the slit spectrometer, analytic samples and measurement data are identically represented by Eqn. (9.5). As described by Eqn. (5.43), the variance of the sample data due both to various additive components and to signal dependent shot noise is

$$\sigma_S^2 = \sigma_r^2 + \kappa_p \bar{S} \tag{9.24}$$

For the coded aperture system, in contrast, the analytic samples S_n are obtained by computational inversion of the measurement samples g_n. The variance of the coded aperture measurements is

$$\sigma_g^2 = \sigma_r^2 + \frac{N}{2}\kappa_p \bar{S} \tag{9.25}$$

where we assume that half of the spectral channels are collected in each measurement. Using the ordinary least-squares estimator, the variance of the analysis samples for the coded aperture is

$$\sigma_S^2 = \frac{4\sigma_r^2}{N} + 4\kappa_p \bar{S} \tag{9.26}$$

where the factor of 4 assumes the use of the S-matrix code. The variance of both the coded aperture and the slit is reduced by averaging. The slit system averages over the spatial extent of the slit, while the independent column code system breaks the slit up into features. As illustrated in Fig. 9.5, however, the independent column code averages along the columns after reconstruction and alignment. Since the impact of the averaging step is the same for both approaches, we do not consider it further here.

On first glance, Eqns. (9.24) and (9.26) indicate that the slit spectrometer is preferred in shot-noise-dominated systems and the coded aperture is preferred for read-noise-dominated systems. Since one expects read noise to dominate at low signal values and shot noise at high signal values, one may expect coded apertures to be useful in the design of sensitive and short-exposure instruments, while conventional slit spectrometers may perform better when exposure time is not an issue.

Careful comparison of slit and coded aperture spectroscopy must also consider the distribution of source features and noise. A source may be said to consist of a single bright feature if only a single value of S_n is nonzero or if the source consists of a fixed pattern of spectral components. In either case, the SNR for estimation of the bright feature with a slit spectrometer is

$$\text{SNR}_{\text{slit}} = \frac{\bar{S}}{\sqrt{\sigma_r^2 + \kappa_p \bar{S}}} \tag{9.27}$$

whereas the SNR for the coded aperture spectrometer is

$$\text{SNR}_{\text{CA}} = \frac{N\bar{S}}{2\sqrt{\sigma_r^2/N + \kappa_p \bar{S}}} \tag{9.28}$$

where $\bar{S}$ corresponds to the mean energy in the target feature.

On the other hand, if the signal consits of a two unknown spectral components $S_+ \gg S_-$, then SNR for the weaker component with a slit spectrometer is

$$\text{SNR}_{\text{slit}} = \frac{\bar{S}_-}{\sqrt{\sigma_r^2 + \kappa_p \bar{S}_-}} \tag{9.29}$$

while the SNR for the coded aperture is

$$\text{SNR}_{\text{CA}} = \frac{N\bar{S}_-}{2\sqrt{\sigma_r^2/N + \kappa_p \bar{S}_+}} \tag{9.30}$$

meaning that the SNR for measurement of this feature will be worse for the coded aperture system than for the slit if the target feature is more than N^2 times dimmer than the strong feature.

The examples of a measurement of single bright feature and the search for a dim background feature in the presense of a bright obscuring feature illustrates a central distinction between slit and coded aperture spectrometers: even when the mean variances are equal the structure of noise is quite different between the two systems. Shot noise in the slit system introduces larger variance in the brightest channels while shot noise in the multiplex system distributes noise uniformly over all reconstructed channels. A slit and a coded aperture spectrometer with equal estimation variance differ in that all signal components with energy above the mean have lower estimation error for the coded aperture, while all signal components with energy below the mean have lower estimation error for the slit spectrometer. Roughly speaking, this suggests that multiplex systems (e.g., coded apertures) have an advantage for radiant features (as in emission or Raman spectroscopy) but are at a disadvantage for absorptive features (where background dominates).

This point is illustrated in Fig. 9.7, which shows estimated spectra for a slit and a coded aperture spectrometer using an $N = 512$ Hadamard S matrix. Both systems are

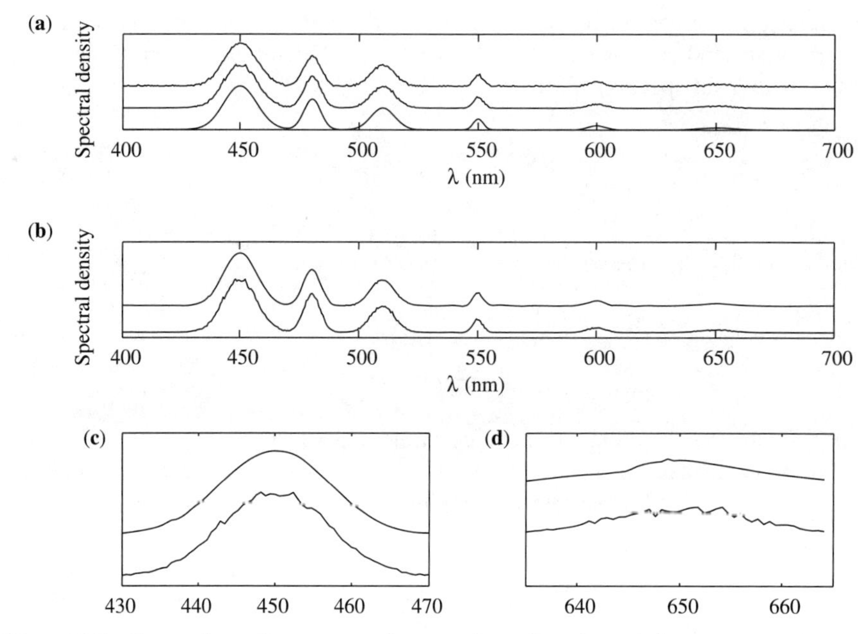

Figure 9.7 Comparison of reconstructed spectra for a slit and a coded aperture spectrometer. Plot (a) (where $\sigma^2_{\text{slit}} = 0.049$, $\sigma^2_{\text{CA}} = 0.056$) shows the reconstructed spectrum for shot-noise-only measurements. The lower trace is the true spectrum; the middle curve is the slit spectrum, assuming a peak measurement count of 1000 quanta per sample. The upper trace is the spectrum reconstructed using nonnegative least squares for $N = 512$ Hadamard S-matrix sampling. Plot (b) (where $\sigma^2_{\text{slit}} = 0.04$, $\sigma^2_{\text{CA}} = 0.022$) shows the wavelet denoised signals of (a). Plots (c) and (d) are detail plots from (b).

achieve similar total signal variance after reconstruction, the square difference between the spectra and the true spectrum for a particular numerical experiment are shown in Fig. 9.7(a). Note, however, that the coded aperture spectrum has noise distributed across all channels of the reconstruction while null values of the slit spectrum are reconstructed as zero. Since the spatial structure of the noise is unrelated to the actual signal structure for the coded aperture system, denoising is much more effective. Figure 9.7(b) shows the estimated spectra of (a) after denoising with the Matlab `wden()` command using minimax thresholding and the order 8 symlet. The lower spectrum is due to a slit, the upper due to the coded aperture. As illustrated in the figure caption, the experimental variance between the estimated and true spectra is reduced by a larger factor for the multiplex system than for the slit spectrometer. Figure 9.7(c) and (d) show details of the plots from (b), illustrating that the CA system is particularly superior near peak spectral values. The multiplex system is less effective in the detection of "noise-like" weak spectral features and is more likely to introduce non-signal-related artifacts to the estimated spectrum.

The two extremes of a slit spectrometer and full S-matrix sampling are endpoints on a continuum wherein diverse coding strategies may be applied to optimally tease out spectral features. In view of the diverse utilities of different coding strategies, some design studies have found dynamically encoded apertures using micromechanical, liquid crystal, and acoustooptic devices to be attractive [55].

As a final comment on dispersive spectroscopy, note that the spectral throughput for both the slit and coded aperture systems is 1 (we have already accounted for the 50% loss in throughput for the coded aperture system in the etendue). The spectral throughput is marginally or substantially reduced by spectrometer designs surveyed in the remainder of the chapter.

9.4 INTERFEROMETRIC SPECTROSCOPY

Dispersion encodes optical signals by directing different components on different spatial paths. Interference encodes optical signals by linearly combining two or more signals on the same path. One may, of course, imagine instruments that combine both dispersion and interference (as we do in Sections 9.6 and 9.8). We first consider the classic instruments of purely interferometric spectroscopy. Interference strategies separate into "two-beam systems," which were introduced in Section 6.3, and "multibeam" resonant systems, which we introduce in Section 9.5. The present section compares the resolving power, etendue, and SNR of two-beam interferometric spectroscopy with the dispersive system metrics derived in the previous two sections.

A Fourier transform (FT) spectrometer based on the Michelson interferometer of Fig. 6.4 gathers serial data by causing interference between a beam and a longitudinally delayed replica of itself. We modeled the interference as a time shift in Eqn. (6.34). Our present goal is to develop a more precise model accounting for a beam spanning a finite solid angle. The interference geometry of a Michelson interferometer is illustrated in Fig. 9.8. An optical field is split into two beams and caused to interfere with itself after a longitudinal delay. Figure 9.8 shows two copies of a field

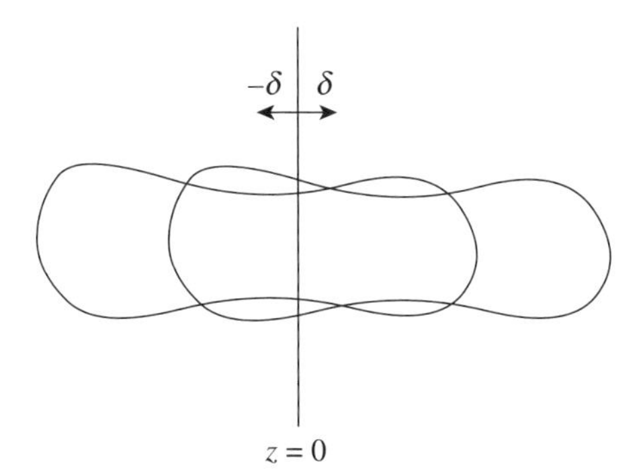

Figure 9.8 A Michelson interferometer creates interference between a light beam and a longitudinally delayed copy of itself. Here the delay is 2δ. The average irradiance is typically measured by integration over the transverse x, y plane at $z = 0$.

with a relative delay of $\Delta z = 2\delta$. An FT instrument measures the irradiance of the interference pattern averaged over a detector in the $z = 0$ plane.

Fourier transform spectrometers collect a range of measurements by either continuously scanning δ or by a step-and-integrate process under which δ changes by discrete values. The continuous motion approach, using micromechanical or piezoelectric positioning stages, is more common for single-detector instruments. Step-and-integrate approaches are necessary in spectral imaging systems where the detector array must acquire 2D frames. Step-and-integrate systems require positioning accuracy to $\lambda/100$ and are extremely sensitive to positioning error [119]. Similarly, scanned systems require uniform and well-characterized motion.

Using the irradiance model of Eqn. (6.34), continuous scanning produces discrete measurements

$$g_n = \frac{A^2\Delta_t}{2}\Gamma(0) + \frac{A^2}{4}\int \Gamma(\Delta_z = -Z + 2vt,\ \tau = 0)p(t - n\Delta_t)\, dt$$
$$+ \frac{A^2}{4}\int \Gamma(\Delta_z = Z - 2vt,\ \tau = 0)p(t - n\Delta_t)\, dt \tag{9.31}$$

where A is the diameter of the detector aperture, v is the velocity of the scan, and $p(t)$ is the temporal sampling function of the detector. One might, for example, assume that $p(t) = \mathrm{rect}(t/\Delta_t)$. We assume that the scan starts at $-Z/2$ and spans $(-Z/2, Z/2)$.

We derive the relationship between g_n and the power spectral density beginning with the cross-spectral density. The cross-spectral density between a light field at $z = -\delta$ and the same field delayed by $z = \delta$ is

$$W(\Delta x = 0, \Delta y = 0, \Delta z = 2\delta, \nu) = \frac{e^{4\pi i(\delta\nu/c)}}{\lambda\delta}\iint$$
$$\times\, W_0(\Delta x', \Delta y', \nu)e^{i(\pi/\lambda\delta)(\Delta x'^2 + \Delta y'^2)}\, d\Delta x'\, d\Delta y' \tag{9.32}$$

$W_0(\Delta x', \Delta y', \nu)$ is the cross spectral density in the plane $z = 0$, and we model diffraction using the Fresnel kernel. We assume a Schell model (spatially stationary) object. Further assuming a Gaussian–Schell model such that

$$W_0(\Delta x, \Delta y, \nu) = S(\nu)\exp\left(-\frac{\pi(\Delta x^2 + \Delta y^2)}{w_0^2}\right) \tag{9.33}$$

yields

$$W(\Delta z = 2\delta, \nu) = e^{4\pi i(\delta\nu/c)}\frac{w_0^2\nu}{w_0^2\nu - ic\delta}S(\nu) \tag{9.34}$$

As indicated by Eqn. (9.34), the cross-spectral density of the longitudinally shifted fields decays inversely in the ratio $\delta\lambda/w_0^2$. It is possible to accurately model both the amplitude and phase of the decay factor, but for the present purposes it is more illustrative to note that the decay factor limits the effective scan range to $|\delta| < w_0^2/\lambda$. This range is specific to the Gaussian–Schell model; it may be possible to extend it by a constant factor through wavefront engineering based on the optical extended depth of field techniques described in Section 10.2 Within the effective scan range we assume

$$W(\Delta z = 2\delta, \nu) \approx e^{4\pi i(\delta\nu/c)} S(\nu) \tag{9.35}$$

Noting that

$$\Gamma(\Delta z, \tau = 0) = \int W(\Delta z, \nu)\, d\nu \tag{9.36}$$

we observe that

$$\begin{aligned}
&\int \Gamma(\Delta_z = 2vt - Z, \tau = 0) p(t - n\Delta_t)\, dt \\
&\quad = \iint \exp\left(2\pi i \frac{(2vt - Z)\nu}{c}\right) S(\nu) p(t - n\Delta_t)\, dt\, d\nu \\
&\quad = \int \hat{p}\left(\frac{2v\nu}{c}\right) \exp\left(-2\pi i \frac{(Z + 2nv\Delta_t)\nu}{c}\right) S(\nu)\, d\nu \\
&\quad = \sum\nolimits_{n'} e^{-2\pi i(nn'/N)} S_{n'}
\end{aligned} \tag{9.37}$$

where

$$\begin{aligned}
S_{n'} = &\int \hat{p}\left(\frac{2v\nu}{c}\right) e^{-2\pi i(Z\nu/c)} \exp\left[\pi i\left(\frac{n'}{N} - \frac{2v}{c}\Delta_t \nu\right)\right] \\
&\times \frac{\sin\{\pi[n' - (2Z/c)v]\}}{\sin\{\pi[(n'/N) - (2Z/Nc)\Delta_t \nu]\}} S(\nu) d\nu
\end{aligned} \tag{9.38}$$

$N = Z/v\Delta_t$ is the number of samples recorded. Equation (9.37) is derived by taking the discrete Fourier transform, as in Eqn. (7.9), to isolate $S_{n'}$ and then taking the inverse transform. Approximating the Dirichlet kernel as in Eqn. (7.10) yields

$$\begin{aligned}
S_{n'} &\approx \sum\nolimits_{n''=-\infty}^{\infty} \int \hat{p}\left(\frac{2v\nu}{c}\right) \operatorname{sinc}\left(\frac{2Z}{c}\nu - n' - n''N\right) S(\nu) d\nu \\
&= \sum_{n''=-\infty}^{\infty} \hat{p}\left[\frac{2v}{Z}(n' + n''N)\right] S\left[\frac{c}{2Z}(n' + n''N)\right]
\end{aligned} \tag{9.39}$$

The values $S_{n'}$ are the analytic samples of the power spectrum that one estimates from the data g_n. We see in Eqn. (9.39) that these samples are spaced by $\Delta\nu = c/2Z$, meaning that the spectral resolution is $\delta\lambda = \lambda^2/2Z$. Applying the constraint from Eqn. (9.34) that $Z < w_0^2/\lambda$, we find that the resolving power of the FT spectrometer is

$$R = \frac{2w_0^2}{\lambda^2} \tag{9.40}$$

Recalling from Eqn. (6.22) that the coherence cross section is related to the angular extent of a beam by $w_0 \approx \lambda/\Delta\theta$, the resolving power becomes

$$R \approx \frac{2}{\Delta\theta^2} \tag{9.41}$$

If the interferometer collects light through a focal system, we may assume that $\Delta\theta \approx 1/f/\#$ and $R = 2(f/\#)^2$.

The etendue of an FT spectrometer with aperture diameter A is $L = A^2/(f/\#)^2$, corresponding to efficiency $E = A^2$. Estimating the volume of the instrument to be $V = f/\#A^3$, the efficiency in terms comparable to the dispersive instruments is

$$E = \frac{V^{2/3}}{(f/\#)^{2/3}} \tag{9.42}$$

Comparing Eqn. (9.42) with Eqns. (9.23) and (9.11), we find that the relative efficiencies of FT and dispersive instruments depend strongly on $f/\#$. The $f/\#$ of an FT instrument is linked to resolving power. R=1000, for example, requires $f/\# \geq 23$. The $f/\#$ of dispersive instruments, in contrast, is determined by the angular bandwidth of the diffractive element, as discussed for volume holograms in Section 9.6.1, and by lens scaling issues discussed in Section 10.4.1. $f/10$ is typical in practical designs. For typical parameters, the spectral efficiency of a slit spectragraph is made worse than FT instruments. Coded aperture systems may achieve efficiencies comparable to or somewhat better than FT systems.

The idea that the *LR* product is a constant of spectrometer design was popularized in the 1950s [125]. At that time, the balance of design favored FT instruments. The emergence of high-quality detector arrays explains the difference between analysis then and now. The 1950s era design assumed that the spectrometer would use a single-detector element. If the detector area is the same for the dispersive and FT instruments, then the resolving power and efficiency are much better for the FT system. FT systems remain the approach of choice at wavelengths, such as the infrared and ultraviolet extremes, where reliable detector arrays are unavailable. Dispersive systems gain substantially as the detector area is reduced and arrays are fabricated. With the emergence of 2D detector arrays, the "throughput advantage" often associated with interferometric instruments has actually swung substantially in favor of dispersive design. Spectrometer design remains remarkably fluid, however. There are many interferometric designs that produce spatial patterns and yield somewhat better efficiency than the Michelson interferometer.

Equation (9.38) indicates aliasing of spectral components separated by the range $cN/2Z = N\delta\lambda$. Aliasing is unlikely to be an issue for the continuous scan–integrate approach, however, because the spectrum is lowpass-filtered by the sampling function. Typically, we assume that $\hat{p}(\nu) = \text{sinc}(\Delta_t \nu)$, meaning that terms in the series are substantially attenuated for $|n' + n''N| > N/2$. Since we are exclusively interested in values of S_n for positive n, this means that we obtain on the order of one spectral value for every two samples (although samples near the band edge will be severely attenuated). In practice, sampling at the rate of $v\Delta_t = \lambda_{\max}/4$ may be advisable. With the multiplex spectrometer, the spectral values $nc/2Z$ span the range from DC ($n = 0$) to $\nu_{\max}$. In practice, of course, DC values are not part of the optical spectrum. If we consider an instrument spanning the range from $\lambda = 2-20\ \mu\text{m}$, then 90% of the spectral range from DC to $\lambda_{\max}$ contains useful information. On the other hand, an instrument in the visible spanning the range $\lambda = 500-700\,\text{nm}$ must collect over six measurements per data value. This potential sampling inefficiency puts interferometric systems at a disadvantage to dispersive systems in using detector arrays.

Substituting Eqn. (9.38) in Eqn. (9.31) and assuming that $\Gamma(0) = \sum_n S_n$, the measurement model for an FT spectrometer becomes

$$g_n = \frac{1}{2}\sum_{n'=0}^{N/2}\left[1 + \cos\left(2\pi\frac{nn'}{N}\right)\right]S_{n'} \tag{9.43}$$

This mapping can be inverted using methods discussed in Chapter 8. Under ordinary least-squares estimation, the Fourier code of Eqn. (9.43) yields an estimate with twice the variance of the Hadamard S matrix, meaning that the variance in estimates of S_n is

$$\begin{aligned}\sigma_S^2 &= 8\frac{\sigma_g^2}{N}\\ &= \frac{8\sigma_r^2}{N} + 4\kappa_p\overline{S}\end{aligned} \tag{9.44}$$

where $\overline{S}$ is the mean of S_n integrated over one sampling period.

The function σ_r^2 represents the variance for a single detector measurement over a fixed time window. An FT instrument is most easily compared to an instrument with a tunable narrowband filter in front of the single detector. As indicated by Eqn. (9.44), the variance of the FT instrument will be a factor of $N/8$ less than the variance of the tunable instrument if read noise dominates. In shot-noise-dominated systems, however, both approaches produce the same variance. Dispersive multiple detector systems, in contrast, integrate each spectral channel over the entire measurement window. In a simple model of such an instrument the read signal variance increases in proportion to the length of the recording window. The mean signal value also increases by the recording time, however. Since the single channel mean signal is a $N\times$ less than the mean signal for the multiplex instrument, the photon noise per measurement is about the same. Summarizing, a detector array based dispersive element measuring the same signal as in Eqn. (9.44) over the same total measurement time window produces estimation variance of $\sigma_S^2 = \sigma_r^2 + \kappa_p N\bar{S}$ on a mean signal value of $N\bar{S}$. The dispersive system therefore achieves approximately $2\sqrt{N}\times$ better SNR than does the FT system. As with the coded aperture system, changes in the relative SNR may arise when denoising and application-specific measurements are considered.

The advantage of FT systems relative to narrowband filters in single-detector systems dominated by read noise is termed the "multiplex advantage" [71,72]. This factor was a primary motivator in the development of FT systems from 1960 through the 1980s. The basic idea is that the spectral throughput of a single-detector FT instrument is approximately $\frac{1}{2}$, while the spectral througput of a single-spectral-channel instrument is $\delta\lambda/\Delta\lambda$. For broadband measurements with additive noise, the multiplex instrument achieves an SNR advantage of $\sqrt{\Delta\lambda/2\delta\lambda}$. The analysis is more complex for signal-dependent noise, as discussed in Section 9.3.

Multiplexing remains attractive when detector arrays are unavailable, when the object is diffuse, and when a spectral image is desired. When 2D detector arrays are available, coded aperture systems have higher etendue, spectral throughput, and mechanical stability than do FT systems. The development of large-scale micromechanical modulator arrays has added further variety to multiplex system design. Using a modulator array, it is possible to make a Hadamard, rather than Fourier, coded system and thereby achieve a modest increase in SNR. Of course, this increase comes at the cost of dramatically more complex micromechanical control requirements. On the other hand, one may use large modulator arrays to dynamically and adaptively sample spectral channels. Increasing integration time on features of interest and decreasing attention to null features could substantially improve SNR. This type of approach was demonstrated, for example, by Maggioni et al. in an adaptive spectral illumination system [163].

9.5 RESONANT SPECTROSCOPY

The introduction to optical elements way back in Section 2.2 lists four classes of devices: refractive, reflective, and diffractive elements and *interferometric* devices. By this point in the text, the reader is generally familiar with the nature and potential utility of the first three categories. However, the two-beam interferometric systems encountered thus far (the Michelson interferometer, the Michelson stellar interferometer, and the rotational shear interferometer) are far from representative of the true capabilities of interferometric devices. Resonant devices introduce qualitatively novel features into optical systems. The next several sections provide a brief introduction to resonant devices in spectroscopy. Design, fabrication, and analysis tools for these systems continue to evolve rapidly and radical system opportunities are emerging. We cannot predict the ultimate nature of these devices, but we hope to motivate their continued development.

The Fabry–Perot (FP) etalon, sketched in Fig. 9.9, is the simplest resonant interferometer. The instrument consists of two partially transmissive/partially reflective surfaces separated by an dielectric gap of thickness d. An incident wave is partially

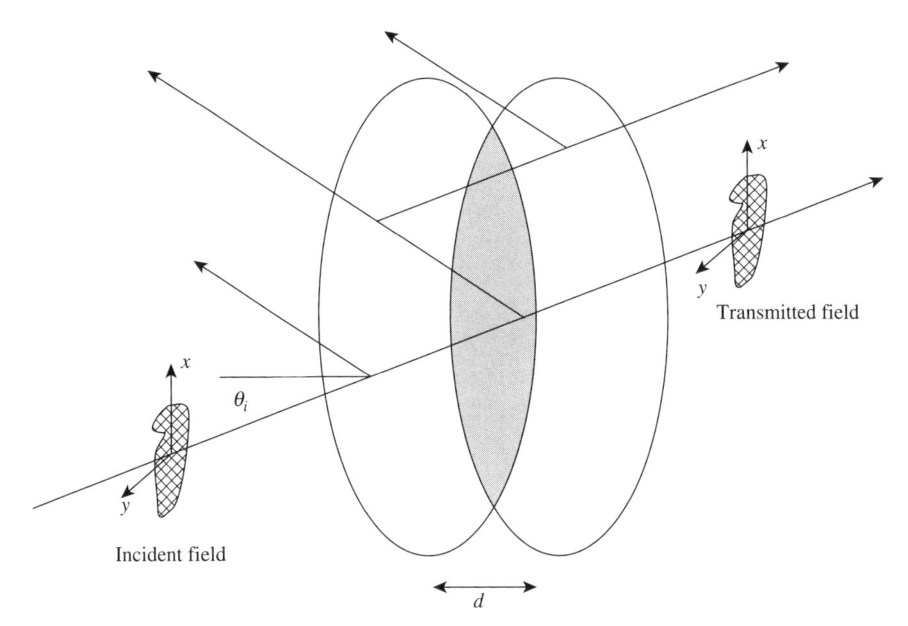

Figure 9.9 A Fabry–Perot etalon consists of a pair of partially transmissive surfaces separated by a gap of thickness d.

reflected by the first surface and partially transmitted into the cavity. Once in the cavity, the wave experiences an infinite series of partial reflection and transmission events at each surface.

As always, we approach analysis of an optical element by first considering the modulation that the device induces on a coherent input field. An FP resonator does not produce a local modulation of the incident field, like a transmission mask or a lens. Rather, the output field is a shift-invariant linear transformation of the input field. As always, shift invariance means that if the field on the input aperture is $E_i(x, y)$, then the field on the output aperture is

$$E_0(x', y', \nu) = \int h(x' - x, y' - y, \nu) E_i(x, y, \nu) \tag{9.45}$$

The coherent transfer function $\hat{h}(u, v, \nu)$ corresponding to the shift-invariant impulse response is derived by considering the transmittance for the incident plane wave $E_i e^{2\pi i(ux+vy)} e^{i2\pi z\sqrt{(1/\lambda^2)-u^2-v^2}}$. Assuming that the plane wave transmittance at each interface is τ and that the reflectance is r, the wave transmitted by the etalon is

$$\begin{aligned} E_t(u, v) &= \tau^2 e^{i\phi(u,v)} E_i + \tau^2 r^2 e^{i3\phi(u,v)} E_i + \tau^2 r^4 e^{i5\phi(u,v)} E_i + \tau^2 r^6 e^{i7\phi(u,v)} E_i + \cdots \\ &= \frac{\tau^2 e^{i\phi(u,v)}}{1 - r^2 e^{i2\phi(u,v)}} E_i \end{aligned} \tag{9.46}$$

where the phase delay in propagating through the etalon is $\phi(u, v) = 2\pi n d\sqrt{1/\lambda^2 - u^2 - v^2}$; n is the index of refraction of the cavity dielectric. We find, therefore, that the transfer function between the field on the input aperture of an etalon and the output aperture is

$$\hat{h}(u, v) = \frac{\tau^2 e^{i\phi(u,v)}}{1 - r^2 e^{i2\phi(u,v)}} \tag{9.47}$$

In deriving Eqn. (9.47) we have neglected the fact that τ and r also depend on (u, v). While it is not difficult to account for this dependence in numerical analysis, our analytic discussion is simpler without it. If the surfaces of the etalon consist of metal films or high-permitivitty dielectrics, then the (u, v) dependence of τ and r is relatively weak. The reflectivities of practical surfaces, as well as additional model parameters such as surface smoothness, scatter, and finite etalon apertures are discussed in Ref. 117.

The coherent impulse response for the etalon is, of course, the inverse Fourier transform of $\hat{h}(u, v)$, and the incoherent impulse response is the squared magnitude of the coherent impulse response. Figure 9.10 shows the incoherent impulse response for various cavity thicknesses. Since the cavity thickness and the spatial scales are given in wavelengths, one may imagine similar plots varying λ rather than d. The point of this exercise is to confirm that one obtains a PSF that is strongly dependent on wavelength and a cavity thickness, although thc structure is not yet particularly promising for spectral analysis. Spectral analysis using the etalon requires insertion of the device in more complex optical systems.

As illustrated in Fig. 9.11, a typical FP spectrograph places an etalon in the aperture plane of a Fourier transform lens. As with the FT spectrometer, we model the response of this instrument to a Schell model object. A spatially stationary object transformed by a linear shift-invariant system remains a Schell model object after transformation. Specifically, if $W_0(\Delta x, \Delta y, v)$ is the cross-spectral density at the input to the etalon, immediately after the etalon

$$W(\Delta x, \Delta y, \nu) = \iiiint W_0[\Delta x - (x_1' - x_2'), \Delta y - (y_1' - y_2'), \nu] \times h^*(x_1', y_1', \nu) h(x_2', y_2', \nu)\, dx_1'\, dy_1'\, dx_2'\, dy_2' \tag{9.48}$$

where $h(x, y, \nu)$ is the coherent impulse response of the etalon. We find the cross-spectral density at the focal plane of the spectrograph by substituting Eqn. (9.48) in by Eqn. (6.56), which yields

$$S(x', y', \nu) = \iint \hat{W}_0\left(u = \frac{\nu x}{cF}, v = \frac{\nu y}{cF}, \nu\right) \left|\hat{h}\left(u = \frac{\nu x}{cF}, v = \frac{\nu y}{cF}, \nu\right)\right|^2 \times h_{ic}(x' - x, y' - y, \nu)\, dx'\, dy' \tag{9.49}$$

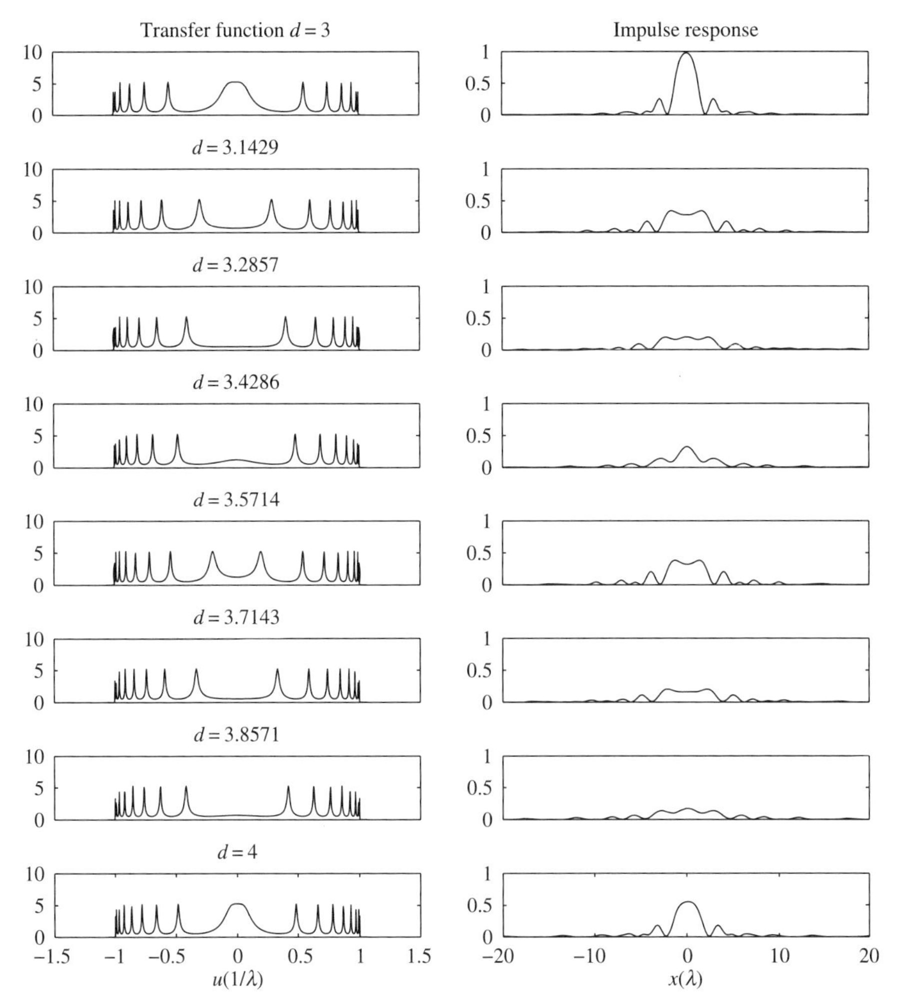

Figure 9.10 Transfer function and incoherent impulse response for a thin Fabry–Perot etalon. The thickness d is given in wavelengths. The plots on the left show $\hat{h}(u, v = 0)$, and the plots on the right show $|h(x, y = 0)|^2$.

where $h_{ic}(x, y)$ is the PSF of the focusing lens. For the Gaussian–Schell model source of Eqn. (9.33), we obtain

$$\hat{W}_0(u, v, \nu) = S_0(\nu)e^{-\pi w_0^2(u^2+v^2)} \tag{9.50}$$

Typically, w_0 will be of order λ and $\hat{W}_0(x/\lambda F, y/\lambda F)$ will be uniform over a region comparable to F. In this case, $S(x, y, \nu)$ in the focal plane is an image of the etalon transfer function blurred by the optical PSF (as illustrated by the ring pattern in the focal plane of Fig. 9.11). Efficient energy transfer from the input aperture to the

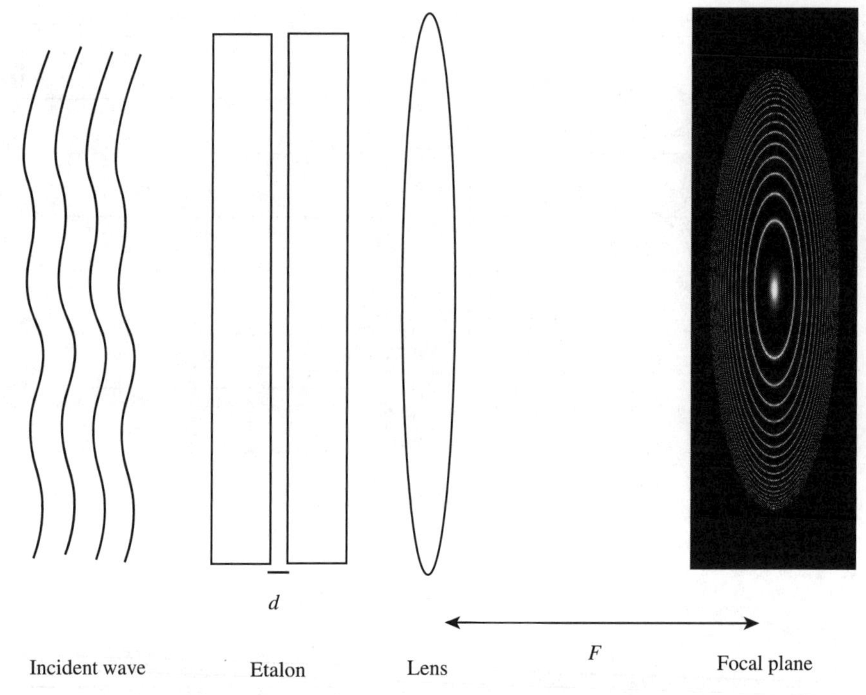

Figure 9.11 A Fabry–Perot spectrograph combines an etalon and a Fourier transform lens.

focal plane is ensured if we select the angular extent of the object to match the numerical aperture of the focal system, which implies $\Delta\theta \approx \lambda/w_0 = 1/f/\#$.

The ring pattern induced by the FP spectrograph is

$$
\begin{aligned}
q(x, y, \nu) &= \left|\hat{h}\left(\frac{\nu x}{cF}, \frac{\nu y}{cF}, \nu\right)\right|^2 \\
&= \frac{(1-|r|^2)^2}{1+|r|^4 - 2|r|^2\cos\left(4\pi(n\nu d/c)\sqrt{1-((x^2+y^2)/F^2)}\right)} \\
&= \frac{1}{1+C\sin^2\left\{2\pi(n\nu d/c)\sqrt{1-[(x^2+y^2)/F^2]}\right\}}
\end{aligned}
\tag{9.51}
$$

where we note that $|\tau|^2 = 1 - |r|^2$ and

$$
C = \frac{4|r|^2}{(1-|r|^2)^2} \tag{9.52}
$$

As illustrated in Fig. 9.12(b), the ring pattern modulates the focal plane power spectral density periodically in ν. The period of these modulations is called the *free spectral*

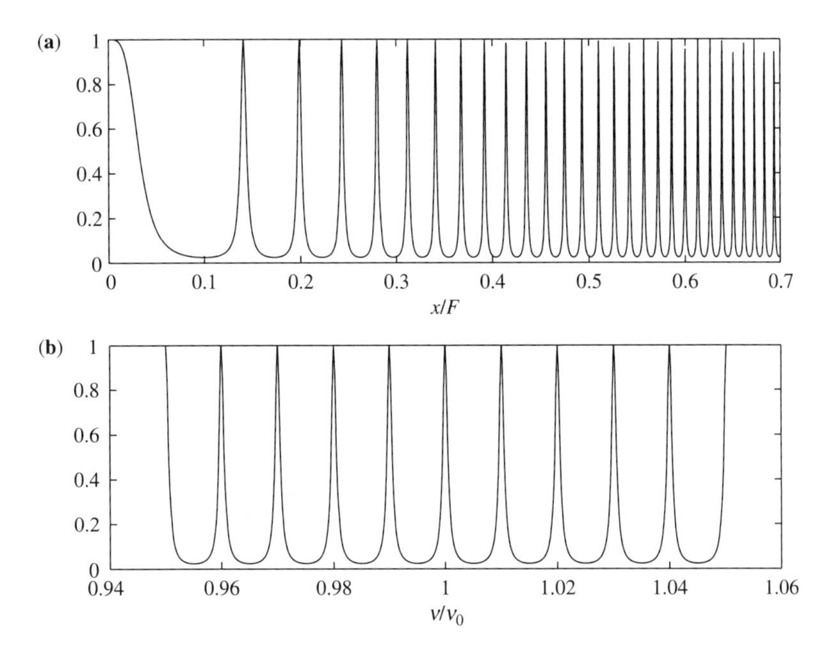

Figure 9.12 Cross sections of the Fabry–Perot ring pattern $q(x, y, \nu)$: (a) plots $q(x, 0, \nu_0)$ as a function of x/F for $nd = 50\lambda_0$; (b) plots $q(0, 0, \nu)$. The finesse is 10.

range (FSR). With reference to Eqn. (9.51), we see that

$$\nu_r(x, y) = \frac{c}{2nd\sqrt{1 - [(x^2 + y^2/F)]}} \tag{9.53}$$

The variation in the FSR across the ring pattern is illustrated in Fig. 9.13, which plots $q(x, 0, \nu)$. Assuming a spatially homogeneous source, one may use the spatial variation in ν_r to remove order ambiguity and reconstruct over a broader spectral range.

The width of spectral and spatial features determines the spectral resolution of FP instruments. The full-width at half-maximum (FWHM) of the peaks along the spectral axis is

$$\begin{aligned} \delta\nu &= \frac{2\nu_r}{\pi}\sin^{-1}\left(\frac{1}{\sqrt{C}}\right) \\ &= \frac{\nu_r}{\mathcal{F}} \end{aligned} \tag{9.54}$$

where we assume that $C \gg 1$ and $\mathcal{F} = \pi|r|/(1 - |r|^2)$ is the *finesse* of the cavity. $\delta\nu$ is the approximate spectral resolution of the FP instrument. The resolving power is

$$R = \frac{\lambda}{\delta\lambda} = \frac{\nu}{\delta\nu} = \frac{\nu\mathcal{F}}{\nu_r} \tag{9.55}$$

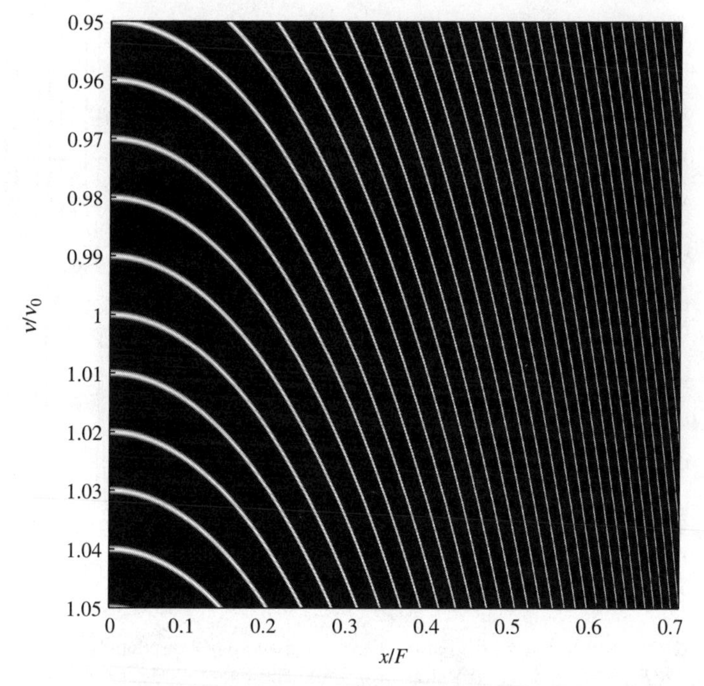

Figure 9.13 Density plot of $q(x, 0, \nu)$ for the etalon of Fig. 9.12.

Along the radial axis, the FWHM of the peaks near the edge of the focal plane is

$$\begin{aligned}\Delta x &\approx \frac{F^2 \nu_r}{x \mathcal{F} \nu} \\ &= \frac{Ff/\#}{R}\end{aligned} \tag{9.56}$$

Effective spatial sampling of the ring pattern thus seems reasonable to resolving powers on the order of 1000, which would correspond to 10 μm features for a 1 cm focal length.

The spectrum $S_0(\nu)$ is inferred from observations of spatial structure of the FP ring pattern [51] and/or from observations of the variation of the ring pattern as the optical thickness *nd* is modulated as a function of time. Over the range such that the optical, pixel and longitudinal bandpass are sufficient to capture the signal, these observations sample the function

$$g(u) = \int \frac{S_0(\nu)}{1 + (4/\pi^2)\mathcal{F}^2 \sin^2(\pi u \nu)} d\nu \tag{9.57}$$

where $u = 1/v_r$. For the Fourier lens FP system of Fig. 9.11, the range of u is

$$u \in \left(\frac{2}{c}(nd)_{\min}\sqrt{1 - \frac{1}{(f/\#)^2}}, \frac{2}{c}(nd)_{\max} \right) \tag{9.58}$$

For a static system (nd constant), the practical range of u might be 10–20% of $1/\nu_r$. In presssure-scanned systems, n may be varied by 1–2%. Mechanically scanned systems may change d by an octave or more. While we recognize that considerable attention must be devoted to the fact that $g(\nu_{\mathrm{FSR}})$ is nonuniformly sampled in the model spectrograph, we choose to neglect this issue for the moment and focus instead on the process of estimating $S_0(\nu)$ from Eqn. (9.57).

The simplest and most commonly adopted approach to FP spectroscopy assumes that the support of $S_0(\nu)$ is limited to a single free spectral range. Suppose that the object illumination is prefiltered such that $S(\nu) = 0$ for $\nu \leq (N - 1/2)\nu_{r0}$ and for $\nu \geq (N + 1/2)\nu_{r0}$, where ν_{r0} is a baseline free spectral range for the etalon. As illustrated in Fig. 9.14, the instrument function is approximately shift-invariant in the over the support of $S_0(\nu)$ for $N \gg 1$. The instrument function is given by

$$h(u, \nu) = \frac{1}{1 + (4/\pi^2)\mathcal{F}^2 \sin^2\{\pi[u_0\nu - N(u/u_0)]\}} \tag{9.59}$$

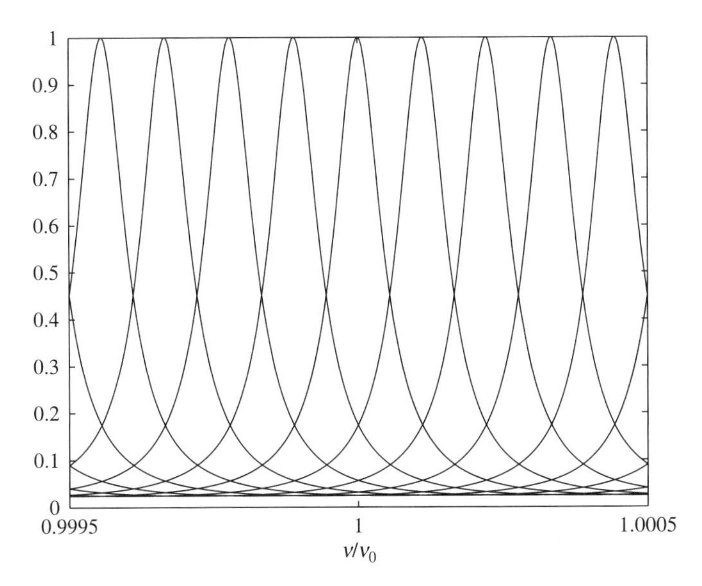

Figure 9.14 Instrument function as a function of ν and u for a Fabry–Perot spectrometer with $\nu_0/\nu_r = 1000$. u is increased by $1/\mathcal{F}\ \nu_r$ in each successive plot.

The sampled spectrum for this instrument is

$$S_n = \int h(n\Delta_u, \nu) S_0(\nu)\, du \tag{9.60}$$

where Δ_u is the sampling period over u. Simple Fourier analysis of the system transfer function confirms the resolving power estimated in Eqn. (9.55) as well as the spectral resolution

$$\delta\lambda = \frac{\lambda^2 \nu_r}{c\mathcal{F}} \tag{9.61}$$

The etendue of the FP instrument, assuming that transverse spatial sampling effectively captures the ring pattern, is $L = \pi A^2/4(f/\#)^2$, where A is the aperture diameter. Substituting the resolving power from Eqn. (9.56), the efficiency

$$E = \frac{\pi V}{4(f/\#)\Delta x} \tag{9.62}$$

where Δx is the minimum feature size used in the focal plane. Comparing with Eqns. (9.42), (9.23), and (9.11), we find that the FP spectrometer efficiency substantially exceeds the slit and FT efficiency. Assuming that the order of the coded aperture is $N \approx V^{1/3}/\Delta x$, the efficiency of the FP and coded aperture instruments is approximately comparable.

As with previous systems, the efficiency does not tell the whole story for the FP spectrometer. Operating over a single free spectral range, the etalon is effectively a narrow-pass spectral filter. Only one spectral channel is detected for each value of u. Thus, the spectrally averaged throughput is reduced by a factor of $1/\mathcal{F}$ relative to dispersive and multiplexed interferometric systems. An FP system thus needs $\mathcal{F}$ times greater efficiency to achieve the same SNR as a dispersive system. This comparison is also not quite fair, however, because the spectral resolution that one can obtain from an FP system is extraordinary, and the spectral range is typically quite limited. The high spectral resolution is typical of resonant systems generally and of systems resonant with modes oscillating along the longitudinal axis specifically.

Overcoming the limited spectral range of the FP spectrometer requires only that we expand our mathematical horizons. *Multiplex Fabry–Perot spectroscopy* solves Eqn. (9.57) for $S_0(\nu)$ spanning multiple free spectral ranges. We consider multiplex instruments here from a somewhat different perspective than in previous studies [52,117,231]. Equation (9.57) is a "Fredholm integral equation of the first kind" with a symmetric kernel. Such equations may be inverted by standard methods [134]. In spectroscopy, where one typically need estimate only a few thousand channels, direct algebraic methods are convenient.

We first suppose that our goal is to estimate $S_0(\nu)$ over two free spectral ranges. Each measurement samples the sum of two spectral channels, for example

$$g(u_1) \approx S\left(\nu_a = \frac{N}{u_1}\right) + S\left(\nu_b = \frac{N+1}{u_1}\right) \tag{9.63}$$

In order to disambiguate $S(\nu_a)$ from $S(\nu_b)$, we make a second measurement

$$g(u_2) \approx S\left(\nu_a = \frac{M}{u_2}\right) + S\left(\nu_c = \frac{M+1}{u_2}\right) \tag{9.64}$$

This measurement produces independent data if $|\nu_c - \nu_b| > \delta\nu = 1/u_2\mathcal{F}$, where we assume that $u_2 > u_1$. A very little algebra yields the constraint

$$\frac{1}{u_1} - \frac{1}{u_2} > \frac{1}{u_2\mathcal{F}} \tag{9.65}$$

or

$$\nu_{r1} - \nu_{r2} > \frac{\nu_{r2}}{\mathcal{F}} \tag{9.66}$$

We find, therefore that the fractional change in the free spectral range between adjacent free spectral ranges must exceed $1/\mathcal{F}$ if one hopes to measure across a span of $2\nu_r$. Similar analysis suggests that the fractional change in the free spectral range must be $M/\mathcal{F}$ if one hopes to estimate the spectrum across a range of $M\nu_r$.

Alternative strategies for extending the spectral range of Fabry–Perot spectroscopy combine spatial dispersion and resonant devices. Historically, the most common strategy uses a slit-based spectrometer as a pre- or postfilter on Fabry–Perot systems, with a goal of using the dispersive system to limit the spectrum to a single free spectral range. Alternatively, one may use a coded aperture in combination with a Fabry–Perot to maintain the naturally high efficiency of the instrument while also obtaining high spectral resolution. It is also possible to dispense with spatial filters altogether. If one images the Fabry–Perot ring pattern through a diffraction grating with dispersion rate $\alpha = F_g c/\Lambda$, the resulting system mapping

$$q(x, y, \nu) = \left[1 + C\sin^2\left(2\pi\frac{n\nu d}{c}\sqrt{1 - \frac{(x-\alpha/\nu)^2 + y^2}{F^2}}\right)\right]^{-1} \tag{9.67}$$

is no longer ambiguous from one free spectral range to the next.

The nonuniformity and redundancy of the spatial distribution of spectral projections in the ring pattern is the primary disadvantage of the Fabry–Perot spectrometer. We have seen in the present section that resonant devices offer extraordinary resolution, resolving power, efficiency, and integration. We have also seen, however, that analysis of these devices is much more complex than simple dispersive or interferometric systems. Even though we have sidestepped most of the complexity of sampling the Fabry–Perot ring pattern, our analysis of signal estimation has been unusually complex and incomplete. On the other hand, the Fabry–Perot is the simplest interferometric device. For better or worse, we turn to systems of greater complexity in Section 9.6.

9.6 SPECTROSCOPIC FILTERS

We have compared spectroscopic instruments based on resolution, resolving power, spectral throughput, and spectral efficiency as a function of volume. We are naturally led to wonder whether the limits that we have derived thus far are close to fundamental physical limits. The answer to this question is "No."

The volume, etendue, resolving power, and SNR of spectral sensors are not linked by fundamental physical law. While readout poses obvious challenges, one can imagine sensors consisting of individual atoms tuned to absorb each spectral line. On the scale of the optical wavelength, such atomic absorbers may be arbitrarily small. For example, the quantum dot spectrometer [127] uses electronic resonators to create single-pixel devices with hundreds of spectral channels. Many other examples of the design of the spectral response of molecular, semiconductor, metal, and dielectric materials may be considered. These systems apply on the micrometer or nanometer scale the same tools in diffraction, interferometry, and resonance as the macroscopic spectrometers that we have thus far considered.

Given that the performance metrics of spectroscopic instruments are not limited by physical law, one may wonder why large and inefficient systems have not been completely displaced by integrated devices. The answer to this puzzle lies in the complexity of the design and fabrication of high-performance metamaterials and optical circuits. Over time, instruments will become increasingly small. For the present it is sufficient to explore the basic nature of structured devices.

Spectroscopic filters use microscopic structure to modulate the power spectral density. Filters are constructed based on the following effects:

- *Atomic and Molecular Resonance*. These filters use the intrinsic spectral sensitivity of quantum transitions. They may consist of semiconducting wafers, inorganic color centers doped in solids or organic dyes in a polymer matrix. Semiconductors yield long-pass filters; wavelengths above the band edge are not absorbed and wavelengths below are. Color center and dyes yield filters with modest spectral responsivity (10–100) due to the broad absorption bands necessary to achieve high quantum efficiency in solids. They are the basis of coarse spectroscopy, as indicated by their inclusion in the Bayer pattern of RGB imaging. As indicated by the quantum dot spectrometer, however, the responsivity of quantum systems can be dramatically increased using artificial nanostructures.

- *Plasmonic Resonance*. Plasmons are optical frequency electronic excitations in conductors or semiconductors. *Plasmons* are quantized quasiparticles and, in structure nanoparticles or nanowires, are limited to discrete states. The color associated with stained glass windows are due to plasmonic resonances in metal nanoparticles.
- *Optical Resonance*. Volume holograms and the Fabry–Perot resonator are just two examples of many devices that filter light on the basis of optical resonances. Filter design and manufacturing based on optical resonance is more advanced than atomic or plasmonic resonance because design and manufacturing tools for devices structured at optical wavelengths (e.g. μm) are much more advanced than tools at electronic wavelengths (e.g., nm). The most common narrowband optical filters are "thin film" filters consisting of layered dielectric and metal films.
- *Polarization Filtering*. Molecular and plasmonic structure modulates the polarization of the light field as well as the spectrum. Spectral filters may be created by exploiting dispersion in the polarization response. Liquid crystal devices enable electrically tunable modulators and filters based on polarization effects.

For brevity, we limit our attention in this section to filters based on optical resonance. We consider polarization filtering in Section 9.7. Filters considered in the present section rely on optical structures modulated along one dimesion consisting of gratings and layered materials. While thin-film and holographic filters have a long history, research in artificially structured materials for spectroscopic filters has accelerated dramatically since the mid-1990s with the development of photonic crystals, nanomaterials, and metamaterials. An understanding of 1D filters is essential to design, but one expects that 2D and 3D metamaterials will be needed to fundamentally advance spectrometer design. Multidimensional filters are briefly considered in Section 9.8.

9.6.1 Volume Holographic Filters

Volume holograms are the simplest optical microstructure-based filters. We have already implicitly assumed the use of holograms in our discussion of dispersive spectroscopy. Volume transmission holograms are attractive as devices for the diffraction gratings illustrated in Figs. 9.2 and 9.4 because they achieve high diffraction efficiencies over broad spatial and spectral bandwidths. Etched or ruled reflection gratings achieve similar efficiencies and spatial bandwidth, but are somewhat more challenging to fabricate. While we do not include a detailed comparison here, it is useful to note that in each case the spatial bandwidth and uniformity of the response are critical factors.

The spectral–spatial response of a hologram may be analyzed using coupled mode theory. For example, Fig. 9.15 shows the plane wave diffraction efficiency of a transmission hologram appropriate for a dispersive spectrometer as a function of angle of incidence and wavelength calculated using Eqns. (4.90) and (4.92). The wavelength dependence of the diffraction efficiency is weak in this geometry, although the center

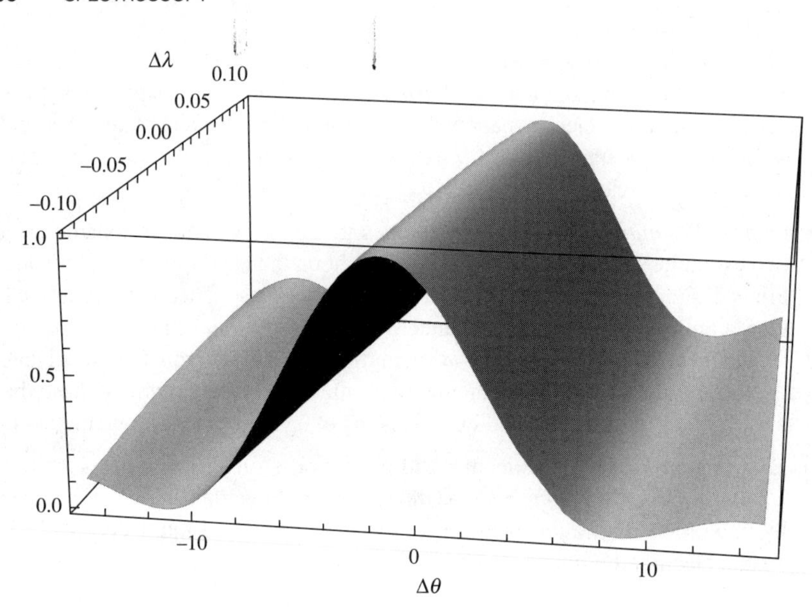

Figure 9.15 Plane wave diffraction efficiency of a transmission volume hologram as a function of angle of incidence and wavelength. The hologram is designed for a center wavelength diffraction efficiency of 100% from a wave incident at 15° below to a diffracted wave 15° above the surface normal, corresponding to $\Lambda = \lambda_0/2\sin(\pi/12)$. $\Delta\theta$ is the deviation from the design angle of incidence in degrees and $\delta\lambda = \lambda/\lambda_0 - 1$. We assume $\Delta\varepsilon/\varepsilon = 5 \times 10^{-2}$, which yields 100% diffraction efficiency for a hologram thickness of 9.6 λ_0.

angle of the hologram shifts with λ. The approximately 10° width of the central diffraction lobe is typical of volume phase gratings used in diffractive instruments. The limited spatial bandpass of the hologram may be a major contributor in determining a spectrograph's PSF. The tightest PSF for system based on a grating with angular bandpass $\Delta\theta$ has an approximate width of $w = \lambda/\Delta\theta$. Substituting in Eqn. (9.7), the grating limited resolution is

$$\delta\lambda = \frac{\lambda_0^2}{2\Delta\theta F \sin\theta} \tag{9.68}$$

where θ is the half-angle of the holographic deflection at the design wavelength and we note that $\Lambda = \lambda_0/2\sin\theta$. For the hologram of Fig. 9.15, this corresponds to a resolution of approximately $11\lambda_0^2/F$. The limited angular bandpass is a greater issue for coded aperture instruments, which can utilize aperture features at resolution approaching the optical limit, than for slit-based instruments, which typically use spatial features on scales much larger than the PSF.

The response of a hologram to wavelength and angular shifts varies as a function of the recording and reconstruction geometry. The holographic diffraction efficiency falls as the Bragg mismatch increases. We first encountered the Bragg mismatch in

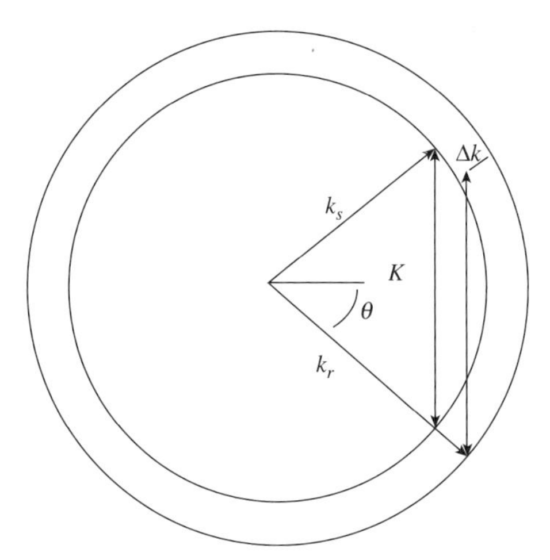

Figure 9.16 Bragg mismatch arising from a change in the reconstruction wavelength. The reconstruction wavevector is matched to the grating wavevector on the inner wave normal surface, but reducing the wavelength such that the reconstruction wavevector lies on the outer sphere produces the mismatch Δk.

Eqn. (4.90). Figure 4.22 illustrates the Bragg mismatch when the angle of incidence of the reconstruction beam differs from the recording angle. In filter applications we are particularly interested in the sensitivity of the hologram to spectral shifts, Fig. 9.16 illustrates the Bragg mismatch arising from a change in the reconstruction wavelength. The sensitivity of the hologram to changes in reconstruction angle and wavelength depend on the reconstruction geometry. Figure 9.16 illustrates Bragg matching when the reconstructing beam makes an angle $\theta_0 = \sin^{-1}(K/k_0)$ with respect to the optical axis, where K is the grating wavenumber and $k_r = 2\pi n/\lambda_0$ is the incident beam wavenumber. If the reconstruction beam is instead incident at an angle θ with wavelength λ, then the Bragg mismatch is

$$\Delta k = \frac{2\pi}{\lambda}\left|1 - \sqrt{\left(\sin\theta - \frac{2\lambda\sin\theta_0}{\lambda_0}\right)^2 + \cos^2\theta}\right| \tag{9.69}$$

A hologram is particularly insensitive to changes in incident angle or wavelength in geometries where the Bragg mismatch changes slowly with respect to such variations. The hologram is sensitive to wavelength if the mismatch changes rapidly with spectral variation. Figure 9.17 plots the gradient of the Bragg mismatch with respect to reconstruction angle and wavelength as a function of the Bragg-matched reconstruction angle θ_0. As illustrated in Fig. 9.17(a), the rate of change of the Bragg mismatch with respect to λ is maximal at $\theta_0 = \pi/2$, which corresponds to a *reflection hologram*. The reconstruction and diffracted beams are counterpropagating in this geometry. As illustrated in Fig. 9.17(b), the rate of change of the Bragg mismatch with respect to angle is maximal for $\theta_0 = \pi/4$, which corresponds to a 90° diffraction

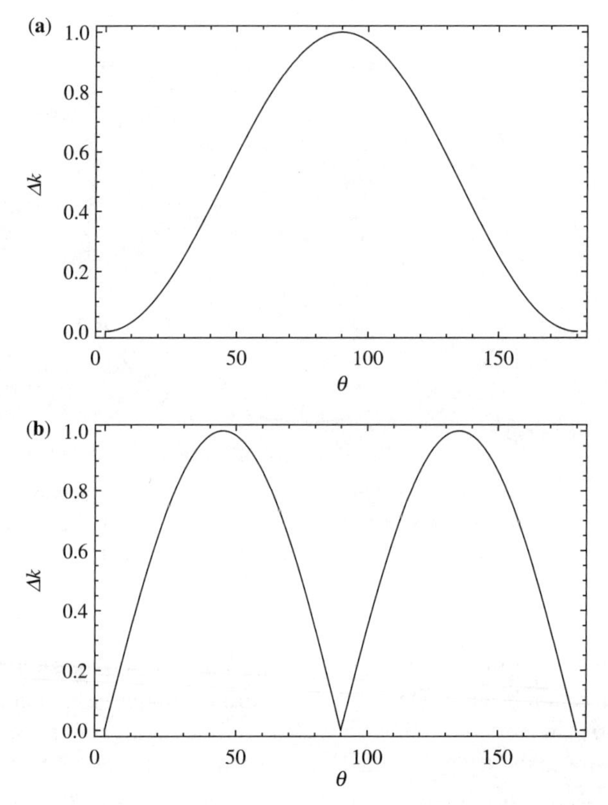

Figure 9.17 (a) Gradient of the Bragg mismatch with respect to wavelength $\partial\Delta k/\partial\lambda$ as a function of the Bragg matched grating half-angle θ_0. The gradient is evaluated at the Bragg-matched angle and wavelength; (b) gradient of the Bragg mismatch with respect to angle $\partial\Delta k/\partial\theta$ versus θ_0.

geometry. The maxima of the angular sensitivity in this geometry guides the design of angularly multiplexed holographic data storage [28]. The maxima of the wavelength sensitivity in the reflection geometry means that reflection holograms are preferred as spectral filters. It is also useful in filter applications that the reflection geometry is a minima of angular sensitivity.

The basic geometry of a reflection hologram is illustrated in Fig. 9.18. The wave equation for this system is

$$\nabla^2 U + \mu\omega^2[\varepsilon + \Delta\varepsilon\cos(Kz)] = 0 \tag{9.70}$$

One attempts a coupled wave solution to this equation under the assumption that $U = R(z)e^{i(k_x x + k_z z)} + S(z)e^{i(k_x x - k_z z)}$, which produces the coupled equations

$$ik_z\frac{dR}{dz} + \frac{k_z^2}{2}\frac{\Delta\varepsilon}{\varepsilon}Se^{i(K-2k_z)z} = 0 \tag{9.71}$$

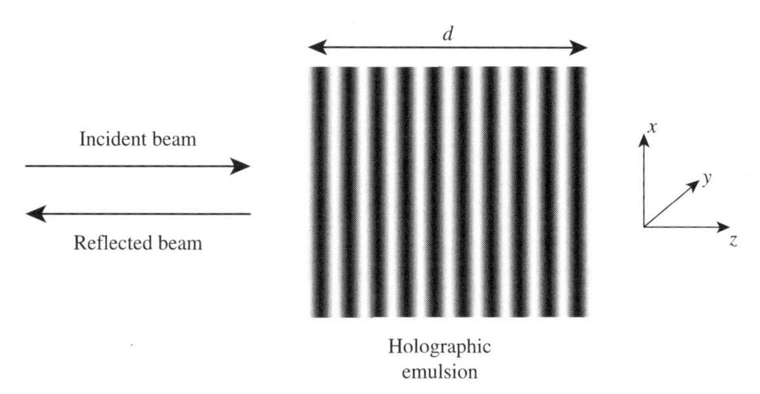

Figure 9.18 Reflection hologram geometry.

$$-ik_z \frac{dS}{dz} + \frac{k_z^2}{2} \frac{\Delta\varepsilon}{\varepsilon} R e^{i(2k_z - K)z} = 0 \tag{9.72}$$

Equations (9.71) and (9.72) are nearly identical to Eqns. (4.86) and (4.87) except for a change in sign due to the counterpropagating diffracted beam. The boundary conditions are also different; we still require that $R(z=0) = R_0$ but the second boundary boundary condition is now $S(z=d) = 0$. The solution to Eqns. (9.71) and (9.72) satisfying these conditions is

$$\begin{aligned} R(z) &= e^{i(\Delta kz/2)} R_0 \left[\frac{\gamma \cosh[\gamma(z-d)] - i(\Delta k/2)\sinh[\gamma(z-d)z]}{\gamma \cosh(\gamma d) + i\frac{\Delta k}{2}\sinh(\gamma d)} \right] \\ S(z) &= \frac{i}{2} \frac{k_z \Delta\varepsilon}{\varepsilon} e^{-i(\Delta kz/2)} R_0 \frac{\sinh[\gamma(z-d)]}{\gamma \cosh(\gamma d) + i(\Delta k/2)\sinh(\gamma d)} \end{aligned} \tag{9.73}$$

where $\Delta k = K - 2k_z$ and $\gamma = \frac{1}{2}\sqrt{k_z^2 \Delta\varepsilon^2/\varepsilon^2 - \Delta k^2}$.

Figure 9.19 plots the diffraction efficiency of a reflection hologram as a function of wavelength and angular detuning. Note that the hologram does not produce significant diffraction at wavelengths longer than the retroreflected wavelength. Shorter wavelengths are Bragg-matched for efficient reconstruction at detuned angles. The critical point at the retroreflection wavelength creates a relatively broad angular range with uniform spectral response. The spectral filtering properties of reflection holograms makes them ideal for display holography. Transmission holograms are difficult to view in white light because angular ambiguity creates a blur of overlapping diffracted orders. Reflection holograms, in contrast, reflect only frequencies above a critical value. Accounting for interface refractive effects and relatively narrow illumination angles, reflection holograms in display applications are sharply colored and easy to view under white light.

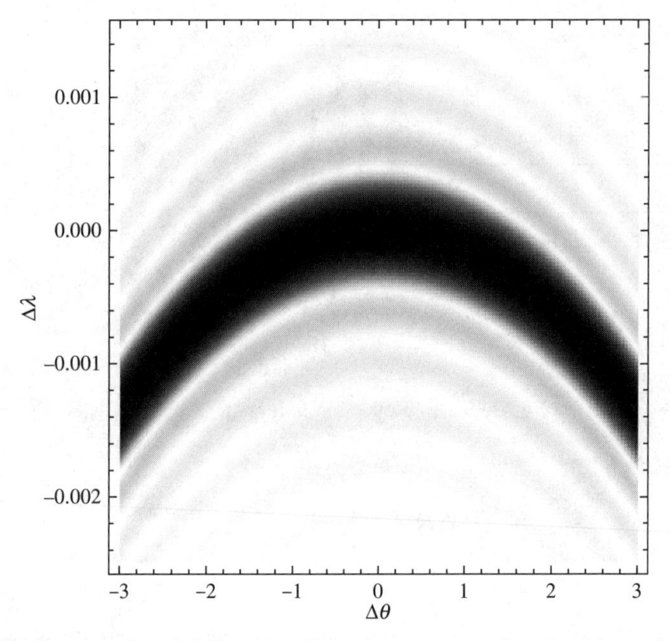

Figure 9.19 Density plot of diffraction efficiency versus wavelength and angular detuning for a reflection volume hologram. The holographic modulation is $\Delta\varepsilon/\varepsilon = 5 \times 10^{-4}$, and the thickness is $d = 4\varepsilon/k_0\Delta\varepsilon = 1273\lambda$. The maximum diffraction efficiency for this thickness is $\eta = 0.93$. ($\Delta\theta$ is in degrees.)

Estimating the angular and spectral resolving power of a reflection hologram based on the region over which γ is real yields

$$\delta\lambda \approx \lambda \frac{\Delta\varepsilon}{\varepsilon} \tag{9.74}$$

and

$$\Delta\theta \approx \sqrt{\frac{2\Delta\varepsilon}{\varepsilon}} \tag{9.75}$$

Given that $\Delta\varepsilon/\varepsilon$ may be 10^{-5} or less, the resolving power of a holographic filter may be extraordinary. The etendue is $\pi A^2 \Delta\theta^2/4 \approx \pi A^2 \Delta\varepsilon/\varepsilon$, where A is again the diameter of the entrance aperture. The R–L product is thus

$$E = \frac{\pi A^2}{2} = \frac{\pi V \Delta\varepsilon}{\lambda\varepsilon} \tag{9.76}$$

where we note that thickness required to achieve high efficiency is $d \approx 2\lambda\varepsilon/\pi\Delta\varepsilon$ and that $V = \pi A^2 d/4$. With E proportional to the aperture area, the spectral efficiency of a holographic filter is approximately equivalent to a slit spectrometer and less than a coded aperture or Fabry–Perot. As with the Fabry–Perot, one might increase the spectral efficiency by using a Fourier lens to disambiguate spectral channels over a wider angular range. It is also possible to increase the spectral efficiency of a holographic filter by creating holograms with focusing beams [182].

Holographic filters are most commonly used in applications that require one to isolate a single narrow spectral line, such as laser line stabilization and mode selection filters. The primary challenge in using holographic filters in other applications is the difficulty associated with creating a filter that responds to more than one wavelength. In principle, one can overcome this challenge by recording multiple-exposure holograms, with each exposure recording a grating for a target wavelength, but materials control issues associated with this approach favor filters fabricated by layered (nonoptical) methods [27].

9.6.2 Thin-Film Filters

As illustrated in Fig. 9.20, a *thin-film filter* consists layers of optical materials. While spectral filtering by a thin film is known to anyone that has observed soap bubbles, the technology of thin-film filter design and fabrication is extraordinarily sophisticated. As discussed by Macleod [162], the modern thin-film filter emerged with the development of advanced deposition technologies in the 1930s. Given that thin-film deposition technologies are central to modern microelectronics as well as optics, the current state of chemical and physical deposition systems is highly advanced for metals and dielectrics. Subnanometer layer thicknesses and smoothness are commonly available. In optical applications filters may consist of over 100 layers.

Thin-film filter analysis follows the same strategy that we applied to the Fabry–Perot etalon; we consider the field reflected and transmitted by the filter when the

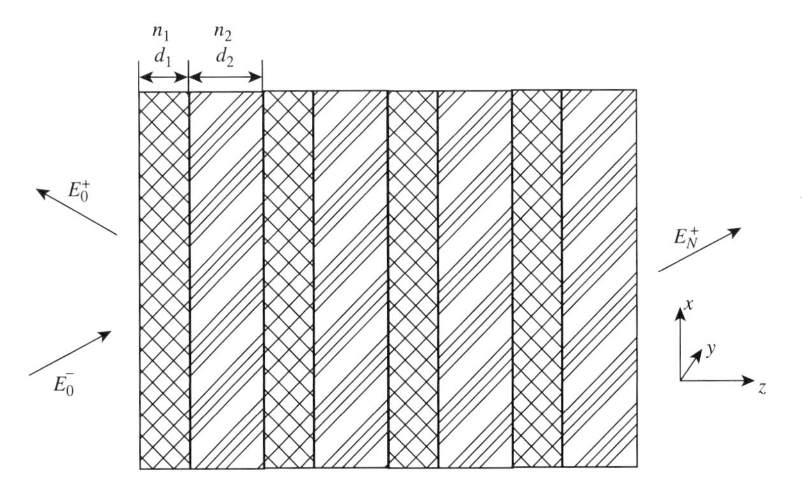

Figure 9.20 A thin-film filter is formed from layers of different optical materials. The (potentially complex) index of refraction of the nth layer is n_i and the thickness is d_i.

plane wave $E_0^+ \exp(ik_x x + ik_z z)$ is incident. The thickness and index of refraction of the lth layer are d_l and n_l. n_l may be complex to account for absorption. The incident field induces plane waves propagating in the positive and negative z directions in each layer. Let the amplitudes of these fields at the left edge of the layer be E_l^+ and E_l^-. The corresponding amplitudes at the right edges are $E_l^+ \exp(ik_{lz}d_l)$ and $E_l^- \exp(-ik_{lz}d_l)$, where $k_{lz} = \sqrt{4\pi^2 n_l^2/\lambda^2 - k_x^2}$.

The response of a thin-film filter depends on the polarization of the incident field. One accounts for this dependence by separately considering the transverse electric (TE) and transverse magnetic (TM) components. The TE (TM) wave is polarized such that **E** (**H**) lies along the y axis. In each case the boundary condition that the transverse components of **E** and **H** must be continuous across the interface is applied to relate E_l^+ and E_l^- and E_{l+1}^+ and E_{l+1}^-. In the TE case, we note from Eqn. (4.10) that $H_x = -i(\omega/\mu_0)\partial E_y/\partial z$. Continuity of E_y and H_x then yields

$$E_l^+ e^{ik_{lz}d_l} + E_l^- e^{-ik_{lz}d_l} = E_{l+1}^+ + E_{l+1}^- \tag{9.77}$$

$$k_{lz}\left(E_l^+ e^{ik_{lz}d_l} - E_l^- e^{-ik_{lz}d_l}\right) = k_{(l+1)z}\left(E_{l+1}^+ - E_{l+1}^-\right) \tag{9.78}$$

Equations (9.77) and (9.78) may be rearranged to form the difference equation [257]

$$\begin{pmatrix} E_{l+1}^+ \\ E_{l+1}^- \end{pmatrix} = \mathbf{M_l} \begin{pmatrix} E_l^+ \\ E_l^- \end{pmatrix} \tag{9.79}$$

where

$$\mathbf{M_l} = \frac{1}{2}\begin{bmatrix} \left(1 + \dfrac{k_{lz}}{k_{(l+1)z}}\right) e^{ik_{lz}d_l} & \left(1 - \dfrac{k_{lz}}{k_{(l+1)z}}\right) e^{-ik_{lz}d_l} \\ \left(1 - \dfrac{k_{lz}}{k_{(l+1)z}}\right) e^{ik_{lz}d_l} & \left(1 + \dfrac{k_{lz}}{k_{(l+1)z}}\right) e^{-ik_{lz}d_l} \end{bmatrix} \tag{9.80}$$

Equation (9.79) corresponds to Eqns. (9.71) and (9.72) and is solved under the same boundary conditions.

Consider, as an example, a periodically layered structure such that $d_{l+2} = d_l$ and $n_{l+2} = n_l$. Defining $\mathbf{M} = \mathbf{M_2 M_1}$, the difference equation in this case becomes

$$\begin{pmatrix} E_{l+2}^+ \\ E_{l+2}^- \end{pmatrix} = \mathbf{M} \begin{pmatrix} E_l^+ \\ E_l^- \end{pmatrix} \tag{9.81}$$

for odd l. **M** is the "characteristic matrix" of the thin-film structure. Among several interesting properties, one may show show that $|\mathbf{M}| = 1$ [23].

Attempting a solution to Eqn. (9.81) of the form

$$\begin{pmatrix} E_{l+2}^+ \\ E_{l+2}^- \end{pmatrix} = \gamma \begin{pmatrix} E_l^+ \\ E_l^- \end{pmatrix} \tag{9.82}$$

yields the characteristic equation $|\mathbf{M} - \gamma\mathbf{I}| = 0$. The characteristic equation is expressed in terms of the elements of $\mathbf{M}$ as

$$M_{11}M_{22} - \gamma(M_{11} + M_{22}) + \gamma^2 - M_{21}M_{12} = 0 \tag{9.83}$$

or, applying the fact that $\mathbf{M}$ is unimodular, as

$$\gamma^2 - \gamma\mathrm{Tr}(\mathbf{M}) + 1 = 0 \tag{9.84}$$

Multiplying $\mathbf{M_2M_1}$ one finds the trace of $\mathbf{M}$

$$\mathrm{Tr}(\mathbf{M}) = 2\cos k_{1z}d_2 \cos k_{2z}d_2 - \sin k_{1z}d_1 \sin k_{2z}d_2 \left(\frac{k_{1z}}{k_{2z}} + \frac{k_{1z}}{k_{2z}}\right) \tag{9.85}$$

If k_{1z} and k_{2z} are real, then $\mathrm{Tr}(\mathbf{M})$ is also real. The eigenvalue of $\mathbf{M}$ is

$$\gamma = \frac{\mathrm{Tr}(\mathbf{M})}{2} \pm \sqrt{\frac{\mathrm{Tr}(\mathbf{M})^2}{4} - 1} \tag{9.86}$$

Note that if $\mathrm{Tr}(\mathbf{M}) < 2$ then $|\gamma| = 1$. When $|\gamma| = 1$ the eigenvectors of $\mathbf{M}$ describe modes that propagate through the layered structure without attenuation.

Values of λ or θ such that $|\gamma| \neq 1$ lie in the stopband of the thin-film filter. The eigenmodes in this region decay exponentially on propagation. The boundaries of the stopband are described by the condition

$$\cos k_{1z}d_2 \cos k_{2z}d_2 - \frac{1}{2}\sin k_{1z}d_1 \sin k_{2z}d_2 \left(\frac{k_{1z}}{k_{2z}} + \frac{k_{1z}}{k_{2z}}\right) = \pm 1 \tag{9.87}$$

At normal incidence with $n_2d_2 = n_1d_1 = d$, for example, the edges of the stopband occur at λ such that

$$\cos^2\left(\frac{2\pi d}{\lambda}\right) - \frac{1}{2}\sin^2\left(\frac{2\pi d}{\lambda}\right)\left(\frac{n_1}{n_2} + \frac{n_2}{n_1}\right) = \pm 1 \tag{9.88}$$

Solutions to Eqn. (9.88) corresponding to a stopband take the form

$$\sin^2\left(\frac{2\pi d}{\lambda}\right)\left[1 + \frac{1}{2}\left(\frac{n_1}{n_2} + \frac{n_2}{n_1}\right)\right] = 2 \tag{9.89}$$

Equation (9.89) may be simplified in the case of weak modulation by assuming that $n_1/n_2 = 1 - \Delta n/n + \Delta n^2/n^2$ and $n_2/n_1 = 1 + \Delta n$, which yields a band edge at

$$\sin^2\left(\frac{2\pi d}{\lambda}\right) \approx 1 - \frac{\Delta n^2}{2n^2} \tag{9.90}$$

Stopbands are centered on the wavelengths

$$\lambda = \frac{4d}{2q+1} \tag{9.91}$$

where q is an integer.

A thin-film filter consisting of alternating layers of thickness $\lambda/4n_i$ is called a *quarter-wave stack*, which is the discrete analog of the reflection volume hologram; both are periodic with period $\lambda/2$. Such a filter reflects wavelengths within the stop-band and passes wavelengths outside the stopband. The approximate width of the stopband is

$$\delta\lambda = \lambda\frac{2\Delta n}{\pi n} \tag{9.92}$$

which is roughly equivalent to the stopband width observed for a volume hologram in Eqn. (9.74). Of course, a quarter-wave stack may achieve much greater index contrast than a volume hologram. For example, Fig. 9.21 plots the stopband as a function of the index ratio n_2/n_1 for normal incidence on a quarter-wave stack.

The angular response of a quarter-wave thin-film reflection filter is also similar to volume reflection filter. As with the holographic filter, the angular range is proportio-nal to the square root of the index contrast. As illustrated in Fig. 9.22, however, the higher-index contrast available to thin-film devices produces a wider angular range (and decreased spectral resolution).

The eigenvalues $\gamma\pm$ corresponding to the positive and negative choices in Eqn. (9.86) correspond to eigenvectors

$$\mathbf{E}_\pm = \begin{pmatrix} M_{12} \\ \gamma_\pm - M_{11} \end{pmatrix} \tag{9.93}$$

The solution to Eqn. (9.81) is

$$\begin{pmatrix} E^+_{2l+1} \\ E^-_{2l+1} \end{pmatrix} = \alpha\gamma^l_+\mathbf{E}_+ + \beta\gamma^l_-\mathbf{E}_- \tag{9.94}$$

where α and β are constants determined from boundary conditions. Considering, for example, a filter with L periods embedded in a substrate of index n_1, the boundary

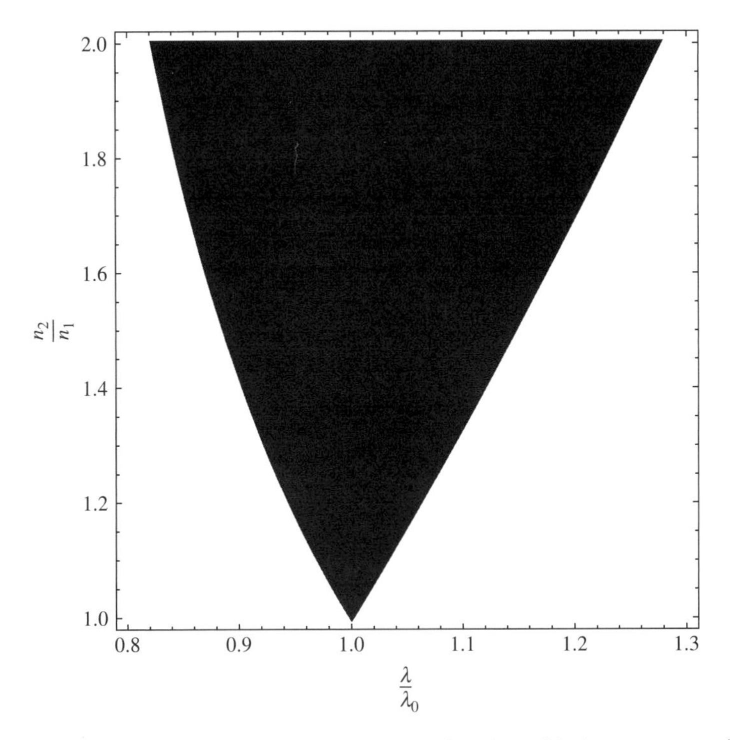

Figure 9.21 Stopband of a quarter-wave stack as a function of index contrast n_2/n_1. The wavelength axis is plotted in units of the quarter-wave resonance wavelength. The dark region corresponds to the stopband.

conditions take the form

$$\alpha\gamma_+^L(\gamma_+ - M_{11}) = -\beta\gamma_-^L(\gamma_- - M_{11}) \tag{9.95}$$

and

$$\begin{pmatrix} E_0^+ \\ E_0^- \end{pmatrix} = \alpha\mathbf{E}_+ + \beta\mathbf{E}_- \tag{9.96}$$

Solving for α and β in terms of E_0^+, one finds that the reflectance and transmittance of the quarter-wave stack are

$$\begin{aligned} r &= \frac{\gamma_- - M_{11}}{M_{12}} \frac{1 - (\gamma_-/\gamma_+)^L}{1 - (\gamma_-/\gamma_+)^L[(\gamma_- - M_{11})/(\gamma_+ - M_{11})]} \\ \tau &= \gamma_-^L \frac{\gamma_+ - \gamma_-}{\gamma_+ - M_{11} - (\gamma_-/\gamma_+)^L(\gamma_- - M_{11})} \end{aligned} \tag{9.97}$$

In practice, of course, the filter is likely to be embedded in air rather than a dielectric of index n_1. This discrepancy is typically resolved by adding antireflection filters at

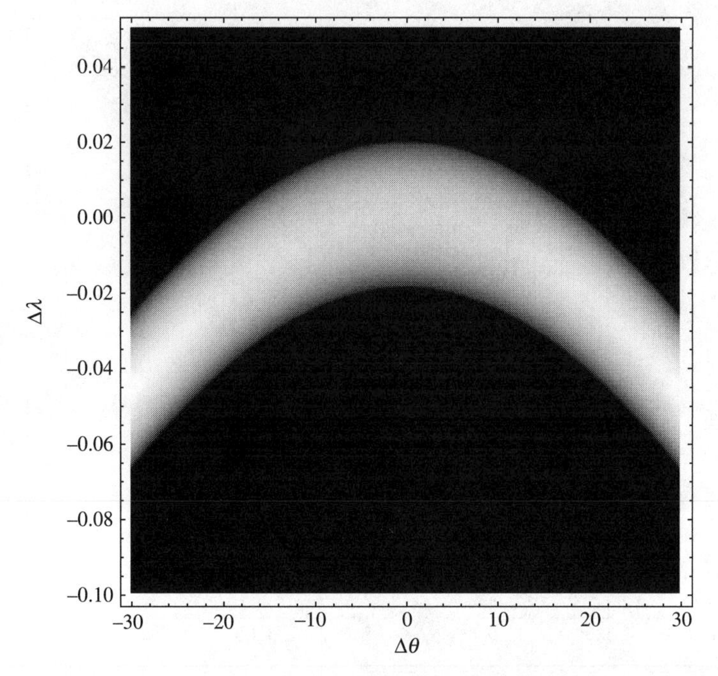

Figure 9.22 $|\gamma|$ as a function of $\delta\lambda = \lambda/\lambda_0$ and $\Delta\Theta$ in degrees from normal for a quarter-wave stack with $n_1 = 1.7$ and $n_2 = 1.65$.

the ends of the periodic layers. The antireflection filters consist of one or more additional layers designed to match the impedance of the filter to the surrounding air.

Within the stopband, one can show that

$$\left|\frac{\gamma_- - M_{11}}{M_{12}}\right| = 1 \tag{9.98}$$

and the ratio $(\gamma_-/\gamma_+)^L$ goes to zero as $L \to \infty$. Finite L produces a finite reflectance, as illustrated in Fig. 9.23 for a 30-period quarter-wave stack. The angular sensitivity of the finite structure is well described by Fig. 9.22. Thin-film filters are generally designed to operate in transmission rather than reflection. The filter of Fig. 9.23 is a band rejection filter, blocking a reasonably broad range of wavelengths centered on λ_0. Much higher index contrast is readily available to create broader band rejection filters.

Narrow-bandpass thin-film filters are created by putting layer dislocations in otherwise periodic structures. For example, Fig. 9.24 illustrates a dislocation consisting of a single $\lambda/2$ layer in a quarter-wave stack. The dislocation creates a localized mode (a "bound state") within the stopband. Tunneling through the bound state creates a sharp spectral feature in the transmission of the filter.

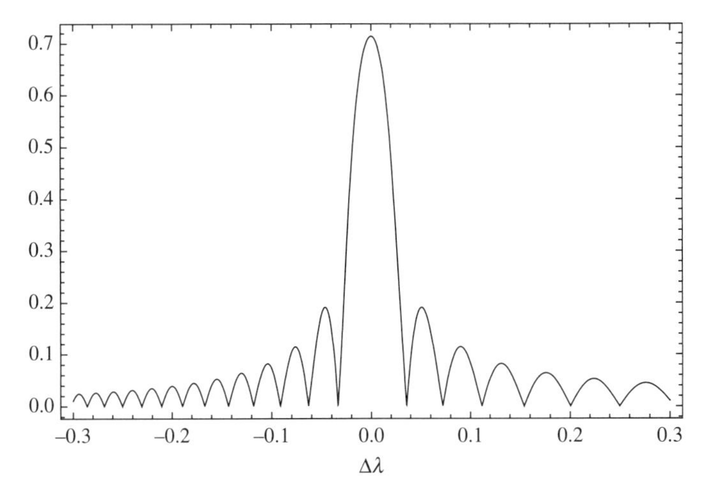

Figure 9.23 Reflectance as a function of wavelength of a 30-period quarter-wave stack with $n_1 = 1.7$ and $n_2 = 1.65$. As in Fig. 9.22, $\delta\lambda$ is in units of the quarter-wave design wavelength.

The Fabry–Perot analysis of Section 9.5 can be applied to obtain the transmission characteristics of a periodic filter with a dislocation. Treating the periodic structure on either side of the dislocation as a cavity mirror with reflectance and transmission described by Eqn. (9.97), we can describe the transmittance of the overall filter by Eqn. (9.47). As an example, Fig. 9.25 plots the transmittance of a resonator formed from two quarter-wave stack dielectric mirrors. The curious aspect of this

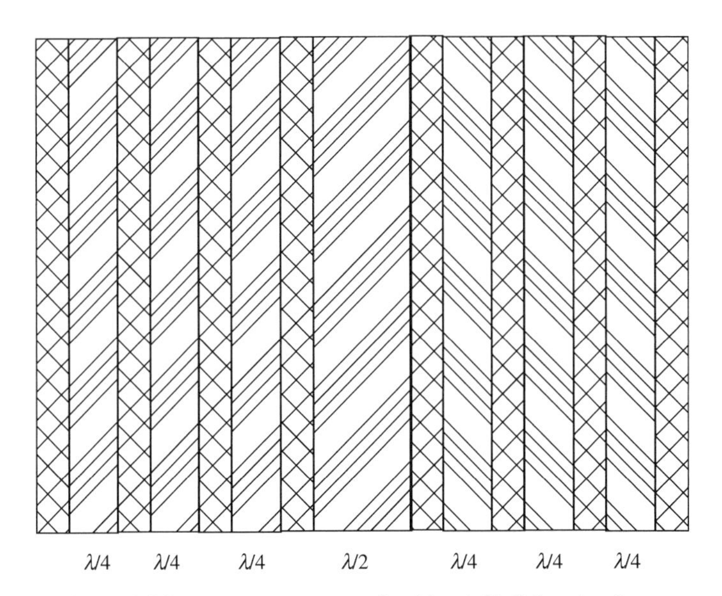

Figure 9.24 A quarter-wave stack with a $\lambda/2$ dislocation layer.

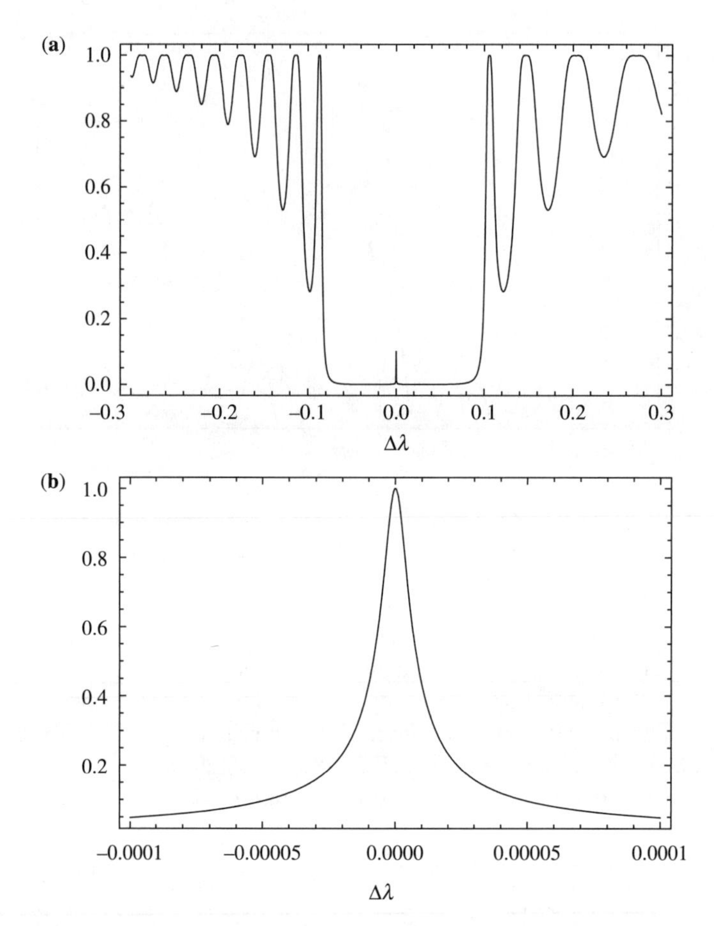

Figure 9.25 (a) Absolute value of the transmittance of a quarter-wave stack resonator with $n_1 = 1.7$ and $n_2 = 2.2$ as a function of wavelength detuning at normal incidence. (b) Magnified plot of (a) focusing on the tunneling resonance. The resonator consists of two 20-period dielectric mirrors sandwiched back to back such that the center layer is of thickness $\lambda/2$ with index n_1.

resonator with respect to our earlier discussion of the Fabry–Perot is that we even though we include no resonant cavity layer between the mirrors, we observe a sharp resonant transmission peak. The peak appears as a blip in full stopband plotted in Fig. 9.25(a), but when we zoom in we see that the transmittance reaches 1 over a narrow spectral range in the center of the stopband.

The resonance occurs because the phase of the reflectance of a dielectric mirror varies as a function of wavelength and angle of incidence. For example, Fig. 9.26 shows the phase of the reflectance as a function of wavelength detuning for the dielectric mirrros used in Fig. 9.25. One finds in general that the phase varies approximately

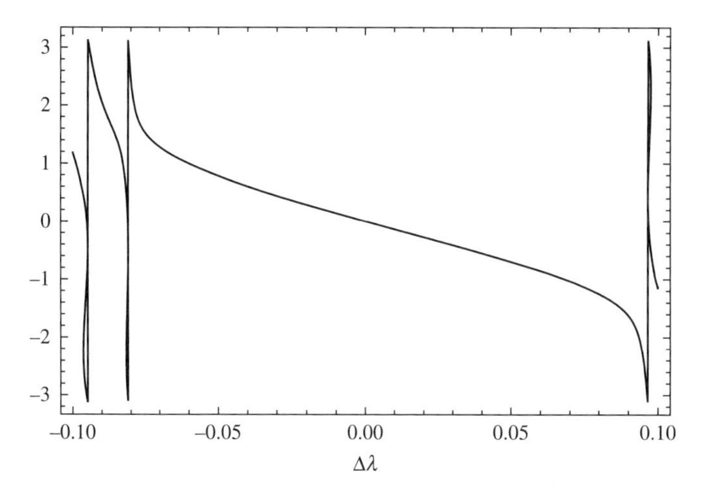

Figure 9.26 Phase of the reflectance at normal incidence for the dielectric mirrors of Fig. 9.25.

linearly between $-\pi/2$ and $\pi/2$ across the stopband. Replacing $1/\nu_r$ with the rate of phase variation $\pi/2\Delta\nu_{sb}$, one substitutes in Eqn. (9.55) to find that the resolving power of the bound-state resonator is

$$R = \frac{\lambda}{\delta\lambda} = \frac{2\mathcal{F}}{\pi}\frac{\lambda}{\delta\lambda_{sb}} \tag{9.99}$$

where $\Delta\nu_{sb}$ and $\delta\lambda_{sb}$ are the widths of the stopband in ν and λ and $\mathcal{F}$ is the finesse. We find, in short, that the resolving power associated with tunneling through the bound state is $2\mathcal{F}/\pi$ times greater than the resolving power of the quarter-wave stack. The finesse for the mirrors of Fig. 9.25 is 2.3×10^4, which corresponds to the approximately 10^4 ratio between the widths of the stopband and the tunneling resonance observed in Fig. 9.25. The tunneling resonance also increases the angular resolving power of the filter, as illustrated in Fig. 9.27.

In comparison with a Fabry–Perot resonator, the tunneling resonance enables thin-film filters to achieve comparable resolving power in a smaller volume and without free spectral range ambiguity. The spectral efficiency limits for thin-film filters are similar to those for a Fabry–Perot. The state of the art of thin film filter design and manufacturing is extremely sophisticated. Filters with multiple complex reflectance or transmittance resonances and wide angular performance are routinely available. Even simple variations, like using asymmetric layers in a periodic stack, introduce rich transmission and reflectance features. As thin-film devices continue to improve, one expects their application in spectroscopy will greatly expand. In particular, two- and three-dimensional resonant filters, as discussed in Section 9.8, greatly increase the power of this technology.

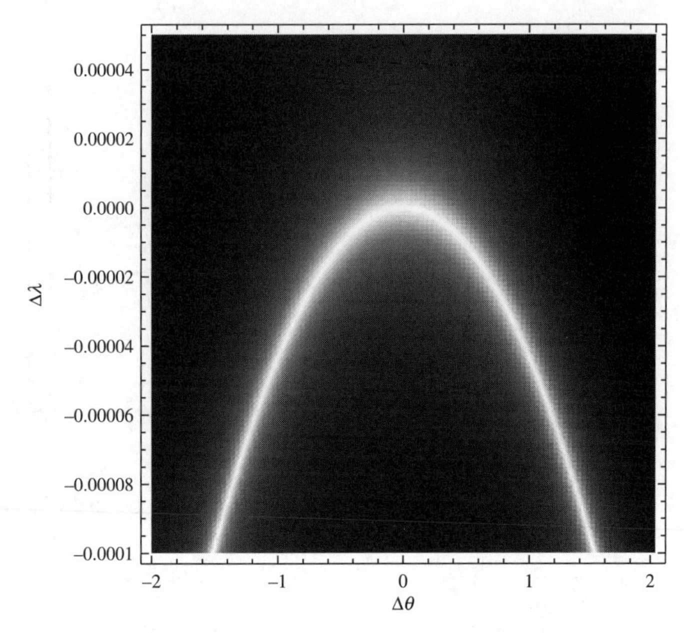

Figure 9.27 Transmittance as a function of angular and wavelength detuning for the thin-film resonator of Fig. 9.25.

9.7 TUNABLE FILTERS

Returning once again to the philosophy with which Girard opened this chapter, a spectrometer is formed by measuring data dispersed over space or time. We have briefly encountered each of these strategies for spectral filters in the form of temporally varying Michelson and Fabry–Perot interferometers and in the form of spatial modulation in the FP ring pattern. The present section and Section 9.8 discuss strategies for temporal and spatial modulation of spectral filters in more detail.

We focus in this section on strategies for creating narrow-bandwidth single-channel tunable filters. Among the systems we have encountered thus far, the Fabry–Perot etalon and the dispersive spectrometer offer the best hope for creating such a device. As we have seen, however, the etalon is effective over only a single free spectral range. The dispersive spectrograph working as a "monochromator" is commonly used as a single-channel spectral filter, but it is an extremely bulky and inelegant solution to this problem.

Polarization-based filters utilizing liquid crystal and acoustooptic devices have emerged as compact and effective tunable filters since the mid-1980s. The *liquid crystal tunable filter* utilizes a stack of polarization analyzers, birefringent crystals, and liquid crystal layers to isolate spectral channels using wavelength-dependent polarization rotation. The *acoustooptic tunable filter* uses polarization-dependent Bragg scattering from acoustic waves. The acoustic grating wavevector selects the scattered wavelength. This section reviews the basic design of these devices and describes their resolving power, etendue, spectral throughput, and spectral efficiency.

9.7.1 Liquid Crystal Tunable Filters

A *liquid crystal* is a homogeneous material formed of asymmetric molecules. The molecular state is defined by orientational and positional order parameters. In a normal liquid the relative orientation and position of molecules decorrelates in just a few molecular spacings. In a liquid crystal, however, the relative order spans macroscopic domains. Figure 9.28 shows a typical visual model for a *nematic* liquid crystal. The liquid crystal is a state of matter, similar to a liquid, solid, and gas, with phase transitions between states as at critical temperatures and pressures. In the nematic phase the long-range orientational order parameter is nonzero. Other phases include the *smectic* phase, in which orientation and 1D translation are ordered, and the *cholesteric* phase, in which orientation and 2D translation and rotation are ordered. The smectic phase consists of layers of ordered molecules. In the cholesteric phase the orientation of successive layers tends to rotate. The direction of the orientational order parameter is often determined by boundary conditions or by an electric field.

The most significant optical implication of the order parameter is that liquid crystals are birefringent. For nematic liquid crystals, the transmittance of a thin liquid crystal layer for light polarized parallel to the order parameter differs from the transmittance for light polarized orthogonally to the order parameter. The effect of a liquid

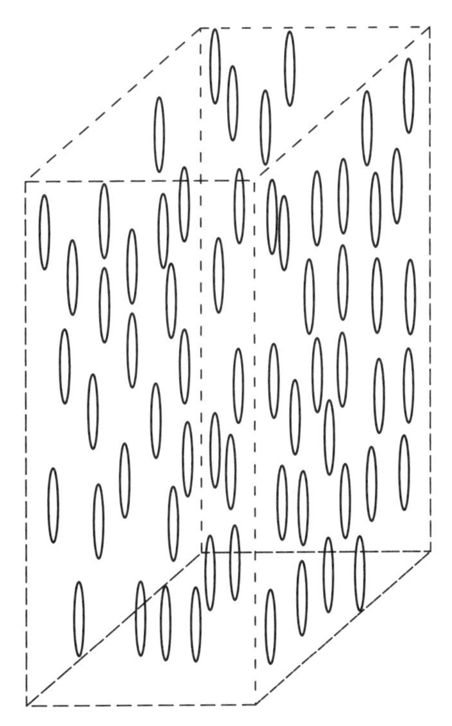

Figure 9.28 In an nematic liquid crystal the long-range order parameter for molecular alignment is nonzero. The molecules themselves must be asymmetric, as with the rod-like molecules illustrated here. Alignment in real materials is not nearly as uniform as illustrated here; one needs only a nonzero statistical correlation in the alignment to obtain long-range order.

crystal layer on the optical polarization is more complex in cholesteric or *twisted nematic* materials. The birefringent axis rotates in such materials and can rotate the optical polarization with it. Twisted nematics and *ferroelectric* liquid crystals are used to make displays, where one typically seeks simply to switch between a transmissive state and a nontransmissive state. Simple birefringent devices suffice for liquid crystal filters, however.

The building blocks of a liquid crystal tunable filter, consisting of linear polarizers and liquid crystal layers, are illustrated in Fig. 9.29. Devices may also include static birefringent crystals. The order parameter of the nematic layer in the liquid crystal cell is described by a 3D vector, the component of this vector in the boundary plane of the liquid crystal layer defines the extraordinary axis. The transmittance of the liquid crystal layer for a plane wave polarized along the extraordinary axis is $\exp\left(2\pi n_e d\sqrt{1/\lambda^2 - u^2 - v^2}\right)$, where d is the layer thickness and n_e is the extraordinary index of refraction. The transmittance for a wave polarized along the ordinary axis is similar with index n_o. In view of the polarization dependence of the device, the transfer function for the liquid crystal layer generalizes from the scalar form of Eqn. (9.47) to the matrix form

$$\hat{\mathbf{H}}(u, v) = \begin{pmatrix} e^{2\pi i n_o d\sqrt{(1/\lambda^2)-u^2-v^2}} & 0 \\ 0 & e^{2\pi i n_e d\sqrt{(1/\lambda^2)-u^2-v^2}} \end{pmatrix} \tag{9.100}$$

where $\hat{\mathbf{H}}$ is the *Jones matrix* for the liquid crystal layer.

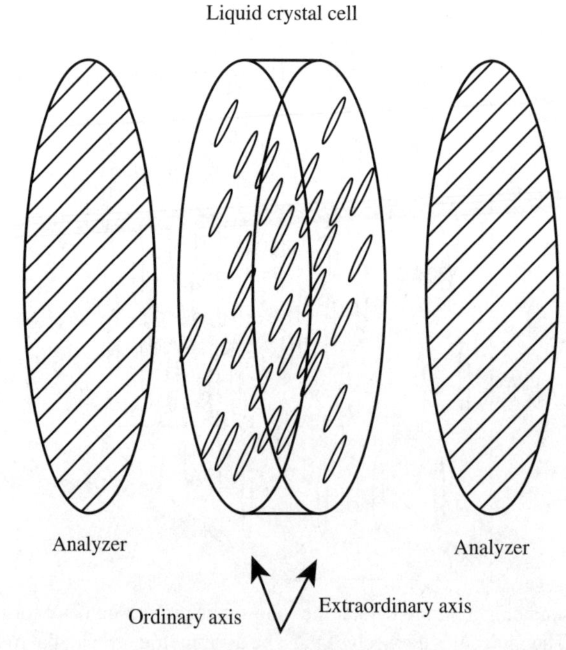

Figure 9.29 A liquid crystal tunable filter consists of electrically modulated birefringent liquid crystal layers sandwiched between polarization analyzers and birefrigent crystals.

Jones matrix calculus is commonly applied to describe the operation of polarization devices. The Jones matrix transforms the incident field **E** into the output field according to $\mathbf{E_o} = \mathbf{HE_i}$. The two components of **E** correspond to the electric fields polarized along the x and y axes. In writing a diagonal Jones matrix for the liquid crystal layer in Eqn. (9.100), we assume that the liquid crystal axes are aligned along x and y.

The Jones matrix for a linear polarizer is

$$\begin{pmatrix} 1 & 0 \\ 0 & 0 \end{pmatrix} \tag{9.101}$$

Propagation through this device blocks the E_y field component. The operation of a Jones matrix depends on the orientation of a device relative to the polarization axes of an incident beam. Rotating a device relative to fixed polarization axes results in a transformed Jones matrix $\mathbf{RHR}^{-1}$. For example, rotation of the linear polarizer by $\pi/2$ produces the Jones matrix

$$\begin{pmatrix} 0 & 0 \\ 0 & 1 \end{pmatrix} \tag{9.102}$$

A liquid crystal tunable filter is designed by judicious application of linear polarization analysis, birefringent filtering, and coordinate rotation. For example, the transmittance for a plane wave normally incident on the simple cell illustrated in Fig. 9.29 with both analyzers oriented at $\pi/4$ with respect to the ordinary axis is

$$\begin{aligned} \hat{\mathbf{H}}(0, 0) &= \frac{1}{4}\begin{pmatrix} 1 & 1 \\ 1 & 1 \end{pmatrix}\begin{pmatrix} e^{2\pi i n_o (d/\lambda)} & 0 \\ 0 & e^{2\pi i n_e (d/\lambda)} \end{pmatrix}\begin{pmatrix} 1 & 1 \\ 1 & 1 \end{pmatrix} \\ &= \frac{1}{2} e^{\pi i (n_o + n_e) d/\lambda} \cos\left(\pi (n_o - n_e)\frac{d}{\lambda}\right)\begin{pmatrix} 1 & 1 \\ 1 & 1 \end{pmatrix} \end{aligned} \tag{9.103}$$

Assuming that the incident light is unpolarized, the fraction of the normally incident irradiance transmitted by this system is $\cos^2(\pi \Delta n d/\lambda)/2$, where $\Delta n = n_o - n_e$. The transmittance as a function of λ for this cell is shown in one of the two curves in the lower half of Fig. 9.30(a). The birefringence available in liquid crystals greatly exceeds that of typical crystals; values of Δn may reach 0.5. A value of 0.2 is selected in Fig. 9.30.

A *Lyot filter* is a cascade of alternating birefringent and analyzer stages [66,161]. One analyzes such a device by multiplying successive matrices as in Eqn. (9.103); the resulting irradiance transmittance is

$$\frac{\prod_i \cos^2[\pi \Delta n (d_i/\lambda)]}{2} \tag{9.104}$$

where d_i is the liquid crystal thickness in the ith stage. The simplest Lyot filter uses layer thickness increasing in powers of 2 (e.g., $d_i = 2^{i-1} d_1$). The thickness of the first

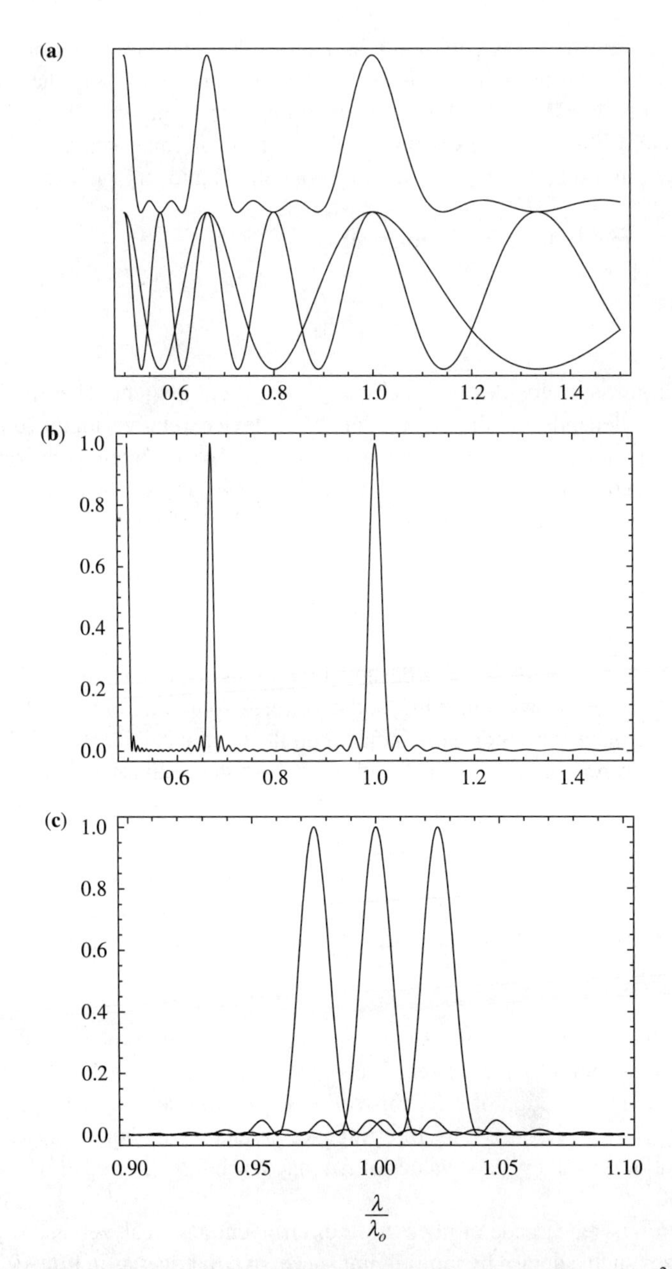

Figure 9.30 Transmittance as a function of wavelength for a Lyot filter: (a) $\cos^2[\pi\Delta n(d/\lambda)]$ and $\cos^2[2\pi\Delta n(d/\lambda)]$ in lower curves and product $\cos^2[\pi\Delta n(d/\lambda)]\cos^2[2\pi\Delta n(d/\lambda)]$ in upper curve, with birefringence $\Delta n = 0.2$ and $d = 10n_o\lambda$; (b) composite transmittance for a stack of d, $2d$, $4d$, $8d$, and $16d$ filters; (c) a detail of the transmittance of (b) with tuning of $\Delta n =$ 0.195, 0.2, 0.205.

stage determines the free spectral range $\Delta\nu_{FSR} = c/\Delta n d_1$, and the thickness of the final state determines the spectral resolving power

$$R = \frac{\lambda}{\delta\lambda} \approx \frac{\Delta n 2^{N-1} d_1}{\lambda} \tag{9.105}$$

Figure 9.30(b) shows the irradiance transmittance of a five-stage Lyot filter. The filter is tuned by changing the birefringence Δn. An applied field moves the molecular order parameter closer to the optical axis, thereby reducing the birefringence observed by a wave propagating along the axis. Assuming that the birefringence of all layers is simultaneously modulated the change required to move the resonance by one resolution element is

$$\delta n = \frac{\Delta n}{R} = \frac{\lambda}{2^{N-1} d_1} \tag{9.106}$$

The number of spectral channels one can resolve is the total range over which one can vary the birefringence divided by δn. For the five-stage system of Fig. 9.30, $R \approx 32$ and the step change in Δn is about 3% of the baseline. A 50% range in birefringence would enable approximately 15 distinct spectral channels.

As illustrated in Fig. 9.31, the angular sensitivity of a Lyot filter is essentially similar to the response of thin-film, holographic, and Fabry–Perot filters and the

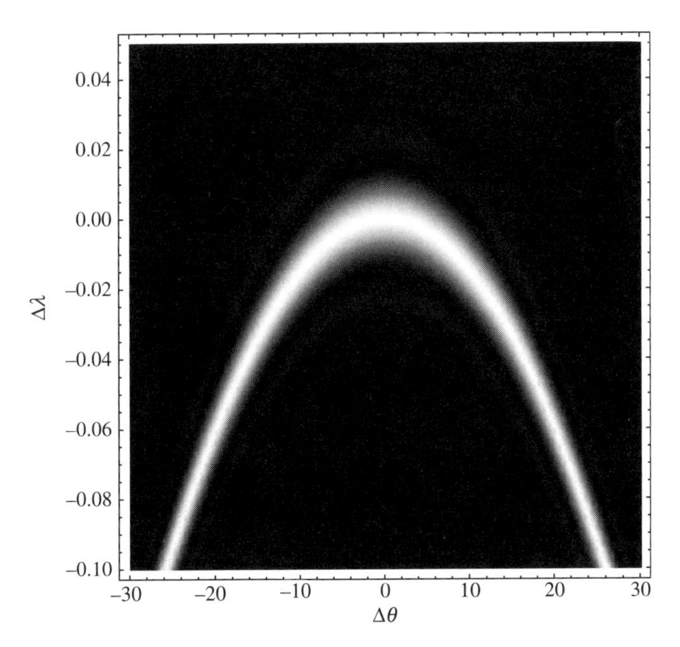

Figure 9.31 Angular sensitivity of a Lyot filter. Compare with Fig. 9.22 for a thin-film filter; note, however, that the present figure plots a transmission resonance while Fig. 9.22 plots a stopband (reflection resonance).

Michelson interferometer. In each case, the physical effect filters on the effective longitudinal wavelength, which is proportional to the cosine of the angle of incidence. The spectral width of the filter $\delta\lambda$ is related to the change in the effective wavelength with angle as

$$\delta\lambda = \lambda(1 - \cos\Delta\theta) \approx \lambda\frac{\Delta\theta^2}{2} \tag{9.107}$$

meaning that for longitudinal filters and interferometers

$$\Delta\theta^2 \approx \frac{2}{R} \tag{9.108}$$

This approximation appears previously in Eqn. (9.75) and Eqn. (9.41). Considerable effort in "wide field" thin-film and birefringent filter design focuses on pushing the limits of this approximation. Indeed, since we are making a Taylor series expansion on small angles, the approximation does not exactly hold, and design does help in pushing the relationship between resolving power and acceptance angle. Such efforts push a constant of proportionality; however, the basic link between R and $\Delta\theta$ is intrinsic to longitudinal devices. We avoided this limit in our discussion of Fabry–Perot spectroscopy by assuming that we could decode the ring pattern over a wide angle rather than just using longitudinal filtering. We return to multidimensional interferometry to avoid this limit in Section 9.8.

The etendue of a liquid crystal tunable filter is $L = \pi A^2/4\Delta\theta^2$, and the spectral efficiency is $E = \pi A^2/2$, the same as for a holographic or thin-film filter. For all of the approaches considered thus far, except for the slit spectrometer, etendue and efficiency increase in proportion to the aperture area rather than the aperture diameter. The spectral efficiency is $\delta\lambda/\Delta\lambda = \lambda/R\Delta\lambda$. Spectral throughput inversely proportional to resolving power is characteristic of narrow-pass spectrographs. For broad-spectral-range systems, this leads to an increase in the signal acquisition time necessary to achieve SNR targets.

The thickness required to obtain a given resolving power using a Lyot filter is approximately $1/\Delta n$ times greater than the thickness required to reach the same resolving power using a hologram or thin-film filter. One ought not discount, however, the incredible utility of real-time tunablity. Liquid crystal tunable filters are wonderful devices for fairly course but rapid single-color tuning. Current systems involve optimized stage design, often using the Solc approach [227] of using only two polarization analyzers in a chain of birefringent elements. System design may include static birefringent plates and thin-film elements in addition linear polarization filtering, liquid crystal layer thickness, and axis rotation. One may also imagine multiplex filter designs using liquid crystal devices.

9.7.2 Acoustooptic Tunable Filters

Acoustooptic tunable filters (AOTFs) also use polarization filtering, although the motivation for polarization effects is much different. Acoustooptic devices are

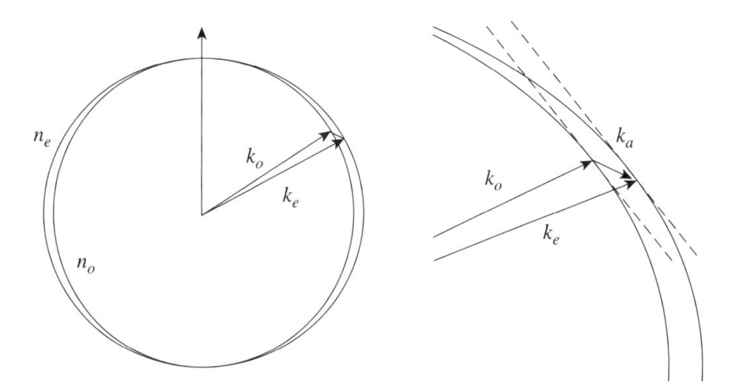

Figure 9.32 Wave normal surfaces for TeO_2. The ordinary index is 2.26; the extraordinary is 2.41. The optical axis is vertical in the figure at right. An acoustooptic tunable filter uses Bragg scattering to diffract light from one surface to the other. Maximal angular bandwidth for the scattering is obtained if the tangents to the ordinary wave sphere and the extraordinary wave ellipsoid are parallel.

based on the scattering of optical fields from permitivitty gratings induced by acoustic waves. Even though acoustic frequencies are much less than optical frequencies, acoustic wavelengths may be small because the acoustic velocity in crystals is about five orders of magnitude less than the speed of light. Unfortunately, maximum acoustic frequencies range from 100 MHz to 1 GHz, about six orders of magnitude less than optical frequencies. This means that the period of an acoustic grating is at least an order of magnitude greater than the optical wavelength. With reference to Eqn. (9.7) or to our plot of spectral selectivity versus Bragg angle in Fig. 9.17(a), we see that long-period acoustic gratings are unlikely to be effective spectral filters.

The solution to this problem is illustrated in Fig. 9.32, which is a wave matching diagram for scattering from the ordinary polarization to the extraordinary polarization in a birefringent crystal. The wave normal surfaces are scaled to the indices of tellurium dioxide (TeO_2), which is commonly used for visible and near IR filters. An acoustic wave scattering light from one polarization to the other must satisfy $\mathbf{k_a} = \mathbf{k_e} - \mathbf{k_o}$, where $\mathbf{k_a}$ is the acoustic wavevector and $\mathbf{k_e}$ and $\mathbf{k_o}$ are the wavevectors of the extraordinary and ordinary optical waves. Assuming collinear beams, the acoustic frequency is

$$f = \frac{(n_e - n_o)v}{\lambda} \tag{9.109}$$

At $\lambda = 633$ nm in TeO_2, $n_e = 2.41$ and $n_o = 2.26$. The acoustic velocity is $v = 620$ m/s [103]. The acoustic frequency for colinear coupling is thus $f = 147$ MHz, which is well within practical limits.

The basic design of an AOTF is illustrated in Fig. 9.33. Incident signals are polarized along the ordinary or extraordinary axis by an input polarizer. A radio frequency

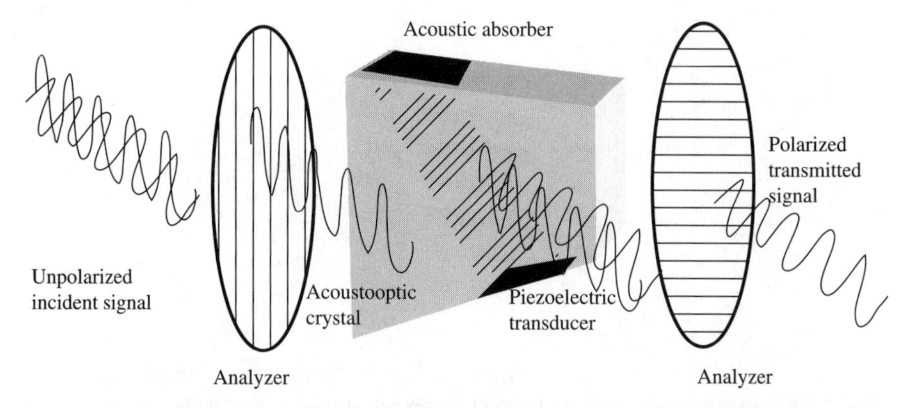

Figure 9.33 Optical system layout for an acoustooptic tunable filter.

acoustic wave launched by a piezoelectric transducer creates a diffraction grating in birefringent crystal. This grating scatters the optical wave under Bragg selective conditions from one polarization wave normal surface to the other. Undiffracted input light is blocked by an output polarizer, which passes the Bragg-filtered diffracted beam. In a noncollinear design, the angle of propagation of the diffracted beam is shifted from the angle of incidence.

The acoustic wave changes the indices of refraction of a crystal via the *photoelastic* effect, under which mechanical stress modulates the permittivity. The change in refractive index is described by a photoelastic tensor. The utility of a material for acoustooptic filtering depends on the structure and amplitude of coefficients in this tensor. TeO_2, for example, has strong photoelastic properties but lacks photoelastic tensor coefficients to mix the ordinary and extraordinary waves copropagating in the plane orthogonal to the optical axis. Off-axis designs would normally have limited field of view, but if the propagation direction of the ordinary and extraordinary beams differ such that the tangent planes to the wave normal surface are parallel, as illustrated in Fig. 9.32, then the coupling will be relatively insensitive to angular Bragg mismatch. For this reason, noncollinear geometries are preferred in AOTF design. A designer generally must use computational methods to combine frequency response data, photoelastic coefficients, and refractive geometries to optimize system performance metrics such as spectral resolving power, spectral range, and field of view.

Equations (4.89) are easily modified to account for collinear polarization switching, for example

$$ik_e \frac{dR}{dz} + \frac{k_e^2}{2} \frac{\Delta\varepsilon}{\varepsilon} \tilde{S} = 0 \tag{9.110}$$

$$ik_o \frac{d\tilde{S}}{dz} - k_o(k_e - k_o - k_a)\tilde{S} + \frac{k_o^2}{2} \frac{\Delta\varepsilon}{\varepsilon} R = 0 \tag{9.111}$$

where we assume that $S(z)$ describes the amplitude of the ordinary wave and $R(z)$ describes the extraordinary. $\Delta\varepsilon$ is a function of tensor coefficients coupling the ordinary and extraordinary waves. Solutions take exactly the form of Eqn. (4.90) with $\Delta k = k_e - k_o + k_a$, where k_a is again the acoustic wavevector. Δk is nonzero owing to detuning of the input wavelength or angle of incidence. For example, the change in Δk for wavelength detuning $\delta\lambda$ is $\Delta k = 2\pi\Delta n\delta\lambda/\lambda^2$, where $\Delta n = (n_e - n_o)$. As in Eqn. (4.91), substantial loss in diffraction efficiency occurs if Δk is comparable to $k\Delta\varepsilon/\varepsilon$. This implies that the spectral resolving power of the AOTF is

$$R = \Delta n \frac{\varepsilon}{\Delta\varepsilon} \tag{9.112}$$

As a practical matter, AOTF resolution is limited by crystal size. One sets the acoustic power entering the device to obtain a grating modulation consistent with high diffraction efficiency (e.g., $L = \pi/\gamma \approx \varepsilon\lambda/2\Delta\varepsilon$). The maximum value of L available is determined by crystal growth and uniformity limitations. In terms of the grating interaction length, the resolving power is [103]

$$R = \frac{2\Delta nL}{\lambda} \tag{9.113}$$

It is interesting to note that the period of the acoustic grating is $\lambda/\Delta n$, so the resolving power of the AOTF is proportional to the number of grating periods in the interaction length. This result, and the similar result obtained for dispersive spectrometers, is, of course, expected from Fourier uncertainty arguments.

As with volume holograms, thin-film filters and liquid crystal tunable filters (LCTFs), the field of view of an AOTF is inversely proportional to the square root of the resolving power and the spectral efficiency is proportional to the input aperture area. AOTFs commonly achieve resolution in the nm range, corresponding to a spectral resolving power of 100–1000. The field of view is typically 5–10°. The spectral throughput shares the $\lambda/R\Delta\lambda$ characteristic of all narrow-pass instruments, although one can imagine implementing multiplex AOTF filtering. Tuning speed and spectral range are particular advantages of AOTF technology, it is not uncommon for an AOTF to tune from one wavelength to the next in $<10\ \mu$s, and crystals are available for filters spanning the ultraviolet to longwave infrared.

9.8 2D SPECTROSCOPY

As we saw in Section 7.5, dimensionality may have many meanings in sensor systems. In most spectroscopy literature, "two-dimensional" refers to molecular analysis based on joint consideration of excitation and relaxation frequencies [6]. "*N*-Dimensional" and "hyperdimensional" strategies have evolved in nuclear magnetic resonance and optical vibrational spectroscopies as mechanisms for joint analysis of molecular harmonics [142]. This sense of dimensionality points to the incredible richness of spectroscopy, which ultimately focuses more on the nature of the object than on the nature of the optical field. While it is fruitful to apply generalized sampling and signal analysis to excitation/relaxation spectroscopy, we limit the present narrative to optical system analysis.

We focus, therefore, on a more prosaic meaning of 2D, specifically, "How can we best take advantage of 2D detector arrays and 2D optical modulation to improve resolving power, etendue, and spectral efficiency?" The most striking similarity between the diffractive, interferometric, and resonant spectrographs that we have discussed is that, with the notable exception of the coded aperture, all rely on one-dimensional dispersion and coding strategies. The resolving power of the diffractive systems is determined by the number grating periods transverse to the optical axis, and the resolving power of the interferometric and resonant systems is determined by the scan range or dielectric modulation along the optical axis. We did consider off-axis modulation in the analysis of Fabry–Perot rings, but even in that case the basic spectroscopic pattern was one-dimensional in the radial coordinate. Given that 2D detector arrays are readily available and that the cost of an optical system rises nonlinearly in the aperture diameter, it is unfortunate that spectroscopic designs do not take advantage of the possibility of multidimensional optical elements.

Although 2D design does not improve the resolving power, etendue, and efficiency limits described earlier in this chapter, it does enable substantial increases in spectral range within a given volume. We are interested in systems, as illustrated in Fig. 9.34, in which a spatially inhomogeneous optical element modulates an incident field. The incident field is typically specified by prior information; for instance, the object may be a diffuse source radiating into a fixed solid angle, a plane wave, a focal image, and so on. The purpose of the optical element is to spatially and

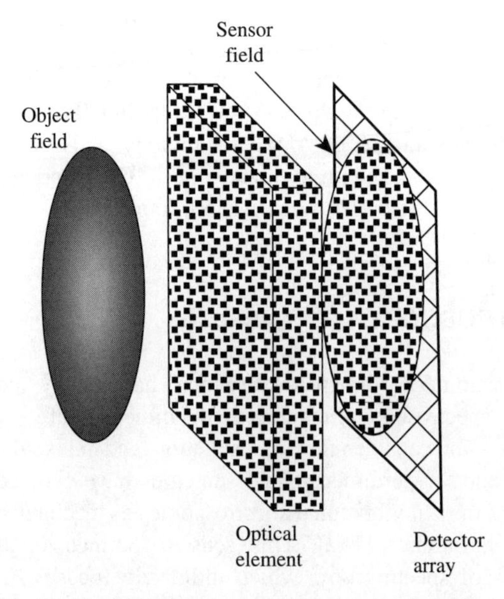

Figure 9.34 Field modulation by a multidimensional dispersive element.

spectrally modulate the optical field such that one can accurately estimate the mean power spectral density of the object.

We may make some general statements regarding optical processing with a multidimensional element. First, the element cannot increase the maximum focal radiance of any targeted spectral channel. We know from Section 6.6.3 that the brightest focal spot one could produce in any color channel has irradiance proportional to the largest coherent mode in that spectral channel. A passive optical element cannot violate this principle. On the other hand, it is possible to imagine an optical element that implements arbitrary independent spectral projections on each object image pixel. By an arbitrary spectral projection we mean a measurement of the form

$$g_i = \int S(x, y, \lambda) h_j(x, y, \lambda)\, d\lambda \tag{9.114}$$

Spectral projections g_i and g_j are independent if h_i and h_j are independent functions.

An optical fiber bundle provides an existence proof for a system capable of implementing an array of independent spectral projections. If we collect the object field into an array of fibers, we are free to implement complex holographic and interferometric filters independently on the light in each fiber. The number of different fields samples that we can process is limited by the fiber spacing, but once the light is in the fiber, processing of each channel becomes completely independent.

Moving from existence proofs to practical design, the following sections describe several strategies for efficient use of 2D detector arrays in spectrometer design. This idea continues to evolve rapidly; Section 9.8.4 offers thoughts on future directions.

9.8.1 Coded Apertures and Digital Superresolution

As we saw in Eqn. (9.12), coded aperture systems produce a 2D sensor pattern. In the model of Eqn. (9.19), a coded aperture implements dimension increasing generalized sampling as the 1D spectrum is mapped onto 2D measurements. Depending on the structure of the 2D code, each measurement may correspond to an independent projection. Assuming the uniformly pixelated aperture of Eqn. (9.14), however, the instrument function of Eqn. (9.15) applies and the number of measurements is typically much greater than the number of spectral channels captured. The purpose of increased dimensions in this case is to enable signal averaging over many pixels in order to increase SNR. An alternative interpretation of this system as a dimension reducing generalized sampling strategy for spectral images is introduced in Section 10.6.

One goal in the present section is to consider strategies whereby the number of measurements in a 2D system is equal to or even less than the number of spectral channels estimated. An example strategy arises in coded aperture systems where the pixel sampling function is larger than the optical PSF and the coded aperture features. We generally assume, of course, that the pixel pitch in a dispersive spectrometer is less than half the slit or aperture feature width. This sampling rate ensures that the object spectrum is sampled at the Nyquist rate. It is possible,

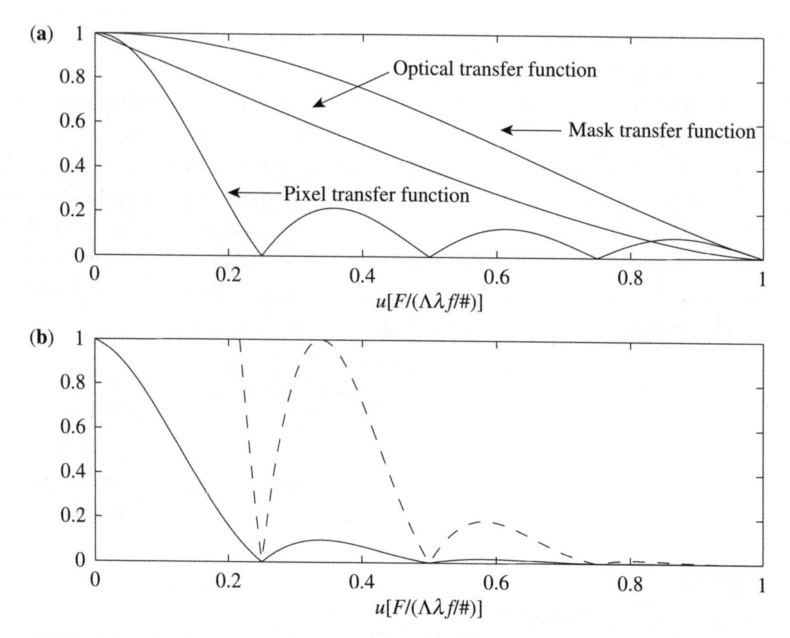

Figure 9.35 Transfer functions for a coded aperture spectrometer with mask features of width $\lambda f/\#$ and focal plane pixels of width $4\lambda f/\#$. As illustrated in (a), the system transfer function is the product of the optical, pixel, and mask transfer functions. The dashed curve in (b) is the system transfer function magnified along the vertical axis by a factor of 10. The aliasing limit for single-channel sampling is $u = 0.125F/\Lambda\lambda f/\#$. The goal of nondegenerate 2D sampling is to extend the effective system transfer function to the optical diffraction limit, increasing the spectral resolution in this case by a factor of 4.

however, to use 2D sampling to antialias an undersampled 1D spectrum and make the effective focal plane pixel size smaller.

Example transfer functions for this case are illustrated in Fig. 9.35. The plots assume that the mask feature size is $4\times$ smaller than the pixel feature size and that the mask features are just at the optical resolution limit. As discussed in Chapters 5 and 7, the pixel pitch in modern visible and shortwave infrared focal planes is usually much larger than the optical Nyquist limit of $\lambda f/\#/2$. Interpreting the pixel pitch as the resolution limiting feature in Fig. 9.35 using the signal inference strategies of Section 9.3, one predicts that the useful bandwidth limit is $u_{\max} = 0.125F/\Lambda\lambda f/\#$. Spectral features beyond this limit would be aliased, meaning both that they would not be resolved and that they would add ambiguity to features within the nonaliased band.

"Pixel superresolution" techniques using multiple nondegenerate samples of a sampled image to overcome the aliasing limit are well established for multiaperture and video imaging systems [195]. We considered a rough version digital superresolution in Section 8.4 and ask the reader to consider them again in Problem 9.14. We discuss these techniques for imaging in Section 10.14. *Pixel superresolution* in the current context consists of repeating rows of the coded aperture with a slight horizontal shift from one row to the next. The sampling model of Eqn. (9.12) applies, but by repeating rows of the code with a 1 code pixel shift from one row to the next, one may collect data such that the sample pitch is $\Delta_l = \Lambda a/qF$, where the code pitch a is less than the FPA pixel pitch Δ.

In considering this sampling strategy, one observes the different impacts of the pixel transfer function and the sampling aliasing limit on instrument design. The pixel pitch in rectangularly sampled systems often leaves substantial system bandpass beyond the aliasing limit, in part because a rectangular pixel has significant bandpass beyond its corresponding Nyquist limit. Multichannel sampling of signals with subpixel shifts enables one to create sampling systems with spatially overlapping pixel sampling functions such that the aliasing limit corresponds to the full pixel bandpass.

The instrument function for the pixel superresolution strategy described here remains $h_{lr}(\lambda)$ as defined in Eqn. (9.15), meaning that the system transfer function is the STF illustrated in Fig. 9.35. As illustrated in the figure, the system transfer function remains potentially useful out to $0.5F/\Lambda\lambda f/\#$, indicating an improvement in resolution by a factor of 3–4 over the conventional limit. Signal improvements arise from both the elimination of aliasing noise and the broadened bandpass.

With pixel superresolution one may imagine dispersive spectrometers with resolving power at the diffraction limit. Substituting $a_x = \lambda f/\#$ in Eqn. (9.8) yields

$$R = \frac{A}{\Lambda} \tag{9.115}$$

where A is the aperture diameter and Λ is the grating period. However impressive this result may be in comparison with many dispersive designs, the fact that the resolving power is proportional to the aperture diameter rather than the aperture area means that we have not achieved truly 2D coding.

The dispersive spectrographs in Figs. 9.2 and 9.4 consist of three major components: the input aperture, the dispersive element, and the focal plane array. We have found great success thus far in considering 2D structures on the input aperture and the FPA. Our next step is to consider 2D dispersion. 2D dispersion requires that we expand our stable of diffractive elements beyond the thin amplitude or phase gratings of Section 4.5 and the volume phase gratings of Section 4.8. Because of limitations of modeling, fabrication, and imagination, there is considerable room for continuing diffractive element innovation. We describe just three examples in this section to give a general idea of the possiblities.

9.8.2 Echelle Spectroscopy

Our first example is the echelle grating. In French, *echelle* refers to a ladder. The echelle grating improved on the earlier concept of an echelon grating, which was a staircase-shape stack of refractive elements, by adding a blaze angle to the stair facets to enhance diffraction into targeted orders [113]. The basic structure of an echelle grating is illustrated in Fig. 9.36. The grating itself consists of a reflective surface into which periodic rulings have been etched. A "blazed grating" is a relief

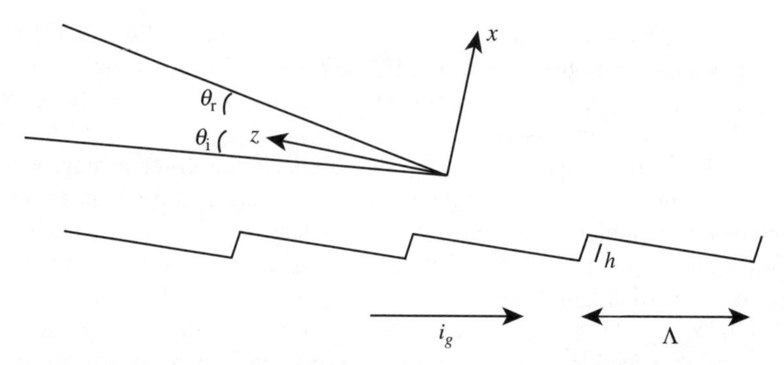

Figure 9.36 Geometry of an echelle grating mounted in a Littrow geometry. The z axis is selected parallel to the blaze normal. The grating lies in a plane with surface normal $\mathbf{i}_g$.

structure consisting of planar patches tilted with respect to the surface normal. The tilt is at the "blaze angle." As discussed momentarily, a blaze dramatically enhances diffraction efficiency into targeted diffraction orders. One may regard a blaze as a form of phase grating; blaze gratings have a longer history than do most diffractive optical elements because they require relatively modest spatial resolution and can be fabricated by diamond turning.

Rigorous analysis of diffraction from echelle gratings is a challenging problem; any device with sharp discontinuities will tend to introduce strong scattering artifacts [156]. As we are satisfied for the present purposes with a general idea of the echelle function, we simply assume that the incident signal wave uniformly illuminates the blaze surfaces, ignoring secondary scattering from the longer nearly horizontal surfaces. We focus on the *Littrow* geometry in which the incident and diffracted beams are nearly parallel to the blaze. In this geometry the diffracted wave is nearly retroreflected and the effective grating period must be near $\lambda/2$. Since echelle gratings consist of rulings of period $\Lambda \gg \lambda$, this means that the Littrow reflection is a high-order diffraction mode.

As illustrated in Fig. 9.36, we select $\mathbf{i_z}$ normal and $\mathbf{i_x}$ parallel to the blaze facets. $\mathbf{i_g} = \cos\theta_g \mathbf{i_z} + \sin\theta_g \mathbf{i_x}$ is a unit vector parallel to the grating wavevector. We consider a plane wave incident on the grating with wavevector $\mathbf{k_i} = -k\cos\theta_i \mathbf{i_z} + k\sin\theta_i \mathbf{i_x}$. The wave field on the nth grating facet is

$$U_n(x) = \mathrm{rect}\left(\frac{x - n\Lambda\sin\theta_g}{h}\right) e^{(2\pi i/\lambda)[(x-n\Lambda\sin\theta_g)\sin\theta_i + n\Lambda\cos\theta_g\cos\theta_i]} \tag{9.116}$$

Observed in under the Fraunhofer diffraction approximation at range $z = R$, we find from Eqn. (4.42) that the wave field reflected from this facet is

$$\begin{aligned} U_n(x') = {} & e^{2\pi i R/\lambda}\mathrm{sinc}\left[\frac{h}{\lambda}\left(\frac{x'}{R} - \sin\theta_i\right)\right] \\ & \times \exp\left[\frac{2\pi i n\Lambda\sin\theta_g}{\lambda}\left(\frac{x'}{R} - \sin\theta_i\right)\right] \exp\left(\frac{4\pi i n\Lambda\cos\theta_g\cos\theta_i}{\lambda}\right) \end{aligned} \tag{9.117}$$

The scattered field radiates in a cone of angular width λ/h centered on $\theta_r = x'/R = \theta_i$. Summing over N grating facets, we may approximate the combined diffracted field as

$$U(\theta_r) = U_0 \text{sinc}\left(\frac{\theta_r - \theta_i}{\theta_o}\right) \times \frac{\sin\left\{\pi(N+1)\left[(\Lambda \sin\theta_g/\lambda)(\theta_r - \theta_i) + (2\Lambda/\lambda)\cos\theta_g \cos\theta_i\right]\right\}}{\sin\left\{\pi\left[(\Lambda \sin\theta_g/\lambda)(\theta_r - \theta_i) + (2\Lambda/\lambda)\cos\theta_g \cos\theta_i\right]\right\}} \tag{9.118}$$

for some constant U_0. We define $\theta_o = \lambda/H$ and we assume $\sin\theta_i \approx \theta_i$.

The diffracted field consists of a series of diffraction orders satisfying the grating equation

$$\theta_r - \theta_i = q\frac{\lambda}{\Lambda \sin\theta_g} - 2\frac{\cos\theta_i}{\tan\theta_g} \tag{9.119}$$

The diffraction orders are modulated by the facet diffraction pattern $\text{sinc}[(\theta_r - \theta_i)/\theta_o]$. For the peak diffraction order corresponding to $\theta_r = \theta_i$, we obtain

$$q = 2\frac{\Lambda}{\lambda}\cos\theta_i \cos\theta_g \tag{9.120}$$

Anticipating that θ_i, $\theta_g \ll 1$ and that $\Lambda \gg \lambda$, this means that the peak diffraction order corresponds to the value of $q \gg 1$. Typical values of q for echelle gratings range from 10 to 100; the actual value is determined in the selection of Λ and θ_g. As expected, in the retroreflection geometry $\Lambda/q \approx \lambda/2$.

The angular width of each diffraction order is $\delta\theta \approx \lambda/Asin\theta_g$, where $A = N\Lambda$ is the aperture of the echelle. $A\sin\theta_g$ is the cross section of the aperture as viewed along the x axis. With λ fixed, we find the angular separation from one order to the next by increasing q by one, which leads to a shift in θ_r of $\lambda/\Lambda\sin\theta_g$. With $\theta_g \ll 1$, this shift may be comparable to $\theta_o = \lambda/h$, meaning that we may observe only one or two diffraction orders for a given λ. The *free spectral range* of the echelle is the change in ν from one order to the next:

$$\Delta\nu = \frac{c}{2\Lambda\cos\theta_i \cos\theta_g} \tag{9.121}$$

The wavelength range corresponding to one free spectral range, $\delta\lambda \approx \lambda^2/2\Lambda$, may be 1–10 nm at visible wavelengths.

The ideal field scattered from an echelle consists of a set of diffraction orders separated in frequency by the free spectral range but all clustered around $\theta_r = \theta_i$. These diffraction orders must be separated to allow spectral analysis. As illustrated in Fig. 9.37, this separation is achieved by including a cross-disperser. The figure shows an echelle spectrograph in simplified form; a practical system includes

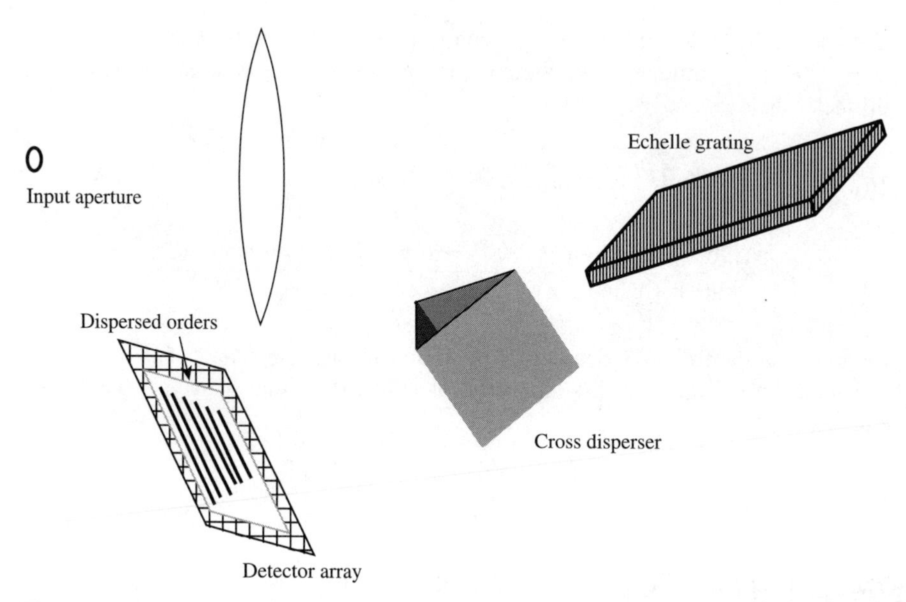

Figure 9.37 An echelle spectrograph that combines fast dispersion on the echelle grating with coarse separation due to a cross-disperser to produce a 2D output mapping.

imaging mirrors and lenses, neglected here for simplicity. The basic concept of the design is identical to a grating spectrometer, where the input aperture is imaged onto a detector array through a dispersion system. The use of different diffraction orders for different free spectral ranges enables true 2D dispersion, however, and allows one to use an effective grating period consistent with high resolving power while also maintaining a broad spectral range. Notice that the figure assumes that the echelle is illuminated at a nonzero azimuthal angle (e.g., $\mathbf{k_i}$ has a $\mathbf{i_y}$ component). We neglected this possibility to simplify our narrative in describing the grating; azimuthal illumination simplifies separation of the incident and diffracted beams in a Littrow system. Figure 9.38 is an image of echelle spectrometer data. Depending on the cross-dispersion aperture, a given spectral channel may appear in more than one diffraction order in an echelle. On this particular instrument (Optomechanics Research SE 200) each spectral feature appears in two diffraction orders.

The spectral resolution of an angularly dispersive instrument is determined by the rate of change of θ_r with respect to λ:

$$\Delta\theta_r = \frac{q}{\Lambda \sin\theta_g}\,\delta\lambda \tag{9.122}$$

For an input aperture of width a imaged through a system with effective focal length F, each spectral channel occupies an angular band of width $\Delta\theta_r = a/F$. We find therefore that the resolving power of an echelle spectrometer is

$$R = \frac{q\lambda F}{a\Lambda \sin\theta_q} \tag{9.123}$$

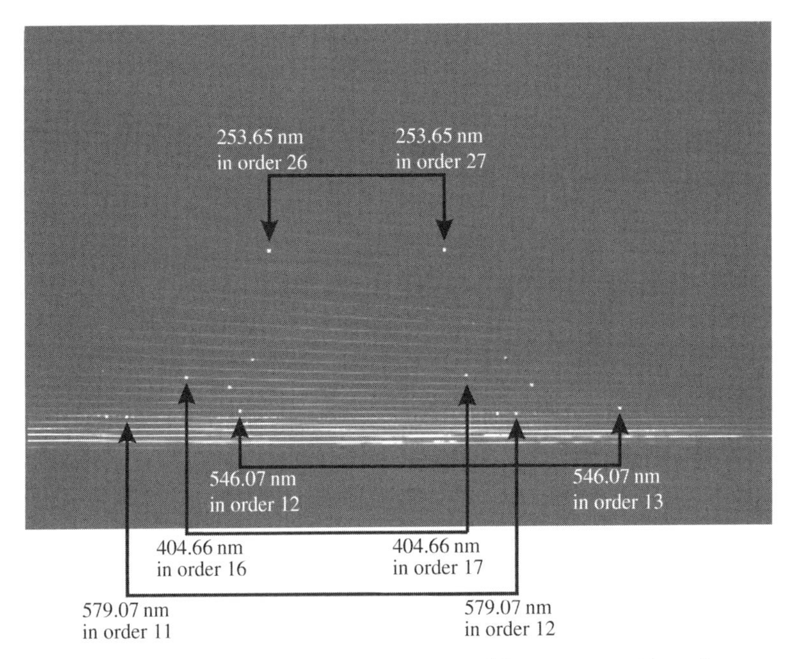

Figure 9.38 A CCD image of the spectrum generated by a superposition of mercury and deuterium–tungsten lamps. The spectrum was collected by the Optomechanics Research SE 200 echelle spectrometer using Catalina Scientific's KestrelSpec software. The image was provided courtesy of Catalina Scientific Instruments.

As with the slit spectrometer, the diffraction limited resolving power is proportional to the number of grating periods. Replacing a with $\lambda f/\#$ and noting that the effective system aperture is the cross-sectional aperture of the echelle, $A \sin\theta_g$, we find the resolving power to be qN_g. As an example, the SE 200 achieves resolving powers ranging from 2000 to 6000, depending on grating geometry and pixel size.

To allow for order sorting by 2D dispersion, echelle instruments generally use pinhole apertures, rather than slits. One might overcome this limitation using a 1D coded aperture along the y axis. Under this strategy an echelle achieves spectral efficiency comparable to a conventional grating spectrometer while also obtaining extraordinary spectral range. Echelle systems are not amenable to 2D coded apertures because the numerical aperture in x is limited. Assuming a pinhole aperture, one replaces the input aperture area Aa in Eqn. (9.9) with a^2 for an echelle instrument. The resulting loss in spectral efficiency is balanced by the much shorter effective grating period for the echelle, however, so the overall spectral efficiency is comparable to that for a slit spectrometer.

In summary, the echelle grating obtains the high spectral range and resolution of true 2D coding but is generally limited to low-etendue input signals. It is interesting to note that one could obtain similar advantages with much higher etendue with using a dielectric mirror consisting of uniformly thick layers. However, the fabrication technology to make a dielectric stack with 20–40-μm layers to submicrometer uniformity is not currently available. When Harrison introduced the echelle grating, the art of fabrication was a critical factor in system performance and design. The art of blazed grating manufacturing has advanced considerably in the intervening half-century, but system constraints due to manufacturing art remain a central theme of spectrometer development.

9.8.3 Multiplex Holograms

Multiplex volume holograms are an example of a physically plausible technology lacking the current manufacturing art for widespread integration in spectroscopy. As discussed in Section 4.8, a hologram may be recorded between fields of arbitrary complexity. For simplicity, we limit our discussion here to holograms recorded between plane waves. Such holograms produce simple diffraction gratings described by a grating wavevector $\mathbf{K}$, as illustrated in Fig. 4.21. In a multiplex hologram, multiple plane wave gratings are recorded in a single material. Two of the various recording strategies for multiple grating holograms are illustrated in Fig. 9.39. Figure 9.39(a) illustrates a strategy for recording grating wavevectors $\mathbf{K}_1$ and $\mathbf{K}_2$ using a single reference beam. As discussed in Ref. 27, recording with a single reference produces higher modulation depth and diffraction efficiency. The process of reading the single-reference hologram at a shorter wavelength is illustrated by the outer wave normal surface in Fig. 9.39(a). At the shorter wavelength, the Bragg-matched reference angle for both recorded gratings changes.

Our goal in using a multiplex grating hologram in spectrometer design is to efficiently map multiple spectral ranges from the angular field of the input aperture on to the angular field of the output aperture, much like an echelle grating. The potential

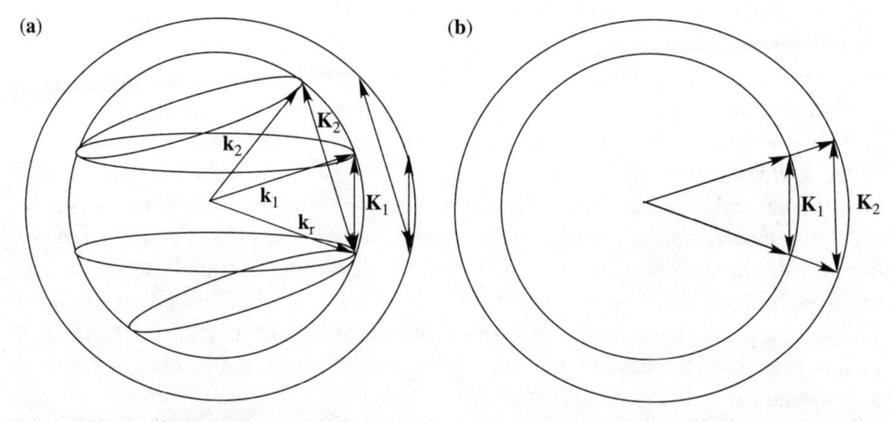

Figure 9.39 Recording geometries on the wavenormal surfaces for multiple-grating holograms: (a) recording of two grating wavevectors using a single reference beam—the cross-grating between $\mathbf{k}_1$ and $\mathbf{k}_2$ is also recorded; (b) recording of two wavevectors between pairs of beams at different wavelengths.

advantage of the multiplex grating approach is that the input field of view will be comparable to conventional instruments, allowing convenient use of coded apertures or slits. Unfortunately, the multiplex grating hologram of Fig. 9.39(a) does not achieve this objective because the recording field is Bragg-matched to both gratings along the input wavevector. The shorter wavelength field is nearly Bragg-matched to both gratings, but at a shifted input angle. If the input field scattered by all of the grating orders, then one must increase the output field of view to capture all of the grating orders, and the system resolving power is simply determined by the largest value of **K**.

In the alternative recording geometry of Fig. 9.39(b), independent gratings are recorded at two different wavelengths. In this case both recording wavelengths are Bragg-matched at the same input and output angles. We also illustrate the Bragg matching condition on the outer wave normal surface of the grating $\mathbf{K}_1$ recorded on the inner wave normal surface and the Bragg matching condition on the inner wave normal surface for the grating $\mathbf{K}_2$ recorded on the outer surface. The shift in the Bragg condition can be used in combination with an angularly filtered input aperture to ensure that a spectral band centered on $\mathbf{k}_1$ scatters from $\mathbf{K}_1$ while a spectral band centered on $\mathbf{k}_2$ scatters from $\mathbf{K}_2$ from the same input aperture into the same output aperture.

To use multiplex volume gratings in a diffractive spectrometer, one decreases $\Delta\varepsilon$ and increases L to increase the angular selectivity of the hologram. In a conventional design, one seeks low angular selectivity to achieve reasonable etendue. In a multiplex design, one maintains etendue by adding grating components. Figure 9.40 shows the increased angular selectivity for a lower-index-modulation hologram. Figure 9.40(a) is a density plot of the data shown in Fig. 9.15, showing the angular selectivity of a high-modulation, thin hologram. Figure 9.40(b) shows the decreased angular range of thicker and lower-modulation hologram. Figure 9.41 shows the diffraction efficiency as a function of angular and wavelength detuning

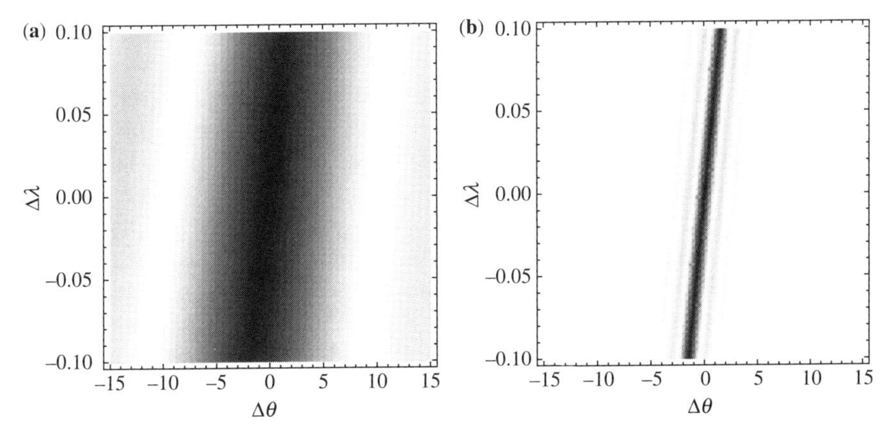

Figure 9.40 Diffraction efficiency versus angular and wavelength detuning for (a) the volume hologram of Fig. 9.15 and (b) the same hologram geometry with $\Delta\varepsilon/\varepsilon = 5 \times 10^{-3}$, which yields 100% diffraction efficiency for a thickness of $96\lambda_0$.

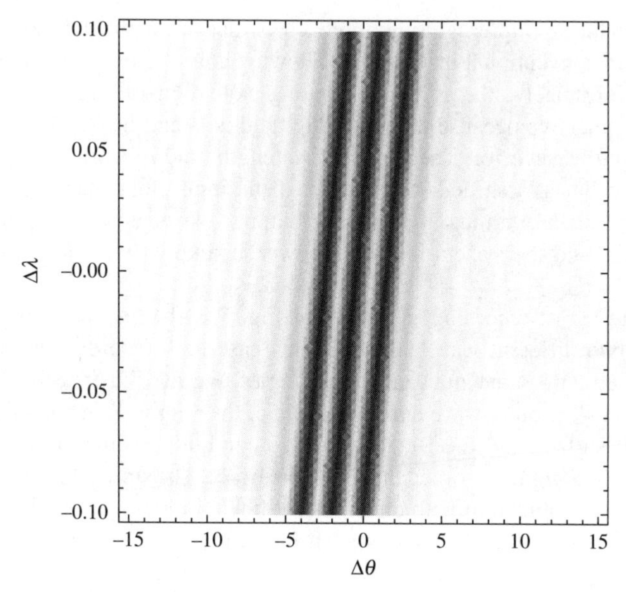

Figure 9.41 Diffraction efficiency versus angular and wavelength detuning for a three-grating multiplex hologram, each grating using the same parameters as in Fig. 9.40(b).

for a three-grating multiplex hologram. The wavevectors of the gratings have been separated enough to permit one to see the three orders; in a practical system one selects the orders such that at each wavelength one can see at least one order within the numerical aperture of the system while also attempting to maximize the overall photon efficiency and etendue.

Feller et al. [70] demonstrated a multiple-order volume holographic 2D spectrometer using a coded aperture to enable overlap in the spectral images on the detector plane. Figure 9.42 shows sensor data for a monochromatic input for this instrument, illustrating the multiple diffraction orders of an order 37 MURA code. The goal of Feller's instrument was to create a 2D spectral mapping such that each position on the plane corresponds to a single wavelength. Just as this mapping restricts throughput for a pinhole camera, such a mapping restricts throughput for a spectrometer if the input source is diffuse. Feller overcomes this issue using the coded aperture strategy discussed in Section 2.5 by placing each spectral channel on a 2D pattern matched to deconvolution. The vertical shift between the diffraction orders in Fig. 9.42 is introduced by a tilt in the recorded grating **K** vectors rather than by a cross-disperser.

In principle, one could imagine the use of highly multiplexed volume holograms to create complex 2D spectral mappings. In practice, however, the art of creating multiplex gratings with even just a few gratings is undeveloped. As discussed in Ref. 27, recording with M reference beams reduces the total diffraction efficiency of a hologram by approximately M^{-2} compared to a hologram recorded with only one reference. At the time of this writing there are no commercial sources for multiplex

Figure 9.42 2D image detector for a monochromatic source using the multiple-order coded aperture spectrometer described by Feller et al. [70].

gratings, although the use of commercial use of multiplex holograms in data storage systems remains promising. Multiplexing could, of course, also be applied to great effect for the volume reflection gratings of Section 9.6 to create multiple band rejection or isolation filters. At present, however, the art of 2D thin film filter and microresonantor fabrication appears likely to dominate multiplex holography in spectroscopic applications.

9.8.4 2D Filter Arrays

The echelle and multiplex volume hologram spectrographs demonstrate that dispersive instruments can utilize a 2D aperture. One may, of course, also encode interferometric and resonant instruments over 2D sensor arrays. Early studies of spatially distributed measurements with such instruments focused on designs that map data that might otherwise be distributed in time onto 1D detector arrays, as in two-beam interferometers with spatially varying phase delay [192,208] and spatially variable thin-film filters [34,85,215]. More recently, technologies have been developed to create 2D instruments using active interferometer arrays based on liquid crystal or micromechanical filters [238] or using spatially patterned 2D arrays of thin-film filters [34,245]. While these approaches are promising in the near term, miniaturized interferometers and filters are still subject to the same resolving power, etendue, and efficiency scaling laws as their macroscopic cousins described earlier in this chapter. As considered in our discussion of volume scaling laws, simply making an instrument smaller often degrades performance. To achieve truly revolutionary performance, one must account for multidimensional structure; for example, it is not enough to make 1D gratings or 1D filters smaller—one must make 3D optical structures.

The seeds of 2D spectrometers with arbitrary etendue and resolving power are apparent in two trends. In the first example, mentioned earlier in the chapter but generally beyond the scope of this text, one creates spectrally sensitive absorbing materials. The Foveon X3 sensor, a visible focal plane consisting of a stack of red, green, and blue photodiode layers, is the most widely known example of this approach [193]. More recently, substantial progress has been made to extend this concept using layered or electrically tunable materials in infrared devices [127,140]. One may view this approach as an attempt to transfer spectral analysis

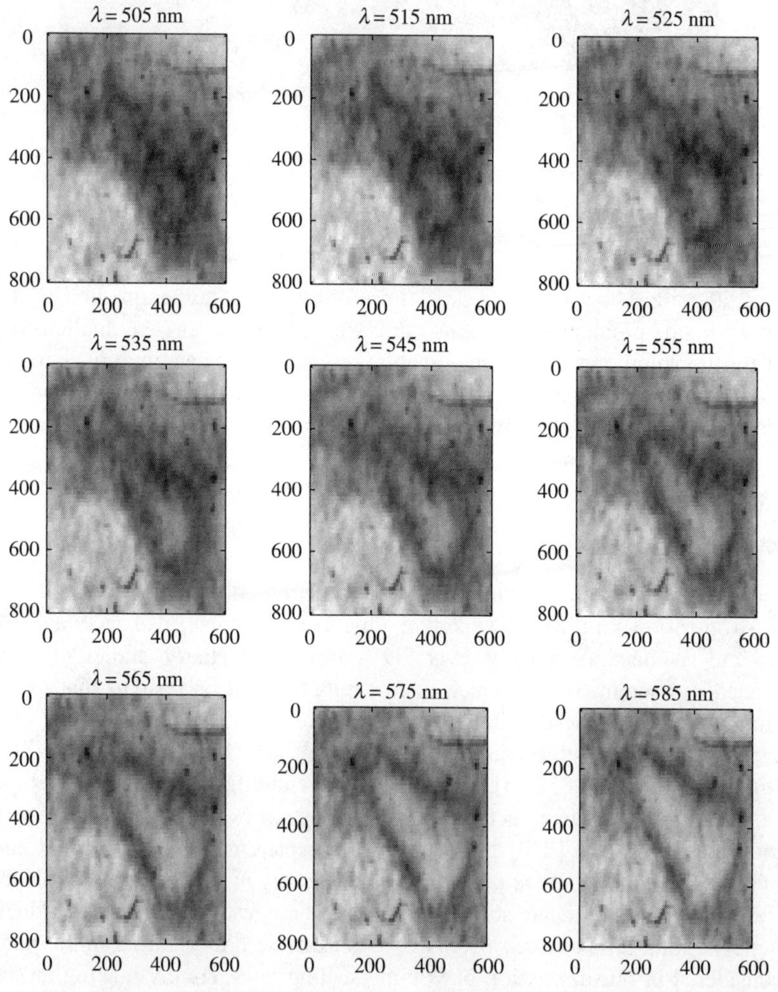

Figure 9.43 Images of a self-assembled opal illuminated by diffuse quasimonochromatic light. The spatial axes are plotted in micrometers.

from the optical domain to the electronic domain and thereby achieve smaller devices based on the smaller electronic wavelength. One may also view it as an attempt to use electronic nanostructures or metamaterials to create novel devices.

The second trend involves the use of 2D and 3D photonic metamaterials in spectroscopic analysis. Although there have been a few interesting demonstrations of this approach [216,255], design and fabrication of spectroscopic instruments based on multidimensional metamaterials is in its infancy. A simple example of the basic idea is illustrated in Fig. 9.43, which shows a 3D photonic crystal illuminated by diffuse quasimonochromatic light at frequencies ranging over the visible spectrum. The photonic crystal is an "opal" formed by self-assembly of nanometer-scale dielectric beads. If the beads were perfectly uniform in size, the transmission pattern would be spatially uniform. Random defects in the crystal structure lead to complex 2D spectral transmission patterns. By calibrating the spatial pattern for each wavelength, one can invert the spatial pattern under broadband illumination to estimate the spectrum. A spectrometer based on this strategy consists simply of a thin photonic crystal layer deposited directly on a 2D detector array.

While early demonstrations of photonic and electronic metamaterial-based instruments are imperfect, the basic concept of nanoscale instruments based on these technologies are sound. It seems clear that the story of spectrometer design and miniaturization is just beginning.

PROBLEMS

9.1 *Space–Bandwidth Product.* Use Fourier uncertainty relationships to explain why the resolving power of dispersive spectrometers and the AOTF is proportional to the number of grating cycles in the instrument. How does this result relate to the resolving power of a volume reflection hologram [Eqn. (9.74)]? Can you make similar arguments regarding the resolving power of a Fabry–Perot or thin-film filter?

9.2 *Dispersive Spectroscopy.* Design a slit spectrometer spanning the spectral range 350–950 nm with a spectral resolution of 5 nm. Your design should specify slit size, the focal lengths of any lenses or mirrors used and the period of the grating used. What are the resolving power, etendue, and volume of your system?

9.3 *Coded Aperture Spectroscopy.* Design and simulate a Hadamard S-matrix spectrometer with $N = 128$. The spectrometer should span the spectral range 350–950 nm with 1 nm resolution. Specify the pixel pitch on the detector array, the coded aperture feature size, and grating and lens parameters.

(a) Use synthetic spectra consisting of 1–10 Lorentzian lines of various widths to simulate data collection and processing. Evalutate the mean-square estimation error under linear least-squares and nonnegative least-squares inversion with various levels of additive Gaussian and Poisson noise.

(b) Substitute a 127×127 random matrix for the S matrix. Compare reconstructions using linear least squares and regularized least squares with the results obtained with the Hadarmard code.

9.4 *Spectroscopic Gratings*

(a) Discuss the impact of grating efficiency on system volume and $f/\#$ for spectrometers based on volume phase gratings and on echelle gratings.

(b) Datasheets for spectroscopic gratings are available from suppliers such as Wasatch Photonics, Richardson Gratings, and Ondax. Compare quantum efficiencies, spectral range, and angular range for available grating types and geometries.

9.5 *Interferometric Spectroscopy.* Design an FT spectrometer spanning the spectral range 2–20 μm with spectral resolution of 0.01 nm. What is the scanning range and scan resolution required? What are the resolving power, etendue, and volume of the system? What optics might be required? What detector would you use?

9.6 *Spectrometer Analysis.* Compare limits on the resolving power, etendue, and volume of spectrometer designs discussed in this chapter. On a graph of resolving power versus etendue, mark the limits of coded aperture, slit, FT, multibeam interferometery, and 2D spectroscopy. Construct a similar graph of resolving power versus volume.

9.7 *Fabry–Perot Estimation.* Design a spectrometer operating over the spectral range 500–520 nm with 0.1 nm resolution using the dispersive Fabry–Perot modeled by Eqn. (9.67). Select a region of the xy plane for sampling and specify resonator thickness and index, grating and lens parameters, and pixel pitch.

9.8 *TM Modes of Thin-Film Filters.* Derive the characteristic matrix $\mathbf{M}$ for TM modes of a quarter-wave stack. Assuming $n_1 = 2.25$ and $n_2 = 2.5$, replicate Fig. 9.22 for the TM modes over the range $\Delta\Theta = \pm 30^\circ$.

9.9 *Thin-Film Eigenvectors.* Prove that $\mathbf{E}_\pm$, as described by Eqn. (9.93), are eigenvectors of the characteristic matrix $\mathbf{M}$.

9.10 *Thin-Film Bandpass Filters.* Replicate Fig. 9.25 using 10 period dielectric mirrors. The center of the resonator for Fig. 9.25 is a $\lambda/2$ dislocation. Demonstrate tuning of the resonance across the stopband by varying the thickness of the center layer.

9.11 *Liquid Crystal Tunable Filters*

(a) Design a Lyot filter with a resolving power exceeding 100 using a liquid crystal with $\Delta n = 0.2$.

(b) Plot representative transmittance curves, similar to Fig. 9.30, for your device as Δn is tuned from 0 to 0.2.

9.12 *Acoustooptic Tunable Filters*. Estimate the range of acoustic frequencies used to drive a TeO_2 AOTF tuned to operate from $\lambda = 0.5$ to 1 μm. What factors determine the spectral range of an AOTF?

9.13 *Echelle Spectroscopy*. Estimate the free spectral range for the echelle spectrograph illustrated in Fig. 9.38.

9.14 *Pixel Superresolution*. Design a "slit" spectrograph with $a_x = \lambda f/\# = \Delta/4$. The goal of the system is to estimate the spectral density with an effective sampling pitch $\Delta_l = \Lambda a_x/F$. One achieves this objective by shifting the center of the slit from one row to the next by $\Delta/4$ and applying the pixel superresolution technique of Section 9.8. Plot the mask code t_{ij} that you would use to achieve this goal. Is it possible to reduce the effective pixel size on both the x and y axes? What is the maximum resolving power per unit volume for this coding approach? Develop code to simulate measurement data on your proposed system. Plot the measurement data for your mask code for the spectral density

$$S(\lambda) = S_0[1 + \cos(2\pi u\lambda)] \tag{9.124}$$

for $u = 0.125u_c$, $0.25u_c$, and $0.375u_c$, where $u_c = F/\Lambda\lambda f/\#$. Use a Wiener filter to estimate the spectrum from each set of simulated data. Plot an estimated spectrum for Poisson noise and for additive Gaussian noise, in each case specifying the noise statistics used.

10

COMPUTATIONAL IMAGING

> Better results should be achieved by a procedure in which the image-gathering system is designed specifically to enhance the performance of the image-restoration algorithm to be used.
>
> —W. T. Cathey, B. R. Frieden, W. T. Rhodes, and C. K. Rushforth [43]

10.1 IMAGING SYSTEMS

Optical sensor design balances performance metrics against system implementation and operation constraints. Interesting performance metrics include angular, spatial, or spectral resolution; depth of field; field of view; zoom capacity; camera volume; sensed data efficiency; spectral or polarization sensitivity; and tomographic fidelity. Constraints include fundamental, practical, and financial limits based on physical, information-theoretic, and data processing issues. Examples include spatial and temporal bandwidth, coherence and statistical properties, system model limitations, and computational complexities.

Because of the complexity of system metrics and constraints and the embrionic state of many design tools, none of the designs or design strategies we discuss in this, or previous, chapters are in any sense optimal. While the pessimist may disdain the ad hoc nature of current digital imaging and spectroscopy design, we find promise in the rapidly evolving design landscape and hope that the student finds frank discussion of design strategies illuminating and suggestive.

We may divide the optical sensor design process into

- *Specification*, consisting of a description of the class of objects and object features that a system must measure, the nature of image and object data that the system must produce, and performance specifications for measurement and image generation. Specification may include the field of view, angular

Optical Imaging and Spectroscopy. By David J. Brady

resolution, spectral range, depth of field, zoom, and spectral resolution. Typically, the system designer begins with an application in microscopy, telescopy, machine vision, or photography and seeks to achieve maximal performance within a certain monetary and system form factor budget. Under this scenario, specifications evolve under feedback from subsequent steps in the design process. Initial system specification generally consumes less than 5% of a design cycle.

- *Architecture*, which consists of broad specification of system sensor and optical components. The system architect decides whether and where to use pixel, convolutional, and implicit coding strategies. The goal of system architecture is to lay out a strategy for matching desired performance specifications with a realistic engineering strategy. Architecture design typically consumes 10% of the design cycle and may include idealized simulation of system performance.
- *Engineering*, consisting of detailed design of optical elements, detector arrays, readout electronics, and signal analysis algorithms. Optical engineering generally accounts for 40% of a design cycle and will include computer simulation of optical components and signal analysis algorithms as well as tolerancing studies.
- *Integration*, which consists of optical component manufacturing, testing, and optoelectronic systems and processing integration. Integration accounts for about 40% of the design cycle.
- *Evaluation*, consisting of testing of prototype designs and confirmation of system performance.

This text focuses exclusively on the architecture component of system design. The skilled system architect will, of course, wish to complement this text with more detailed studies in lens design, image processing, and optoelectronics. A system architect uses high-level design concepts to make systems perform better than naive design might predict. While an architect will in practice seek to balance diverse performance specifications, we illustrate the design process in this chapter by singly optimizing particular performance metrics. Subsequent sections consider design under the constraints that we wish to optimize depth of field, spatial resolution, field of view, camera volume, and 3D spatial or spatiospectral data cube acquisition.

10.2 DEPTH OF FIELD

Focal imaging occurs only for object and image geometries satisfying the image condition [Eqn. (2.17)]. As an object is displaced from the plane $z_o = z_i F/(z_i - F)$, the image blurs owing to a broader PSF and narrower OTF. The range of distances z_o over which the object may be displaced without unacceptable loss of image fidelity is called the *depth of field*. Section 6.4.3 described the defocus transfer function and considered Hopkins' criterion limiting the defocus parameter w_{20}.

Given a maximum acceptable value for w_{20}, the object field is the range of z_o such that

$$-\frac{2w_{20}}{A^2} \leq \frac{1}{z_o} + \frac{1}{z_i} - \frac{1}{F} \leq \frac{2w_{20}}{A^2} \tag{10.1}$$

For simplicity, we limit our discussion to object fields extending from some *near point* to $z_o = \infty$. We set the distance between the lens system and the focal plane z_i, such that a point at infinity is defocused to the maximum acceptable blur. This yields

$$z_i = \frac{FA^2}{A^2 - 2w_{20}F} \tag{10.2}$$

Moving in from infinity, the defocus decreases until the thin-lens imaging law is satisfied at $z_H = A^2/2w_{20}$, which is called the *hyperfocal distance*. Moving in from the hyperfocal distance, the defocus increases up to the near point for acceptable focus (e.g., the point such that $1/z_o + 1/z_i - 1/F = 2w_{20}/A^2$). The near point for a lens focused on the hyperfocal distance is $z_o = z_H/2$.

Figure 10.1 illustrates a system imaging the plane at the hyperfocal distance. The point at infinity focuses at the lens system focal point and is blurred at the sensor plane, which is displaced approximately F^2/z_H from the focal plane. Using the similarity of the triangle between the lens and the focal point at the bottom of Fig. 10.1

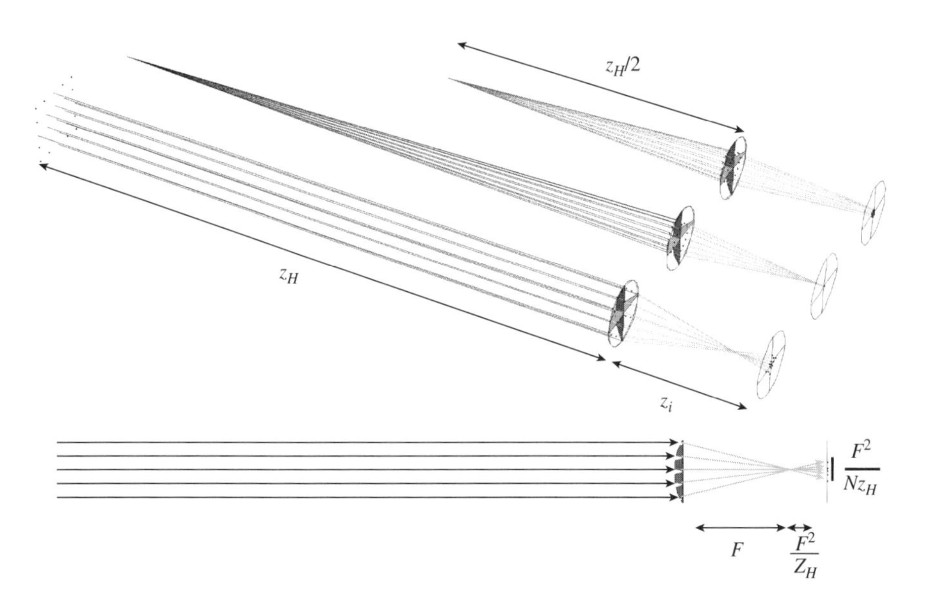

Figure 10.1 Geometry for imaging at the hyperfocal distance. Images formed from a point source at $z_H/2$ (top) or from a point source at infinity (bottom) are blurred. A well-formed image is formed for a point source at the hyperfocal distance (center).

and the triangle between the focal point at the sensor plane, one can see that $A/F = Cz_H/F^2$, where C is the extent of the blur spot for a point at infinity. C is called the *circle of confusion*. In terms of the circle of confusion

$$z_H = \frac{F^2}{Cf/\#} \tag{10.3}$$

The conventional understanding of imaging systems observing from a near point to infinity without dynamic refocusing is thus that the near point is $z_H/2$, where z_H is as given by Eqn. (10.3). In conventional systems, one increases the depth of field (e.g., reduces the range to the near point) by decreasing z_H. One achieves this objective by increasing $f/\#$ or decreasing F. One increases $f/\#$ by *stopping down* an imaging system with a pupil. This strategy sacrifices resolution, sensitivity, and SNR, but is effective in increasing the depth of field.

Alternative strategies for increasing the depth of field by PSF engineering have emerged since the early 1980s. In considering these strategies, one must draw a distinction between *lens design* and "wavefront engineering." The art of lens design plays an enormous role in practical imaging systems. A lens typically consists of multiple materials, coatings, and surfaces designed with a goal of obtaining an aberration-free field with an approximately shift-invariant PSF. One may distinguish the lens design, however, from the wavefront that the lens produces on its exit pupil for an incident plane wave. In *diffraction-limited* systems this wavefront is parabolic in phase and uniform in amplitude, as in Eqn. (4.64). In practical systems the pupil function $P(x', y')$ does not reflect the transmittance of any single lens surface; rather, it is the distortion from uniform phase and amplitude on the exit aperture of the lens. The remainder of this section reviews design strategies for $P(x', y')$ aimed at extending the depth of field. We do not consider lens design strategies to produce the target pupil function.

Two pupil design strategies are particularly popular for systems with extended depth of field (EDOF). The first strategy, referred to here as *optical EDOF*, emphasizes optical system design with a goal of jointly minimizing the extent of the PSF and the rate of blur as a function of object range. The second approach, *digital EDOF*, emphasizes codesign of the PSF and computational postprocessing to enable EDOF in digital estimated images. The remainder of this section considers and compares these strategies. Alternative strategies based on multiple aperture and spectral coding are discussed in Sections 10.4 and 10.6.

10.2.1 Optical Extended Depth of Field (EDOF)

Optical EDOF aims to extend depth of field by designing optical beams with large focal range. To this point we have explicitly considered four types of beams:

1. The plane wave
2. The 3D focal response, defined by Eqn. (6.74)

3. Hermite–Gaussian and Lagurre–Gaussian beams, as described in Eqn. (4.39) and Problem 4.2
4. Bessel beams, as described in Problem 4.1

Each type of beam is associated with a depth of focus and a focal concentration. The *depth of focus*, which describes the range over which the image sensor can be displaced while maintaining acceptable focus, is complementary to the *depth of field*, which describes the range over which an object can be displaced while remaining in acceptable focus. Since the transverse distribution of a plane wave does not change on propagation, one might consider that plane waves have infinite depth of focus. On the other hand, since the plane wave does not have a focal spot, one might say that it has zero depth of focus. The Bessel beam, with localized maxima, is more interesting but also fails to localize signal power in a finite spot.

An imaging system transforms light diverging from an object into a focusing beam. In our discussion so far, the object beam has generally consisted of plane waves, and the focusing beam has consisted of the clear aperture diffraction limited Airy beam. One can imagine, however, optical systems that implement transformations between more general beam patterns. Prior to considering such systems, it is useful to consider whether the structure of the focusing beam makes a difference, specifically, *whether it is possible to focus light such that the rate of defocus differs from conventional optical designs*.

Referring to Eqn. (10.1), we can see that the depth of focus for the Airy beam is $\Delta z_i = 4w_{20}(f/\#)^2$. Recalling that the Airy spot size is approximately $\Delta x = 1.2\lambda f/\#$, the relationship between depth of focus and focal spot size is

$$\Delta z_i = \frac{2.78 w_{20} \Delta x^2}{\lambda^2} \tag{10.4}$$

Figure 10.2 shows cross sections of the Airy focal intensity for various focal spot sizes. As expected, the depth of focus grows as the square of the focal spot cross section.

In the case of the Hermite–Gaussian beam, reference to Eqn. (4.40) yields the beam waist as a function of defocus, $w(\Delta z_i) = \Delta x\sqrt{1 + \lambda^2 \Delta z_i^2/\Delta x^4}$. Figure 10.3 shows the fundamental Gaussian beam irradiance distribution as a function of the focal spot width. A tighter focus defocuses more rapidly than a defocused spot. Assuming that defocus corresponds to an increase in the focal spot diameter by a factor of N, the depth of focus for a Gaussian mode is

$$\Delta z_i = 2\sqrt{N^2 - 1}\left(\frac{\Delta x^2}{\lambda}\right) \tag{10.5}$$

In comparing Eqns. (10.4) and (10.5) and Figs. 10.2 and 10.3, one finds that while the depth of focus for the Airy beam is comparable within a constant factor to the depth

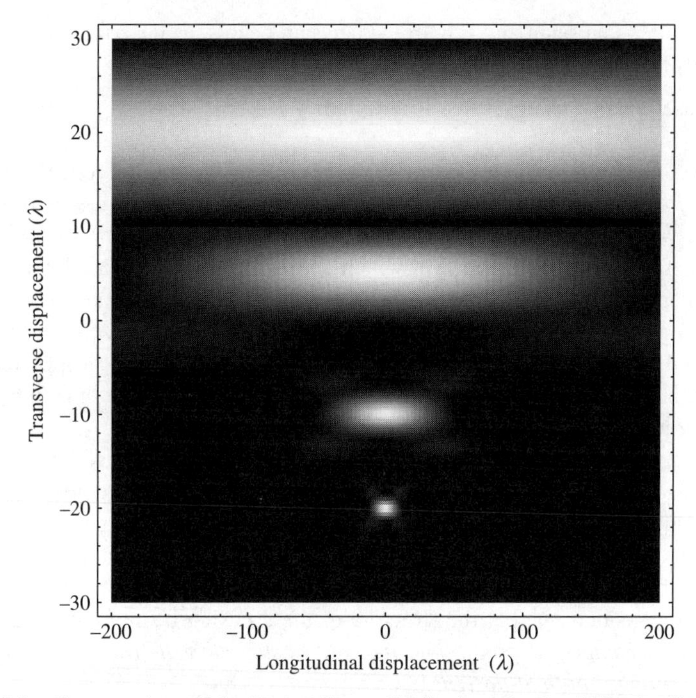

Figure 10.2 Cross sections of the 3D irradiance distributions for the diffraction limited Airy beam with focused beam waists of Δx of 2λ, 4λ, 8λ, and 16λ. The horizontal axis corresponds to the longitudinal focal direction; the vertical axis is transverse to the focal plane.

of focus for the Gaussian beam, the structure and rate of blurring near the focus is substantially different for the two-beam patterns and that the depth of focus for the Airy pattern exceeds the depth of focus for the Gaussian with similar waist size.

An increase in the depth of focus by just a few micrometers can lead to dramatic increases in the depth of field. Given that the Airy beam outperforms the Gaussian beam in certain circumstances, one may reasonably ask whether there exist beams that outperform the Airy beam by a useful factor. Optical EDOF seeks to create such beams by coding $P(x, y)$ to balance depth of focus and resolution. Diverse amplitude and phase modulations of the pupil function have been considered over the long history of optical EDOF. The aperture stop is the simplest amplitude modulation for EDOF; more sophisticated amplitude filters were pioneered by Welford [247], Mino and Okano [178], and Ojeda-Castaneda et al. [189]. As optical fabrication and analysis systems have improved, phase modulation has become increasingly popular. The potential advantage of phase modulation is that it does not sacrifice optical throughput. In practice, of course, one may choose to use both phase and amplitude modulation.

Suppose, as an example, that we wish to extend the depth of field using a radially symmetric phase modulation of the pupil function. With reference to Eqns. (4.66) and (6.24), the incoherent impulse response for a defocused imaging system with

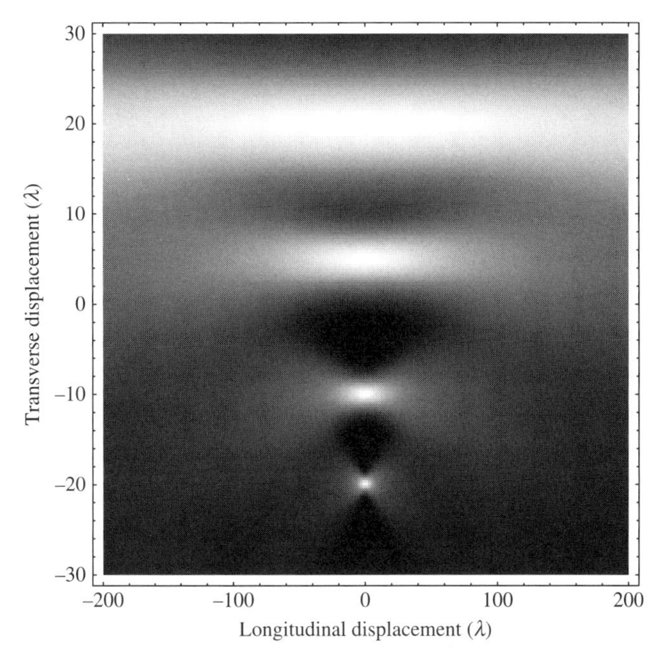

Figure 10.3 Cross sections of the 3D irradiance distributions for fundamental Gaussian beams with focused beam waists of 2λ, 4λ, and 16λ. The horizontal axis corresponds to the longitudinal focal direction; the vertical axis is transverse to the focal plane.

phase modulation $\phi(\rho)$ is

$$\begin{aligned} h_{\theta_z}(\rho, \phi) &= \left| \frac{1}{\lambda^2 z_1} \int_0^{\infty} \int_{-\pi}^{\pi} P(\rho') e^{i\phi(\rho)} e^{i[(\pi\rho'^2 \theta_z)/\lambda]} \right. \\ &\quad \left. \times \exp i2\pi \frac{\rho\rho'}{\lambda d_i} \cos(\phi - \phi') \frac{\rho'}{\sqrt{\rho'^2 + d_i^2}} d\rho' d\phi' \right|^2 \\ &= \left| \frac{2\pi}{\lambda^2 z_1} \int_0^{\infty} P(\rho') e^{i\phi(\rho')} e^{i[(\pi\rho'^2 \theta_z)/\lambda]} \right. \\ &\quad \left. \times \frac{\rho'}{\sqrt{\rho'^2 + d_i^2}} J_0\left(2\pi \frac{\rho\rho'}{\lambda di}\right) d\rho' \right|^2 \end{aligned} \tag{10.6}$$

where we apply the Bessel identity from Problem 4.1 and, consistent with Chi and George [46], we do not approximate the distance term in the denominator of the Fresnel kernel. Assuming that the phase of the defocus and modulation terms are rapidly varying over the aperture, we may evaluate Eqn. (10.6) using the *method of stationary phase* [23], which yields

$$h_{\theta_z}(\rho) = \frac{\rho_o^2}{|\phi''(\rho_o) + (2\pi\theta_z/\lambda)|(\rho_o^2 + d_i^2)} J_0^2\left(2\pi\frac{\rho_o\rho}{\lambda di}\right) \tag{10.7}$$

where ρ_o is the stationary point of the integrand phase corresponding to

$$\phi'(\rho_o) = \frac{-2\pi\theta_z\rho_o}{\lambda} \tag{10.8}$$

and we have neglected nonessential factors.

Various studies have adopted the design goal of making the on-axis PSF invariant with respect to defocus, for example, rendering $h_{\theta_z}(0)$ independent of θ_z. To achieve this goal, we select $\phi(\rho)$ such that $\rho_o^2/|\phi''(\rho_o) + 2\pi\theta_z/\lambda|(\rho_o^2 + d_i^2)$ is independent of θ_z. We use Eqn. (10.7) to eliminate θ_z from this ratio, but since ρ_o varies as a function of θ_z, the ratio must also be invariant with respect to ρ_o to achieve our objective. Selecting

$$\phi''(\rho) - \frac{\phi'(\rho)}{\rho} = \frac{\alpha\rho^2}{(\rho^2 + d_i^2)} \tag{10.9}$$

yields a solution

$$\phi(\rho) = \frac{\alpha(\rho^2 + d_i^2)}{4}\log[\beta(\rho^2 + d_i^2)] \tag{10.10}$$

where α and β are constants. This solution is a variation on the "logarithmic asphere" lens derived by Koronkevitch and Palchikova [139]. Figure 10.4 compares the PSF as a function of defocus for this phase modulation with a conventional diffraction-limited lens. As expected, the phase aberration produces a blurred PSF but is much less sensitive to defocus than the conventional system. The lens in Fig. 10.4(a) is an $f/2$ aperture with $F = 1000\,\lambda$. The defocus varies from $\theta_z = -0.0125/F$ to $\theta_z = 0.0175/F$ in steps of $0.0025/F$ from the bottom curve to the top. The best focus is for the curve starting at 2.5 on the vertical axis. For the lens in Fig. 10.4(b), $\beta = 4 \times 10^{-6}/\lambda^2$, $\beta = 4 \times 10^{-6}/\lambda^2$, and $F = 10^5\lambda$. The phase function of Eqn. (10.10) includes a quadratic modulation such that best focus occurs approximately at 1000λ for these parameters.

A second perspective of the depth of focus of the logarithmic asphere is illustrated in Fig. 10.5, which plots a cross section of the 3D PSF using the design of Chi and George [46]. The lens parameters (in terms of the Chi–George design) are radius $a = 16{,}000\lambda$, $f = 64{,}0000\,\lambda$, and $s_1 = 4 \times 10^7\,\lambda$. The PSF produces non-negligible sidelobes, but considerably greater depth of focus in comparison to Figs. 10.2 and 10.3.

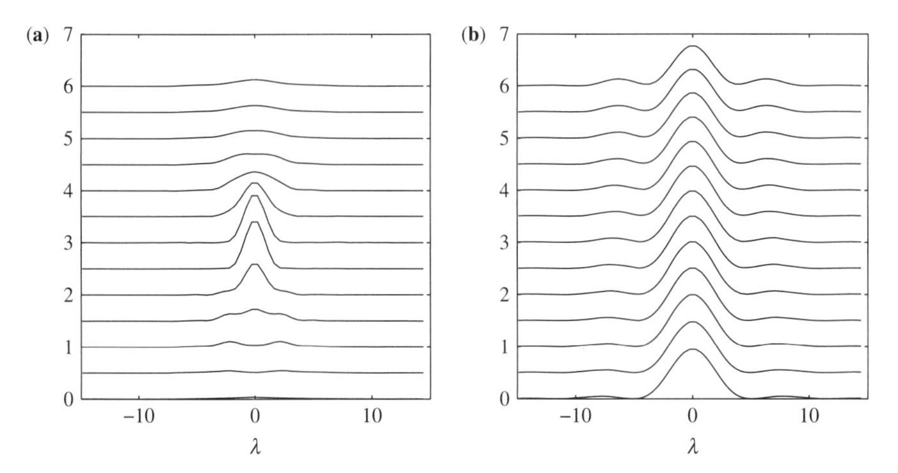

Figure 10.4 PSF versus defocus for (a) a diffraction-limited lens and (b) the logarithmic aspherical lens using the phase modulation of Eqn. (10.10). The range of defocus parameters is the same in (b) as in (a). The PSF was calculated in each case by using the Fresnel kernel and the fast Fourier transform.

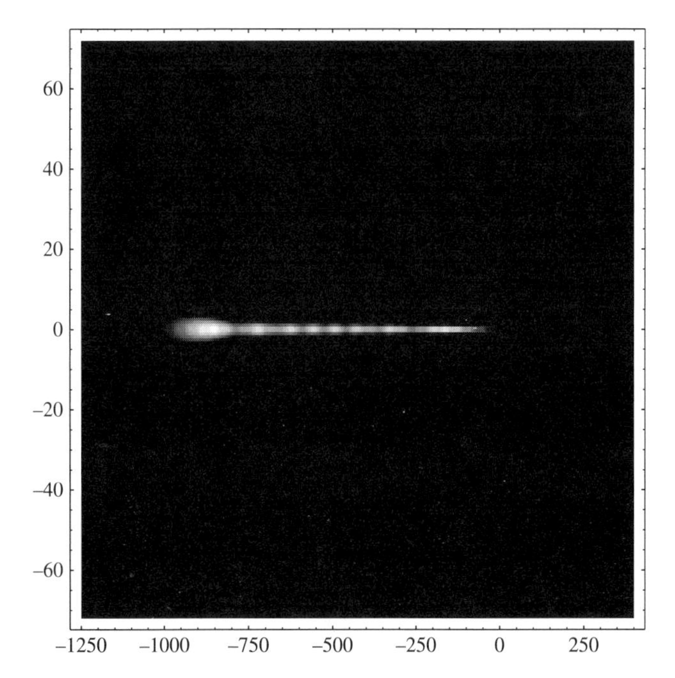

Figure 10.5 Cross sections of the 3D PSF for a point at infinity for a logarithmic aspherical lens. The irradiance was calculated using numerical integration of Eqn. (10.7) by Nan Zheng of Duke University. The horizontal and vertical axes are both in units of λ.

The logarithmic asphere is effectively a lens with a radially varying focal length. One may imagine the aspheric lens as effectively consisting of a parallel set of annular lenses, each with a slightly different focal length. The reduced aperture of the effective lenses produces a blur, but the net effect of all the focal lengths is to extend the depth of field. While the log asphere is an interesting Fourier optics design for this lens, one ought not to consider this solution ideal. In practice, lens design involves optimization over multiple surfaces and thick optical components. One may expect that computational design will yield substantially better results, particularly with regard to off-axis and multispectral performance.

Note that in attempting to keep the on-axis PSF constant, we have not attempted to optimize spatial localization. Serious attempts at optical EDOF must address the general nonlinear optimization problem of localizing the PSF and implementing a 3D lens design. Nonlinear optimization approaches are described, for example, in Refs. 201 and 17. Our discussion to this point, however, should be sufficient to convince the reader that optimization of the pupil transmittance and lens design to balance resolution and depth of field is a rewarding component of system design.

10.2.2 Digital EDOF

While a very early study by Hausler combined PSF shaping with analog processing [123], the first approach to digital EDOF focused on removing the blur induced by a PSF designed for optical EDOF [190]. This approach then evolved into the more radical idea that the defocus PSF should be deliberately designed (e.g., coded) for digital deconvolution [60]. In general, an imaging system maps the 3D object spectral density onto the 2D measurement plane according to

$$g(x, y) = \iiiint S(\theta_x, \theta_y, \theta_z, \lambda) h(\theta_x, \theta_y, \theta_z, \lambda, x, y)\, d\theta_z\, d\theta_x\, d\theta_y\, d\lambda \tag{10.11}$$

In designing an EDOF system, one hopes that $h(\theta_x, \theta_y, \theta_z, \lambda, x, y)$ can be designed such that after digital processing one can estimate the projected image

$$f(\theta_x, \theta_y) = \iint S(\theta_x, \theta_y, \theta_z, \lambda)\, d\theta_z\, d\lambda \tag{10.12}$$

from Eqn. (10.11). With optical EDOF, we have attempted to make a physical system that isomorphically captures $f(\theta_x, \theta_y)$. The goal of digital EDOF, in contrast, is to enable computational estimation of $f(\theta_x, \theta_y)$ from $g(x, y)$. If the processing is based on linear inversion methods, one must assume that all point sources along a ray corresponding to a specific value of θ_x, θ_y produce the same measurement distribution. This is equivalent to assuming that the principal components of the measurement operator (10.11) can be rotated onto the ray projections of Eqn. (10.12). One need not make this assumption with nonlinear inversion methods; we comment briefly on nonlinear digital EDOF at the end of this section.

In general, it is not physically reasonable to expect an imaging system to assign an arbitrary class of radiation to principal components. For example, one could desire a sensor that would produce pattern A from light scattered from any part of "Alice," but produce pattern B from light scattered from any part of "Bob." While the logical distinction between the radiation is clear, in most cases it is not possible to design an optical system that distinguishes A and B light. However, we have previously encountered systems that assign the ray integrals $f(\theta_x, \theta_y)$ to independent components in pinhole and coded aperture imaging [see Eqn. (2.31)] and interferometric imaging [see Eqn. (6.72)].

In the case of the rotational shear interferometer, for example, according to Eqn. (6.46) all sources along the ray (θ_x, θ_y) produce the pattern

$$\frac{1}{2}\left[1+\cos\left(\frac{2\pi\kappa}{\lambda}(y\theta_x + x\theta_y)\right)\right] \tag{10.13}$$

The RSI is thus an existence proof that an optical system can group light radiated from anywhere along a ray onto a common pattern. The disadvantage of the coded aperture and RSI systems is that the system response is everywhere nonnegative and that the support of the response on the measurement space is large. As discussed in Sections 2.5 and 6.3.3, this means that reconstruction SNR is poor for complex objects. The wavefront coding approach of Dowski and Cathey attempts to overcome this problem via a patterned range invariant PSF with more compact support.

Dowski and Cathey [60] propose a "cubic phase" modulation of the pupil function such that the modified pupil function is

$$\tilde{P}(x, y) = e^{i(\alpha/\lambda)x^3} e^{i(\alpha/\lambda)y^3} \operatorname{rect}\left(\frac{x}{A}\right)\operatorname{rect}\left(\frac{y}{A}\right) \tag{10.14}$$

Referring to Eqn. (6.83), the rectangular cubic phase leads to the defocus transfer function $H_{\theta_z}(u, v, \lambda) = H_{r\theta_z}(u, \lambda)H_{r\theta_z}(v, \lambda)$ where

$$\begin{aligned} H_{r\theta_z}(u, \lambda) &= \int e^{i2\pi\theta_z d_i x u} \exp\left[-i\frac{\alpha}{\lambda}\left(x - \frac{\lambda d_i u}{2}\right)^3\right] \exp\left[i\frac{\alpha}{\lambda}\left(x + \frac{\lambda d_i u}{2}\right)^3\right] \\ &\quad \times \operatorname{rect}\left(\frac{2x - \lambda d_i u}{2A}\right)\operatorname{rect}\left(\frac{2x + \lambda d_i u}{2A}\right) d\bar{x} \\ &= e^{i\alpha\frac{\lambda^2 d_i^3 u^3}{4}} \int e^{i2\pi\theta_z d_i x u} e^{i3\alpha d_i u x^2} \operatorname{rect}\left(\frac{2x - \lambda d_i u}{2A}\right)\operatorname{rect}\left(\frac{2x + \lambda d_i u}{2A}\right) d\bar{x} \end{aligned} \tag{10.15}$$

Equation (10.15) can be integrated by again applying the method of stationary phase, which yields

$$H_{r\theta_z}(u,\lambda) \approx \sqrt{\frac{\pi}{12\alpha d_i u}}\, e^{-i(\pi/4)} \times \exp\left(i\alpha\frac{\lambda^2 d_i^3 u^3}{4}\right)\exp\left(-i\frac{\pi^2\theta_z^2 d_i u}{3\alpha}\right) \tag{10.16}$$

for $\theta_z < \alpha A/4\pi$. Figure 10.6 shows the modulation transfer function for a cubic phase distortion for various values of defocus. As expected, the MTF is relatively insensitive to defocus.

With reference to Eqns. (10.14), (6.85), and (4.73), the one-dimensional coherent impulse response for a cubic phase modulation is

$$h_r(\theta_x) = \frac{1}{\lambda d_i}\mathcal{F}\left\{e^{i\pi(\theta_z/\lambda)x^2} e^{i(\alpha/\lambda)x^3}\,\mathrm{rect}\left(\frac{x}{A}\right)\right\}\Big|_{u=(\theta_x/\lambda)} \tag{10.17}$$

Neglecting the effect of the finite aperture and using the identity $\mathcal{F}\{\exp(iax^3)\} = 1/\sqrt[3]{3a}\mathrm{Ai}(-2\pi u/\sqrt[3]{3a})$, where $\mathrm{Ai}(x)$ is the *Airy Ai function*, $h_r(\theta_x)$ reduces to [171]

$$h_r(\theta_x) = \frac{1}{\sqrt[3]{3\alpha\lambda^2 d_i}}\,\mathrm{Ai}\left(-\frac{2\pi\theta_x + \pi^2\theta_z^2/3\alpha}{\sqrt[3]{3\alpha\lambda^2}}\right) \tag{10.18}$$

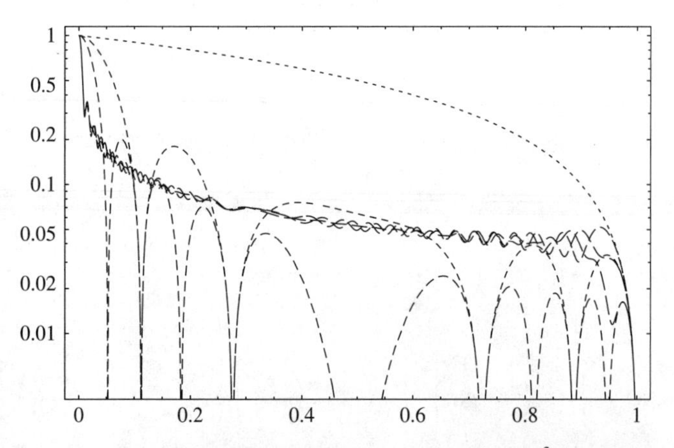

Figure 10.6 MTF for a cubic phase distortion with $\alpha = 750\lambda/A^3$. The top, middle, and bottom plots show the MTF for a conventional diffraction-limited imaging system with $w_{20} = 0\lambda$, 5λ, and 10λ. The three plots grouped across the center of the figure show the MTF for the cubic phase system with the same defocus values. The cubic phase distortion has MTF inferior to that of the well-focused system, but superior to that of the $w_{20} = 5\lambda$ system near the zeros of the conventional MTF and superior to that of the strongly defocused system over most of the passband. The horizontal axis is in units of $A/\lambda d_i$.

Since $h_r(\theta_x)$ includes a range-dependent shift, the cubic phase code does not actually succeed in obtaining a range invariant PSF. However, if α corresponds to N wavelengths of distortion across the aperture, the defocus must reach $w_{20} = \sqrt{\pi \lambda A N}$ for the range-dependent shift to reach one. Clearly, the cubic phase obtains substantial PSF invariance for modest values of α. The impulse response for $\alpha = 10^{-5}\lambda^2$ is presented in Fig. 10.7.

Our analysis of PSF coding has been monochromatic to this point. For optical EDOF the impact of a finite spectral bandwidth is modest except to the extent that materials dispersion may be more significant in aggressive optical designs. For digital EDOF, spectral variation in the PSF, especially spectral scaling of high-frequency features, may substantially degrade deconvolution performance. The

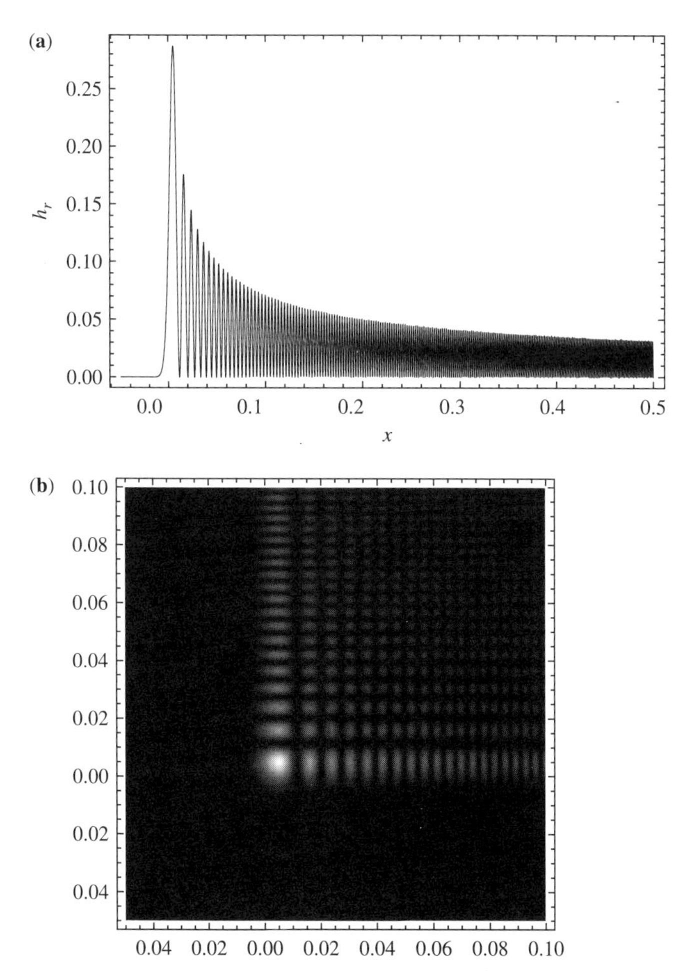

Figure 10.7 (a) Cross section of the cubic phase PSF $h_r(\theta_x)^2$ and (b) density plot of the PSF $h_r(\theta_x)^2\, h_r(\theta_y x)^2$ for $\alpha = 10^{-5}\lambda^2$.

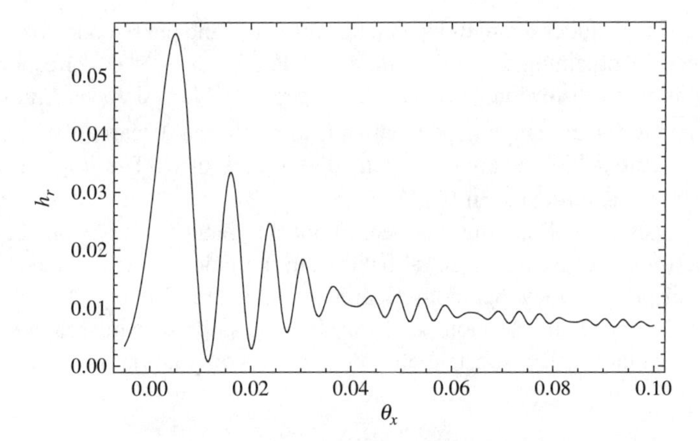

Figure 10.8 Cross section of the polychromatic cubic phase PSF corresponding to Fig. 10.7, $\int h_r(\theta_x)^2 d\lambda$, averaged over one octave of spectral bandwidth.

impact is not as bad as might be expected for the cubic phase pattern since the PSF scales as $\lambda^{2/3}$ rather than linearly in λ, but the effect is nevertheless significant. Figure 10.8 shows the PSF as averaged over a 50% spectral bandwidth. The identifiable structure of the multispectral PSF is more localized [the long oscillation tail of Fig. 10.7(a) is not present in Fig. 10.8], but the multispectral PSF does not average to zero in the tail. The diffuse background scattering from spectral averaging at high frequencies is prejudicial to the system MTF. This effect might be mitigated in color imaging systems where the spectral bandwidth in each channel is reduced. Spectral averaging also has the effect of blurring nulls in the MTF of imaging systems, enabling more accurate deconvolution in some cases.

Figure 10.9 compares the image acquired in an experimental cubic phase camera with the image acquired by a conventional focal camera. The object in this case consists of two targets, the in focus target on the right is 2.5 m from the camera, and the out-of-focus target on the left is 1.75 m away. For the conventional camera, one plane is in focus and one is out of focus. Both targets are blurred by approximately the same PSF for the cubic phase camera. The target was illuminated in this experiment by white incandescent light, so the monochromatic PSF of Eqn. (10.18) is not directly relevant. The broadband PSF was experimentally calibrated and used to construct a slightly shift-variant digital filter [239]. The deconvolved EDOF image is shown in Fig. 10.9(c).

While the highly distributed structure of the cubic phase PSF leads to problems with the magnitude of the MTF, there are some particular advantages to this approach: (1) the MTF has no zeros, which enables effective Wiener filtering for reconstruction across the full system bandwidth—of course, this advantage is less significant if one chooses nonlinear inversion methods; and (2) the PSF is separable in Cartesian coordinates. This enables separable deconvolution and significant reductions in computational complexity. As with other distributed PSF systems (such as the RSI and the coded aperture), the reduced MTF associated with

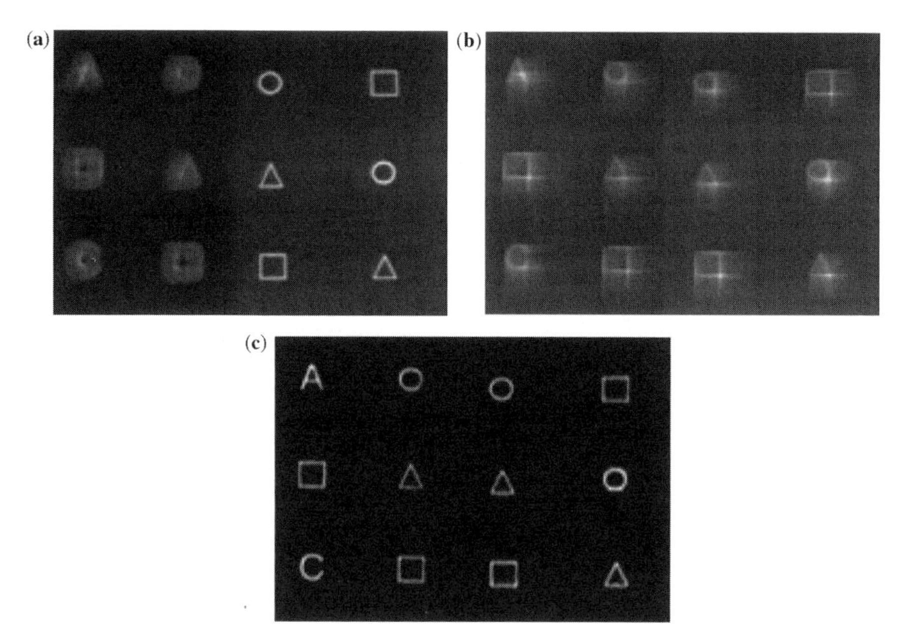

Figure 10.9 Images acquired by (a) a conventional clear aperture imaging system and (b) a cubic phase modulated system with no postprocessing. Panel (c) is the restored image generated by digital deconvolution. The images on the right and left are at the same range for the conventional and cubic phase systems. The right image is at the conventional focus. Both images are recovered by the cubic phase system after deconvolution. (From van der Gracht et al. [239] © 1996 Optical Society of America. Reprinted with permission.)

multiplexing reduces SNR. Since the PSF is not global, however, this loss is less than that for an RSI. As with most multiplexed systems, the cubic phase camera benefits from nonlinear postprocessing.

To summarize discussion thus far, extended depth of field using PSF design is a good idea. None of the PSFs described in this section are ideal, but they do show that PSF design makes a difference in system performance. From this perspective, joint optimization of defocus invariance and digital processing becomes a detailed process of computer-aided lens and materials design, algorithm development, and testing.

Before getting too caught up in design optimization, however, one does well to consider whether one has selected the best design goals. Conventional EDOF design focus simultaneously on three design objectives:

1. The PSF should be range-invariant.
2. The MTF should be broad and flat.
3. The PSF should be well suited to digital deconvolution, meaning that image quality metrics (SNR, MSE, resolution) in the digital processed image should be "good."

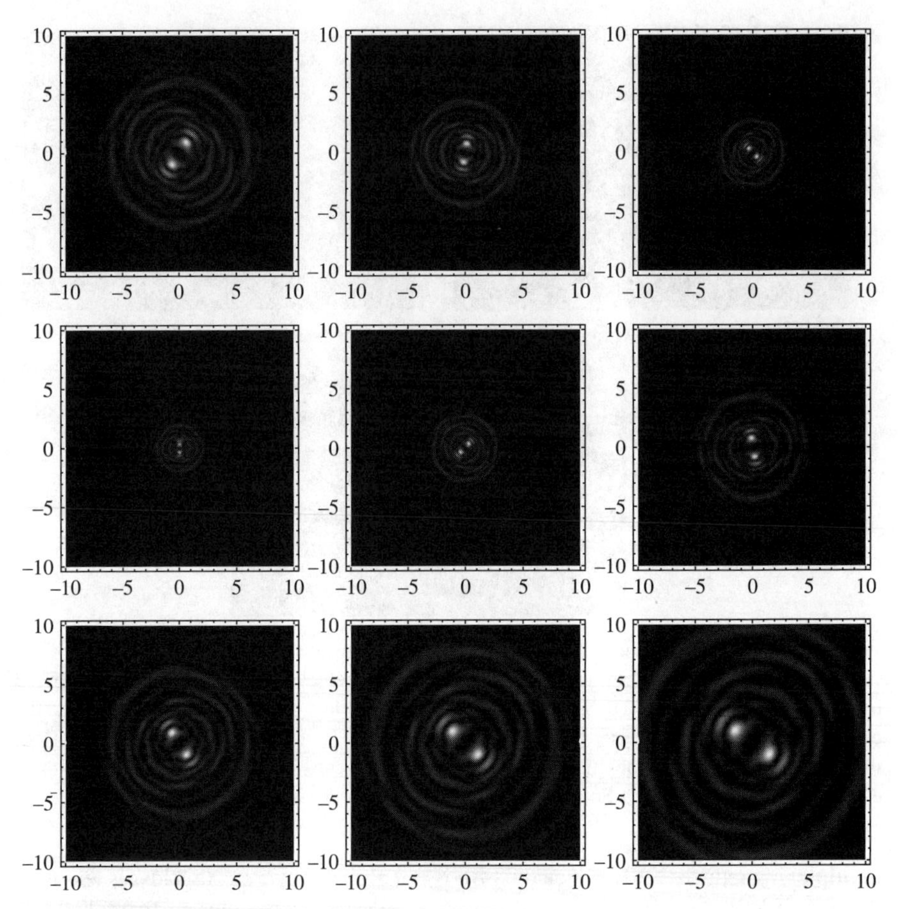

Figure 10.10 Cross sections of the 3D irradiance produced by the field $\Psi(\rho, \phi) = 10\tilde{\psi}_{1,1}/3 + 2\tilde{\psi}_{5,3} + \tilde{\psi}_{9,5}/16 + \tilde{\psi}_{13,7}/312 + \tilde{\psi}_{17,9}/6950$ as a function of defocus. The horizontal axes are in units of w_0. Frames correspond to uniform defocus steps over the range $z = [-3w_0^2/\lambda, 5w_0^2/\lambda]$. $\tilde{\psi}_{mn\tau}$ is described in Eqn. (3.38) and Problem 4.2.

Of course, objective 3 is not particularly precise; research into the meaning of this and other aspects of the problem continues. One may imagine many alternative optimization criteria for the defocus PSF. For example, Sherif and Cathey [222] reverse Cathey's earlier work in attempting to maximize the defocus variance of the PSF to enable passive ranging. Alternatively, one might consider relaxing objective 1 while attempting to maintain objectives 2 and 3. Such a strategy might enable both EDOF and computational ranging.

The challenges under this strategy are to design a lens that maintains MTF over a wide defocus range and to design an image estimation algorithm that effectively combines knowledge of the range-dependent PSF with object priors, such as smoothness or sparsity. We do not attempt to resolve these open challenges here, but we do suggest range-dependent PSFs that could serve as a starting point. Schechner et al. describe a compact range variant PSF based on interference of Laguerre–Gaussian modes [200,214]. Figure 10.10 illustrates an example of a "rotating PSF" produced by a particular example of such a mode. In comparing this PSF with the zeroth-order Gaussian (Fig. 10.11), one observes that while the support of the rotating PSF is larger than the fundamental mode, the rotating version contains more compact features than does the fundamental. Greengard et al. used a similar PSF in an optical ranging system [105]. The PSF was encoded using a computer generated hologram (see Problem 4.13). By deconvolving with the range-dependent PSF, Greengard et al. "digitally focused" the reconstructed image to find both range and the sharpened image.

More generally, a range-variant PSF with higher-frequency defocus MTF than the clear aperture provides a mechanism for inversion of the 3D imaging transformation

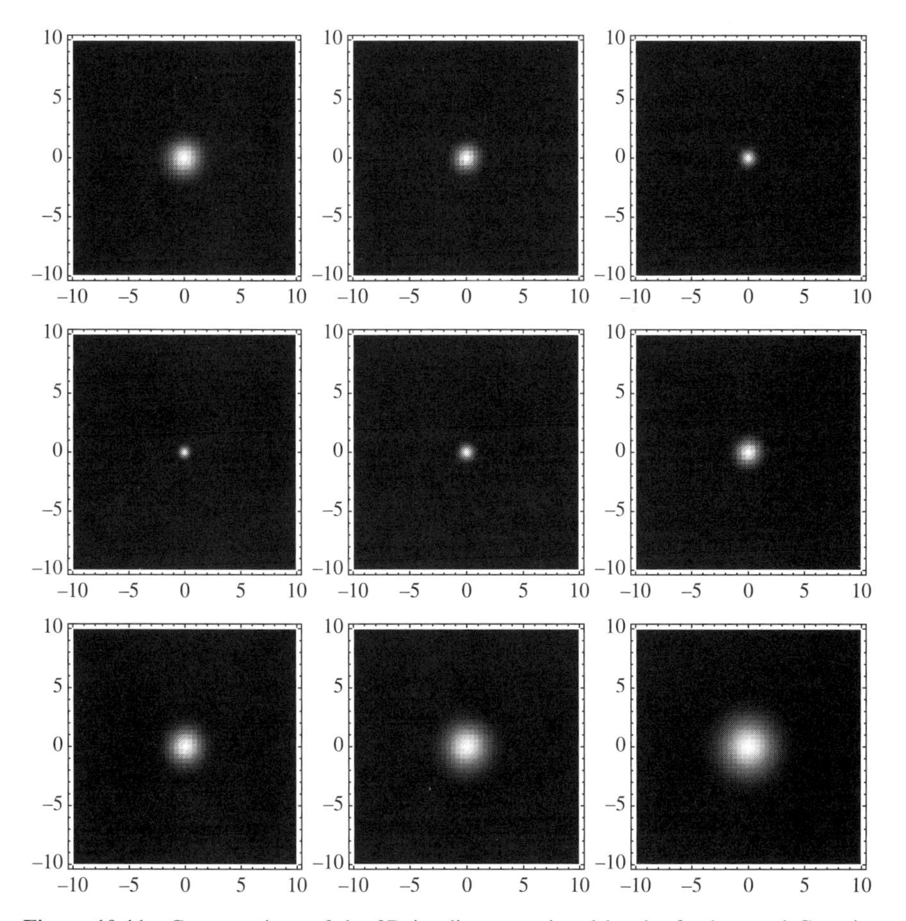

Figure 10.11 Cross sections of the 3D irradiance produced by the fundamental Gaussian mode over the same range of defocus and plotted on the same scale as in Fig. 10.10.

[Eqn. (6.73)]. Of course, the 3D–2D mapping is compressive, but given the 3D PSF, one may attempt inversion using EM algorithms as illustrated in Fig. 7.18 or may utilize algorithms similar to the spectral data cube reconstructions described in Section 10.6.

10.3 RESOLUTION

Imagers and spectrometers are bandlimited measurement systems. One generally assumes that the resolution of such systems is determined by the Fourier uncertainty relationship, meaning that the resolution is inversely proportional to the bandpass. The bandpass is the width of the system transfer function (STF). As discussed in Section 7.1, the STF is determined jointly by the optical transfer function and by electronic sampling and processing. Of course, STF limited resolution is achieved only if the sampling rate is sufficient to avoid aliasing. Aliasing has the effect of both reducing the effective bandpass and introducing noise from aliased frequencies. We discuss the use of multichannel sampling to recover aliased signals in Section 10.4. Antialiasing using multiple apertures or exposures is called *digital superresolution*.

Estimation of images at resolution beyond the Fourier uncertainty limit is called *optical superresolution*. The limits of optical resolution are based on the relationship between aperture size and system bandpass, which is expressed in its most basic form by Eqn. (6.71). We repeat the equation using slightly different variables here

$$W(\Delta x, \Delta y, q, \lambda) = \iiint S(\theta_x, \theta_y, \theta_z, \lambda) \frac{\theta_z^2}{\lambda^2} \times e^{-i(2\pi/\lambda)(\theta_x \Delta x + \theta_y \Delta y + q\theta_z)} d\theta_x \, d\theta_y \, d\theta_z \tag{10.19}$$

where $\theta_x = x/z$ and $\theta_z = 1/z$. The cross-spectral density across an aperture is the 3D Fourier transform of the power spectral density of a remote object. According to this equation, the support over which one samples the Fourier space of the object is proportional to the system entrance aperture. The sampled frequencies along the transverse components are $u = \Delta x/\lambda$ and $v = \Delta y/\lambda$. The longitudinal frequency is $w = q/\lambda$. The bandpass is determined by the limits of Δx, Δy, and q within the aperture and cannot be increased by optical or electronic processing after the field has passed through the aperture. For a circular aperture of diameter A, $|u|_{\max}$, $|v|_{\max} = A/\lambda$, and $|w|_{\max} = A^2/8\lambda$. The band volume covers the disk $\sqrt{u^2 + v^2} \leq A/\lambda$ in the $w = 0$ plane. The extent along w depends on u and v. The structure of the bandpass is discussed in Section 6.3, and the limits of the band volume are sketched in Fig. 6.15.

On the basis of Fourier uncertainty, the bandlimits imply resolution in θ_x and θ_y of approximately λ/A and in θ_z of approximately $8\lambda/A^2$. The corresponding resolutions in object space x, y, z are $\lambda z_o/A$ in the transverse coordinates and $8\lambda z_o^2/A^2$ in the longitudinal coordinate, where z_o is the object range. These values are termed the *diffraction limits* because one need only assume that the Fresnel kernel applies to derive them. The spatial resolution of most imaging systems is worse than the diffraction limit owing to suboptimal sampling and processing of $W(\Delta x, \Delta y, q, \lambda)$.

Optical superresolution is a complex and profound subject with a long history of mixed success and failure. Concomitant with the microprocessor revolution, work since the mid-1990s has demonstrated modest success in computational resolution enhancement, and there are suggestive ideas that future improvements are possible. To date, however, the impact of computational processing is far greater in enabling systems to achieve metrics approaching the diffraction limit over wider fields, with wider depth of focus, and with greater specificity. There are, however, indications that over the long-term systems violating the conventional diffraction limit may be developed. For researchers at the limits of system performance there are many current opportunities for superresolution studies. These opportunities may be grouped into the following categories:

1. Strategies to increase the resolution for signals measured over bandlimited channels. Examples mentioned below include channel coding and estimation algorithms and multispectral encodings.
2. Strategies to increase the bandpass of optical systems. Examples mentioned below include anomalous diffraction and nonlinear detection.

We briefly review these strategies in Sections 10.3.1 and 10.3.2.

10.3.1 Bandlimited Functions Sampled over Finite Support

The relationship between bandpass and resolution is widely accepted in many disciplines and applications. It is the basis of the Heisenberg uncertainty relationship in quantum mechanics, information transmission limits in communication systems, and diverse resolution limits in imaging and measurement theory. Despite its popularity, however, the relationship is not inviolable. The Whittaker–Shannon sampling theorem is a more precise statement of the link between bandwidth and resolution than is the uncertainty relationship. As discussed in Section 3.6, the sampling theorem tells us that the number of samples necessary to characterize a signal restricted to the frequency band $[-B, B]$ and the spatial support $[-X, X]$ is $4BX$. The significance of this statement for the resolution of a continuous signal is not entirely clear. Assuming that these samples are uniformly distributed, the sampling theorem may suggest that the spatial resolution is equal to the sample spacing, $1/(2B)$. As discussed in Section 7.1, however, the sampling theorem also provides a prescription for interpolation between samples, meaning that the effective spatial resolution may be less than the sample period.

We address this paradox by considering more carefully the measurement model

$$g(x) = \int \text{sinc}[2B(x - y)] f(y) dy + n(x) \tag{10.20}$$

with noise $n(x)$ under the assumption that $g(x)$ is measured with arbitrary precision over $[-X, X]$. Our use of the sinc$(2Bx)$ sampling function means that the transformation from $f(y)$ to $g(x)$ is limited to the bandpass $[-B, B]$. In the following analysis we argue that measurement of $g(x)$ with arbitrary spatial precision on $[-X, X]$ is equivalent to the measurement of $c = 4BX$ discrete coefficients in an expansion of $f(x)$ in "prolate spheroidal wavefunctions." c is termed the *Shannon number* or *space–bandwidth product*. We relate this result to three resolution measures:

1. The information capacity for data transfer from $f(x)$ to $g(x)$
2. The maximum spatial frequency $u_{\max}$ that can be reliably estimated in the Fourier transform $\hat{f}(u)$
3. The minimum resolvable separation d between two point objects $f(x) = \delta(x - d/2)$ and $f(x) = \delta(x + d/2)$

It is helpful to emphasize the relationship between Eqn. (10.20) and optical systems. We saw in Eqn. (4.75) that coherent imaging systems are described by a similar model in 2D. Of course, the focal model of incoherent imaging is more complex, but we may regard the structure of the optical transfer function as an artifact of analog processing. According to Eqn. (10.19), incoherent images could be measured with uniform bandpass by direct characterization of $W(\Delta x, \Delta y, q, \lambda)$ over an aperture. In view of these relationships, a sound understanding of optical superresolution is obtained by simply considering Eqn. (10.20).

Our first step in analyzing continuous forward models, such as Eqn. (10.20), has been to transform them into discrete models by expanding $f(x)$ on a basis. As discussed in Section 7.5, the measurement operator defines a linear space V_H. In the case of Eqn. (10.20), this space is V_B. We have seen that V_B is spanned by the Shannon scaling function, such as sinc$(2Bx)$. One may, however, choose different bases for V_B. The prolate spheroidal wavefunctions $\psi_n(x)$ form the basis of greatest interest in analyzing bandlimited systems over finite spatial support. According to Frieden [81], signal processing interest in $\psi_n(x)$ originated with a 1959 visit to Bell Laboratories by C. E. Shannon. Shannon posed the question "What function $\phi(x) \in V_B$ is most concentrated in the interval $[-X, X]$?" Bell Researchers Pollak, Landau, and Slepian used $\psi_n(x)$ to answer this question [143,144,225].

As implied by the name, *prolate spheroidal wavefunctions* originate in the solution of the 3D wave equation in prolate spheroidal coordinates. The 3D solutions are separable in spheroidal coordinates. The functions of interest in signal analysis are the angular components of the separated solution. Our interest in these functions arises from three facts:

1. $\psi_n(x)$ are orthogonal and complete over V_B.
2. $\psi_n(x)$ are eigenfunctions, with eigenvalues λ_n, of Eqn. (10.20).
3. Expansion of $g(x)$ in terms of $\psi_n(x)$ yields an approximately finite series, rather than the infinite series on the Shannon basis.

The prolate spheroidal wavefunction $\psi_n(x)$ is a real function of $x \in \mathbb{R}$ defined by the eigenvalue relation [81]

$$\int_{-X}^{X} \psi_n(x) e^{2\pi i u x} dx = i^n \sqrt{\frac{X\lambda_n}{B}} \psi_n\left(\frac{uX}{B}\right) \tag{10.21}$$

The eigenvalues λ_n are functions of both the order n and the Shannon number. The right-hand constant is selected to simplify the Fourier transform of Eqn. (10.21) over the bandlimit, which produces

$$2B \int_{-X}^{X} \psi_n(x) \operatorname{sinc}[2B(x-y)]\, dx = \lambda_n \psi_n(y) \tag{10.22}$$

The functions $\psi_n(x)$ are orthogonal and complete over $[-X, X]$ with the weighting factor λ_n such that

$$\int_{-X}^{X} \psi_m(x)\psi_n(x)\, dx = \lambda_n \delta_{mn} \tag{10.23}$$

ψ_n are also orthogonal over $[-\infty, \infty]$ with unit weighting; for instance

$$\int_{-\infty}^{\infty} \psi_m(x)\psi_n(x)\, dx = \delta_{mn} \tag{10.24}$$

Taking the ratio of Eqns. (10.23) and (10.24), one finds that the eigenvalue λ_n is a measure of the concentration of $\psi_n(x)$ in the interval $[-X, X]$ in the sense that

$$\lambda_n = \frac{\int_{-X}^{X} |\psi_n(x)|^2 dx}{\int_{-\infty}^{\infty} |\psi_n(x)|^2 dx} \tag{10.25}$$

Since they are real and positive, the eigenvalues may be arranged in descending order such that

$$1 \geq \lambda_0 > \lambda_1 > \lambda_2 > \cdots > 0 \tag{10.26}$$

The ordering of the eigenvalues means that the bandlimited function that achieves maximal concentration on $[-X, X]$ is $\psi_0(x)$. As illustrated in the plots of $\psi_0(x)$ for low Shannon numbers shown in Fig. 10.12, ψ_0 is a unit area spike centered on the origin. The spike becomes increasingly sharp as c increases.

The function $\psi_1(x)$ satisfies three constraints; it is the (1) eigenfunction of Eqn. (10.21) that is (2) orthogonal to $\psi_0(x)$ that is (3) most concentrated on $[-X, X]$.

Then, $\psi_2(x)$ is the function satisfying ψ_1 constraints that is also orthogonal to ψ_1, and so on. As the area available for concentration is occupied for each successive

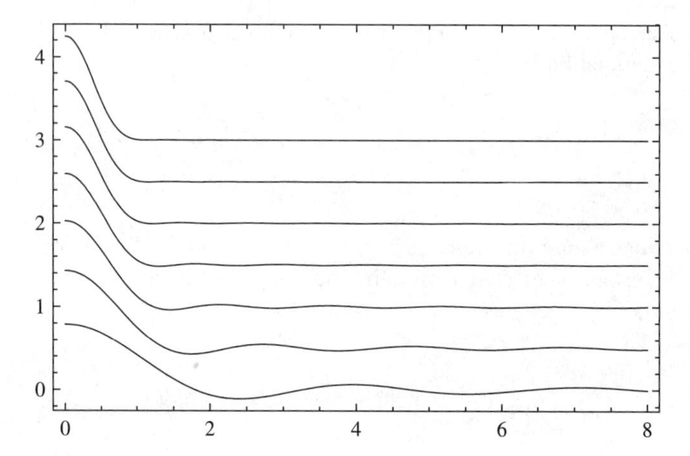

Figure 10.12 Plots of $\psi_0(x)$ for $c = 2$ to $c = 8$. Successive plots are shifted vertically by 0.5. The horizontal axis is in units of X. Only the positive axis is shown; ψ_0 is an even function.

value of n, λ_n must decrease. A series of ψ_n from $n = 0$ to $n = 8$ for $c = 5$ is shown in Fig. 10.13. As n increases, nonvanishing components move away from the origin. For $n > c$, the component of $\psi_n(x)$ within the $[-X, X]$ support flattens and vanishes. This effect is illustrated by the plot of $\psi_{16}(x)$ for $c = 2$ in Fig. 10.14. The wavefunction is very nearly zero over the range $[-X, X]$, corresponding to $\lambda_n = 1.62 \times 10^{-36}$ [81].

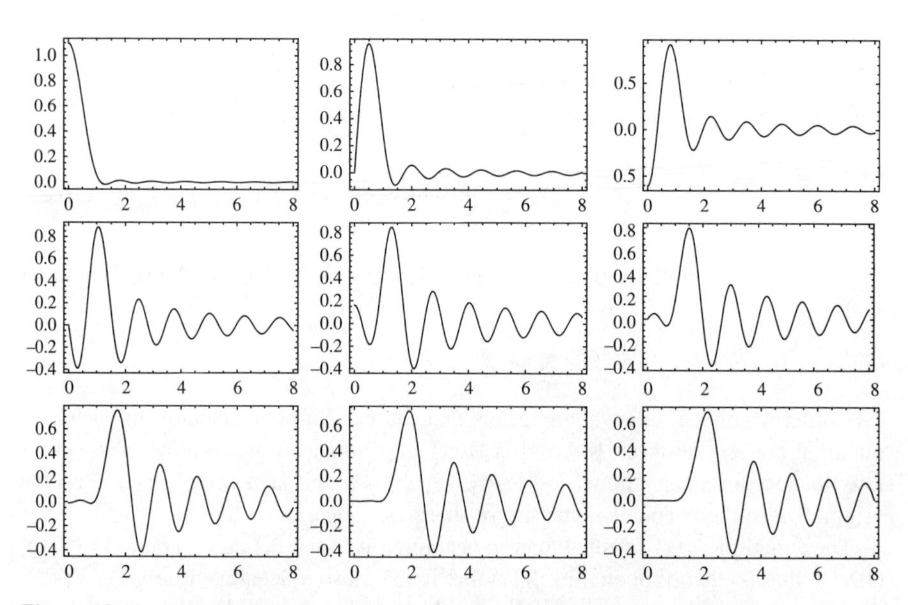

Figure 10.13 Plots of $\psi_n(x)$ for $c = 5$. Values of n are rastered left to right from $n = 0$ to 8. As in Fig. 10.12, we show only the positive axis. ψ_n has odd parity for n odd.

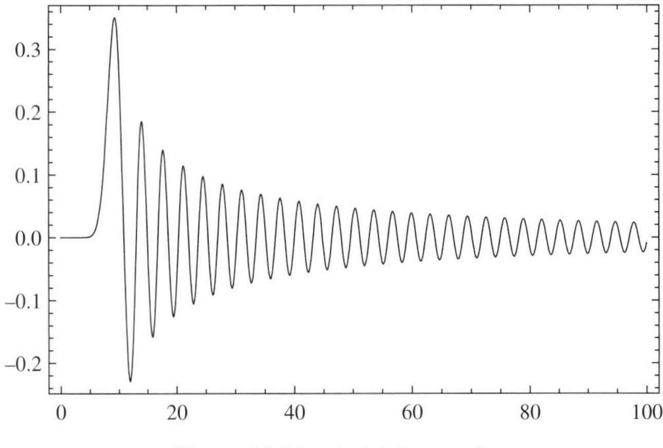

Figure 10.14 $\psi_{16}(x)$ for $c = 2$.

Calculation of λ_n is somewhat more elegant using a normalized form of Eqn. (10.21):

$$\int_{-1}^{1} \psi_n(x) e^{(i\pi cxy)/2} dx = 2i^n \sqrt{\frac{\lambda_n}{c}} \psi_n(y) \tag{10.27}$$

A particularly straightforward strategy for calculation of λ_n integrates Eqn. (10.27) by Gauss–Legendre quadrature [145]. This approach reduces Eqn. (10.27) to the discrete form

$$\sum_l \psi_n(x_l) e^{(i\pi cx_l y_m)/2} w_l = 2i^n \sqrt{\frac{\lambda_n}{c}} \psi_n(y_m) \tag{10.28}$$

where w_l are the Gauss–Legendre weights and x_l and y_m are zeros of the Legendre polynomials as specified by the Gaussian quadrature algorithm. The weights and zeros are accessible in `Mathematica` via the `GaussianQuadratureWeights` function. The prolate spheroidal wavefunction eigenvalues are proportional to the eigenvalues of the homogeneous linear Eqn. (10.28). As illustrated in plots of eigenvalues calculated by this technique for various values of c shown in Fig. 10.15, $\lambda_n \approx 1$ for $n \leq c$ and λ_n approaches 0 very rapidly for $n > c$.

Completeness over V_B means that any bandlimited function can be represented on the $\psi_n(x)$ basis. In particular

$$\text{sinc}(2B(y - x)) = \frac{1}{2B} \sum_{n=0}^{\infty} \psi_n(y) \psi_n(x) \tag{10.29}$$

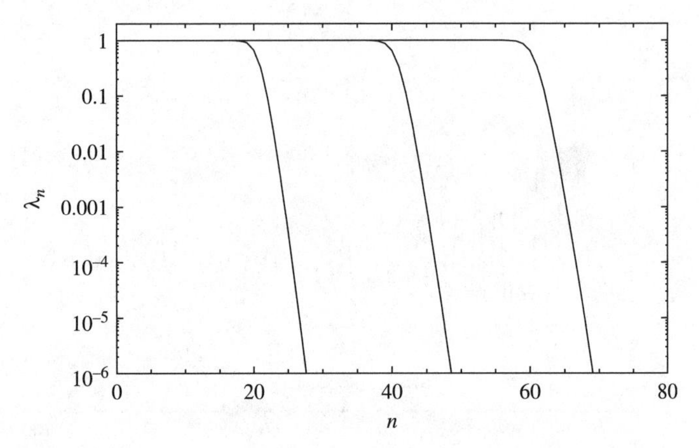

Figure 10.15 Eigenvalues λ_n of $\psi_n(x)$ for $c = 20$, 40, and 60.

for $x, y \in \mathbb{R}$. Given that for $f(x) \in V_B$

$$f(x) = 2B \int f(y)\,\mathrm{sinc}[2B(x - y)]\,dy \tag{10.30}$$

we may substitute Eqn. (10.29) in Eqn. (10.30) to obtain the expansion of $f(x)$ in $\psi_n(x)$

$$f(x) = \sum_{n=0}^{\infty} f_n \psi_n(x) \tag{10.31}$$

where

$$f_n(x) = \int_{-\infty}^{\infty} f(x)\psi_n(x)\,dx \tag{10.32}$$

Alternatively, we may integrate Eqn. (10.31) over $[-X, X]$ using Eqn. (10.23) to obtain

$$f_n(x) = \frac{1}{\lambda_n} \int_{-X}^{X} f(x)\psi_n(x)\,dx \tag{10.33}$$

Having completed a brief tour of $\psi_n(x)$, we are now ready to consider the implications of these beautiful functions for bandlimited measurement. Returning to Eqn. (10.20), we note that the portion of $g(x)$ due to the signal is bandlimited regardless of whether the input signal $f(x)$ is bandlimited. Accordingly, we may project $g(x)$

on to the $\psi_n(x)$ basis by calculating the coefficients

$$g_n = \frac{1}{\lambda_n} \int_{-X}^{X} g(x)\psi_n(x)\,dx$$

$$= f_n + \frac{1}{\lambda_n} \int_{-X}^{X} n(x)\psi_n(x)\,dx$$

$$= f_n + \frac{n_n}{\lambda_n} \tag{10.34}$$

where f_n is the expansion coefficient on the $\psi_n(x)$ basis of the projection of $f(x)$ on V_B. In choosing the finite window of integration to estimate $g(x)$, we recall that the measurement system only measures over $[-X, X]$. According to Eqn. (10.34), the signal-to-noise ratio for estimation of f_n is $\lambda_n f_n/n_n$. If the signal is stronger than the noise and λ_n is 1, then reliable estimation may be expected. For high-order coefficients, however, the signal must be at least $1/\lambda_n$ times stronger than the noise to obtain meaningful data. Since, as illustrated in Fig. 10.15, $\lambda_n \to 0$ for $n > c$, the Shannon number may be rigorously regarded as the maximum number of coefficients that one may extract from a bandlimited measurement.

The Shannon number is often termed the *number of degrees of freedom* of a bandlimited signal. If the value of each degree of freedom is uniformly and independently distributed, then the number of degrees of freedom is a measure of the information in the measurement. As one expects, the number of degrees of freedom is exactly equal to the number of measurements that one would record under Nyquist sampling. Toraldo di Francia analyzes the number of degrees of freedom for various coherent and incoherent imaging systems [232]. In situations where data are uniformly and independently distributed over the image support, one may reasonably argue that the resolution is $2X/c = 1/(2B)$.

Although our analysis of bandlimited sampling has been 1D, extension to multiple dimensions is straightforward. The most interesting difference in multidimensional systems is that the support regions need not be rectangular. $\psi_n(x)$ are termed "linear" prolate spheroidal wavefunctions in this context; circular prolate functions were developed by Slepian shortly after the introduction of the linear functions [224]. The circular functions are well matched to the circular bandlimits associated with lens systems. As in the linear case, the number of degrees of freedom is equal to the space–bandwidth product.

Turning now to more direct links between the degrees of freedom and resolution, we note that analysis in terms of the prolate spheroidal functions is informative in two distinct ways:

1. As we have noted, expansion of $g(x)$ in terms of $\psi_n(x)$ involves c terms.
2. The Fourier spectra of the lowest c order terms $\hat{\psi}_n(u)$ are strongly concentrated in the region $|u| < B$.

Although $\psi_n(x)$ is defined over a finite band, the function is analytic and may be continued over all space. Of course, a bandlimited function cannot have finite support. We have already seen that for $n > c$, most of the signal energy is in the region $|x| > X$. Given that ψ_n is an eigenfunction, the continuation beyond the defining band applies in both real space and Fourier space. One may derive from this continuation an interpolation relationship that reintroduces frequencies beyond the bandlimit and enables arbitrary resolution over the spatial support. However, estimation of frequencies substantially above the bandlimit requires estimation of λ_n for $n > c$.

The fact that $\hat{\psi}_n(u)$ is concentrated within the band $|u| < B$ for $n \leq c$ and within the band $|u| > B$ for $n > c$ tells us that B is generally the greatest frequency that one may estimate in $f(x)$ from measurements $g(x)$. Thus, the prolate spheroidal functions are central to both estimation of the information capacity of the measurement system and to estimation of the limit of postcomputational system transfer function. While the prolate spheroidal analysis greatly clarifies these limits, it also enables one to explore the extent to which estimation of λ_n for n slightly bigger than c and/or extrapolation into the range $|u| > B$ using $\psi_n(u)$ for n slightly less than c might enable superresolution for very high-SNR systems.

Matson and Tyler present a recent and thorough review of "superresolution by data inversion," meaning extrapolation of measured Fourier data to regions outside the measurement bandwidth [175]. They find that for reasonable SNR levels that the inclusion of higher order terms in signal extrapolation may increase the mean bandpass across the support region by a few percent for modest values of c. They also find, however, that the effective bandpass near the edges of the support region may increase by 10–30% of B. The literature on this topic is large, and demonstrable progress is modest.

Our third resolution metric, the separation at which distinct point objects are recognized in an image as distinct, is the oldest and most commonly cited measure. It is called "the Rayleigh criterion" after its originator. The Rayleigh criterion is not directly addressed by prolate spheroidal analysis. The Rayleigh resolution falls into the category of constrained statistical inference problems discussed in Sections 7.5 and 8.5. We are given a constraint that the image consists of one or more point sources and seek to infer parameters, such as the number of point targets and their positions, from the measured data. Since the natural measurement basis (e.g., the prolate functions) and the model basis (sparse spikes) are not strongly correlated, one finds reasonable advantages for nonlinear image inference. Point target images are particularly common in astronomical star field images, and substantial success has resulted from nonlinear resolution enhancement, particularly for systems limited by atmospheric rather than diffractive blur. Most typically, resolution enhancement relies on iterative deconvolution methods [159]. The limits of the Rayleigh criterion are more likely to be decided by statistical decision theory than deconvolution, however. In a review of decision theory-based point target resolution, Shahram and Milanfar argue that point target descrimination an order of magnitude or more below the nominal diffractive resolution limit is achieved with 10–100 dB SNR; scaling as the fourth root of the object separation [217]. A typical plot of minimum detectable point object separation as a function of SNR is shown in

Fig. 10.16. Point target estimation may be regarded as an example of the application of generalized measurement theory. One may similarly imagine that prior constraints to other model bases will yield images that exceed naive resolution limits. One may also imagine that PSF coding might be jointly applied with nonlinear estimation theory to further improve the Rayleigh resolution. Ashok and Neifeld describe the use of coded PSFs for digital superresolution [5]. Similar codings in combination with decision-theoretic estimators could yield optical superresolution, although the need for extremely accurate physical PSF models would likely limit the practicality of such methods.

In the vast majority of imaging applications that do not involve highly constrainted objects and extremely carefully characterized physical systems, the frequency and degree of freedom limits resulting from the prolate spheroidal analysis are hard limits on the image resolution. A little thought leads to an immediate objection, however. Most imaging systems transmit many more than c degrees of freedom; one is allowed c degrees of freedom *per resolvable spectral channel*! The true limit on the degrees of freedom that an imaging system can detect is the product of the space–bandwith product and the time–bandwidth product. If one can imagine a mechanism for sending independent information in diverse temporal and color

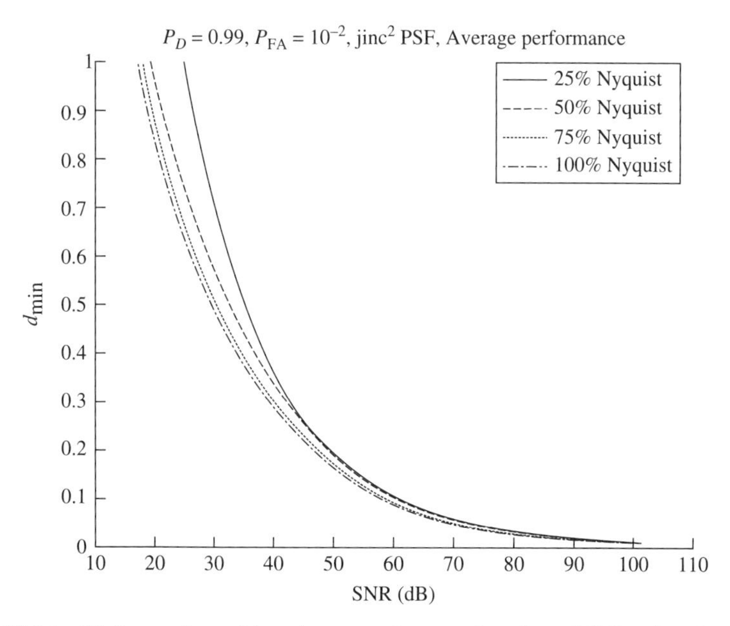

Figure 10.16 Minimum detectable point separation as a function of SNR using the generalized likelihoood ratio test. P_D is the probability that the target is identified as two points; P_{FA} is the false alarm rate. The plots show the performance for the sampling rates indicated averaged over sampling phases. The noise variance is assumed as prior knowledge. The system response is the incoherent Airy PSF, which produces a conventional Rayleigh resolution of 1.22. (From Shahram and Milanfar [217] © 2006 IEEE. Reprinted with permission.)

channels, these degrees of freedom could be used to dramatically improve system resolution.

Most imaging systems are designed either as though all spectral channels are completely correlated (as in black and white imaging) or as though all spectral channels were completely independent (typical color imaging). The real structure of natural images lies between these extremes; images in different color planes are not independent but are also not identical. In most cases, differences between color planes reflect differences in object composition or illumination rather than spatial structure.

Our current focus is on how to use the bounty of spectral degrees of freedom to improve spatial resolution. Polychromatic superresolution strategies are illustrated in Fig. 10.17. Multichannel encoding, as illustrated in Fig. 10.17(a), places optical elements, such as diffraction gratings or microlenses, in the near field of an object. The optical elements encode nonredundant object components on diverse spectrally or temporally modulated channels. As a simple example, one might imagine that the encoding element consists of a microlens array imaging different regions of the object on different color channels. These image channels could then be multiplexed and transmitted over an integrated channel. The basic idea of encoding diverse image features on different spectral channels using optical encoders was first explored by

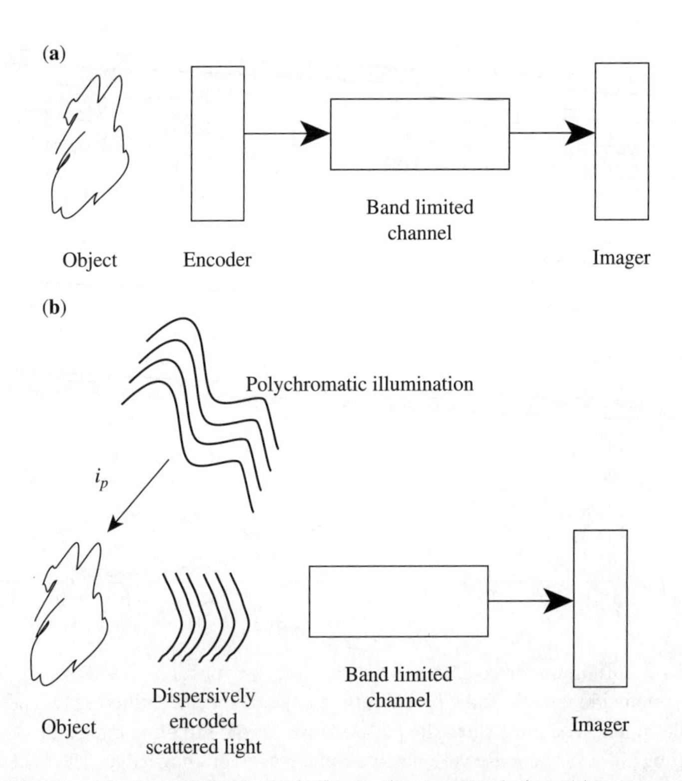

Figure 10.17 Encoding strategies for polychromatic superresolution: (a) superresolution via optical multichannel encoding; (b) superresolution via dispersive scattering.

Lukosz [160]. In an extreme version of this concept, Naulleau and Leith demonstrated image transmission through optical fiber with polychromatic encoding [184]. Of course, the same effect is achieved using optoelectronic sensors to capture images that are then digitally transmitted over fibers. A thorough review of superresolution via strategy (a) is presented by Zalevsky and Mendlovic [260].

Many alternative strategies arise if, as in Fig. 10.17(a), one is allowed to place optical components in the near field of the object. Near-field scanning optical microscopy is the most developed strategy in this class, but diverse schemes in which known objects in the near field are used to encode unknown objects remain unexplored. Rather than develop this approach, however, our focus in the remainder of this subsection turns to the strategy illustrated in Fig. 10.17(b).

Systems based on Fig. 10.17(b) achieve superresolution without placing components on the object side of the bandlimited channel. These systems are based on the modulation of a known illumination signal on scattering from the object. The illumination is encoded with spatial and/or spectral structure. In the case of spatially structured illumination, the spatial pattern changes as a function of time to increase the degrees of freedom beyond the static limit. A simple example of superresolution arises from the illumination of the object with a high-frequency pattern, as illustrated in Fig. 10.18. Limiting our discussion to 1D for simplicity, we consider object distribution $f(x)$ and illumination $t(x)$. We assume that the object irradiance is the product $f(x)t(x)$. The spatial spectrum of the measured object is $\hat{f}(u) * \hat{t}(u)$. Analysis of the measured data is particularly simple if $t(x)$ is periodic with period Λ, in which case $\hat{f}(u) * \hat{t}(u)$ takes the form illustrated in Fig. 10.19. $\hat{f}(u)$ is replicated with period $1/\Lambda$ across the Fourier space. While only the component of $\hat{f} * \hat{t}$ within the passband of the imaging system is observed, this region contains aliased components of $\hat{f}(u)$. From one observation it may be impossible to dealias these signals, but by varying Λ one can obtain sufficient data for unambiguous reconstruction of $f(u)$. The maximum frequency that one can estimate by this method is $B + B_t$, where B_t is the useful bandwidth of $t(x)$ and B is the channel bandwidth. With reference to our previous analysis of the object in angular coordinates the maximum frequency is $u_{\max} = z_o(B + B_t) = A/\lambda + z_o B_t$.

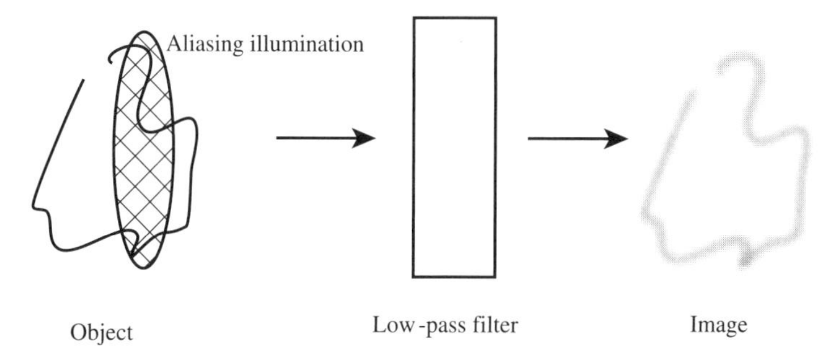

Figure 10.18 Encoding system for superresolution by object plane aliasing.

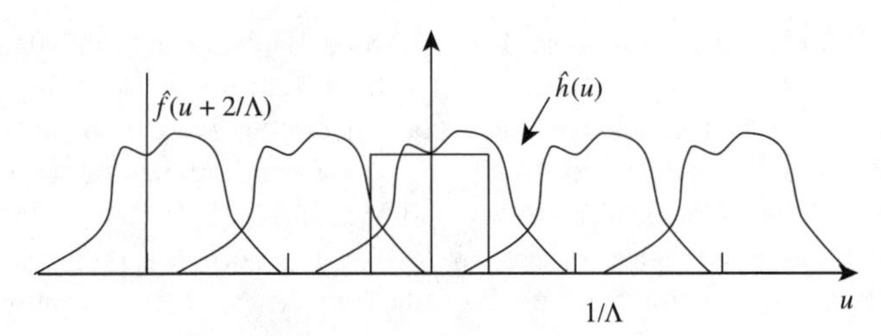

Figure 10.19 Fourier space $\hat{f}(u) * \hat{t}(u)$ for a perioidic aliasing optic.

The challenge with this approach is, of course, how to create the structured illumination. The bandwidth that one can optically project from an aperture is equal to the bandwidth that one measures through the aperture. This means that active illumination from an aperture A can recover object frequencies $u_{\max} = 2A/\lambda$, but to more than double resolution the effective illumination aperture must be larger than the imaging aperture. One may also create effective object space modulation by imaging through resonant devices, such as a Fabry–Perot filter. A resonator placed in between the object and the imaging aperture or in the illumination system superimposes a ring pattern on the captured image. The effective system aperture after dealiasing in this case can equal the resonator aperture, which may exceed the objective entrance aperture.

Particularly dramatic resolution is available in systems that combine nonlinear optical effects with structured illumination, as in stimulated emission depletion (STED) microscopy [116] and nonlinear structured illumination microscopy [107]. Of course, high-resolution illumination presents its own challenges. Where one may reasonably assume the ability to project precise illumination patterns in microscopy, illumination for remote sensing is limited by the same resolution limits as image collection.

Turning to systems using spectrally structured illumination, we note that one can project spectral or coherence patterns at arbitrary ranges independent of aperture size. For simplicity we focus here on passive objects uniformly illuminated by spatially coherent light with known power spectral density. Two types of optical superresolution may be achieved by spectrally dispersive scattering under such illumination

- *Microscopic superresolution* is achieved when sub-wavelength-scale object features are resolved.
- *Remote superresolution* is achieved when subdiffraction limit angular features are resolved.

A primary difference between these two cases lies in the sophistication of the scattering model required for analysis. Subwavelength features cannot be modeled by the multiplicative transmittance functions that have formed the basis of secondary source models in this text.

Basinger et al. consider dispersive encoding of subwavelength features in nondispersive conducting and dielectric materials using rigorous scattering models [9]. By modeling scattering from diverse perfectly conducting and dielectric structures, Basinger is able to characterize polychromatic principal components in the scattered field, abstractly corresponding to the monochromatic eigenfunctions $\psi_n(x)$. As in Fig. 10.15, one can plot the eigenvalues of these scattering modes. Figure 10.20 plots the eigenvalues assuming that one measures one to six frequencies distributed over a 10% bandwidth. The single-frequency curve corresponds to the eigenvalues for $c \approx 10$ for the prolate spheroidal system. As one increases the number of wavelengths observed, the number of degrees of freedom in the scattered light increases. The "theoretical limit" curve plots the eigenvalues that one would obtain if each frequency carried completely independent information, in which case one would replicate the prolate spheroidal eigenvalues 6 times. As illustrated in the figure, while the eigenvalues do not reach the theoretical limit, one polychromatic measurement does obtain a substantial increase in the number of degrees of freedom relative to the monochromatic case. In a subsequent experimental study of dielectric object analysis using polychromatic principal components Basinger et al. characterized object position with $\lambda/10$ resolution [10].

With regard to remote object superresolution, recall that we assumed Schell model illumination in briefly considering objects illuminated by sunlight in Section 6.2 and

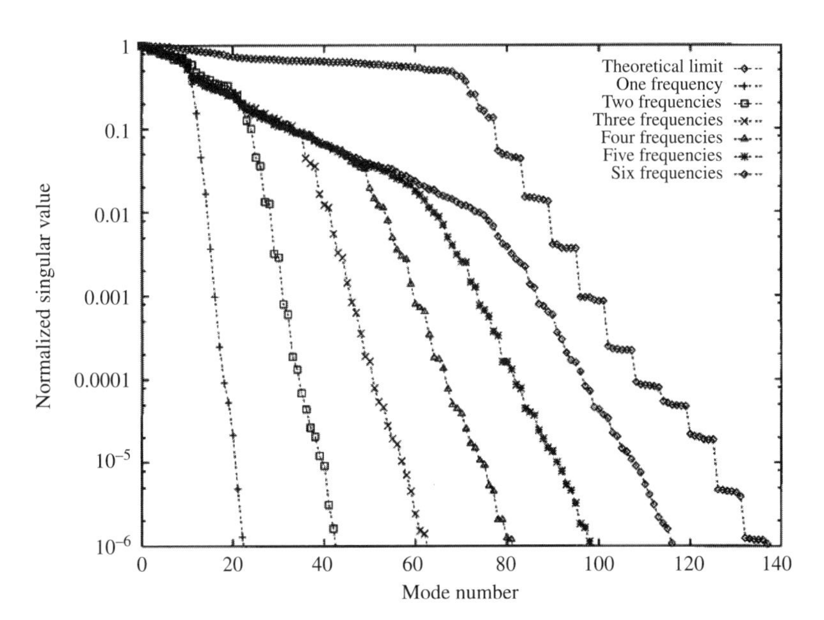

Figure 10.20 Eigenvalues of the polychromatic principal components for scattering from a 1D patterned dielectric. The dielectric is 5 μm long and is illuminated by $\lambda = 1$ μm light, corresponding to an approximate Shannon number of 10. (From Basinger et al. [9] © 1995 Optical Society of America. Reprinted with permission.)

in Problem 6.3. To simplify our current discussion, we assume that the illumination is a spatially coherent plane wave incident along the $\mathbf{i}_p$ direction with spectral density $S(\nu)$. The illuminating cross-spectral density is

$$W(\mathbf{x}_1, \mathbf{x}_2, \nu) = S(\nu)e^{2\pi i(\nu/c)\mathbf{i}_p\cdot(\mathbf{x}_1-\mathbf{x}_2)} \tag{10.35}$$

where $\mathbf{x}_1 \in \mathbb{R}^3$. Generalizing Eqn. (6.25) to a 3D object with scattering density $\sigma(x, y, z)$ and substituting in Eqn. (6.17) using the Fraunhofer impulse response, we find that the cross-spectral density in the far field of the object is

$$\begin{aligned} W(x_1', y_1', x_2', y_2', \nu) = \kappa S(\nu) \int\int & e^{2\pi i(\nu/c)\mathbf{i}_p\cdot(\mathbf{x}_1-\mathbf{x}_2)} \\ & \times \exp\left(2\pi i \nu \frac{x_1x_1' + y_1y_1'}{cz_1}\right)\exp\left(-2\pi i \nu \frac{x_2x_2' + y_2y_2'}{cz_2}\right) \\ & \times \sigma^*(x_1, y_1, z_1)\sigma(x_2, y_2, z_2)\, d\mathbf{x}_1\, d\mathbf{x}_2 \\ = \kappa S(\nu)\Phi^*(x_1', y_1', \nu)\Phi(x_2', y_2', \nu) & \end{aligned} \tag{10.36}$$

where we group nonillustrative factors (including quadratic phase factors) in κ, and we note for the special case of spatially coherent illumination that the far field naturally separates in the coherent mode

$$\Phi(x, y, \nu) = \mathcal{F}(\sigma(x, y, z))|_{u=i_{px}/\lambda+x/\lambda z,\ v=i_{py}/\lambda+y/\lambda z,\ w=i_{pz}/\lambda} \tag{10.37}$$

The coherent modes correspond to the singular vectors of W and may be isolated by standard methods. $\Phi(x, y, \nu)$ corresponds to a particular point in the Fourier space of the object. In practice, variation in x and y is helpful in isolating the phase of $\Phi(x, y, \nu)$, but the range of x/z is much less than the range of i_{px}. As illustrated in Fig. 10.17, i_p is determined by the relative illumination or observation angles. In the example of an object illuminated by sunlight, i_p might change as the Sun transits the sky. While a large range of i_p corresponds to a large effective aperture, under coherent or partially coherent illumination one can isolate $\Phi(x, y, \nu)$ without interference across the full effective aperture. The process of determining $\Phi(x, y, \nu)$ as i_p varies is a form of *synthetic aperture* imaging.

Measurement of $\Phi(x, y, \nu)$ over the full range of i_p at a single wavelength captures the sphere in the object Fourier space corresponding to $u = i_p/\lambda$. Varying λ allows one to fill out the volume of the Fourier space. Figure 10.21 shows that the band volume captured when the illumination angles vary from $\pi/6$ to $\pi/3$ relative to i_z and from $-\pi/6$ to $\pi/6$ in the xy plane. The range of λ is $(\lambda_0, 2\lambda_0)$. With this illumination range, one highpass-filters the object within the band volume illustrated. One often finds that highpass images may be restored by convex optimization and regularization [33].

Measurement of $\Phi(x, y, \nu)$ over a broad range of incidence wavevectors $\mathbf{i}_p$ without focal apertures is the basis of x-ray crystallography. In such systems, one is able

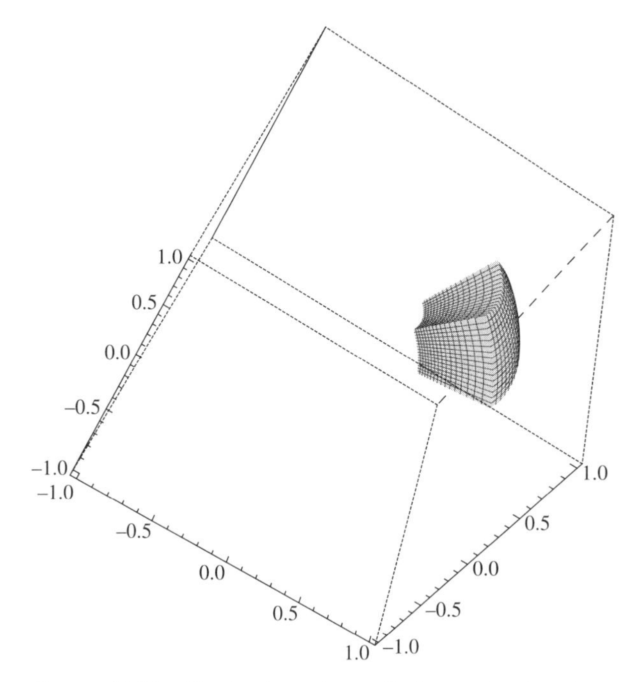

Figure 10.21 Band volume for synthetic aperture imaging.

to achieve wavelength-scale near-field resolution from small-aperture far-field measurements. The ability of focal interferometry to capture the phase of W provides a potential advantage to optical synthetic aperture imaging relative to x-ray systems, which more often must settle for measurement of $|\Phi|$.

In the general case of an object illuminated with partially coherent light, one might choose to analyze the scattered light by finding the independent coherent modes, as discussed in Ref. 172. It is interesting to compare this process with optical coherence tomography; one may consider OCT as a method for measuring $\Phi(\mathbf{x})$ using a reference wave. Synthetic aperture OCT is discussed in Ref. 209. Measurement of the cross-spectral density in Eqn. (10.36) might similarly be considered as a form of self-referencing OCT.

10.3.2 Anomalous Diffraction and Nonlinear Detection

To this point we have accepted Eqn. (10.19) as a valid description of field propagation from an object to a sensor system. There are situations, however, that violate Eqn. (10.19). The fundamental assumptions of Eqn. (10.19) are that (1) the Fresnel transformation describes optical diffraction and (2) objects of interest radiate incoherent power spectra characterized by irradiance sensors. This section considers systems that violate these assumptions.

Field propagation in inhomogeneous media is not described by the Fresnel transform. We are particularly interested in situations where the object field propagates through an intermediate medium prior to the imaging system aperture. Of course, the greatest benefit from an imaging perspective is derived if the intermediate medium happens to be a large-aperture lens, but benefits may be derived from less structured media. Various studies have suggested, for example, that propagation through atmospheric turbulence might be used to increase imaging system bandwidth and enable superresolution [44,90,257,261]. The greatest benefit might be expected in cases where the atmosphere acts as a graded-index material or waveguide, as might be expected for observations across a surface, but some benefit may be expected from observations through random fluctuations.

Figure 10.22 illustrates a simple model of inhomogeneous propagation due to a thin disturbance layer. The effect is similar to modulating the object plane as in Fig. 10.18, although the increase in bandpass decreases as the modulation moves from the object plane to the pupil plane. Diffraction through inhomogeneous or turbulent media is often modeled using a random phase transmittance in an intermediate plane [67,256]. We take an unusual approach here in assuming that the intermediate transmittance is deterministically characterized. Assuming a 1D incoherent imaging system for simplicity, the cross-spectral density at the phase distortion is $W(\Delta x, \nu) = \hat{S}(u = \Delta x/\lambda, \nu)$. The cross-spectral density at the imaging system aperture is then

$$W(x_1, x_2, \nu) = \iint \hat{S}\left(u = \frac{x_1' - x_2'}{\lambda}, \nu\right) t^*(x_1')t(x_2') \times \exp\left(i\pi \frac{(x_1 - x_1')^2 - (x_2 - x_2')^2}{\lambda d}\right) dx_1'\, dx_2' \tag{10.38}$$

where d is the range from the system aperture to the phase distortion. Our immediate goal is to compare the maximum frequency $u_{\max}$ in the Fourier transform of $S(\theta, \nu)$ observable in this system with the maximum frequency observable from Eqn. (10.19). In pursuit of this goal we consider the response to the Fourier

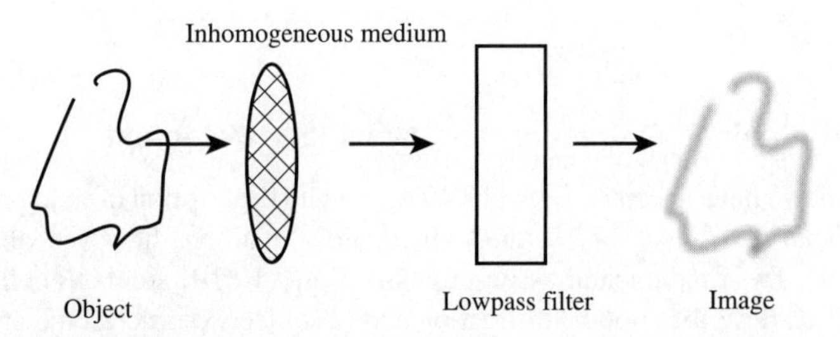

Figure 10.22 Encoding system for superresolution by intermediate turbulence: (a) object; (b) inhomogeneous medium; (c) lowpass fiter; (d) image.

impulse $\hat{S}(u) = \delta(u - u_0)$, which yields

$$W_{u_0}(x_1, x_2, \nu) = \int t^*(x_1')t(x_1' - \lambda u_0) \times \exp\left(i\pi \frac{(x_1 - x_1')^2 - (x_2 - x_2' + \lambda u_0)^2}{\lambda d}\right) dx_1' \quad (10.39)$$

$W_{u_0}(x_1, x_2, \nu)$ is related to the ambiguity function of $t(x)$ [see Eqn. (6.84)]. The ambiguity function is used in radar and sonar imaging to characterize the resolving power of pulses. Design of $t(x)$ to maximize resolution in radar is related to the optical problem addressed here. The student may find it particularly instructive to consider the case $t(x) = \exp(-i\pi x^2/\lambda F)$.

For the case of a homogenous medium $t(x) = 1$ and

$$W_{u_0}(x_1, x_2, \nu) = \delta\left(u_0 - \frac{x_1 - x_2}{\lambda}\right) \quad (10.40)$$

Since $(x_1 - x_2)_{\max} = A$, $u_{\max} = A/\lambda$. As another example, for $t(x) = e^{2\pi i u_t x} t^*(x_1')$ $t(x_1' - \lambda u_0) = e^{2\pi i \lambda u_t u_0}$ and W_{u_0} is again described by Eqn. (10.40). Thus, a single-order diffraction grating in the path does not increase the resolution of an imaging system. If, however, $t(x)$ pans a finite bandwidth B_t, we may estimate the impact on the ambiguity by replacing $t^*(x_1')t(x_1' - \lambda u_0)$ with $e^{2\pi i B_t x_1'}$, which yields

$$W_{u_0}(x_1, x_2, \nu) = \delta\left(u_0 - \frac{x_1 - x_2}{\lambda} - B_t d\right) \quad (10.41)$$

As with the object illumination example, we see therefore that the intermediate modulation can alias signal frequencies in to the detection band, in this case enabling

$$u_{\max} = \frac{A}{\lambda} + B_t d \quad (10.42)$$

In typical remote sensing applications A/λ may range from 10^4 to 10^6. For a modulation layer at a range of 1 km with 1 cm feature sizes $B_t d \approx 10^5$. The potential increase in system resolution must be balanced against the substantial difficulty of determining $t(x)$ from measured data and decreased SNR due to multiplexing noise. As with superresolution by structured illumination, one must assume that $t(x)$ varies as a function of time or wavelength to obtain sufficient data for dealiasing. In turbulent systems $t(x)$ will generally be highly dispersive such that the degeneracy of the spectral channels will be broken. One may expect that multispectral sampling of $W(x_1, x_2, \nu)$ will enable joint estimation of $t(x)$ and superresolved $S(\theta)$. The model of $t(x)$ as a phase screen is unlikely to be satisfactory in highly turbulent environments, however.

We turn finally to the use of nonlinear detectors to increase the system bandwidth. We have seen that irradiance detectors achieve twice the spatial frequency bandpass of Maxwell field detectors. Could detectors sensitive to higher-order field correlations similarly achieve higher bandpass?

The greater bandpass of incoherent imaging systems relative to coherent systems is due to the fact that the object information is contained in the irradiance rather than the fact that detectors detect the irradiance. To understand this claim, consider a simple measurement system

$$g(x) = \int f(x')h(x - x')\, dx' \tag{10.43}$$

Simply squaring measurement samples, for example, measuring

$$|g(x)|^2 = \int\int f^*(x'')f(x')h(x - x')h^*(x - x'')\, dx\, dx'' \tag{10.44}$$

does not increase the bandpass of this system. If, when one takes an ensemble average, Eqn. (10.44) reduces to

$$\langle |g(x)|^2 \rangle = \int \langle |f(x')|^2 \rangle |h(x - x')|^2\, dx' \tag{10.45}$$

then, as we have seen (particularly in Problem 6.8), the system bandpass is increased. The incoherence of the field is a property of the object unrelated to the fact that the detector measures the time average of the square of the field.

A similar effect for higher-order field correlations would require that the irradiance or power spectral density also be δ-correlated. To a first approximation for most objects, however, $\langle S(x_1, y_1, \nu)S(x_2, y_2, \nu)\rangle \approx S(x_1, y_1, \nu)S(x_2, y_2, \nu)$, meaning that detectors that measure higher-order field correlations will not easily obtain higher bandpass. There are objects and illumination strategies where higher-order correlations play a significant role, however, and one might indeed expect to achieve optical superresolution in these systems. The emerging discipline of *quantum imaging* seeks to create and exploit higher order field correlations to obtain optical superresolution [138].

10.4 MULTIAPERTURE IMAGING

Multiaperture imaging systems were of limited utility in the analog age because photochemistry provides no mechanism for fusing two images into one, but the combination of multiaperture imaging and digital image fusion is extremely attractive. Multiaperture imaging is common in living systems, which use multiple apertures to ensure against the failure of a single aperture, to provide parallax for 3D scene analysis and to increase field of view. As an example, binocular imaging enables the extraordinary human visual field of approximately 180° in the horizontal plane and approximately 120° in the vertical plane. Emerging computational imaging systems achieve the performance advantages of biological systems while also using multiple apertures to acquire multispectral, polarization, and multiscale data. Multiaperture imaging also enables novel SNR, dynamic range, system power management, and physical packaging strategies.

As will become clear over the course of this section, there are many open research and development issues in multiaperture systems. In considering the move from cyclops to multiaperture design, one must answer the questions "Why more than one?" and "How many apertures?" Sections 10.4.1–10.4.3 consider the following three answers to "why."

1. *Image Quality and Field of View.* As aperture size grows, one must either decrease the effective $f/\#$ or increase optical system complexity to avoid aberration and maintain image quality. For fixed complexity and system volume, multiple apertures achieve superior effective Shannon numbers and larger field of view.
2. *Generalized Sampling.* Multiple apertures enable multiplex image sampling. Generalized sampling strategies provide design freedom for camera form factor, focal plane pixel layout, and power management.
3. *Multidimensional Sampling.* As discussed in Section 7.3, the optical data cube is multidimensional. Focal planes are unfortunately 2D. Multiaperture systems enable intelligent sampling of three spatial dimensions as well as color, time, polarization, coherence, and turbulence.

"How many" apertures is more subtle than "why." We provide some guidance and limits in subsequent discussion but anticipate that the answer to this question will continue to evolve.

10.4.1 Aperture Scaling and Field of View

Depth of field, resolution, and field of view are the primary metrics of monochromatic imager performance. Nominally, Section 10.2 reviewed depth of field, Section 10.3 reviewed resolution, and the present section covers field of view. The *field of view* (FOV) is the angular range observable to the system. For a spatially incoherent source described by $S(\theta_x, \theta_y, \nu)$, the field of view is the range of angles $\text{FOV} = \theta_{\max} - \theta_{\min}$ over which S is observed. To complete our acronym stew, we note that the angular resolution is sometimes termed the *instantaneous field of view* (ifov). The ifov is the field of view for a single pixel; the FOV is the field of view for the entire image.

As we have already seen, DOF, FOV, and resolution are not independent. While the relationship between depth of field and resolution discussed in Section 10.2 (an aperture stop increases depth of field but cuts resolution) is particularly direct and easy to understand, every aspect of optic design impacts all three metrics.

To understand the limits of these metrics, we briefly return to the cross-spectral density on an aperture. If we could directly measure W as described by Eqn. (10.19), the the depth of field of the system would be infinite (the image at any range can be calculated from W); the resolution is determined by the aperture size according to $\Delta\theta = \lambda/A$, and the field of view is determined by the sampling rate in the Δx, Δy space. Specifically, if one samples W to spatial resolution d, then the field of view is approximately λ/d. From this analysis, there is no fundamental relationship between depth of field, resolution, and field of view.

In practice, however, optical system performance is limited by both aperture size and the capabilities of optical and optoelectronic processing to condition and extract data from the field passing through the aperture. For example, optical processing to maximize resolution by focusing to the tightest possible PSF naturally leads to poor depth of field. Field of view is ultimately reduced in high-resolution systems because it is easier to optimize the PSF for resolution over a smaller field. Section 10.2 considered optical design strategies to increase depth of field by advanced optical and digital processing. Similarly, one may imagine codesign of lens systems, sampling, and processing strategies to balance increased field of view against loss of resolution. Currently, however, digital image synthesis from multiple apertures is the most promising approach to wide-field design.

The basic challenge of FOV and resolution in lens design is illustrated in Fig.10.23, which is a ZEMAX ray trace of a simple lens. The lens consists of borosilicate crown glass with spherical surfaces of 1 cm radius. The lens is aligned to produce a sharp focus at the center of the field. The *spot diagram* shows the locus of ray crossings in the focal plane for a given field angle (see Problem 2.7). Ray tracing programs use the spot size as a measure of the focal quality of the lens. The spot size is a good measure of the focal spot area unless it is smaller than $\lambda f/\#$, in which case diffractive analysis is necessary. As indicated by Fig. 10.23, the spot size for this lens is far from diffraction-limited even at the center of the field, and the spot size rapidly degrades as field angle increases.

Sophisticated lens designs are necessary to achieve diffraction limited resolution across a reasonable field of view. As an example, Fig. 10.24 shows ray tracing and spot diagrams for a lithographic lens assembly. Lithographic systems uniquely require very high-resolution imaging over a very large field. As discussed in Section 10.3.1, the number of degrees of freedom of a bandlimited system is $c = 4BX$. The Shannon number may also be interpreted as the ratio of field of view and angular resolution. Since the resolution is inversely proportional to aperture size and since there is no fundamental relationship between field of view and aperture size, one naturally hopes that the Shannon number will grow linearly in the aperture. This does not happen in practice, however, because the field of view for a given lens design must be reduced as aperture size increases.

Simple analysis of the ray tracing diagrams in Figs. 10.23 and 10.24 explains why this is the case. Ray tracing is "scale-invariant." The diagrams in these figures were sketched in units of millimeters, but if one changes the units to centimeters or meters, the diagram is unchanged. Only the scale of the spot diagrams must be revised. The 77-μm center spot for the simple lens becomes a 77-mm spot when ray tracing units are changed to meters, and the 255-μm spot becomes 0.25 m. Similarly, the 0.3-μm spot for the lithographic system becomes a 0.3-mm spot, much larger than the

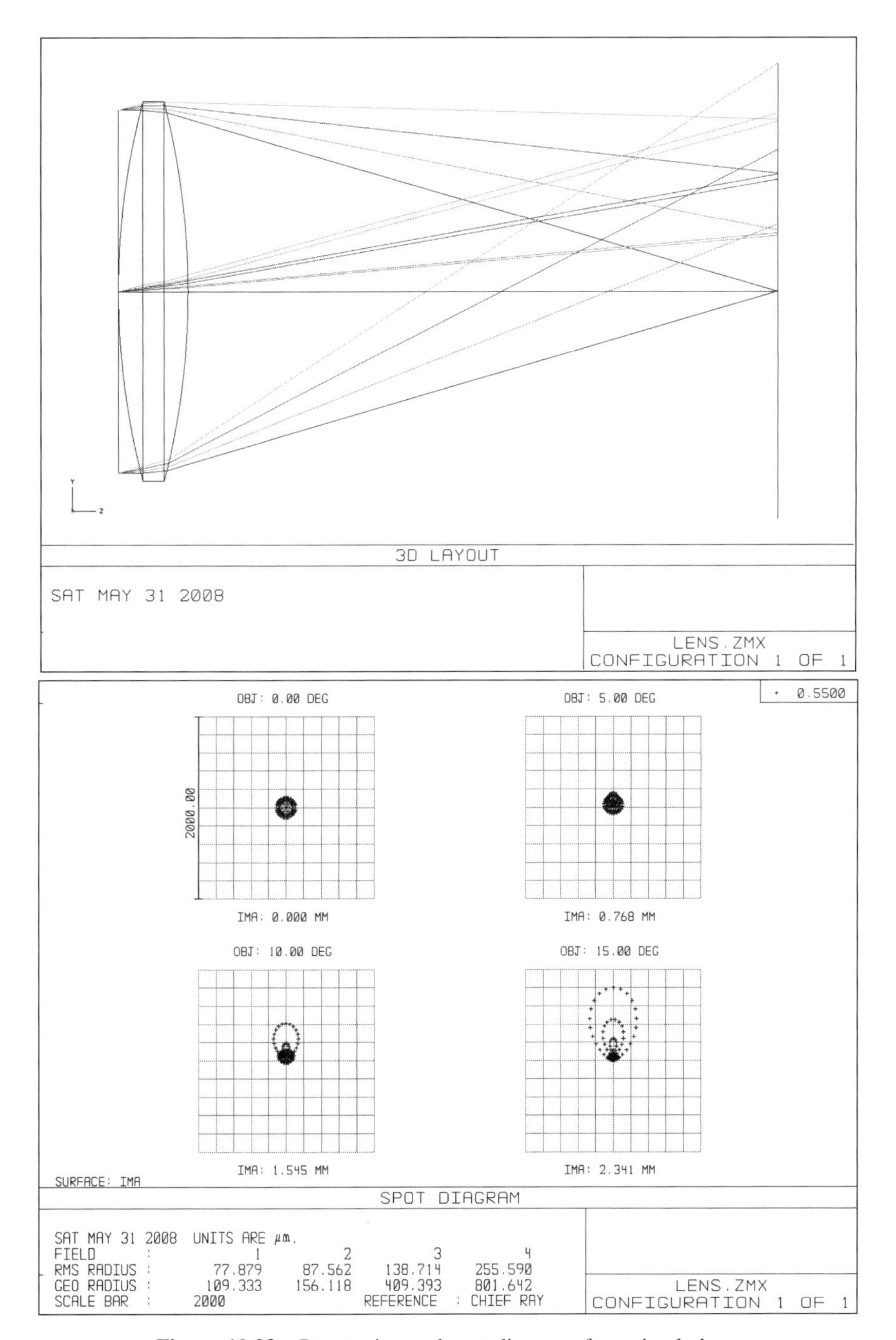

Figure 10.23 Ray tracing and spot diagrams for a simple lens.

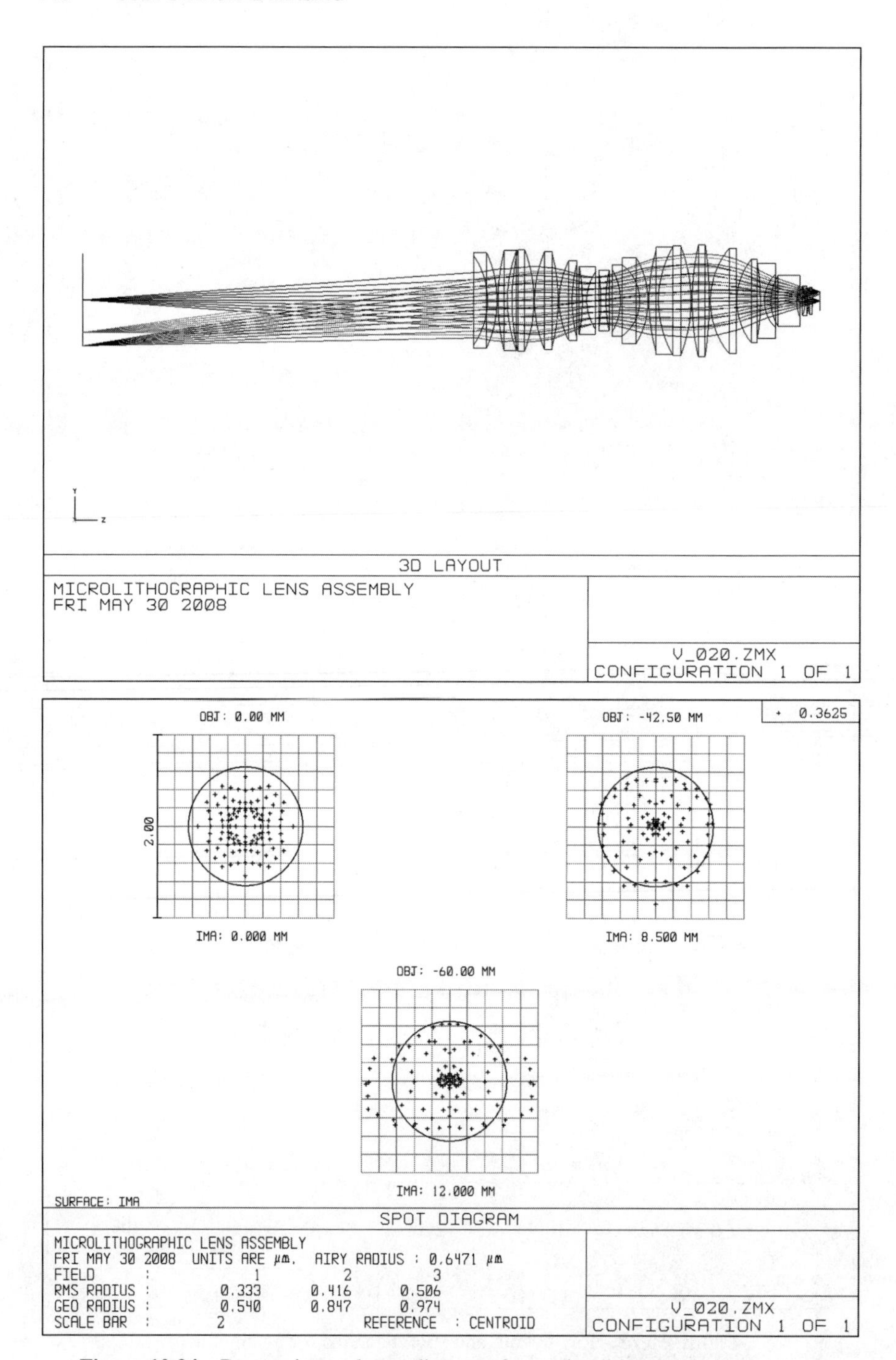

Figure 10.24 Ray tracing and spot diagrams for a microlithographic lens system.

diffraction limit. To counter this scaling problem, one must either increase lens complexity or reduce the field of view when scaling to larger optical systems. In considering these scaling issues, Lohmann presents an empirical observation stating that the effective $f/\#$ tends to increase as $f^{1/3}$, where f is the focal length in millimeters [157]. Under this rule, reasonable design for an $f/1$ system at 1 mm is reduced to $f/3$ at 1 cm aperture and $f/10$ at 1 m aperture.

The alternate strategy of increasing system complexity is adopted to the maximum extent in microlithographic systems. The development of microlithography is driven by *Moore's law*, which predicts that the number of transistors per integrated circuit should double every 2 years. Moore's law has been satisfied over many years by increasing circuit area and by decreasing lithographic feature sizes. Figure 10.25 plots the normalized information capacity of state-of-the-art lithographic lens systems since the early 1980s. The *normalized information capacity* is a measure of the degrees of freedom encoded by the system [176]. The upper curve is Moore's law, doubling system performance every 2 years; the lower curves show the information capacity actually achieved. TTI_{wave} is the total transmitted information relative to the 1980 baseline based on optical improvements alone. TTI_{k1} is the improvement incorporating nonlinear processing strategies in photoresist and exposure. TTI_{tool} is the improvement incorporating system factors in the integrated lithographic tool. The break between the tool and $k1$ curves is due to the implementation of spatially scanned exposure strategies in 1995, effectively corresponding to the introduction of multiaperture lithography. Factors driving improvements in lithographic tools are illustrated in Fig. 10.26. Reduction in the wavelength from >500 nm to <200 nm is of obvious benefit to the Shannon number, as are advances in $k1$. After substantial initial improvements the exposure field size has remained constant over many years, suggesting that a technological or economic limit may be reached in current systems. H is the overall effective etendue, including aperture translation. The dramatic improvement in H since 1995 is due primarily to increased numerical aperture (NA) and to aperture translation. The increased NA corresponds to an extraordinarily high effective FOV.

The optical system improvements illustrated by Figs. 10.25 and 10.26 were enabled by heroic optical design efforts. As illustrated in Fig. 10.27, the optical layout of microlithographic systems from 1980 through 2004 involved massive increases in lens size and complexity. Through these systems, one finds that it is possible to maintain Shannon number as system aperture grows, but only at the expense of substantial increases in volume and manufacturing complexity. Surprisingly little is known regarding the fundamental limits of the relationship between Shannon number and system aperture, however.

In summary, one observes that the number of degrees of freedom predicted as a function of aperture size in Section 10.3.1 is actually achieved for only very small aperture sizes. It is not uncommon for a microscope objective with a submillimeter aperture to have an $f/\#$ less than one and to achieve a Shannon number exceeding 100 in each dimension; 1-mm-aperture systems with $f/\#$ around 1 are reasonable. As aperture size increases, however, $f/\#$ drops and the Shannon number increases sublinearly in A (see Problem 10.8).

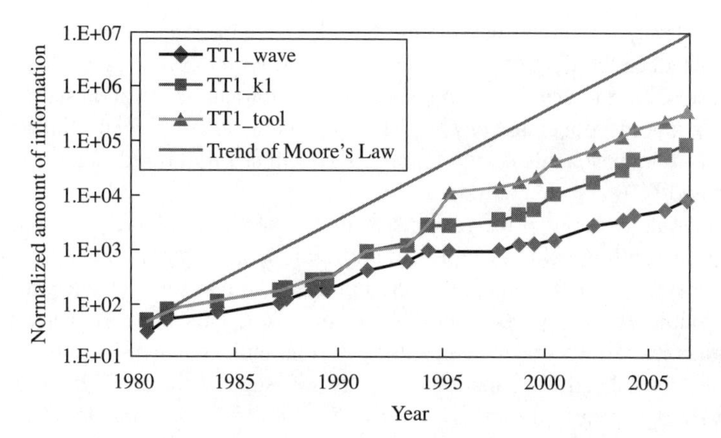

Figure 10.25 Information capacity of lithographic lens systems as a function of time. (From Matsuyama et al. [176] © 2006 SPIE. Reprinted with permission.)

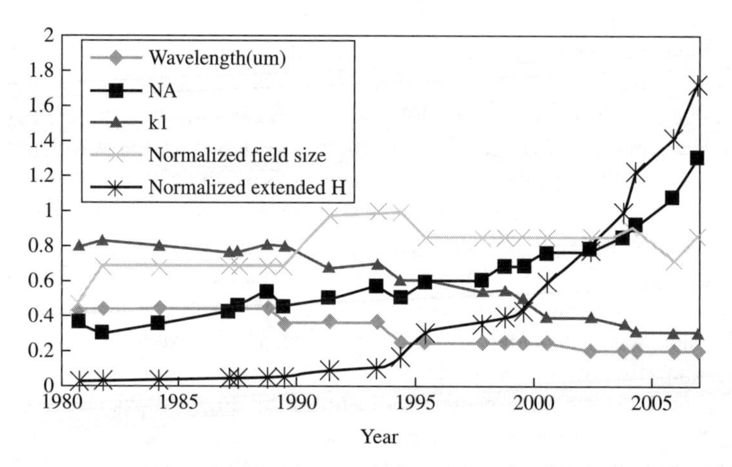

Figure 10.26 Evolution of factors determining the "extended" etendue of microlithographic lens systems. (From Matsuyama et al. [176] © 2006 SPIE. Reprinted with permission.)

Detailed analysis of the origin of Lohmann's scaling law and the limits of lens performance versus aperture size would take us much further into lens design and aberration theory. As we are nearing the end of this text, we leave that analysis to future work. We note, however, that one may increase the degrees of freedom of an imaging system by adding more apertures until the field of view is fully covered. Typical design selects aperture size to achieve resolution targets. Single-aperture field of view is determined by the capabilities of reasonable optics and the size of available focal planes. Additional apertures are added to fill the targeted field of view.

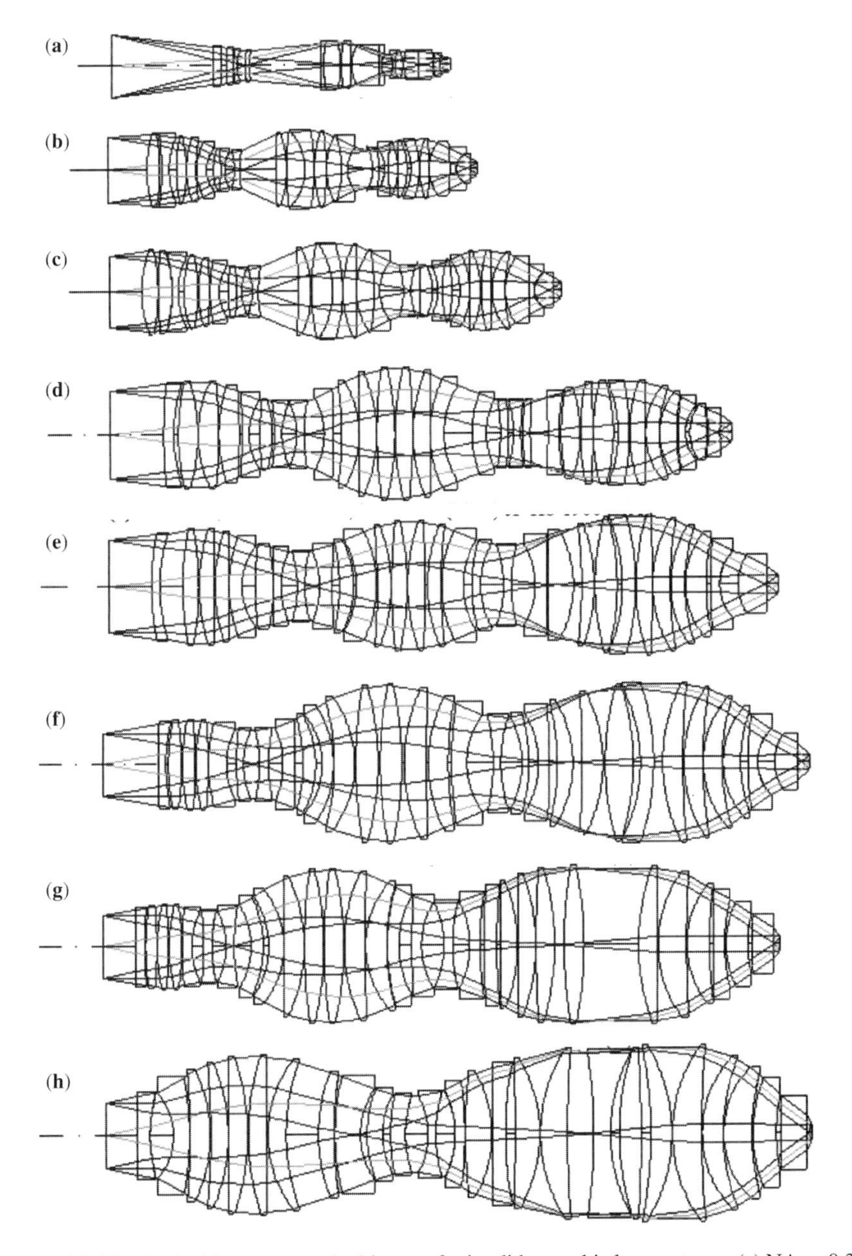

Figure 10.27 Optical layouts over the history of microlithographic lens systems: (a) NA = 0.3, $y_{i,\max} = 10.6$ mm, $\lambda = 436$ nm (g-line); (b) NA = 0.54, $y_{i,\max} = 10.6$ mm, $\lambda = 436$ nm (g-line); (c) NA = 0.54, $y_{i,\max} = 12.4$ mm, $\lambda = 365$ nm (i-line); (d) NA = 0.57, $y_{i,\max} = 15.6$ mm, $\lambda =$ 365 nm (i-line); JP-H8-190047(A); (e) NA = 0.55, $y_{i,\max} = 15.6$ mm, $\lambda = 248$ nm (KrF) JP-2000-56218(A); (f) NA = 0.68, $y_{i,\max} = 13.2$ mm, $\lambda = 248$ nm (KrF) JP-2000-121933(A); (g) NA = 0.75, $y_{i,\max} = 13.2$ mm, $\lambda = 248$ nm (KrF) JP-2000-231058(A); (h) NA = 0.85, $y_{i,\max} =$ 13.8 mm, $\lambda = 193$ nm (ArF) JP-2004-252119(A). (From Matsuyama et al. [176] © 2006 SPIE. Reprinted with permission.)

10.4.2 Digital Superresolution

Multiaperture systems implement generalized sampling when the same object element is observed through more than one aperture. Numerous studies have focused on 3D imaging and resolution enhancement using multiaperture data. The concept of fusing multiple images to increase resolution emerged from diverse sources since the early 1980s. Park et al. present a relatively recent review of digital superresolution [195]. Most historical interest has focused on "images of opportunity" collected as a sequence of video frames from a single aperture. With the introduction of the "thin observation module by bound optics" (TOMBO) microlens array imaging sytem in 2001 [229], however, interest has increasingly focused on computational imagers deliberately designed for multiaperture processing. Of course, biology already had a long history of multiaperture processing, and several previous studies had fabricated compound optical systems based on biological analogies [188,213]. TOMBO-style systems have been implemented by several groups [64,133,179,218].

The original TOMBO system, consisting of an array of microlenses integrated on a single focal plane array, is shown in Fig. 10.28. As illustrated in Fig. 10.28(b) all the

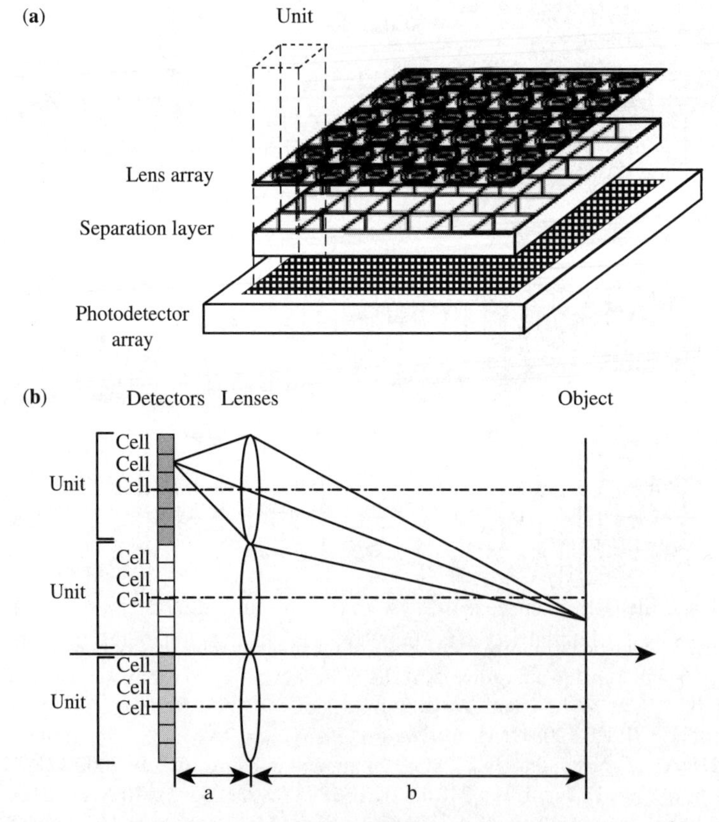

Figure 10.28 TOMBO architecture: (a) system structure; (b) ray tracing. (From Tanida et al. [229] © 2001 Optical Society of America. Reprinted with permission.)

subimagers observe the same object. In practice, parallax leads to variation in the relative object position on the subimagers as a function of range. Compensation of this effect requires scene-dependent registration. Parallax may be neglected for distant objects, however, in which case each camera samples the same image. For distant objects the TOMBO sampling model comparable to Eqn. (7.4) is

$$g_{nmk} = \int_{-\infty}^{\infty}\int_{-\infty}^{\infty}\int_{-\frac{X}{2}}^{\frac{X}{2}}\int_{-\frac{Y}{2}}^{\frac{Y}{2}} f(x, y) h_k(x' - x, y' - y) \times p(x' - n\Delta, y' - m\Delta)\, dx'\, dy'\, dx\, dy \tag{10.46}$$

where $h_k(x, y)$ is indexed by aperture number k and we anticipate variations in the optical PSF from one aperture to the next. In the simplest case, the only difference in the PSF from one subaperture to the next is a shift in the sampling phase, such as $h_k(x, y) = h(x - \Delta_{xk}, y - \Delta_{yk})$. As discussed in Section 7.1, however, shifts in sampling phase do not affect the overall system transfer function (STF), which in this case is $\hat{h}(u, v)\hat{p}(u, v)$. What advantage, then, is obtained by the use of multiple apertures? The answer, of course, is that a diversity of sampling phases changes the aliasing limit and the multitude of apertures increases sensitivity. Prior to considering these points, however, we consider the STF in more detail.

System magnification is a central parameter in multiaperture design analysis. Systems that choose to have a large number of short focal length apertures, like TOMBO, will observe the object with low magnification where a single-aperture system observing the same object will have higher magnification. The effect of the different magnifications and multiple apertures is accounted for by modeling the STF as

$$\mathrm{STF}(u, v) = KM^2 \hat{h}\left(\frac{u}{M}, \frac{v}{M}\right)\hat{p}\left(\frac{u}{M}, \frac{v}{M}\right) \tag{10.47}$$

where M is the relative system magnification and K is the number of apertures. We assume that aperture size is proportional to M, in which case etendue will grow as M^2.

Figure 10.29 plots STF(u, v) for this model for $M = 1$ and for $M = 0.25$, $K = 16$ under the assumption that the focal plane pixel pitch is $\Delta = 4\lambda f/\#$. As discussed in Section 7.1, this pitch undersamples by a factor of 8 relative to Nyquist. We assume for the moment that $f/\#$ is independent of M. The topmost curve in Fig. 10.29(a) and (b) is the optical modulation transfer function, the middle curve is the pixel transfer function, and the bottom curve is the system transfer function (the product of the MTF and PTF). The horizontal (u) axis plots the STF in the unit magnification Fourier space.

With our assumption that $KM^2 = 1$, the systems in Fig. 10.29 have the same light collection efficiency and identical STFs at low frequency. Because of multiplexing noise, however, the STF of the multiaperture system in Fig. 10.29(b) degrades faster than the isomorphic STF in Fig. 10.29(a). Figure 10.29 plots the "excess noise factor" (e.g., the ratio of the multiaperture and single-aperture MSE) based on the Wiener filter MSE described in Eqn. (8.22). We assume that the signal and

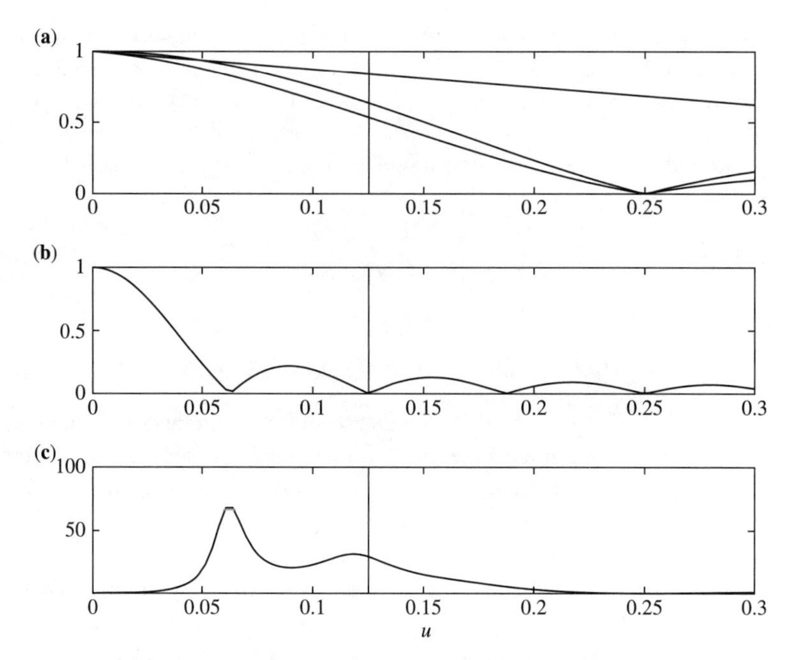

Figure 10.29 System transfer function for $\Delta = 4\lambda f/\#$ with (a) unit magnification; (b) $M =$ 0.25; and (c) excess noise factor for (b) assuming SNR = 20 dB.

noise power spectra are flat and that the SNR $= S_f(u, v)/S_n(u, v) = 20$ dB. At frequencies near the nulls of the pixel sampling function, the MSE of the multiaperture system is much worse than in the single-aperture system. In this particular case, the multiaperture system has competitive SNR up to approximately 25% of the single aperture bandpass, meaning that signal averaging over the apertures achieves reasonable SNR at low frequency but little gain in system resolution is obtained by combining data from the $M = 0.25$ systems.

The vertical lines in Fig. 10.29 represent the aliasing boundaries for each sampling strategy. This boundary is easily calculated for the single-aperture system as $u_{\text{alias}} = 1/(2\Delta)$. The aliasing limit for the multiaperture system is determined by shifts Δ_{xk}. With $\sqrt{K} = M$, the multiaperture and single-aperture systems achieve the same aliasing limit for $\Delta_{xk} = k\Delta/\sqrt{K}$. We assume that this is the case in Fig. 10.29.

We observed in Section 7.1 that pixel pitch, $f/\#$, aliasing, and SNR create a design space from which no single magic design emerges. The focal plane designer has motivations for maintaining a relatively large pixel pitch. For example, small pixels may be difficult to manufacture and may produce excess crosstalk and noise in comparison with larger pixels. The richness of the optical and optoelectronic design space is greatly enhanced by multiaperture designs. A second example design is illustrated in Fig. 10.30. The focal plane pixel pitch is the same as in Fig. 10.29, but we use a 2×2 array of $M = 0.5$ imagers rather than a 4×4 array. We assume a larger SNR of 40 dB. This system achieves reasonable SNR to well

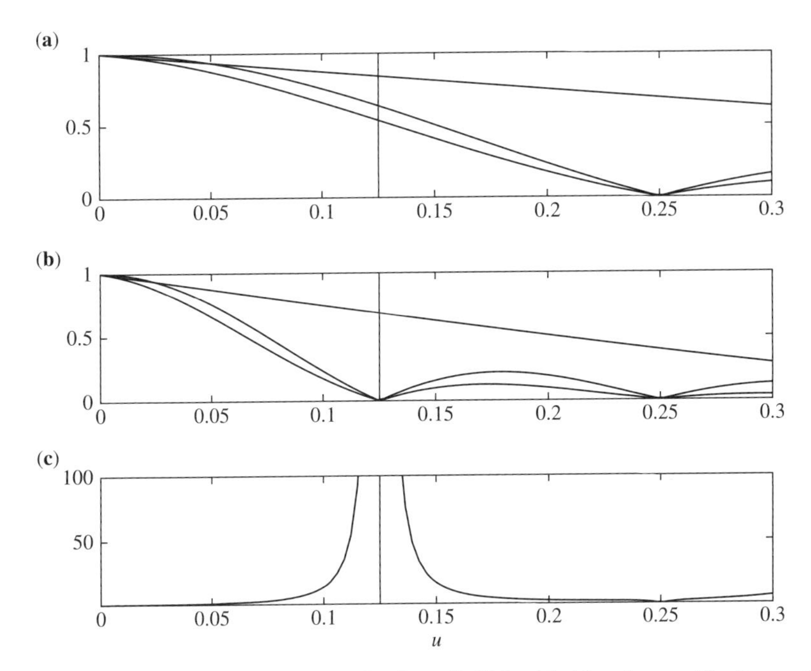

Figure 10.30 System transfer function for $\Delta = 4\lambda f/\#$ with (a) unit magnification and (b) $M = 0.5$; (c) excess noise factor for (b) assuming SNR = 40 dB.

over half of the frequency range of the single-aperture analog using optics with half the focal length of the single aperture system.

Numerous challenges and opportunities remain unexplored in our discussion of multiaperture systems to this point. Ever the optimists, let's begin by considering opportunities. First, as noted in Section 10.4.1, it is not really fair to scale M and $f/\#$ independently. For optical systems on the millimeter–centimeter aperture scale, however, the significance of this coupling impacts mass and complexity of the optics rather than the STF. A factor of much greater significance may arise from aliasing noise. In the worst case, the statistical power spectrum of the signal is flat across the full range of imager sensitivity. Signal components at frequencies above the aliasing limit must be added to the noise spectrum. The Wiener filter MSE accounting for aliasing noise is

$$\varepsilon(u, v) = \frac{S_f(u, v)}{1 + |\hat{h}(u, v)|^2 \{S_f(u, v)/S_n(u, v) + |\hat{h}_a(u, v)|^2 [S_a(u, v)]\}} \tag{10.48}$$

where $h_a(u, v)$ is the STF for frequencies aliased into measured frequency (u, v). Figure 10.31 compares MSE including aliasing noise for the systems of Fig. 10.30(a) and (b). Recalling that the SNR for these systems is 40 dB, aliasing is the dominant noise factor. The curve beginning near the origin in Fig. 10.31 corresponds to the MSE for the 2×2 multiaperture imager. Low-frequency aliasing

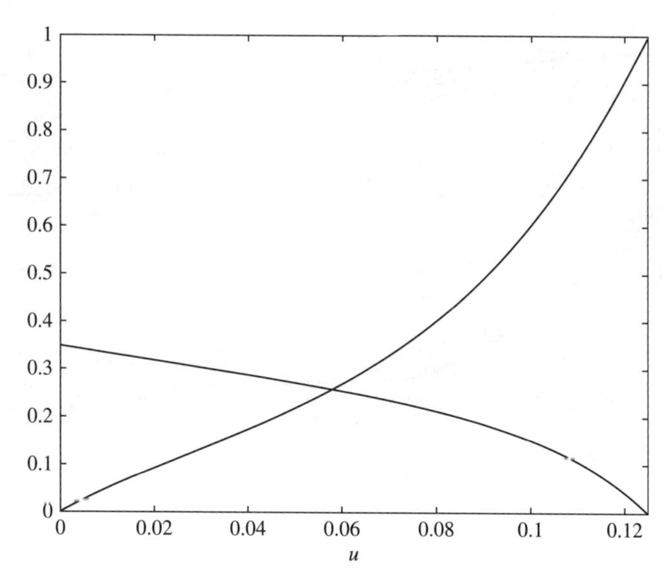

Figure 10.31 Wiener filter MSE based on Eqn. (10.48). The MSE value on the ordinate is relative to the signal spectral density.

noise is weak for this system because the STF passes through a null at the aliasing boundary. MSE increases monotonically to the boundary, where the null in the transfer function makes the error equal to the expected signal value. The MSE for the single-aperture system, in contrast, is high at low frequencies because of the high STF at the aliasing boundary and falls to zero at the aliasing boundary owing to the STF null at $2u_{\text{alias}}$. With this model, the MSE for the multiaperture system is substantially better than for the single-aperture system at frequencies below the crossing point in Fig. 10.31. One typically counters aliasing noise in three ways: by assuming that the object spectral density is not flat, by blurring the optical PSF, and by applying denoising algorithms. In the first approach, compressive coding strategies in multiaperture systems may be considered. The second approach reduces the STF of the single-aperture system to be more comparable with the multiplex multiaperture system. To understand the third approach, we must consider image estimation from multiple aperture data in more detail.

Tanida et al. [229] originally inverted TOMBO data using a truncated SVD algorithm. Experimental results illustrated in Fig. 10.32 used 10×10 apertures with 250-μm-diameter, 650-μm-focal-length lenses. The pixel size was 11 μm, meaning that each aperture spanned a 22.7×22.7-pixel grid. The system response was estimated by experimental characterization. The SVD was truncated to singular values $\lambda > \lambda_1/7$. While the reconstructed image is modestly improved relative to the subaperture image, it is not clear that the result is superior to simple interpolation and smoothing. Better results have been achieved by the Osaka University group and others in subsequent studies using diverse linear, convex optimization, and expectation maximization strategies [47,133,187,218].

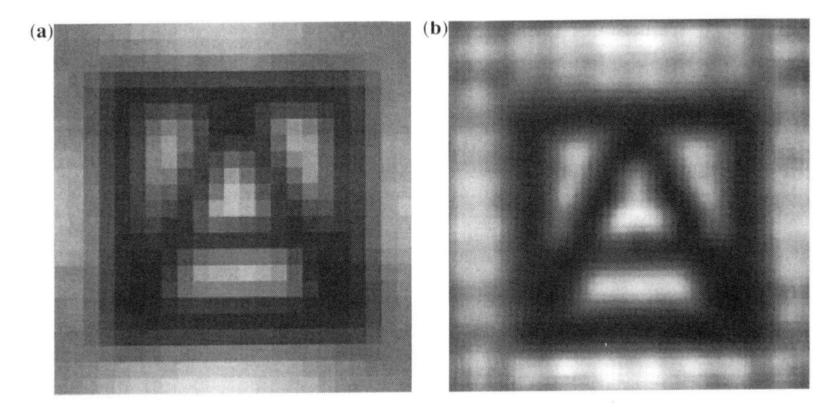

Figure 10.32 Images reconstructed from TOMBO data by truncated SVD: (a) a subaperture image; (b) the reconstructed image. (From Tanida et al. [229] © 2001 Optical Society of America. Reprinted with permission.)

It is important to understand the SVD approach, however, as a baseline for challenges and opportunities in multiaperture systems. A central problem is that highly accurate forward models are critically enabling but are relatively difficult to obtain. Analyses of the sensitivity of multiaperture systems to model error are presented by Prasad [207] and Wood et al. [254]. As an example of the system characterization challenge, we consider data from the Phase II longwave infrared (LWIR) cameras developed at Duke University through the Compressive Optical MONTAGE Photography Initiative (COMP-I). The COMP-I imagers used a 3×3 array of compound germanium and silicon lenses with a 5-mm center-to-center pitch. The effective focal length was 5.8 mm corresponding to $f/1.16$. The lenslet array was positioned over a 640×480 vanadium oxide focal plane array with square 25-μm pixels. The field of view of the image was limited to 20° such that each lenslet actively utilized an 80×80 pixel grid. The ifov of the subapertures was $20/80 = 0.25° = 4.4$ mrad. Reconstructed images up sample by a factor of 3 to create a 240×240 image with 8.3 μm effective pixel size and an ifov of 1.5 mrad.

Image estimation requires a forward model and an inversion strategy. Experimental characterization of the forward model is an attractive first step, but completely empirical forward models are rarely satisfactory. Accurate collection of an empirical forward model requires precise knowledge and control of test targets, compensation for background sources, and extremely stable image collection systems. As an example, Fig. 10.34(a) shows the forward model for COMP-I imagers over a subset of the image field. A point object, consisting of a pinhole in a copper plate, illuminated the imager through a collimation system. A 50×50 grid of object positions evenly distributed over a 40.4-mrad field was sampled by rotating the imager using precision stages.

A 15×15 grid of pixels from each subaperture was used as the output data for each of the 50×50 input images. The resulting mapping can be presented as $\boldsymbol{g} = \boldsymbol{Hf}$ with $\boldsymbol{g}$ a $9 \times 15 \times 15 = 2025$-element measurement vector and $\boldsymbol{f}$ a $50 \times 50 = 2500$-element object vector. The subset of the 2025 × 2500 matrix $\boldsymbol{H}$ shown in Fig. 10.34(a) corresponds to 500 measurement points and 1000 object points. Each column shows nine points corresponding to the object point response over the nine subapertures. The periodic banding in $\boldsymbol{H}$ is due to pixel nonuniformity and uncorrected background. Since microbolometers measure the total thermal flux, there is considerable background in uncooled IR imagery. Substantial nonuniformity correction and background substration is necessary to form images from these systems. Data used to generate Fig. 10.34(a) have already been processed for background subtraction, but, as indicated by the figure, it is difficult to achieve absolutely uniform response from all subapertures.

The singular values of the measurement, illustrated for the first 300 values in Fig. 10.33(a), further illustrate the nature of this problem. The largest singular value is much greater than one would expect from our previous analysis of shift-coded

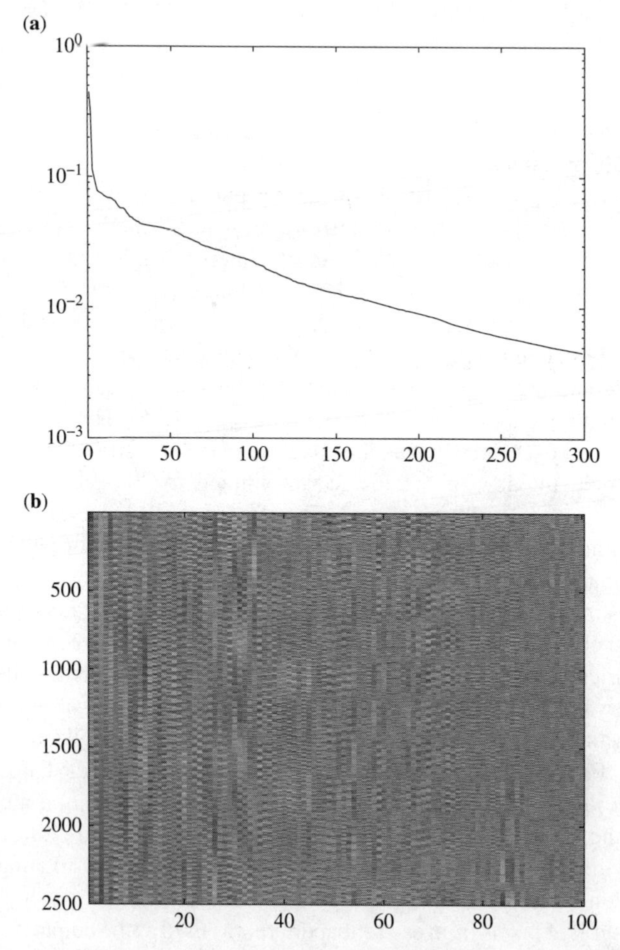

Figure 10.33 Singular values (a) and object space singular vectors (b) for point target characterization of the COMP-I phase II LWIR imaging system.

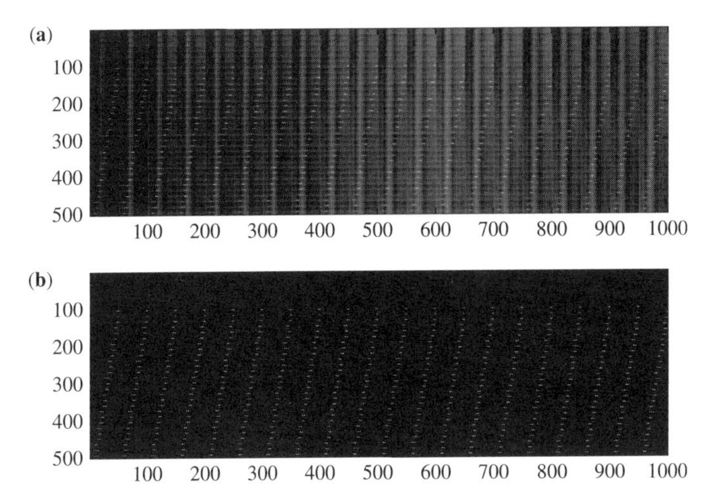

Figure 10.34 Forward model (a) (**H**) and reduced forward model (b) ($\mathbf{H}_r$) for point target characterization of the COMP-I phase II LWIR imaging system.

downsampling due to the substantial static background in the measurements. The first 100 object space singular vectors are illustrated in Fig. 10.33(b). In contrast with previous results, the lowest-order singular vector contains relatively high-frequency components corresponding to static nonuniformity. To compensate for this effect, one may choose to form a reduced forward model truncating both high-order singular vectors and a few low-order singular vectors to eliminate static bias. Figure 10.34(b) shows a reduced system operator created using singular vectors 5–200. Systematic banding is largely eliminated in this operator.

Having characterized the forward model, one may now attempt to form an image by any of the methods discussed in Chapter 8. Figure 10.35 shows least-squares

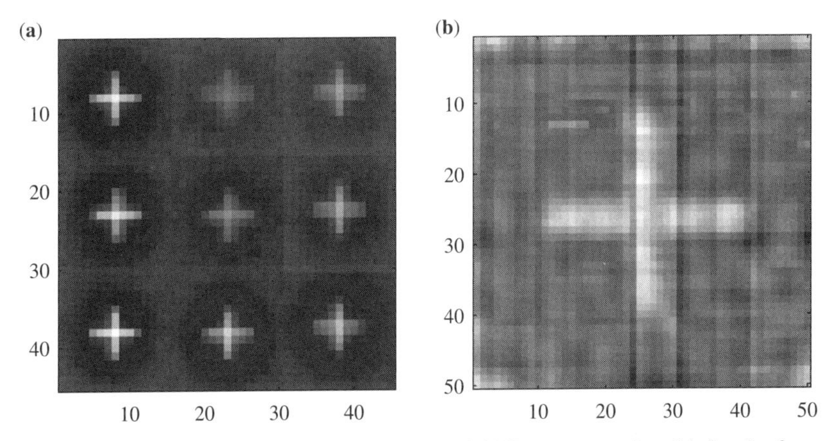

Figure 10.35 Measurement data (a) and truncated SVD reconstruction (b) for the forward model of Fig. 10.33.

reconstruction using the truncated SVD forward model of Fig. 10.33. While the image quality is poor, it is, of course, vastly superior to direct least squares. Important lessons of this exercise include the difficulty of actually measuring the forward model and an appreciation of the scale of the problem. This experiment covered only a very small subset of the system aperture; characterization and algebraic estimation based on the full-aperture system response is numerically intractable. Although experimental forward model characterization is particularly challenging for thermal imagers, these issues are significant for all computational imaging systems. Accurate forward models are essential to virtually all of the spectrometer and imager designs discussed in this text.

Needless to say, given the quality of the subaperture images in Fig. 10.35, the quality of the synthesized image is extremely disappointing. More attractive results are obtained using parameterized physical models rather than fully characterized forward models. For the COMP-I cameras, parameterized models assume that the same image is sampled in each subaperture with an unknown aperture to aperture shift. Assuming the sampling model given by Eqn. (10.46), image synthesis is relatively straightforward. Somewhat more detailed interpolation strategies are necessary for irregular sampling, but if the sampling positions are well known, one may expect to obtain STF-limited performance. The sampling phase for the COMP-I imager was characterized by center-of-mass registration (although developing algorithms for larger-scale imaging systems use dimensionality reduction-based registration [115]). Samples from the registered subimages are then combined on an interpolation grid to reconstruct the full-frame image using subpixel sample spacings determined by the registration algorithm. For objects at infinity, registration data need not be characterized for every image [132]. The final image is smoothed using the least-gradient algorithm based on the null space of the shift coding operator over a finite image window [218].

Figure 10.36 illustrates COMP-I imagery reconstructed by linear interpolation with least-gradient smoothing. One expects the image reconstructed by this algorithm to be subject to the STF described by Eqn. (10.47) for $M = 0.333$ and $\Delta \approx 2.5\lambda f/\#$. A baseline image with $M = 1$ is also illustrated. As expected, the zero-spatial-frequency NEDT is approximately the same for the baseline and multiaperture systems. More detailed experimental analysis yields an ifov equal to approximately $1.5\times$ the baseline value but, as illustrated in the figure, is substantially better than any of the individual lenslets. It is important to emphasize that the improvement in image quality is due to high optical quality of the lenslets and to antialiasing as well as digital superresolution. It is also interesting to note the dramatically improved depth of field of the multiaperture system relative to baseline. According to Eqn. (10.3), one anticipates a factor of 9 reduction in the near point due to the shorter focal length lenses. This effect is illustrated in Fig. 10.36 by simultaneous imaging of a hand inside the near point of the baseline system and a person at a well-focused range.

Sampling on a 3×3 multiaperture system is essentially equivalent to 3×3 down-sample shift operator. As mentioned in Sections 8.4 and 8.5, the structure of the

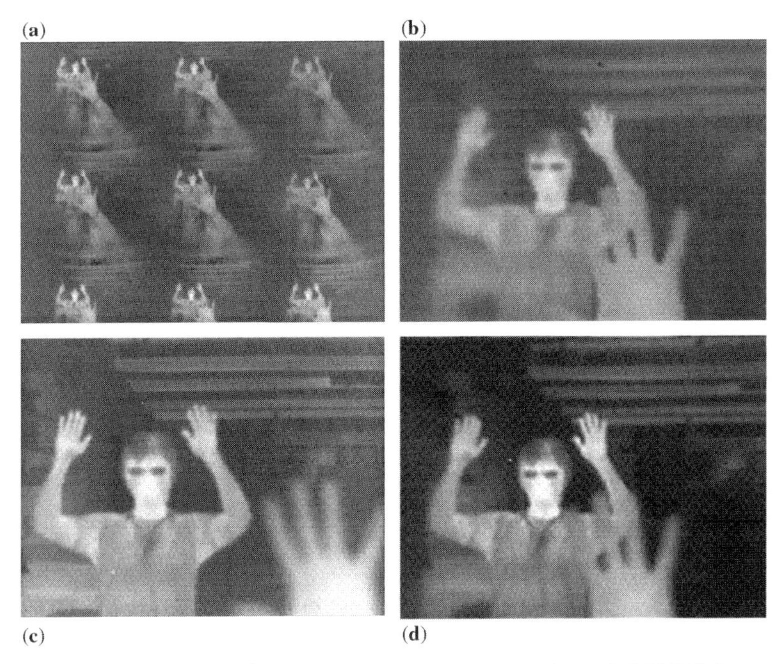

Figure 10.36 Least gradient/linear interpolation reconstruction of COMP-I image data: (a) raw image; (b) single-lenslet image; (c) baseline image; (d) reconstructed image. The person is 3 m from the imagers; the hand in the near field is 0.7 m away. The images are captured simultaneously; the relative shift in the position of the hand and the person is due to parallax between the imagers. (Images collected by Andrew Portnoy and Mohan Shankar.)

sampling function substantially impacts the reconstruction performance. Relatively simple studies have considered modifications to the optical PSF and the pixel sampling function to improve multiaperture imaging performance [204,218], but coding for multiaperture sampling system design remains an active area of research. This topic is closely related to registration and system characterization. In particular, our assumption that registration is range-invariant is incorrect. The effective sampling phase is sensitive to parallax in 3D scenes. Of course, one may consider range-dependent PSF coding to enable 3D image formation without scene-dependent registration. One may also consider localized registration or range-dependent principal component analysis. As always, such strategies require accurate forward models.

10.4.3 Optical Projection Tomography

Sections 10.5 and 10.6 discuss emerging computational imager designs with a particular focus on strategies for fully characterizing the spatial and spectral optical data cube. We begin by considering systems designed to characterize the radiance. As discussed in Section 6.7.1, the spectral radiance is the power density of the field per unit solid angle and wavelength. The radiance is well-defined for quasi-homogeneous fields as the Fourier transform of the cross-spectral density:

$$B(\mathbf{x}, s, \nu) = \iint W(\Delta x, \boldsymbol{x}, \nu) e^{(2\pi i \nu)/cs \cdot \Delta x} d\,\Delta x \tag{10.49}$$

Under this approximation, measurement of the radiance on a surface is equivalent to measuring W. Of course, we observe in Eqn. (6.52) that if $W(\Delta x, \Delta y, \bar{x}, \bar{y}, \nu)$ is invariant with respect to $\bar{x}, \bar{y}$ over the aperture of a lens, then the power spectral density in the focal plane is

$$S(x, y, \nu) = \frac{4\nu^2}{c^2 F^2} \iint W(\Delta x, \Delta y, \bar{x}, \bar{y}, \nu) \mathcal{H}\left(\frac{\nu \Delta x}{2cF}, \frac{\nu \Delta y}{2cF}\right) \times \exp\left(2i\pi\nu \frac{x\Delta x + y\Delta y}{cF}\right) d\,\Delta x\, d\,\Delta y \tag{10.50}$$

where we set $z = F$, $\mathcal{H}(u, v)$ is the optical transfer function and $\bar{x}, \bar{y}$ can be taken as the transverse position of the optical axis. Neglecting the OTF for a moment, we find therefore that the power spectral density at the focus of a lens illuminated by quasi-homogeneous source approximates the radiance, specifically

$$\mathcal{B}\left(\mathbf{x}, s_x = \frac{x}{F}, s_y = \frac{y}{F}, \nu\right) \approx S(x, y, \nu) \tag{10.51}$$

The radiance emitted by a translucent 3D object is effectively the x-ray projection described, for example, by Eqn. (10.12). We have encountered such projections in diverse contexts throughout the text. One may ray-trace the radiance to propagate the field from one plane to the next to construct perspective views from diverse vantage points or apply computed tomography to radiance data to reconstruct 3D objects. As mentioned in our discussion of tomographic reconstruction in Section 2.6, the 4D radiance over a surface containing a 3D object overconstraints the tomographic inverse problem. Reconstruction may be achieved over a 3D projection space satisfying Tuy's condition. Computed tomography from focal images [74,168], from RSI EDOF images [170], and from cubic phase EDOF images [72] are discussed by Marks et al. More recently, optical projection tomography has been widely applied in the analysis of translucent biological samples [220,221].

Optical projection microscopy commonly applies full solid angle sampling to obtain diffraction limited 3D reconstruction. Remote sampling using projection tomography, in contrast, relies on a more limited angular sampling range. Projection tomography using a camera array is illustrated in Fig. 10.37. We assume in Fig. 10.37 that the aperture of each camera is A, that the camera optical axes are dispersed over range D in the transverse plane, and that the range to the object is z_o. The band volume for tomographic reconstruction from this camera array is determined by the angular range

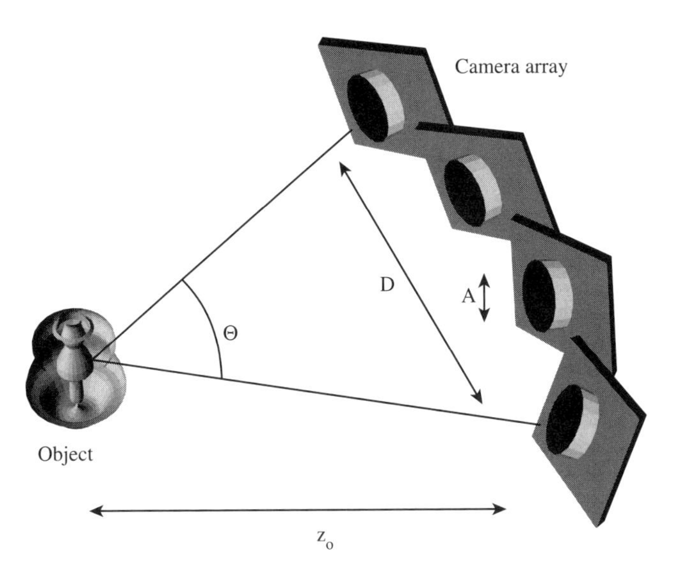

Figure 10.37 Projection tomography geometry. An object is observed by cameras of aperture A at range z_o. The range of camera positions is D. The angular observation range is $\Theta \approx D/z_o$.

$\Theta = D/z_o$. The sampling structure within this bandvolume is determined by the camera-to-camera displacement and camera focal parameters.

Assuming that projections at angle θ are uniformly sampled in l, one may identify the projections illustrated in Fig. 10.38 from radiance measurements by the camera array. The displacement Δl from one projection to the next corresponds to the transverse resolution $z_o\lambda/A$. According to Eqn. (2.52), the Fourier transform of the radiance with respect to l for fixed $s(\theta)$ yields an estimate of the Fourier transform of the object along the ray at angle θ illustrated in Fig. 10.39. The maximum spatial frequency for this ray is determined by Δl such that $u_{l,\max} = A/z_o\lambda$. The spatial frequency w along the z axis is $u_l \sin\Theta$. Assuming that the angular range D/z_o sampled by the camera array along the x and y axes is the same, the band volume sampled by the array is illustrated in Fig. 10.40. The lack of z bandpass at low transverse frequencies corresponds to the "missing cone" that we have encountered in several other contexts. The z resolution obtained on tomographic reconstruction is proportional to the transverse bandwidth of the object. For a point object, the maximum spatial frequency $w_{\max} = u_{\max} \sin\Theta = AD/z_o^2\lambda$ occurs at the edge of the band volume. The longitudinal resolution for tomographic reconstruction is

$$\Delta z = \frac{1}{w_{\max}} = \frac{z_o^2\lambda}{AD} \tag{10.52}$$

Comparing with previous analyses in Sections 10.3 and 6.4, we see that the longitudinal resolution is improved relative to a single aperture by the ratio $8D/A$. The factor of 8 improvement arises from the fact that the tomographic band volume is maximal at

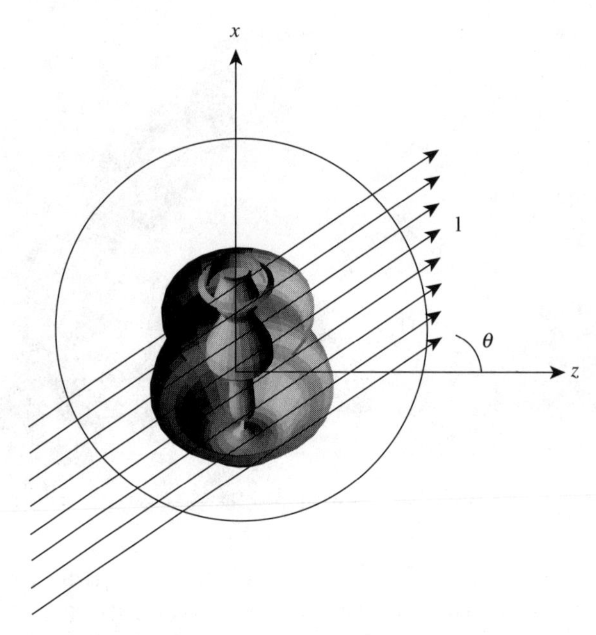

Figure 10.38 Sampling of x-ray projections along angle θ.

the edge of the transverse bandpass, while the 3D focal bandvolume falls to zero at the limits of the transverse OTF. A multiple-camera array "synthesizes" an aperture of radius D for improved longitudinal resolution.

Realistic objects are not translucent radiators such that the observed radiance is the x-ray projection of the object density. As discussed by Marks et al. [168], occlusion

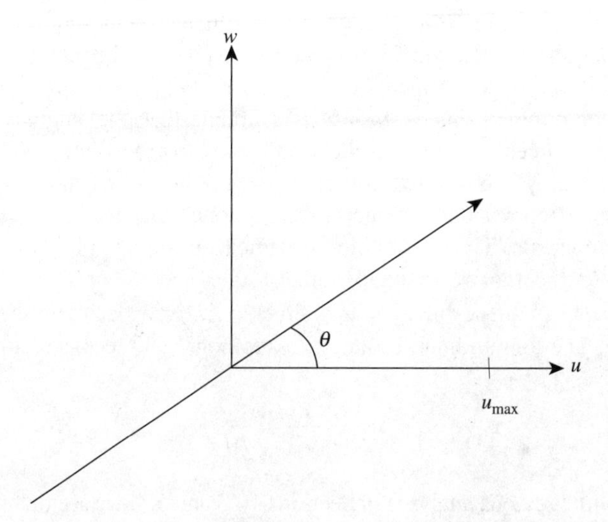

Figure 10.39 Fourier space recovered via the projection slice theorem from the samples of Fig. 10.38.

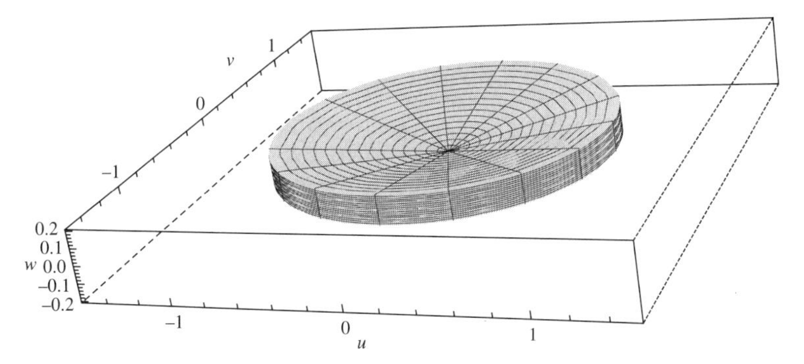

Figure 10.40 Band volume covered by sampling over angular range $D/z_o = 0.175$ in units of $u_{\max}$.

and opaque surfaces may lead to unresolvable ambiguities in radiance measurements. In some cases, more camera perspectives than naive Radon analysis may be needed to see around obscuring surfaces. In other cases, such as a uniformly radiating 3D surface, somewhat fewer observations may suffice.

The assumption that the cross spectral density is spatially stationary (homogeneous) across each subaperture is central to the association of radiance and focal spectral density or irradiance. With reference to Eqn. (6.71), this assumption is equivalent to assuming that $\Delta q/\lambda z \ll 1$ over the range of the aperture and the depth of the object. $\Delta q = A^2/2$ is the variation in q over the aperture. Thus, the quasihomogeneous assumption holds if $A \ll \sqrt{2z\lambda}$. Simple projection tomography requires one to restrict A to this limit. Of course, this strategy is unfortunate in that it also limits transverse spatial resolution to $\lambda z/A \approx \sqrt{\lambda z}$.

Radiance-based computer vision is also based on Eqn. (10.51). For example, *light field photography* uses an array of apertures to sample the radiance across an aperture [151]. A basic light field camera, consisting of a 2D array of subapertures, samples the radiance across a plane. The radiance may then be projected by ray tracing to estimate the radiance in any other plane or may be processed by projection tomography or data-dependent algorithms to estimate the object state from the field radiance. While the full 4D radiance is redundant for translucent 3D objects, some advantages in processing or scene fidelity may be obtained for opaque objects under structured illumination. 4D sampling is important when $W(\Delta x, \Delta y, \bar{x}, \bar{y}, \nu)$ cannot be reduced to $W(\Delta x, \Delta y, q, v)$. In such situations, however, one may find a camera array with a diversity of focal and spectral sampling characteristics more useful than a 2D array of identical imagers.

The *plenoptic camera* extends the light field approach to optical systems with non-vanishing longitudinal resolution [1,153]. As illustrated in Fig. 10.41, a plenoptic camera consists of an objective lens focusing on a microlens array coupled to a 2D detector array. Each microlens covers an $n \times n$ block of pixels. Assuming that the field is quasihomogeneous over each microlens aperture, the plenoptic camera returns the radiance for n^2 angular values at each microlens position. Recalling

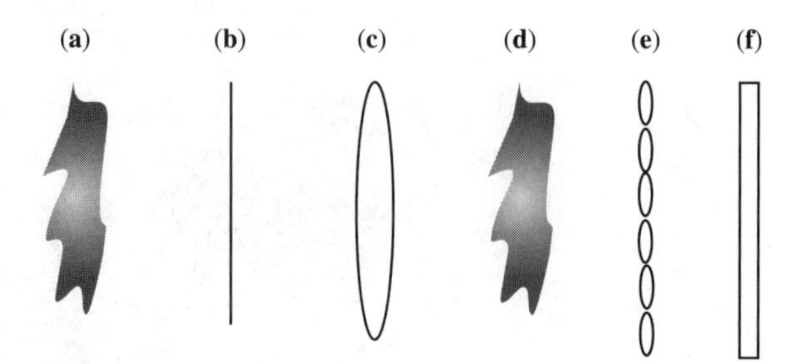

Figure 10.41 Optical system for a plenoptic camera: (a) object; (b) blur filter; (c) objective lens; (d) image; (e) microlens array; (f) detector array.

from Section 6.2 that the coherence cross section of an incoherent field focused through a lens aperture A is approximately $\lambda f/\#$, we find that the assumption that the field is quasihomogeneous corresponds to assuming that the image is slowly varying on the scale of the transverse resolution. This assumption is, of course, generally violated by imaging systems. In the original plenoptic camera, a pupil plane distortion is added to blur the image to obtain a quasihomogeneous field at the focal plane. Alternatively, one could defocus the microlenses from the image plane to blur the image into a quasihomogeneous state. The net effect of this approach is that the system resolution is determined by the microlens aperture rather than the objective aperture and the resolution advantages of the objective are lost. In view of scaling issues in lens design and the advantages of projection tomography discussed earlier in this section, the plenoptic camera may be expected to be inferior to an array of smaller objectives covering the same overall system aperture if one's goal is radiance measurement.

This does not imply, however, that the plenoptic camera or related multiaperture sampling schemes are not useful in system design. The limited transverse resolution is due to an inadequate forward model rather than physical limitation. In particular, the need to restrict aperture size and object feature size is due the radiance field approximation. With a more accurate physical model, one might attempt to simultaneously maximize transverse and longitudinal focal resolution. This approach requires novel coding and estimation strategies; a conventional imaging system with high longitudinal resolution cannot simultaneously focus on all ranges of interest.

The plenoptic camera may be regarded as a system that uses an objective to create a compact 3D focal space and then uses a diversity of lenses to sample this space. Many coding and analytical tools could be applied in such a system. For example, a reference structure could be placed in the focal volume to encode 3D features prior to lowpass filtering in the lenslets, pupil functions could be made to structure the lenslet PSFs and encode points in the image volume, or filters could encode diverse spectral projections in the lenslet images.

The idea of sampling the volume using diverse apertures is of particular interest in microscope design. As discussed in Section 2.4, conventional microscope design seeks to increase the angular extent of object features. In modern systems, however, focal plane features may be of nearly the same size as the target object features. Thus, the goal of a modern microscope may be simply to code and transfer micrometer-scale object features to a focal plane. Object magnification is then implemented electronically.

Transfer of high-resolution features from one plane to another can be implemented effectively using lenslet arrays. As an example, document scanners often exploit lenslets to reduce system volume [3]. The potential of lenslet image transfer is dramatically increased in computational imaging systems, which may tolerate or even take advantage of ghost imaging (overlapping image fields). A conventional camera or microscope objective may be viewed in this context as an image transfer device with a goal of adjusting the spatial scale of the image volume for multiple aperture processing. The light field microscope is an example of this approach [153].

Tomographic imaging relies on multiplex sensing by necessity; there is no physical means of isomorphically mapping a volume field onto a plane. As we have seen, data from multiple apertures observing overlapping volumes can be inverted by projection tomography. We further propose that tomographic inversion is possible in systems that cannot be modeled by geometric rays. The next challenge is to design the sampling strategy, optical system, and inversion strategy to achieve this objective. While we do not have time or space to review a complete system, we do provide some "big picture" guidance with regard to coding strategy in the next section.

10.5 GENERALIZED SAMPLING REVISITED

By this point in the text, it is assumed that the reader is familiar with diverse multiplex sampling schemes. The present section revisits three particular strategies in light of the lessons of the past several chapters. Our goal is to provide the system designer with a framework for comparative evaluation of coding and sampling strategy. An optical sensor may be evaluated based on physical (resolution, FOV, and depth of field), signal fidelity (SNR and MSE), and information-theoretic (feature sensitivity and transinformation) metrics. While detailed discussion of the information theory of imaging is beyond the scope of this text, our approach in this section leans toward this perspective.

We focus in particular on SVD analysis of measurement systems. As discussed in Section 8.4, the singular vectors of a measurement system represent the basic structure of sensed image components, and the singular values provide a measure of how many components are measured and the fidelity with which they can be estimated. When two different measurement strategies are used to estimate the same object features, SVD analysis provides a simple mechanism for comparison. Assuming similar detector noise characteristics, the system with the larger eigenvalue for estimating a particular component will achieve better performance in estimating that component.

While joint design of coding, sampling, and image estimation algorithms is central to computational imager design, SVD analysis provides a basis for comparison that is relatively independent of estimation algorithm. Evaluation of system performance using the singular value spectrum is a generalization of STF analysis. Signal Fourier components are eigenvectors of shift-invariant systems, with eigenvalues represented by the transfer function. SVD analysis extends this perspective to shift-variant systems with the singular vectors playing the role of signal components and the singular values playing the role of the transfer function.

Singular vector structure is central to the image estimation utility of measurements for both shift-variant and shift-invariant systems. Where the singular vector structure of two measurement schemes is different, the strategy with the "better" singular vectors may provide superior performance even if it produces fewer or weaker singular values. "Better" in this context may mean that the strongest sensor singular vectors are matched to the most informative object features or that the singular vectors are likely to enable accurate object estimation or object feature recognition under nonlinear optimization. If a statistical model for the object is available, one may apply the restricted isometry property [Eqn. (7.40)] to compare singular vector bases.

Multiaperture sampling schemes for digital superresolution provide a simple example of comparative SVD analysis. As discussed in Sections 8.4 and 10.4.2, the singular values and singular vectors for shift-coded systems provide useful low-frequency response but do not produce the flat singular value spectrum of isomorphic focal measurement. Of course, the structure of the singular vectors actually provides benefits in lowpass filtering for antialiasing.

The basic shift-coded multiaperture system is modeled as an $N\times$ downsampling operator with variable sampling phase. The alternative shift codes suggested in Section 8.4 could be implemented by PSF coding, with potential advantages in the SVD spectrum as discussed previously. Portnoy, et al. propose an alternative focal plane coding strategy based on pixel masking [204]. The basic idea is to alias high resolution image features into the measurement passband by creating high-resolution features on the pixels.

Portnoy implemented focal plane coding by affixing a patterned chrome mask to a visible spectrum CCD with 5.2 μm pixel pitch. Figure 10.42(a) shows a micrograph of a chrome mask used in the experiments. The subpixel response of the focal-plane-coded

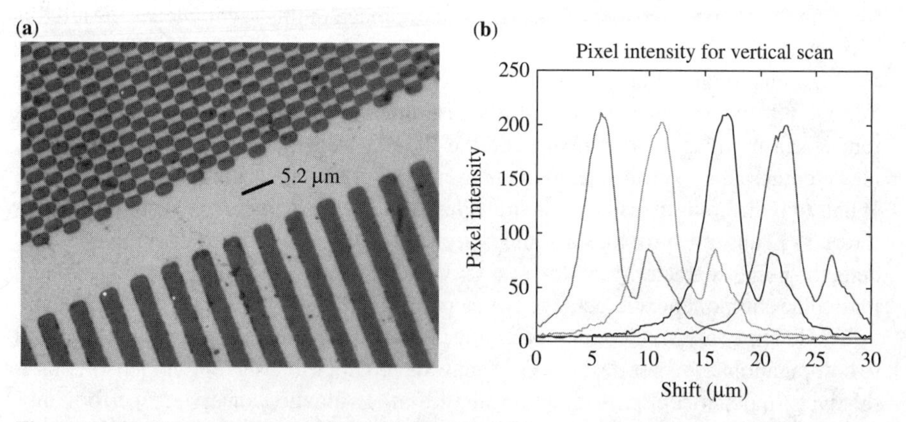

Figure 10.42 Mask for pixel coding (a) and point object response measurement (b) for four adjacent pixels.

system is illustrated by the pixel response curves in Fig. 10.42(b). These curves were obtained by focusing a white point target on the coded focal plane. The output of adjacent pixels is plotted as the target is scanned across the column. The extent of the pixel response is somewhat greater than 5.2 μm because of the finite extent of the target. The period of the pixel response curves is 5.2 μm. Although the mask pattern was not precisely registered to the pixels in this experiment, subpixel modulation of the response is indicated by the twin lobe structure of the pixel response.

We analyze pixel mask-based focal plane coding by modeling each detector as an $n \times n$ block of subpixels. The output of the ith detector is

$$g_i = \sum h_{ij} f_j \tag{10.53}$$

where f_j is the irradiance in the jth subpixel and h_{ij} is 1 if the mask is transparent over the (ij)th subpixel and zero otherwise. A vector of measurements of the subpixels is collected by measuring diverse coding masks over several apertures. As with the shift-coded system, the irradiance available to each pixel in a K aperture imaging system with each aperture observing the same scene is $1/K$ the single-aperture value. Accordingly, the measurement model for binary focal plane coding is

$$\mathbf{g} = \frac{1}{K}\mathbf{H}\mathbf{f} \tag{10.54}$$

with $h_{ij} \in [0,1]$.

For fixed K and independently and identically distributed noise in each measurement, we know from Section 8.2.2 that $\mathbf{H} = \mathbf{S}_K$, where $\mathbf{S}_K$ is the Kth-order Hadamard S matrix, yields minimal variance on estimation of $\mathbf{f}$ from Eqn. (10.54). One may be tempted, therefore, to replace the shift code commonly used for digital superresolution with Hadamard sampling implemented by appropriately masking pixels in each subaperture. Under this approach, one assumes that the sampling phase is identical in each subaperture. Each detector pixel may be regarded as a block of Hadamard sampled subpixels. Figure 10.43 compares the singular value spectrum of $1 - \mathbf{S}_4$ sampling with the shift codes of Fig. 8.9 (we use $1 - \mathbf{S}_4$ rather than $\mathbf{S}_4$ to achieve four-element codes). As illustrated in Fig. 10.43(a), pixel block sampling produces localized singular vectors. Hadamard coding dramatically improves the singular values for the weakest singular vectors, but over most of the spectrum Hadamard singular values are substantially less than the shift code singular values (recognizing that the S-matrix throughput is half the 100% throughput of the shift codes).

On the basis of our discussion of regularized and nonlinear image estimation as well as aliasing noise (and experimental results), it is clear that the increase in the singular values at the right side of the S-matrix spectrum does not justify the reduction shown in Fig. 10.43(b). Part of the greater utility of the shift code arises from implicit priority of low frequencies in image sampling. In assuming that image pixel values are locally correlated, we are essentially assuming that low/moderate-frequency

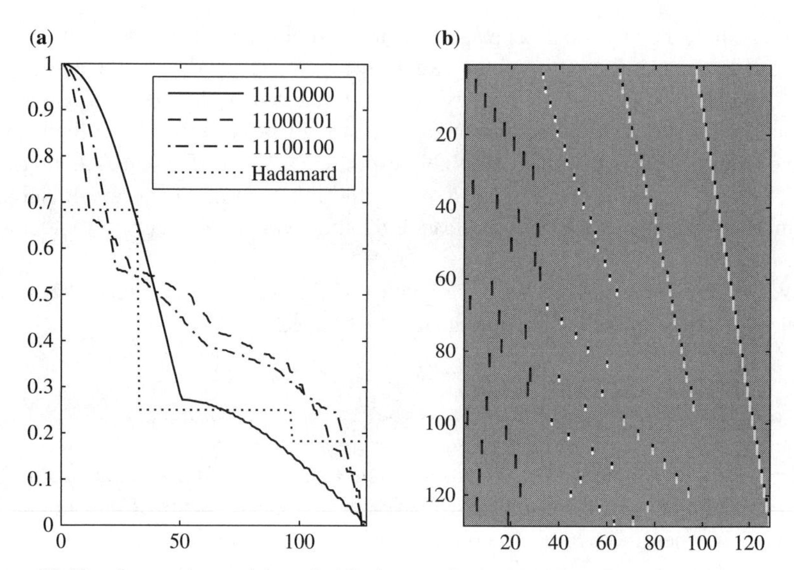

Figure 10.43 Comparison of $1 - S_4$ block sampling with the shift codes of Fig. 8.9: (a) singular value spectra; (b) Hadamard singular vectors.

features may be more informative than features near the aliasing limit. Thus we are generally satisfied with moderate lowpass filtering.

In an analysis of scaling laws for multiple aperture systems, Haney suggests that for fixed integration time the mean-square error of estimated images scales linearly in K for $K \times K$ downsampling [110]. This result is consistent with linear least-squares estimation for S-matrix sampling, but it neglects lens scaling, aliasing noise, and alternative coding and estimation strategies discussed in Section 10.4 Our comparison of STF and aliasing noise in Section 10.4.2 suggests, in fact, that in the balance of passband shaping for resolution, field of view, and antialiasing, multiaperture systems are competitive with cyclops strategies while also providing dramatic improvements in system volume and depth of field.

Expanding on Eqn. (7.37), aliasing arises in a measurement system when the inner product of two object features that one would like to distinguish (such as harmonic frequencies) both produce the same distribution when projected on the object space singular vectors. Design to avoid aliasing noise accordingly attempts to limit the range of the measurement vector to an unambiguous set of object features. Ideal codes must capture targeted features without ambiguity. As discussed in Section 8.4, variations in shift codes may modestly improve image estimation. Continuing research in this area will balance physical implementation, object feature sensitivity, and antialiasing.

Singular value decomposition analysis may also be used to compare spectrometer aperture codes. Figure 10.44, for example, compares a mask with binary elements randomly selected from $t_{ij} \in [0,1]$ with uniform probability with the S matrix $\mathbf{S}_{512}$ using the signal and the noise model of Fig. 9.7. While the first singular value is

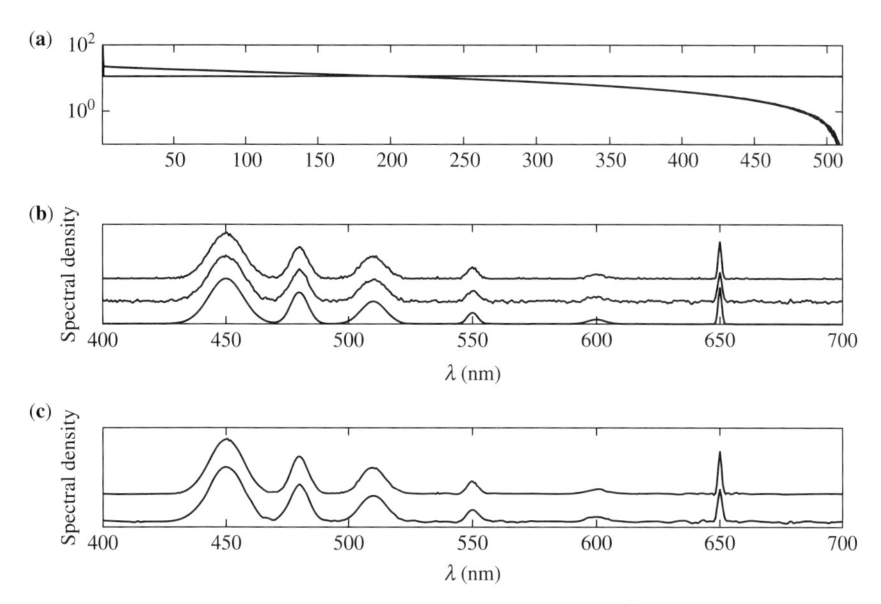

Figure 10.44 Singular values and reconstructed signal spectra for $N = 511$ random and Hadamard coded aperture spectroscopy: (a) singular value spectrum; (b) $\sigma^2_{\text{random}} = 0.24$, $\sigma^2_{\text{Hadamard}} = 0.067$; (c) $\sigma^2_{\text{random}} = 0.18$, $\sigma^2_{\text{Hadamard}} = 0.046$.

256 for both systems, the random measurement produces larger singular vectors over the first half of the band and lower values in the second half. The Hadamard system, by design, produces a flat singular value spectrum.

Figure 10.44(b) compares signal reconstruction from Hadamard and random codes. The bottom curve is the true spectrum, the middle curve is the spectrum estimated from a random code, and the upper curve is the Hadamard code spectrum. The Hadamard spectrum is estimated using nonnegative least squares. The random code spectrum is reconstructed by truncated least-squares estimation using the first 300 singular vector expansion coefficients. The random data are then smoothed using the remaining 211 singular vectors as the null space for least-gradient estimation. Figure 10.44(c) denotes the (b) spectra as in Fig. 9.7. The spectral feature at 650 nm is sharpened relative to Fig. 9.7 to test the resolution of the truncated random SVD and to illustrate artifacts in the reconstruction. While the random code returns inferior SNR in both the initial and denoised spectra, the discrepancy is much less than least-squares analysis would suggest. There is also a strong possibility that the random reconstruction could be substantially improved using nonlinear optimization and/or code optimization. The fact that the random data spectrum improves less under denoising is due to bias in the truncated SVD: this bias could be reduced by shaping the singular vectors in code design and by enforcing l_1, $\mathcal{F}$, total variation (TV), or similar constraints.

In view of this analysis, quasirandom (non-Hadamard) codes are extremely attractive in spectrometer design. Where the standard coded aperture design and inference algorithm assumes that the aperture is uniformly illuminated by a homogeneous power spectral density, quasirandom codes can accommodate spatially localized reconstruction strategies that allow spatial variation in the input signal. This approach is discussed in more detail in Section 10.6.2.

Singular value decomposition analysis may also be applied in considering compressive imaging. As an example, we consider an compressive sampling system under the following constraints:

- The image consists of N pixels.
- The image is sparse such that at most K pixels are nonzero.
- M_s measurements are recorded in M_t timesteps to produce $M = M_s M_t$ total data points.
- The signal power is uniform during the recording process, meaning that the signal energy available in one timestep is $1/M_t$ of the total recording energy.
- Pixels are measured in linear combinations with binary weights drawn from [0,1]. Nonnegative weighting is, of course, required for optical irradiance measurements.

As a first strategy, we consider a single-detector camera such that $M_s = 1$ and $M_t = M$. As in Ref. 63, measurement weights h_{ij}, where the index j refers to pixel number and i to measurement number, are randomly selected from [0,1]. To maintain power conservation, the measurement matrix must be normalized by $1/M$ such that $\sum_j h_{ij} \leq 1$. For uniformly distributed weights, the quantum efficiency of this sampling strategy is $\frac{1}{2}$.

As a second strategy, we consider an M_s-detector camera. Each image pixel is randomly assigned to one of the detectors in each of M_t timesteps. To maintain energy conservation, the measurement matrix is normalized by M_t. While the total image energy available is the same under strategies 1 and 2, the second strategy has a quantum efficiency of 1.

Figure 10.45(a) shows the singular value spectra for these sampling strategies with $N = 1024$ and $M = 128$. The upper curve shows the singular values for strategy 2 with $M_t = 8$ (eight measurement times) using $M_t = 16$ detectors. The lower curve shows the singular values for the single-pixel detector. As illustrated in the figure, the singular values under strategy 2 are $8\times$ larger than those for strategy 1 over most of the spectral range. As illustrated in Fig. 10.45(b), both sampling strategies are effective, in the absence of noise, in reconstructing a sparse signal using l_1 minimization. The signal in this case consists of $K = 30$ values randomly distributed over [0,1] (again consistent with the nonnegativity of optical signals). The true signal is shown at the bottom of Fig. 10.45(b); the middle curve is the strategy 1 reconstruction, and the upper curve is the strategy 2 reconstruction. Reconstruction was implemented using Candes and Romberg's `l1eq_example.m` code distributed at `www.acm.caltech.edu/l1magic` [38].

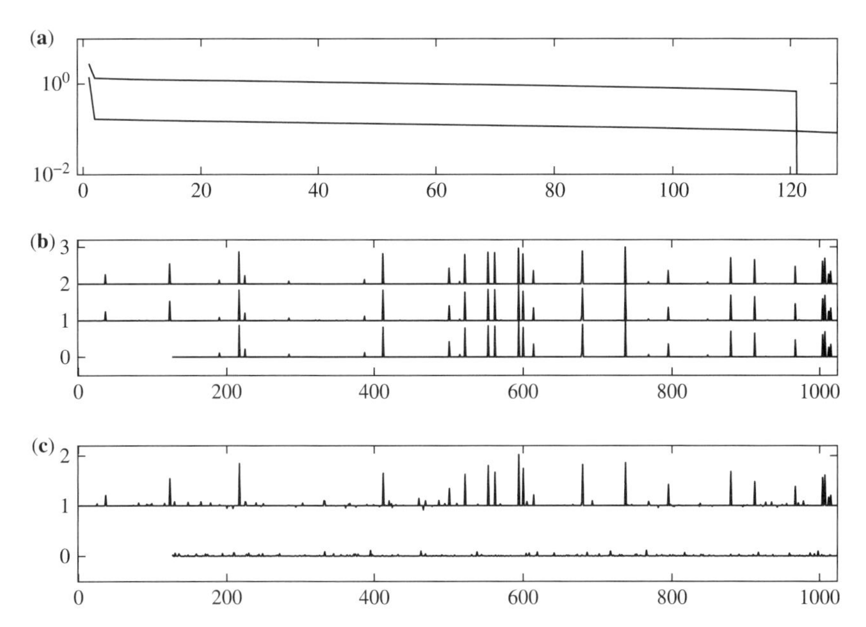

Figure 10.45 Singular value spectra and sparse signal reconstructions for two optical compressive sampling strategies: (a) singular value spectra; (b) noise-free reconstructions; (c) reconstructions with noise variance $\sigma^2 = 10^{-4}$.

As a result of a flatter singular value spectrum, strategy 2 is much less susceptible to noise than strategy 1. This effect is illustrated in Fig. 10.45(c), where zero mean normally distributed noise with $\sigma = 0.01$ is added to each measurement. The lower curve, corresponding to strategy 1, fails to capture the sparse signal. While the upper strategy 2 reconstruction contains numerous noise features, the basic structure of the signal is faithfully reproduced. While the $2\times$ improvement in photon efficiency of strategy 2 is partially responsible for this improvement, the primary improvement comes from the superior singular value distribution.

Fresh from this success, one might push strategy 2 even further by setting $M_t = 1$ and $M_s += 128$. This approach produces orthogonal measurement vectors and a completely flat singular value spectrum (each measurement records an orthogonal set of image pixels). Unfortunately, this approach also fails to map different sparse signals onto different measurements and thereby fails to satisfy the restricted isometry property. If N/M pixels are captured in only one measurement, then each of those pixels will produce the same measurement data no matter which is excited. The goal of optical measurement design is to jointly optimize the structure of the singular vectors to enable unambiguous signal reconstruction while also optimizing the singular value spectrum. In the present case, $M_t = 8$ and $M_s = 16$ appears to balance the singular value advantages of a compact sampling kernel against the compressive sampling advantages of multiplex measurement. None of the example strategies is ideal, however. Continuing research in sampling code and strategy design is highly likely to produce improvements.

10.6 SPECTRAL IMAGING

A *spectral image* is a map of the power spectral density $S(\boldsymbol{x}, \nu)$ over a range of spatial positions. We have assumed throughout this chapter that $S(\boldsymbol{x}, \nu)$ is a measurable quantity. The present section reviews optical systems for characterizing $S(\boldsymbol{x}, \nu)$ with a particular focus on emerging generalized sampling strategies. We focus on systems that characterize $S(x, y, \nu)$ over an image plane; spectral images $S(x, y, z, \nu)$ covering three spatial dimensions may be formed from 2D spatial images using projection tomography [167].

Spectral imagers are obviously useful for their declared purpose: forming a spatial map of the spectral density in an image. Spectral imaging is commonly used in environmental analysis and mineral detection in remote sensing and for molecular imaging in biological and chemical research [22,104]. Beyond the obvious applications, however, spectral imagers are important sensor engines for improving diverse imaging system metrics. We have already discussed several examples of spectral encoding for superresolution in Section 10.3 and have assumed at all points in the text that $S(x, y, \nu)$ is a measurable function.

Spectral imaging may be used to extend depth of field by combining a spectral imaging backplane with a chromatic objective lens. The wavelength-dependent focal length of a chromatic lens may be programmed using materials dispersion or diffractive structures. A spectral imager in combination with such a lens zooms in on a particular focus by simply selecting the appropriate reconstruction wavelength. Some of the most intriguing opportunities for spectral imaging combine focal coherence sensing similar to the astigmatic coherence sensor [172] with emerging trends in generalized sampling theory.

10.6.1 Full Data Cube Spectral Imaging

While one expects that spectral sensors based on the internal quantum dynamics of engineered materials may eventually impact design [92,127,140], current spectral imaging systems rely on optical filtering. Optical filters may be easily designed with essentially arbitrary spectral response and may be adapted to diverse spectral ranges and applications. Each spectrographic measurement strategy described in Chapter 9 may be adapted to spectral imaging applications. The relative merits of each approach are determined by resolving power and etendue as well as imager specific metrics such as resolution, field of view, frame rate, and feature specificity.

Figure 10.46(a) shows the basic structure of the the spectral data cube, while (b)–(e) illustrate common data cube sampling strategies. *Pushbroom scanning*, illustrated in Fig. 10.46(b), captures spectral data along one spatial dimension in each timestep. A typical pushbroom spectrometer relies on a dispersive slit spectrometer. A particular slice of the object is imaged on the input slit and spectrally characterized. The full data cube is captured by translating the image across the slit. The system sampling model for a pushbroom system adds spatial variation to Eqn. (9.5) to obtain

$$g_{nmk} = \iiint S(x, y, \lambda)\, h_r(x - n\Delta_x, y - m\Delta, \lambda - k\Delta_l)\, dx\, dy\, d\lambda \qquad (10.55)$$

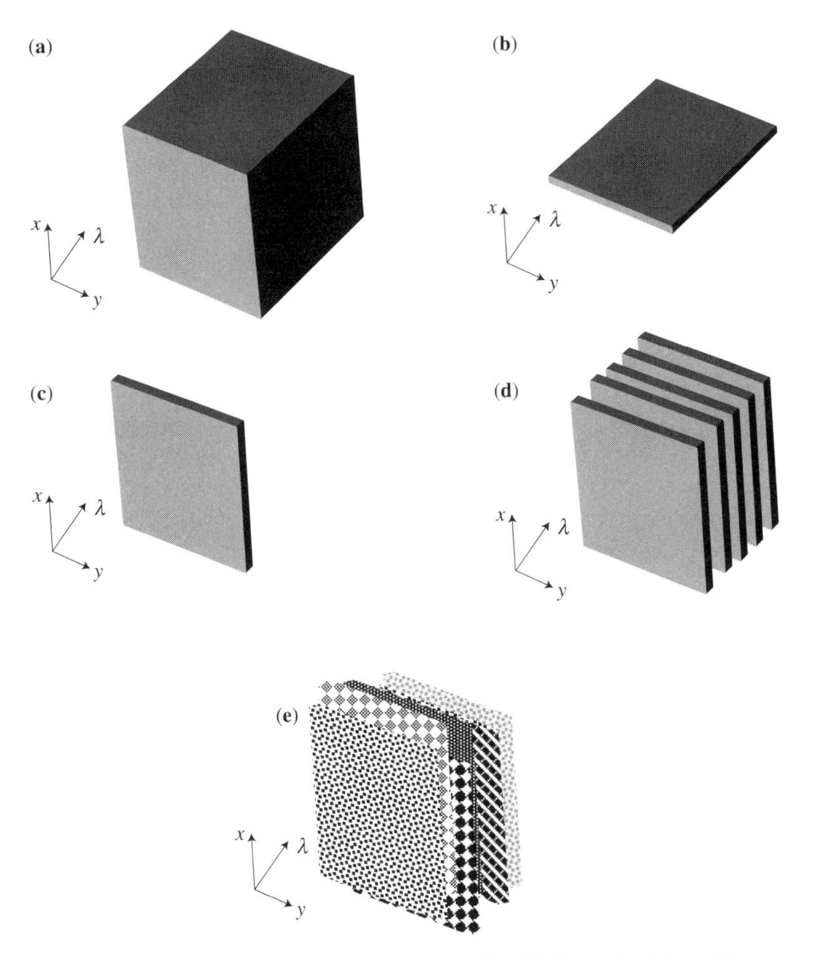

Figure 10.46 Spectral data cube measurement strategies: (a) data cube; (b) pushbroom scanning; (c) tunable filter; (d) interferometric filtering; (e) pixel filters.

where, as in Section 9.2, Δ is the pixel pitch and $\Delta_1 = \Lambda\Delta/F$. Δ_x is the displacement of the slit relative to the image from one recoding step to the next. The system response derived from Eqn. (9.4) is

$$h_r(x, y, \lambda) = t(x) \int\int h(x' - x, y' - y)\, p\left(x' - \frac{F\lambda}{\Lambda}, y'\right) dx' dy' \tag{10.56}$$

and the system transfer function is

$$\hat{h}_r(u, v, w) = \hat{t}\left(u + \frac{\Lambda w}{F}\right) \hat{h}\left(\frac{\Lambda w}{F}, v\right) \hat{p}\left(\frac{\Lambda w}{F}, v\right) \tag{10.57}$$

where we assume for simplicity that the optical impulse response $h(x, y)$ is independent of λ.

While the y resolution of the pushbroom spectrometer is determined by the standard imaging STF for the optics and the focal plane array, the x and λ resolutions are coupled through the slit scanning process. A hard limit on the spectral resolution is set by the pixel size, but spatiospectral resolution may exceed the static limit of the slit in a scanned system. One may regard this process as a form of digital superresolution. This effect is illustrated in the cross section of the STF in the uw plane shown in Fig. 10.47. System parameters in this example are $a_x = 100\,\mu\text{m}$, $\Lambda/F = 10^{-4}$, $\Delta = 10\,\mu\text{m}$, and $\lambda = 1\,\mu\text{m}$. Since the slit width is $10\times$ the pixel width, substantially higher spatial frequencies are achieved in the scanned system.

The limits of the STF are achieved, of course, only if the scan is sampled without aliasing. In most cases, one is likely to scan the pushbroom system such that $\Delta_x \approx a$, in which case the slit width a determines both the x and λ resolutions. Pushbroom systems are often deployed on moving platforms, such as aircraft, to take advantage of natural scanning.

The significance of etendue is slightly different for an imaging system when compared to a spectrometer. The etendue is a measure of the phase space of the instrument and is proportional to the number of degrees of freedom or pixels that the imager can abstract from the field. The etendue of a pushbroom instrument is the same as for the

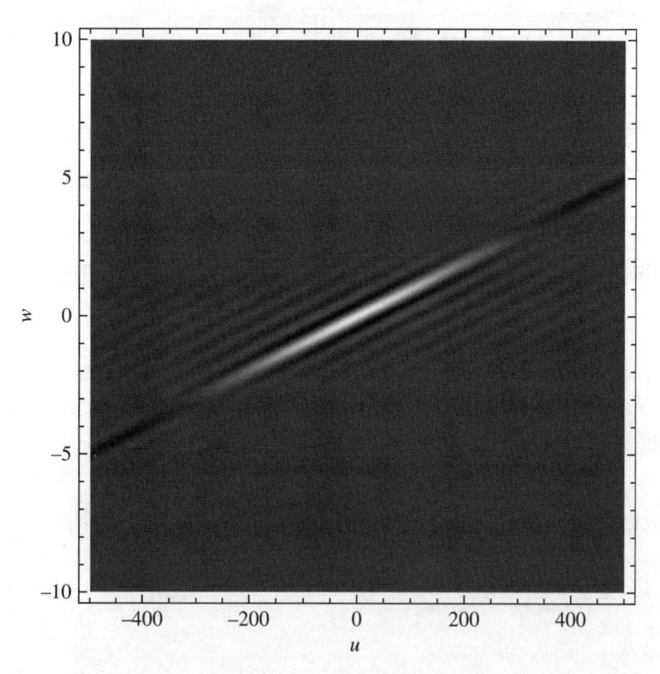

Figure 10.47 Cross section in the uw plane of the STF for a $f/2$ slit-based pushbroom spectrometer. The u axis is in units of mm^{-1} and the w axis is in units of nm^{-1}.

underlying slit spectrometer [Eqn. (9.9)]. Expressed in terms of the resolving power, we find

$$L \approx \frac{\pi^2 A^2 \lambda}{\Lambda R f/\#} \tag{10.58}$$

For an ideal imager, L should approach A^2. The reduction in L in proportion to the resolving power reduces the optical signal collected by the instrument and requires temporal scanning to achieve SNR and FOV objectives.

Coded aperture imaging spectrometers may be constructed using diverse filtering and scanning strategies [86]. For example, the coded aperture design of Section 9.3 actually already functions as 1D spatiospectral imager. Of course, a slit spectrometer is a 1D spatiospectral imager along the y (nondispersed) axis. An independent column code spectrometer, in contrast, images along the x (dispersion) axis. The aligned spectral reconstructions from each column of the aperture code, shown at the center right of Fig. 9.4, correspond to independent spectra for each column of the input aperture. To understand the operation of this instrument, imagine that the power spectral density $S(x, \lambda)$ is uniform as a function of y but varies along the dispersion axis x. Under this assumption, we generalize Eqn. (9.17) to define

$$S_{nj} = \int\int\int S(x + a_x j, \lambda + n\Delta_l)\tau_x(x)h_x(x' - x)p_x\left(x' + \frac{\lambda F}{\Lambda}\right)dx\, dx' d\lambda \tag{10.59}$$

Equation (9.19) generalizes in turn to

$$g_{nm} = \sum_i t_{im} S_{n-\alpha_x i, i} \tag{10.60}$$

For an independent column code there exists $\breve{t}$ such that $\sum_m \breve{t}_{jm} t_{im} = \delta_{ij}$, in which case

$$\sum_m \breve{t}_{jm} g_{nm} = S_{n-\alpha_x j, j} \tag{10.61}$$

A map of S_{nl} for integer values of $n - \alpha_x j$ and j is a 1D spatiospectral image of $S(x, \lambda)$. Fourier analysis of Eqn. (10.60) with respect to n and i yields the STF

$$\hat{h}_r(u, w) = \hat{\tau}\left(u + \frac{\Lambda w}{F}\right)\hat{h}\left(\frac{\Lambda w}{F}\right)\hat{p}\left(\frac{\Lambda w}{F}\right) \tag{10.62}$$

As in Chapter 9, the system response of the coded aperture system mimics that of the slit spectrometer, with the code feature size playing the role of slit width.

A 1D spectral image for an $N = 32$ Hadamard S-matrix spectrometer is shown in Fig. 10.48. The object is an ethanol solution encasing an acetaminophen solution in a cuvette 1 mm below the ethanol surface. The Raman spectrum of the return signal is plotted on the horizontal axis, while the vertical axis corresponds to position (mask column number). The ethanol Raman spectrum is evident at low wavenumbers on the

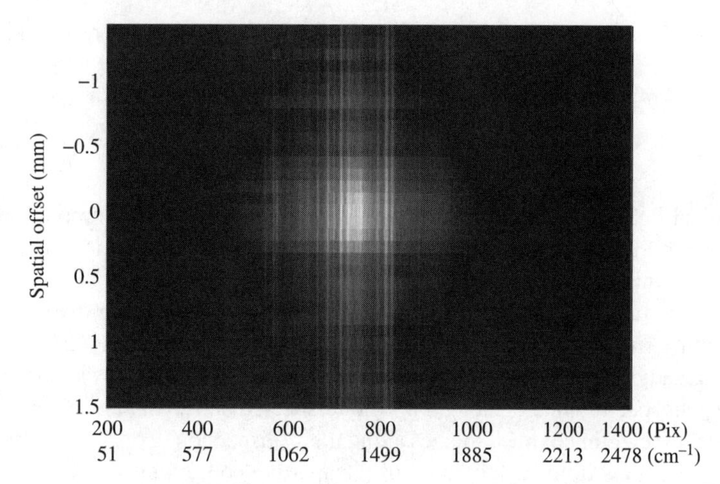

Figure 10.48 Spatiospectral image for a coded aperture spectrometer.

left of the image. While the acetaminophen spectrum is also strong near $x = 0$, it grows in strength relative to ethanol at the edge of the field. This effect is known as "spatially offset Raman spectroscopy" [174].

A potential problem in using a coded aperture system as a 1D spectral imager arises when the power spectral density is not uniform as a function of y. This problem also affects nonimaging coded aperture spectroscopy. For diffuse sources, a Fourier transform lens or diffuser is typically added in front of the coded aperture to ensure field uniformity. For the 1D imaging case, a cylindrical lens assembly may be used to image along x while diffusing along y. Coded aperture pushbroom instruments take another approach to achieving uniform y illumination [88]. For $S(x, y, \lambda)$ we define

$$S_{njm} = \int\int\int\int\int S(x + a_x j, y - m\Delta, \lambda + n\Delta_l)\tau_x(x)\tau_y(y)h(x' - x, y' - y) \times p_x\left(x' - \frac{\lambda F}{\Lambda}\right)p_y(y')\,dx\,dx'\,dy\,dy'\,d\lambda \quad (10.63)$$

and the discrete measurement model [Eqn. (9.19)] becomes

$$g_{nm} = \sum_i t_{im} S_{n-\alpha_x i,i,m} \quad (10.64)$$

In contrast with the slit pushbroom, which sweeps along the dispersion axis, a coded aperture pushbroom may sweep along the y axis. A sequence of measurements taken as the image shifts along the y axis produces the data cube

$$g_{nm'm} = \sum_i t_{im'} S_{n-\alpha_x i,i,m} \quad (10.65)$$

for $m' = 1$ to M. The data plane $g_{nm'm}$ for variable n and m' and fixed m is identical to the data that one would obtain for spectral density uniform with respect to y. Equation (10.65) can be inverted using independent column coding to estimate $S_{n-\alpha_x i,i,m}$ for integer $n - \alpha_x\ i,i,m$. Fourier analysis of Eqn. (10.63) yields an approximate transfer function

$$\hat{h}_r(u, w) = \hat{\tau}_x\left(u + \frac{\Lambda w}{F}\right)\hat{h}\left(\frac{\Lambda w}{F}, v\right)\hat{p}\left(\frac{\Lambda w}{F}, v\right) \tag{10.66}$$

where, as in Section 9.3, we have neglected diffractive crosstalk along y.

The reader may find a graphical recap useful in understanding coded aperture imaging spectroscopy. As illustrated in Fig. 10.49, the basic function of a dispersive spectrometer is to image the input object while shifting the color planes of the data cube. S_{nj} in Eqn. (10.59) refers to the nth color in the jth input column. As illustrated in Fig. 10.49, a dispersive spectrometer images the nth spectral channel from the jth input object column onto column $(j - 1) + n$ in the spectrally dispersed image.

The Nth column of the output image, therefore, consists of superimposed images of λ_N from the first column of object, λ_{N-1} from the second column, λ_{N-2} from the third column, and so on. For the coded aperture spectrograph of Section 9.3, one assumes that the input object is spatially uniform in each column. The coded aperture modulates each column with a unique spatial code such that the contribution of each object column to the signal measured in each output column can be computationally isolated. After decoding, the contributions $S(\lambda_N)$ from the first column, $S(\lambda_{N-1})$ the second column, and so forth in the Nth output column are determined independently.

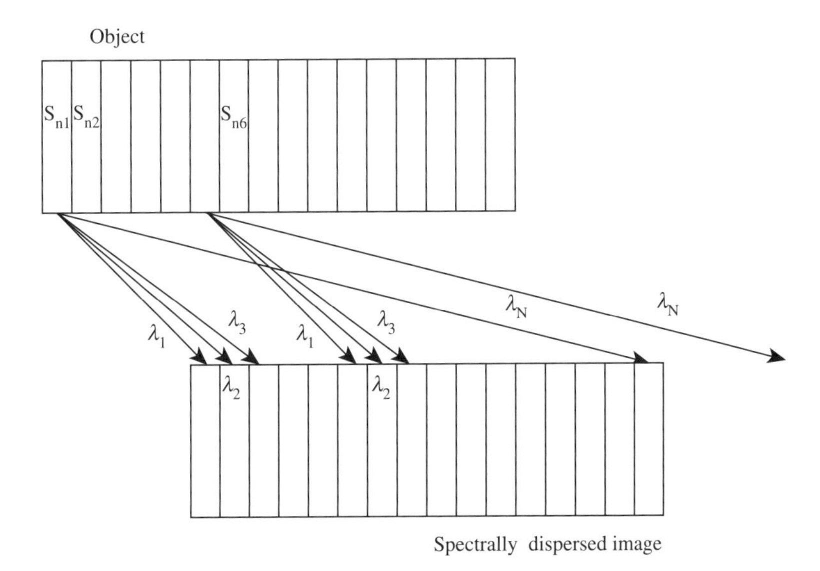

Figure 10.49 Dispersive imaging geometry.

This process transforms the raw image in Fig. 9.5(a) into the 1D spectral image in Fig. 9.5(d). In this particular case, the spectrum was uniform in all object columns, and one may average the column spectra to produce the mean spectrum in Fig. 9.5(e). If the spectra of the columns are different, one produces a 1D image, as in Fig. 10.48.

Suppose, however, that the object varies as a function of y such that S_{njm} must be indexed by row as well as column. In principle, this means that we can no longer use independent column coding to isolate the column spectra. As discussed above, however, one may sweep a pushbroom along the column to retrieve data consistent with illuminating the entire column with S_{njm}. Since each row is recorded independently, sampling S_{njm} when this object row illuminates the first row of the mask, then the second, and so on until a particular row has scanned the entire mask produces a data plane that enables imaging of the mth row.

A Hadamard coded aperture imager increases throughput relative to a slit with similar resolving power by the factor $N/2$, where N is the order of the aperture mask. Increased throughput may in turn enable more rapid scanning or improved spectral resolution. If N is increased in proportion to R, then the throughput is independent of resolving power. Noise tradeoffs with increasing throughput and multiplexing are similar to those discussed in Section 9.3.

Spectral imaging using a *tunable filter* is illustrated in Fig. 10.46(c). The system model for tunable filters, introduced in Eqn. (7.18), is extremely simple

$$g_{nmk} = \iiint S(x, y, \lambda) h_r(x - n\Delta, y - m\Delta) t(\lambda - k\Delta_l)\, dx\, dy\, d\lambda \tag{10.67}$$

where $h_r(x, y)$ accounts for both optical blur and pixel sampling. Tunable filters are typically implemented using liquid crystal and acoustooptic devices as discussed in Section 9.7. The relationship between angular field of view and resolving power is the primary limitation of these devices in imaging applications. The relationship between $\Delta\theta$ and R derived in Chapter 9 limits the $f/\#$ through the filter to

$$f/\# < \sqrt{\frac{R}{2}} \tag{10.68}$$

For this reason, tunable filters are most popular in microscopy, where the image space $f/\#$ is naturally large. The etendue for tunable filters is

$$L = \frac{\pi A^2}{2R} \tag{10.69}$$

where N_l is the number of wavelength channels. Tunable filters, especially acousto-optic devices, may achieve scan rates in the microsecond range. In addition to the relatively limited field of view, tunable filters also suffer from poor spectral throughput. Depending on the inversion algorithm and the structure of the object, this may not be an issue in shot-noise-limited systems, but it is a substantial drawback in systems dominated by additive noise. As discussed in Section 9.7, the spectral throughput is $\lambda/R\Delta\lambda$, meaning that the spectral throughput–efficiency product is inversely proportional to R^2.

An *interferometric filter* captures linear combinations of the spectral data planes in each timestep, as illustrated in Fig. 10.46(d). Interferometric imagers based on scanning Michelson interferometers are common. The system model for an interferometric system is identical to Eqn. (9.31) with spatially dependent mutual coherence $\Gamma(x, y, \Delta z)$. The etendue is similar to the tunable filter result of Eqn. (10.69), but the spectral throughput is $\frac{1}{2}$.

Pixel filters, as illustrated in Fig. 10.46(e), capture different linear combinations of the spectral channels at each spatial pixel. The RGB sampling strategies discussed in Section 7.3 are a form of pixel filtering, but this strategy may be extended to more than three colors. Pixel filtering may be implemented using the spectroscopic filter technologies described in Section 9.6. As discussed in Section 9.8.4, patterned diachronic filters have become available for spectral imaging [35,245]. One expects that such devices as well as continuing advances in photonic crystal and metamaterial filters will eventually enable pixel filter integration directly on focal plane arrays. At present, however, coded apertures provide the most direct and easily programmed pixel-level filtering platform.

10.6.2 Coded Aperture Snapshot Spectral Imaging

The spectral data cube is generally "highly compressible," meaning that image data in different spectral bands tend to be redundant. Digital compression and feature extraction algorithms may be reasonably expected to compress spectral data cubes by several orders of magnitude with little or no loss. One expects, therefore, that compressive sampling will be effective in spectral imaging.

One uses compressive sampling in spectral imaging to reduce image acquisition time, to increase throughput and sensitivity, and to simplify image acquisition hardware. Full-data-cube sampling strategies discussed to this point each rely on temporal scanning to fill in the 3D data cube using 2D detector arrays. These strategies fall in the "conventional measurement" category discussed in Section 7.5.1. Compressive measurement systems, on the other hand, need not preserve the dimensionality of the object embedding space in measurement data. A compressive spectral imaging system, in particular, can characterize the 3D data cube in a single snapshot on a 2D detector array.

One expects a successful compressive spectral imager design to have the following characteristics:

- *Multiplex Sampling*. The goal of the system is to estimate a signal $\boldsymbol{f} \in \mathbb{R}^N$ using M measurements. While one might achieve this objective without measuring many of the signal pixels, doing so would decrease quantum efficiency. One supposes, therefore, that measurements will consist of linear combinations of data cube voxels.

- *Flat SVD Spectrum.* As discussed in Section 10.5, multiplex sampling strategies may be evaluated according to the structure of the singular value spectrum. The effective number of measurements in a generalized sampling system is more reasonably related to the number of singular values above a noise floor than to the number of optoelectronic detector values recorded.
- *Restricted Isometry Property* (*RIP*). Where the singular value spectrum reflects on the quantity of measurements recorded, the structure of the singular vectors reflects on the quality of the data. Measurement systems must separate distinct signals into distinct measurement data consistent with the RIP discussed in Section 7.5.

With these principles in mind, the coded aperture spectrometer illustrated in Fig. 9.4 and discussed as a spectrometer in Section 9.3 and as a pushbroom imager in Section 10.6.1 is also an excellent candidate for compressive spectral imaging. To use the system as a compressive imager, one need only reinterpret Eqn. (10.63). We term instruments based on this approach as *coded aperture snapshot spectral imagers* (CASSIs) [88,242]. The "snapshot" capability, such as the ability to estimate the full spectral data cube from a single 2D frame of measurements, is the primary distinction of CASSI instruments relative to full-data-cube spectral imagers.

A CASSI instrument based on the spectrograph of Fig. 9.4 is most simply described as a 2D imager in the x–λ plane. As discussed above, this instrument is a simple imager along the y (undispersed) axis with y image pixels mapping isomorphically to object pixels indexed by m in Eqn. (10.63). Accordingly, we focus on a 2D version of Eqn. (10.60)

$$g_n = \sum_i t_i S_{n-i,i} \tag{10.70}$$

where we assume for simplicity that $\alpha_x = 1$. Defining $i' = n - i$, Eqn. (10.70) can be rewritten

$$g_n = \sum_i t_{n-i'} S_{i',n-i'} \tag{10.71}$$

Measurement based on Eqn. (10.73) is simulated in Fig. 10.50. While, as discussed above, the sampling system mixes spatial and spectral structures, one may roughly associate the n axis in S_{nj} with the color spectrum and the j axis with spatial position. Figure 10.50(a) plots an example slice S_{nj}. To illustrate the coding structure, we assume that the image consists of a rectangle in the nj plane. Figure 10.50(b) plots $t_{n-i'} S_{i',n-i'}$ when a pseudorandom binary code t_i uniformly drawn from [0,1] modulates the spectral image of Fig. 10.50(a). A CASSI system integrates Fig. 10.50(b) along the vertical axis to produce the measurement data shown in Fig. 10.50(c). The baseline is shifted to allow the sampling code to be plotted beneath the measurement data.

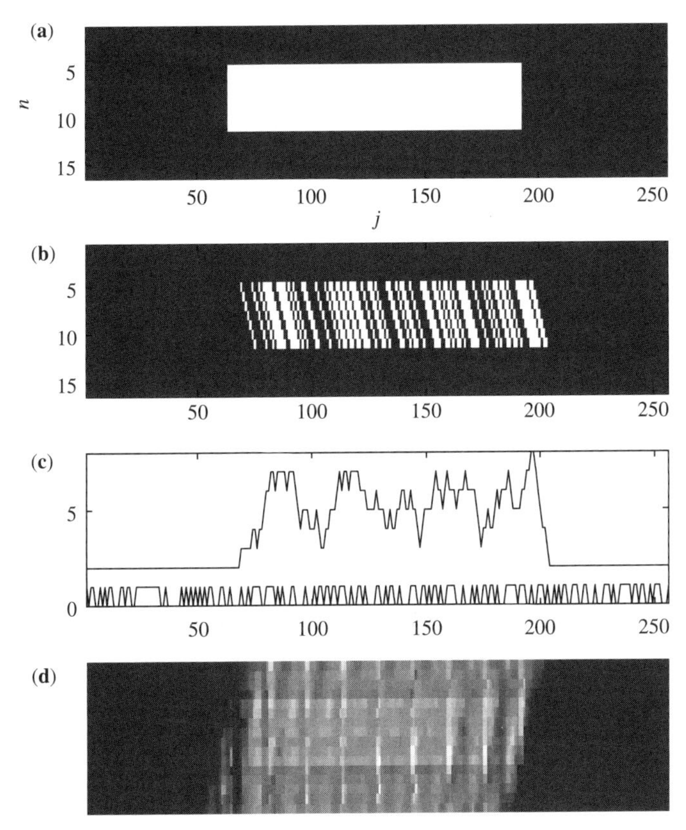

Figure 10.50 Simulated measurement data based on Eqn. (10.73): (a) x–λ spectral data plane; (b) data plane after punch and shear operations; (c) "smashing" of the modulated data plane to produce the measurement vector; (d) estimation of the true data plane from the measurements using Bioucas-Dias and Figueiredo's TWIST algorithm [20] with $\tau = 0.1$.

A CASSI system seeks to estimate the full spectral image of Fig. 10.50(a) from the data plotted in Fig. 10.50(c). The basic idea is that the high-frequency code structure will be uncorrelated with natural image features. The code modulation is abstracted to jointly estimate the spatial and spectral structure. Diverse code patterns may be considered. While the code of Fig. 10.50 is a multiplex code in the sense that multiple voxels are added in each data point, single-channel CASSI codes are also possible. Since each spatiospectral voxel is assigned to a measurement with weight 1 or 0 CASSI codes are orthogonal. While this produces an absolutely flat singular value spectrum the primary question is, of course; does the CASSI code separate distinct objects into distinct measurement data?

Figure 10.50(d) is a reconstruction of the full (a) data plane using the measurements of (c) and constrained total variation optimization implemented via the two-step iterative shrinkage/thresholding (TWIST) algorithm [20]. The reconstruction quality is poor, but one may alternatively choose to be amazed by the similarity to the original image when one considers that the reconstruction is based on $16\times$ compressed data using a single projection. The basic reconstruction problem is very similar to Radon reconstruction from limited projections, although the sampling code is used to create a data prior and higher-frequency response. Results are somewhat better for the sparser data plane illustrated in Fig. 10.51.

While the 2D CASSI projection would not be effective in reconstructing complex S_{nj} data planes, spectral data cubes are highly correlated. Reconstruction algorithms enforcing wavelet sparsity [242] and total variation constraints in the xy plane for a modest number of spectral channels are effective full-data-cube estimation from CASSI data. For example, Fig. 10.52 shows a spectral data cube reconstructed in an experimental CASSI system from the measurements illustrated in Fig. 10.53. The object consists of several plastic models illuminated by standard

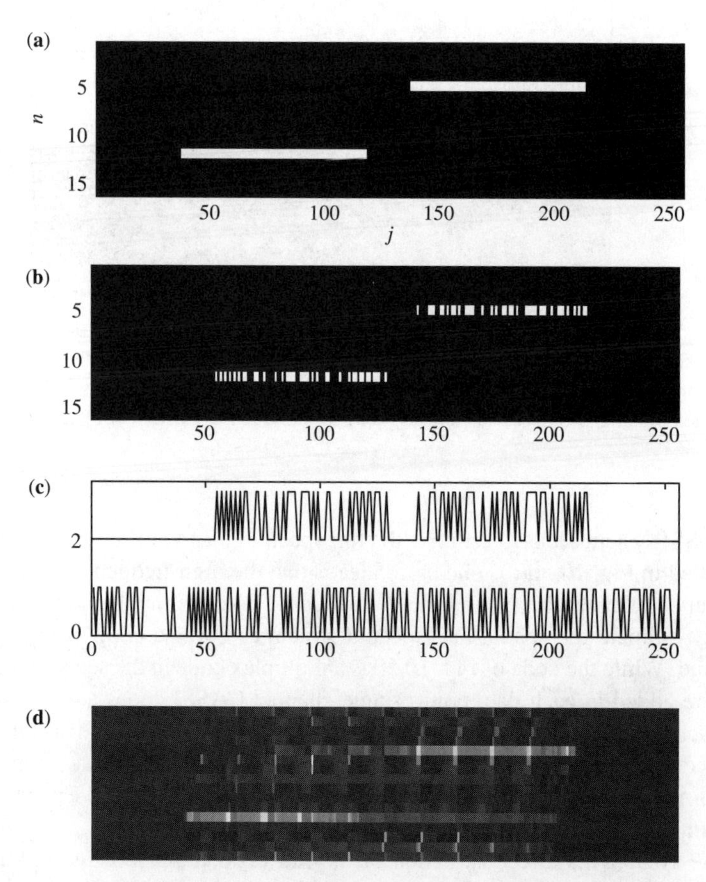

Figure 10.51 Data plane (a), punch-and-shear plane (b), measurements (c), and reconstruction (d), using the the same code and estimation algorithm as in Fig. 10.50.

fluorescent lighting. The illumination spectrum includes a blue band at 495 nm, green at 550, yellow at 590, and red above 600. This spectrum is modulated in turn by the color of the objects, including a blue stapler and a variety of plastic fruit (a yellow banana, red apple, and green pineapple). One also observes both specular and diffuse reflection from the objects. As expected, the banana and pineapple are more apparent in the yellow and green bands, and the apple is clear in the red band. This experiment used a 2D pseudorandom code with reconstruction with the TWIST algorithm under a *TV* constraint with regularization parameter $\tau = 0.1$ [20].

Diverse reconstruction and coding approaches may be applied to the basic CASSI architecture. A monochromatic object illuminating a CASSI system produces a clean image of the object shifted in proportion to the wavelength and modulated by the code. In this case, simple correlation may be used to find the shift and identify the

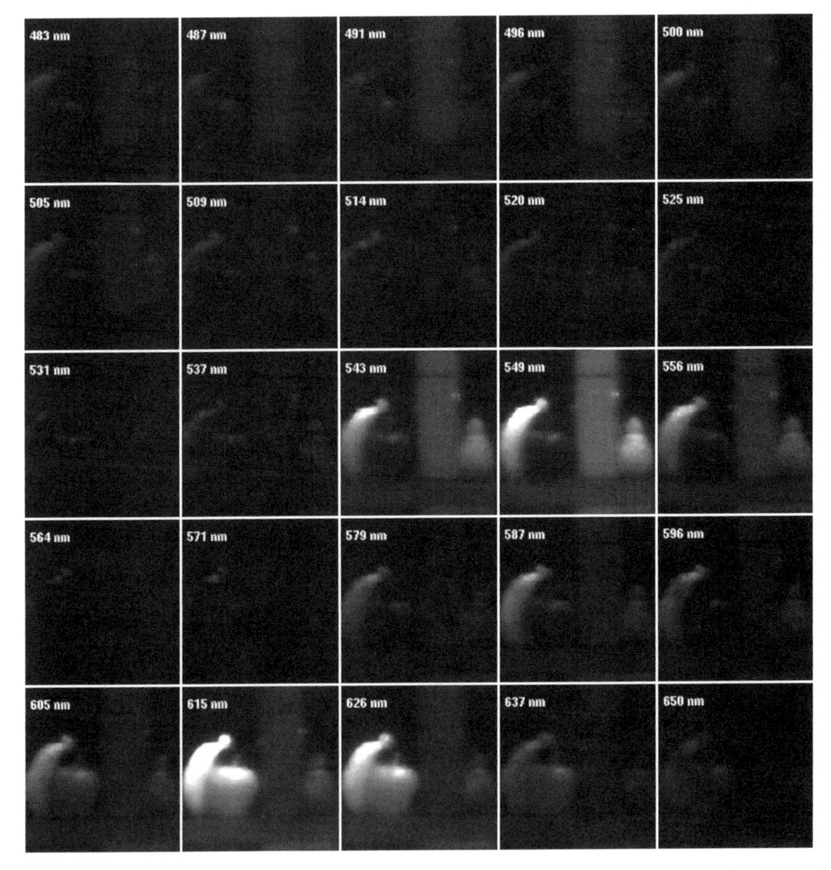

Figure 10.52 Spectral data cube reconstructed from CASSI measurements using TV minimization with the TWIST algorithm. (Figure generated by Ashwin Wagadarikar.)

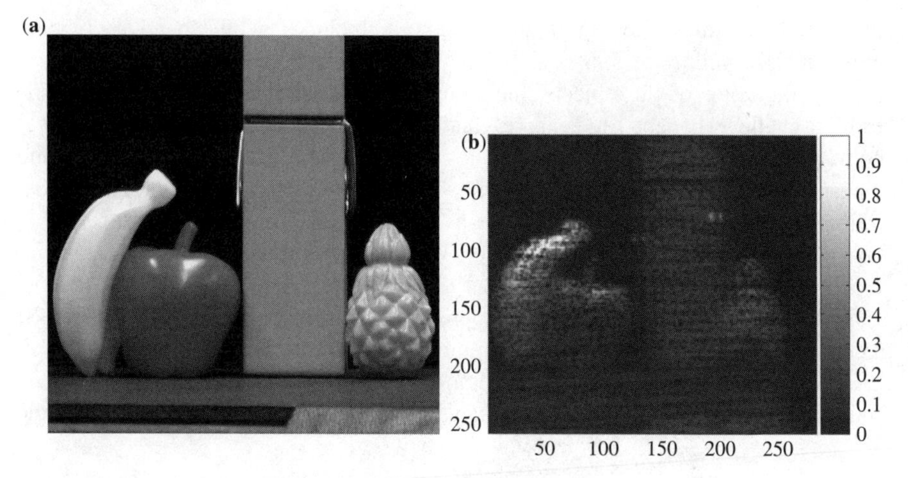

Figure 10.53 Black-and-white image (a) and CASSI data plane (b) for the object of Fig. 10.52.

wavelength. The code modulation might then be removed by denoising. More generally, one finds that images of natural scenes captured at different wavelengths tend to look very similar, in which case a separable model is appropriate. The separable model $S(x, \lambda) = f(x)S(\lambda)$ may be parameterized with 1D coefficients such that Eqn. (10.73) admits algebraic solutions. One anticipates, however, that most scenes are not fully separable. More likely, a sconce consists of a sparse array of locally separable features.

To this point we have focused on pseudorandom aperture codes. In practice, one is likely to optimize the CASSI code t_i to eliminate long sequences of 1 or 0 and to improve object feature separation. The question naturally arises, however, "Why use codes at all?" CASSI systems incorporate

- "Punch" operations in which voxels are removed from the object data cube by a transmission mask or modulator array
- "Shear" operations in which spatial dispersion is used to translate spectral data planes relative to each other [as in Fig. 10.50(b)]
- "Smash" operations in which a detector array is used to integrate the signal along the spectral axis

The smash operation is intrinsic to the detector operation, shear and punch operations are added to enable data cube estimation. The simplest strategy is conventional black-and-white imaging. A black-and-white image is formed by integrating Fig. 10.50(a) along the spectral axis. The black-and-white code also has a flat singular value spectrum with singular vectors consisting of columns of 1s at each spatial pixel. One may well ponder why the black-and-white measurement is less effective than the CASSI

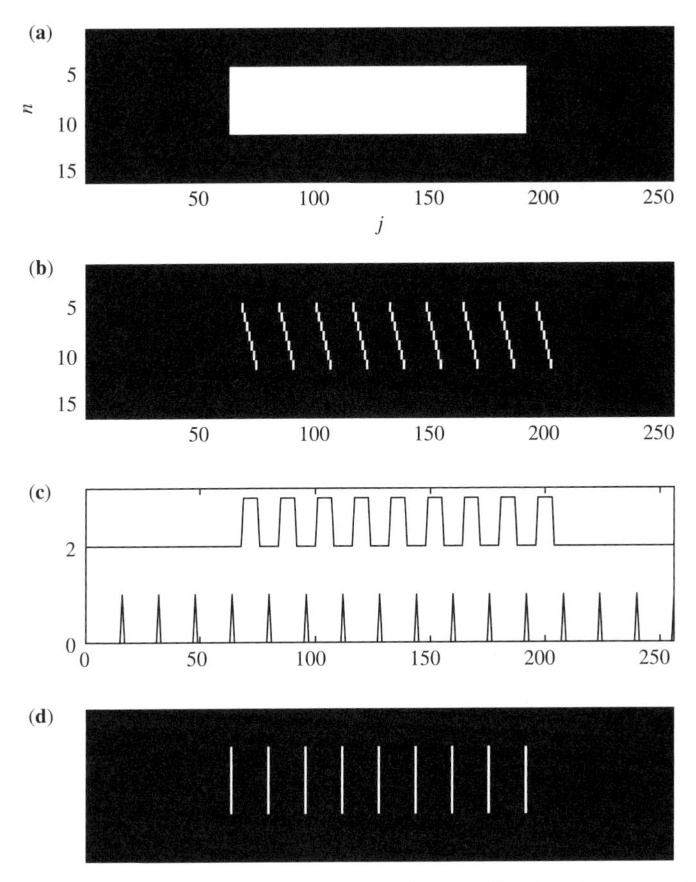

Figure 10.54 CASSI sampling with a sparse periodic code. The data plane (a) is the same as in Fig. 10.50 and is again reconstructed using the TWIST algorithm, although in this case each measurement corresponds to a single (nonmultiplexed) voxel in the data cube. The full data cube might be reconstructed by interpolation within aliasing limits similar to those discussed in Section 7.3.

system in producing the full data cube. The answer, of course, is that the black-and-white system does not produce different measurement data for images with similar spatial features but different spectra. The idea of CASSI coding is to create noise-like features in the measurement data to more effectively exploit the information capacity of the detector array.

Multiple punch and shear operations may be implemented in various orders to vary the system model and sampling strategy. One might, for example, apply the sparse periodic CASSI code

$$t_i = \begin{cases} 1 & \text{if } \mathrm{mod}(i, N_l) = 1 \\ 0 & \text{otherwise} \end{cases} \tag{10.72}$$

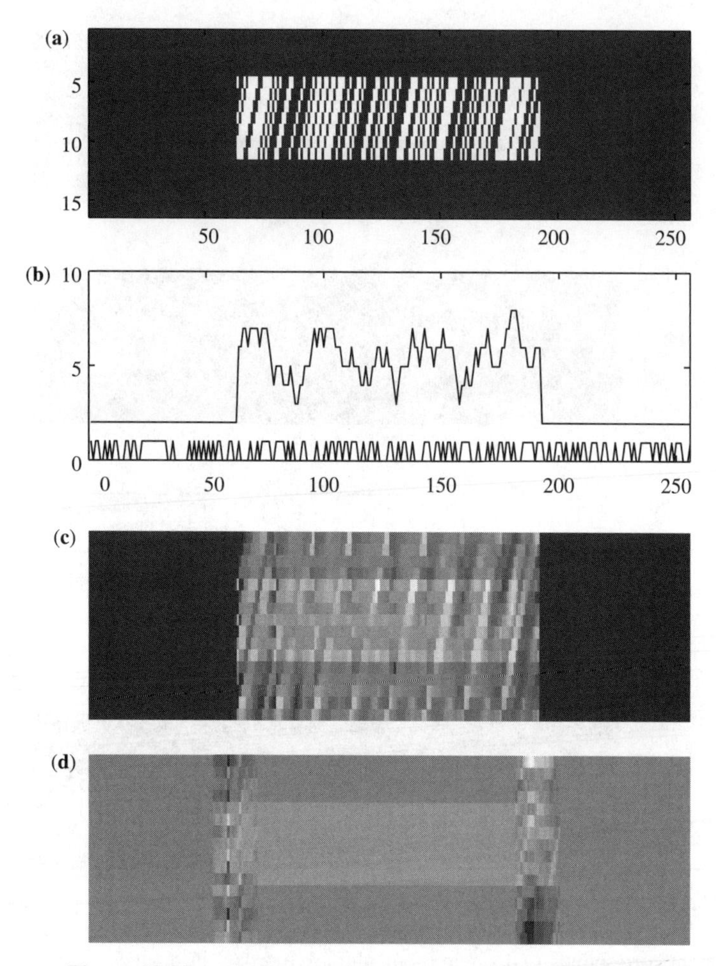

Figure 10.55 CASSI sampling with a dual-disperser design.

With this code, each measurement g_n returns the power spectral density S_{ni} for a single value of n and i. The full data cube is then reconstructed by interpolation from the known voxels. Figure 10.54 samples and reconstructs the data of Fig. 10.50 using a sparse code. A sparse code CASSI system is effectively an array of slit spectrometers, although better results are obtained if the input features are dispersed from row to row (e.g., along y) rather than aligned in slits. This approach enables spatial sampling of N_l color channels with period $\sqrt{N_l}$ in the $x-y$ plane. Analysis of sparse sampling is similar to the RGB color sampling of Section 7.3 with N_l color channels rather than just three. This sampling strategy may be preferred in shot-noise-dominated systems imaging weak spectral features, as discussed in Section 9.3.

Another approach uses a shear–punch–shear operation in which the spectral data planes are sheared prior to modulation and then restored to original relative positions. This "dual disperser" design produces well-registered spectral data planes at the detector [87]. The dual-disperser sampling model is

$$g_n = \sum_i t_{n-i} S_{i,n} \tag{10.73}$$

Dual-disperser sampling with using the same random code as in Fig. 10.50 is illustrated in Fig. 10.55. The data plane is the same rectangle as in Figs. 10.50(a) and 10.54(a). The code modulation comparable to Fig. 10.50(b) is shown in Fig. 10.55(a). Figure 10.55(b) shows the measured data corresponding to (a) and (c) as the TWIST reconstruction comparable to Fig. 10.50(d). A dual disperser produces periodic spectral projections on the measured image similar to Bayer pattern RGB sampling and sparse periodic CASSI. Figure 10.55(d) is a least-squares reconstruction of the (b) data generated using a neighborhood of 20 points around each x position in the measurement vector. Of course, a more compact neighborhood could be used in a fully 3D spectral imager. While image interpolation is similar in each case, the dual disperser produces a multiplex code with $N_l/2$ greater throughput than Bayer or sparse coded aperture sampling strategies.

As of this writing, code, sampling, inference algorithm, and system design for CASSI architectures and more general 2D spectral imaging filter arrays is evolving rapidly. The reader may be better advised to focus on the opportunity of generalized and compressive measurement rather than the details of current implementations. One may reasonably expect that many of the spectral modulation technologies described in Chapter 9 will eventually be integrated in snapshot spectral imaging systems. Some imagination will be necessary, however, as emerging designs expand the range of spectral filtering components beyond those covered in Chapter 9. For example, since the number of spectral channels is modest and high image quality and quantum efficiency is paramount, current CASSI designs use prism-based dispersion elements rather than diffractive gratings [242]. One anticipates that novel materials, mathematics, and manufacturing technologies will continue to drive rapid innovation.

PROBLEMS

10.1 *Depth of Field and Depth of Focus.* A lens system with effective focal length F and aperture A is aligned to image objects at range z_o.

(a) For circle of confusion C, what is the depth of focus Δz for this system?

(b) Assuming a circle of confusion $C = 10\ \mu\text{m}$, compare numerical values for the depth of focus for Airy and Gaussian PSFs with $\Delta x = 3\ \mu\text{m}$ and $\lambda = 1\ \mu\text{m}$.

10.2 *Logarithmic Aspherical Lenses*

(a) Assuming that the logarithmic aspherical lens pupil modulation of Fig. 10.4 is implemented using refractive components, plot the lens SAG.

(b) Numerically calculate and plot the PSF for this lens for θ_z varying from $-0.05/F$ to $0.05/F$ [e.g., calculate the plots of Fig.10.4(b) over a wider range of defocus].

(c) The Strehl ratio for a lens is the ratio between the peak value of the incoherent PSF and the peak value for a diffraction-limited PSF using the same aperture. Plot the Strehl ratio as a function of defocus for lens of Fig.10.4(a) and for the lens of Fig.10.4(b). (*Hint*: Use Fresnel propagation and FFT analysis to numerically calculate the PSFs.)

10.3 *Cubic Phase Coding for Extended Depth of Field.* Consider the model PSF for a cubic phase EDOF system with finite spectral bandwidth

$$h_r(\theta_x) = \frac{1}{\lambda d_i} \mathcal{F}\left\{ e^{i\pi(\theta_z/\lambda)x^2} e^{i(\alpha/\lambda)x^3} \operatorname{rect}\left(\frac{x}{A}\right) \right\}\Big|_{u=(\theta_x/\lambda)} e^{-1000\theta_x^2} \tag{10.74}$$

(a) Simulate the detected image for this PSF for an image of your choosing. Describe your sampling parameters and resolution. Specify a reasonable value for α and discuss physical implementation of a cubic phase modulation with the targeted α.

(b) Assuming normally distributed additive noise with SNR = 20 dB, simulate Wiener filter and Richardson–Lucy image estimation with this PSF.

10.4 *Range-Variant PSFs.* Plot the MTF as a function of defocus for the fundamental Gaussian mode and for the rotating PSF of Fig. 10.10. Compare the rotating PSF with the PSFs of the constituent higher-order Laguerre–Gaussian modes.

10.5 *Prolate Spheroidal Wavefunctions*

(a) Use Gauss–Legendre integration to calculate the eigenvalues of $\psi_n(x)$ for $c = 100$. Compare your result with the plots shown in Fig. 10.15.

(b) Use the `Mathematica` function `SpheroidalPS[n, 0, c, x]` to plot $\psi_5(x)$ for $c = 2$, $c = 5$, and $c = 10$.

10.6 *Polychromatic Superresolution in Remote Imaging*. Suppose that one illuminates a remote object over a 0.1 steradian (sr) cone offset by 1 radian from the optical axis. The optical system collects 10^{-4} sr. Assuming one octave spectral bandwdith, describe the band volume for measurement of $\sigma(\mathbf{x})$ based on Eqn. (10.37). Estimate the object space imaging resolution.

10.7 *The Rayleigh Criterion.* Two closely spaced point sources are imaged by a diffraction-limited imaging system. Assume that the angular separation is

$\lambda/4A$. Develop a sensor model, assuming Nyquist or better optoelectronic sampling. Implement and compare Wiener filter, Richardson–Lucy, and l_1 minimization strategies for deconvolution and object estimation. Assume normally distributed additive noise with SNR varying from 100 to 10,000.

10.8 *Degrees of Freedom.* Accepting Lohmann's [157] observation that $f/\# \propto f^{1/3}$, derive scaling laws for imager volume and image degrees of freedom as a function of f. Compare your scaling laws with currently available imagers.

10.9 *Longitudinal Resolution.* On the basis of the bandpass limits for focal imaging, plot the aperture size A as a function of range z_o necessary to obtain longitudinal resolution $\Delta z = z_o/100$ for $\lambda = 1\,\mu\text{m}$ for $100\ \mu\text{m} < z_0 < 100$ m. Assuming that $A = 100\ \mu\text{m}$, plot the aperture displacement necessary to achieve similar longitudinal resolution using optical projection tomography.

10.10 *Shower Curtains.* An observer (Bob) standing close to, although not immediately against, a phase diffuser has trouble seeing objects through it. An observer (Alice) at some distance on the other side of the diffuser sees Bob relatively clearly. The ability of Alice to observe Bob and the inability of Bob to observe Alice is called the "shower curtain effect."

(a) With reference to the discussion of superresolution through phase modulation in Section 10.3.2, explain this effect.

(b) Suppose that Alice and Bob each know the transmittance of the curtain. Who finds it more useful for superresolved imaging? How useful is it?

10.11 *Digital Superresolution.* Consider a 1D signal $f(x) = [(1 + \cos(20\pi x^{1.25})]/2$ on the range $x \in (0, 1)$.

(a) Plot the measurement data when $f(x)$ is sampled uniformly with period $\Delta = 0.1$ with pixel sampling function $p(x) = \text{rect}(x/\Delta)$. Interpolate a display signal using cubic spline and Shannon sampling.

(b) Repeat **(a)** using the same pixel function with $\Delta = 0.05$.

(c) Repeat **(a)** using $p(x) = \text{rect}(x/2\Delta)$ and $\Delta = 0.05$.

(d) Discuss the relationship between this problem and multiple aperture digital superresolution.

10.12 *Optical Projection Tomography.* A digital camera with 40° FOV observes an object at a range of 1 m. The camera aperture is 1 cm, and the focal length is 2 cm. The object rotates by 360° around an axis normal to the optical axis of the camera. Plot the band volume observed by the camera, labeling all axes in SI units. Assuming diffraction-limited resolution, estimate the resolution with which the object can be estimated in all three spatial dimensions.

10.13 *Pixel Coding.* Consider a 256 × 256 image, such as the "cameraman" image distributed with Matlab. Develop model measurements for 4× downsampling with perfectly registered subpixel shifts and for Hadamard pixel sampling as described in Section 10.5. Assume normally distributed additive noise in each measurement with $\sigma = 0.01\mu$ (where μ is the mean measurement value). Compare least squares and Tikhonov regularized reconstruction for each measurement model.

10.14 *Noise and Compressive Sampling.* You are given 10 detectors. Your goal is to characterize a signal consisting of 128 elements on the canonical basis. At most 10 of these elements are nonzero. The mean value of the nonzero elements is μ. You must make your measurement in time T. Each measurement produces standard deviation $\sigma = 0.01\mu\delta t/T$, where δt is the duration of the measurement. Use numerical simulation to compare measurement fidelity for the following strategies:

(a) Measure each channel individually. This strategy makes 128 measurements, each of duration $10T/128$.

(b) Make 40 measurements, each of duration $T/4$, using random multiplex projections. In each time period, the energy from each signal channel is assigned to one of the 10 detectors. Estimate the signal using l_l minimization.

(c) Divide the signal into two groups of 64 elements. Make 20 measurements on each group, each measurement of duration $T/4$. In each time period, the energy from each signal channel is assigned to one of the five detectors assigned to the corresponding group. Estimate the signal using l_l minimization.

Can you devise a measurement strategy with signal estimation fidelity better than that of any of the three strategies listed here?

10.15 *Pushbroom Coded Aperture Spectroscopy*

(a) Prove that Eqn. (10.66) accurately describes the STF of the 1D coded aperture imaging system.

(b) Plot the cross section of the STF in the uw plane for $a_x = 30$ μm, $\Lambda/F = 10^{-3}$, $\Delta = 10$ μm and $\lambda = 0.5$ μm.

(c) Suppose that

$$S(x, \lambda) = \exp\frac{(x - 20\Delta)^2}{400\Delta^2}S_1(\lambda) + \exp\frac{(x + 20\Delta)^2}{400\Delta^2}S_2(\lambda) \qquad (10.75)$$

where, as in Problem 9.3, $S_1(\lambda)$ and $S_2(\lambda)$ consist of 1–10 Lorentzian lines. Simulate the forward model and image reconstruction for a 127 × 127 1D spectral image of $S(x, \lambda)$ using a Hadamard S-matrix code.

10.16 *2D CASSI.* Consider a CASSI system imaging in the x–λ plane.

(a) Design a CASSI system with 1 nm spectral resolution and 1 mrad angular spatial resolution operating over the 500–700 nm spectral band. Specify mask feature sizes, focal lengths, grating or prism dispersion rates, and focal plane pixel pitch.

(b) The signal

$$S(\theta_x, \lambda) = \phi_0 \left(\frac{\theta_x}{\sigma_x} \right) S(\lambda) \tag{10.76}$$

is imaged using this system. σ_x spans 15 mrad and $S(\lambda)$ is a Lorentizan line of width 2 nm centered on 575 nm. Replicate Figs. 10.50, 10.54, and 10.55 for this image using the system design of (a).

10.17 *3D CASSI.* Use a 128×128 RGB image of your choosing to generate a synthetic 16-spectral-channel image by modeling the red channel with a Gaussian spectrum centered on 600 nm of width 75 nm, the green channel with a Lorentizian spectrum centered on 550 nm, and width 50 nm, and the blue channel with a Lorentzian spectrum centered on 500 nm and width 10 nm.

(a) Generate synthetic CASSI measurements for your image on the bases of random sampling in the single disperser and dual-disperser architectures and 16-channel sparse sampling.

(b) Use *TV* minimization in the *xy* plane to estimate full data cubes from your measurements. For the dual-disperser and sparse sampling systems, compare your result with interpolated estimation.

REFERENCES

1. E. H. Adelson and J. Y. A. Wang, Single lens stereo with a plenoptic camera, *IEEE Trans. Pattern Anal. Mach. Intell.* **14**:99–106, 1992.
2. D. K. Agrawal, D. J. Brady, and J. Matusek, Segmenting object space by geometric reference structures, *ACM Trans. Sensor Networks* **2**(4):455–465, 2006.
3. R. H. Anderson, Close-up imaging of documents and displays with lens arrays, *Appl. Opt.* **18**(4):477, 1979.
4. A. Ashok and M. Neifeld, Information-based analysis of simple incoherent imaging systems, *Opt. Express* **11**(18):2153–2162, 2003.
5. A. Ashok and M. A. Neifeld, Pseudorandom phase masks for superresolution imaging from subpixel shifting, *Appl. Opt.* **46**(12):2256–2268, 2007.
6. R. Aster, B. Borchers, and C. Thurber, *Parameter Estimation and Inverse Problems*, Elsevier, 2004.
7. W. P. Aue, E. Bartholdi, and R. R. Ernst, Two-dimensional spectroscopy: Application to nuclear magnetic resonance, *J. Chem. Phys.* **66**:2229–2246, 1976.
8. H. H. Barrett and K. J. Myers, *Foundations of Image Science*, Wiley-Interscience, 2004.
9. S. A. Basinger, E. Michielssen, and D. J. Brady, Degrees of freedom of polychromatic images, *J. Opt. Soc. Am. A (Opt. Image Sci. Vision)* **12**(4):704, 1995.
10. S. A. Basinger, R. A. Stack, K. B. Hill, and D. J. Brady, Superresolved optical scanning using polychromatic light, *J. Opt. Soc. Am. A* **14**(12):3242–3250, 1997.
11. M. J. Bastiaans, Wigner distribution function applied to optical signals and systems, *Opt. Commun.* **25**(1):26–30, 1978.
12. M. J. Bastiaans, Application of the Wigner distribution function to partially coherent light, *J. Opt. Soc. Am. A* **3**(8):1227–1238, 1986.
13. M. J. Bastiaans, *Gabor Analysis and Algorithms: Theory and Applications*, Birkhäuser, 1998, Chapter 14, pp. 427–451.
14. S. Basty, M. A. Neifeld, D. Brady, and S. Kraut, Nonlinear estimation for interferometric imaging, *Opt. Commun.* **228**(4–6):249–261, 2003.
15. G. Battle, A block spin construction of ondelettes. 1. Lemarie functions, *Commun. Math. Phys.* **110**(4):601–615, 1987.
16. B. Bayer, *Color Imaging Array*, US Patent 3,971,065, 1975.
17. E. Ben-Eliezer, E. Marom, N. Konforti, and Z. Zalevsky, Radial mask for imaging systems that exhibit high resolution and extended depths of field, *Appl. Opt.* **45**:2001–2013, 2006.

18. H. Beneking, Gain and bandwidth of fast near-infrared photodetectors—a comparison of diodes, phototransistors, and photoconductive devices, *IEEE Trans. Electron. Devices* **29**(9):1420–1431, 1982.

19. M. Bertero and P. Boccacci, *Introduction to Inverse Problems in Imaging*, Taylor and Francis, 1998.

20. J. M. Bioucas-Dias and M. A. T. Figueiredo, A new twist: Two-step iterative shrinkage/thresholding algorithms for image restoration, *IEEE Trans. Image Process.* **16**:2992–3004, 2007.

21. R. E. Blahut, *Theory of Remote Image Formation*, Cambridge Univ. Press, Cambridge, UK, 2004.

22. M. Borengasser, W. S. Hungate, and R. L. Watkins, *Hyperspectral Remote Sensing: Principles and Applications*, CRC Press, Boca Raton, FL, 2008.

23. M. Born and E. Wolf, *Principles of Optics*, 6th ed., Cambridge Univ. Press, 1980.

24. S. P. Boyd and L. Vandenberghe, *Convex Optimization*, Cambridge Univ. Press, 2004.

25. W. S. Boyle and G. E. Smith, Charge-coupled semiconductor devices, *Bell Syst. Tech. J.* **49**:587, 1970.

26. R. N. Bracewell, *Fourier Analysis and Imaging*, Springer, 2004.

27. D. Brady and D. Psaltis, Control of volume holograms, *J. Opt. Soc. Am. A* **9**(7):1167–1182, 1992.

28. D. Brady and D. Psaltis, Information capacity of 3-d holographic data-storage, *Opt. Quantum Electron.* **25**(9):S597–S610, 1993.

29. D. J. Brady, N. Pitisianis, X. Sun, and P. Potuluri, *Compressive Sampling and Signal Inference*, US Patent 7,283,231, 2007.

30. D. J. Brady, N. P. Pitsianis, and X. Sun, Reference structure tomography, *J. Opt. Soc. Am. A* **21**(7):1140–1147, 2004.

31. D. J. Brady, M. E. Gehm, N. Pitsianis, and X. Sun, Compressive sampling strategies for integrated microspectrometers, *SPIE (Soc. Photoopt. Instrum. Eng.)* **6232**:6232OC, 2006.

32. W. H. Bragg and W. L. Bragg, The reflection of x-rays by crystals, *Proc. Roy. Soc. Lond. Ser. A* **88**:428–438, 1913.

33. A. K. Brodzik and J. M. Mooney, Convex projections algorithm for restoration of limited-angle chromatomographic images, *J. Opt. Soc. Am. A* **16**(2):246–257, 1999.

34. A. Bruststing, *Simultaneous Multiple Wavelength Photometer*, US Patent 4,657,398, 1985.

35. P. Buchsbaum and M. Morris, *Method for Making Monolithic Patterned Dichroic Filter Detector Arrays for Spectroscopic Imaging*, US Patent 6,638,668, 2003.

36. C. B. Burchardt, Diffraction of a plane wave at a sinusoidally stratified dielectric grating, *J. Opt. Soc. Am.* **56**:1502–1509, 1966.

37. T. M. Buzug, *Computed Tomography*, Springer, 2008.

38. E. Candes and J. Romberg, Sparsity and incoherence in compressive sampling, *Inverse Problems* **23**(3):969–985, 2007.

39. E. J. Candes, J. K. Romberg, and T. Tao, Stable signal recovery from incomplete and inaccurate measurements, *Commun. Pure Appl. Math.* **59**(8):1207–1223, 2006.

40. E. J. Candes and T. Tao, Near-optimal signal recovery from random projections: Universal encoding strategies? *IEEE Trans. Inform. Theory* **52**(12):5406–5425, Dec. 2006.

41. E. J. Candes, J. Romberg, and T. Tao, Robust uncertainty principles: Exact signal reconstruction from highly incomplete frequency information, *IEEE Trans. Inform. Theory* **52**(2):489–509, Feb. 2006.
42. E. J. Candes and T. Tao, Decoding by linear programming, *IEEE Trans. Inform. Theory* **51**(12):4203–4215, Dec. 2005.
43. W. T. Cathey, B. R. Frieden, W. T. Rhodes, and C. K. Rushforth, Image gathering and processing for enhanced resolution, *J. Opt. Soc. Am. A* **1**(3):241–250, March 1984.
44. V. A. Myakinin, M. I. Charnotskii, and V. U. Zavorotnyy, Observation of superresolution in nonisoplanatic imaging through turbulence, *J. Opt. Soc. Am. A* **7**(8):1345–1350, 1990.
45. S. S. Chen, D. L. Donoho, and M. A. Saunders, Atomic decomposition by basis pursuit, *SIAM J. Sci. Comput.* **20**(1):33–61, 1999.
46. W. Chi and N. George, Computational imaging with a logarithmic asphere: Theory, *J. Opt. Soc. Am. A* **20**:2260–2273, 2003.
47. K. Choi and T. J. Schulz, Signal-processing approaches for image-resolution restoration for TOMBO imagery, *Appl. Opt.* **47**(10):B104–B116, 2008.
48. M. A. Choma, M. V. Sarunic, C. H. Yang, and J. A. Izatt, Sensitivity advantage of swept source and Fourier domain optical coherence tomography, *Opt. Express* **11**:2183–2189, 2003.
49. A. Cohen, I. Daubechies, and J.-C. Feauveau, Biorthogonal bases of compactly supported wavelets, *Commun. Pure Appl. Math.* **45**(5):485–560, 1992.
50. R. Coifman, F. Geshwind, and Y. Meyer, Noiselets, *Appl. Comput. Harmonic Anal.* **10**(1):27–44, 2001.
51. M. Conde, Deriving wavelength spectra from fringe images from a fixed-gap single-etalon Fabry–Perot spectrometer, *Appl. Opt.* **41**(14):2672–2678, 2002.
52. W. B. Cook, H. E. Snell, and P. B. Hays, Multiplex Fabry–Perot interferometer 1. Theory, *Appl. Opt.* **34**(24):5263–5267, 1995.
53. I. Daubechies, *Ten Lectures on Wavelets*, SIAM, Philadelphia, 1992.
54. J. F. de Boer, B. Cense, B. H. Park, M. C. Pierce, G. J. Tearney, and B. E. Bouma, Improved signal-to-noise ratio in spectral-domain compared with time-domain optical coherence tomography, *Opt. Lett.* **28**:2067–2069, 2003.
55. R. A. de Verse, R. M. Hammaker, and W. G. Fateley, Realization of the Hadamard multiplex advantage using a programmable optical mask in a dispersive flat-field near infrared spectrometer, *Appl. Spectrosc.* **54**:1751–1758, 2000.
56. A. P. Dempster, N. M. Laird, and D. B. Rubin, Maximum likelihood from incomplete data via the EM algorithm, *J. Roy. Stat. Soc.* **39**(1):1–38, 1977.
57. E. L. Dereniak and G. D. Boreman, *Infrared Detectors and Systems*, Wiley, 1996.
58. D. H. Devorkin, Michelson and the problem of stellar diameters, *J. Hist. Astron.* **6**:1–18, 1975.
59. D. L. Donoho, Compressed sensing, *IEEE Trans. Inform. Theory* **52**(4):1289–1306, 2006.
60. E. R. Dowski and W. T. Cathey, Extended depth of field through wave-front coding, *Appl. Opt.* **34**(11):1859–1866, 1995.
61. R. G. Driggers, P. Cox, and T. Edwards, *Introduction to Infrared and Electro-optical Systems*, Artech House, Boston, 1999.
62. D. Z. Du and F. K. Hwang, *Combinatorial Group Testing and Its Applications*, World Scientific, 2000.

63. M. F. Duarte, M. A. Davenport, D. Takbar, J. N. Laska, T. Sun, K. F. Kelly, and R. G. Baraniuk, Single-pixel imaging via compressive sampling [building simpler, smaller, and less-expensive digital cameras], *IEEE Signal Process. Mag.* **25**(2):83–91, March 2008.
64. J. Duparre, P. Dannberg, P. Schreiber, A. Brauer, and A. Tunnermann, Thin compound-eye camera, *Appl. Opt.* **44**:2949–2956, 2005.
65. J. Enderlein and F. Pampaloni, Unified operator approach for deriving Hermite–Gaussian and Laguerre–Gaussian laser modes, *J. Opt. Soc. Am. A* **21**(8):1553–1558, 2004.
66. J. W. Evans, The birefringent filter, *J. Opt. Soc. Am.* **39**(3):229, 1949.
67. R. L. Fante, Imaging of an object behind a random phase screen using light of arbitrary coherence, *J. Opt. Soc. Am. A* **2**(12):2318, 1985.
68. W. G. Fateley, R. M. Hammaker, and R. A. DeVerse, Modulations used to transmit information in spectrometry and imaging, *J. Molec. Struct.* **550**:117–122, 2000.
69. L. A. Feldkamp, L. C. Davis, and J. W. Kress, Practical cone-beam algorithm, *J. Opt. Soc. Am. A* **1**:612–619, 1984.
70. S. D. Feller, H. Chen, D. J. Brady, M. E. Gehm, C. Hsieh, O. Momtahan, and A. Adibi, Multiple order coded aperture spectrometer, *Opt. Express* **15**(9):5625–5630, 2007.
71. P. Fellgett, A propos de la theorie du spectrometre interferentiel multiplex, *J. Phys. Radium* **19**:187–191, 1958.
72. P. B. Fellgett, *The Multiplex Advantage*, PhD thesis, Univ. Cambridge, Cambridge, UK, 1951.
73. E. E. Fenimore and T. M. Cannon, Coded aperture imaging with uniformly redundant arrays, *Appl. Opt.* **17**(3):337–347, 1978.
74. M. R. Fetterman, E. Tan, L. Ying, R. A. Stack, D. L. Marks, S. Feller, E. Cull, J. M. Sullivan, D. C. Munson, S. T. Thoroddsen, and D. J. Brady, Tomographic imaging of foam, *Opt. Express* **7**(5):186–197, 2000.
75. R. D. Fiete, Image quality and lambda fn/p for remote sensing systems, *Opt. Eng.* **38**(7):1229–1240, 1999.
76. M. A. T. Figueiredo and R. D. Nowak, An EM algorithm for wavelet-based image restoration, *IEEE Trans. Image Process.* **12**:906–916, Aug. 2003.
77. A. R. FitzGerrell, E. R. Dowski, and W. T. Cathey, Defocus transfer function for circularly symmetric pupils, *Appl. Opt.* **36**(23):5796–5804, 1997.
78. E. R. Fossum, CMOS image sensors: Electronic camera-on-a-chip, *IEEE Trans. Electron. Devices* **44**(10):1689–1698, Oct. 1997.
79. A. T. Friberg, Existence of a radiance function for finite planar sources of arbitrary states of coherence, *J. Opt. Soc. Am.* **69**(1):192–198, 1979.
80. B. R. Frieden, Optical transfer of a three-dimensional object, *J. Opt. Soc. Am.* **57**:56–66, 1967.
81. B. R. Frieden, Evaluation, design and extrapolation methods for optical signals, based on use of the prolate functions, *Progress Opt.* **IX**:311–407, 1971.
82. D. Gabor, Theory of communication, *J. IEE (Inst. Electrical Engineers) Radio Commun. Eng.* **93**(26, Part III):429–457, 1946.
83. D. Gabor, A new microscopic principle, *Nature* **161**:777–778, 1948.
84. D. Gabor, Guest editorial, *IEEE Trans. Inform. Theory* **5**(3):97, Sept. 1959.
85. N. Gat, *Spectrometer Apparatus*, US Patent 5,166,755, 1992.

86. M. E. Gehm and D. J. Brady, High-throughput hyperspectral microscopy, *SPIE* **6090**:13–21, 2006.
87. M. E. Gehm, R. John, D. J. Brady, R. M. Willett, and T. J. Schulz, Single-shot compressive spectral imaging with a dual-disperser architecture, *Opt. Express* **15**(21): 14013–14027, 2007.
88. M. E. Gehm, M. S. Kim, C. Fernandez, and D. J. Brady, High-throughput, multiplexed pushbroom hyperspectral microscopy, *Opt. Express* **16**(15):11032–11043, 2008.
89. M. E. Gehm, S. T. McCain, N. P. Pitsianis, D. J. Brady, P. Potuluri, and M. E. Sullivan, Static two-dimensional aperture coding for multimodal, multiplex spectroscopy, *Appl. Opt.* **45**(13):2965–2974, 2006.
90. D. R. Gerwe and M. A. Plonus, Superresolved image reconstruction of images taken through the turbulent atmosphere, *J. Opt. Soc. Am. A* **15**(10):2620–2628, 1998.
91. A. C. Gilbert, M. J. Strauss, and J. A. Tropp, A tutorial on fast Fourier sampling [how to apply it to problems], *IEEE Signal Process. Mag.* **25**(2):57–66, March 2008.
92. D. L. Gilblom, S. K. Yoo, and P. Ventura, Operation and performance of a color image sensor with layered photodiodes, *SPIE*, **5074**:318–331, 2003.
93. A. Girard, Spectrometre a grilles, *Appl. Opt.* **2**(1):79–88, Jan. 1963. [Translated as "Among spectral analysis systems, it is now classic to distinguish on the one hand between spectrometers that explore the spectrum serially in time, element by element, and on the other hand, spectrographs that simultaneously capture all the elements of the spectrum."]
94. M. J. E. Golay, Multislit spectroscopy. *J. Opt. Soc. Am.* **39**:437–444, 1949.
95. M. J. E. Golay, Anent codes, priorities, patents, etc., *Proc. IEEE* **64**(4): 572, 1976.
96. T. Goldstein and S. Osher, *The Split Bregman Algorithm for l1 Regularized Problems*, Computational and Applied Mathematics Reports 08–29, UCLA, April 2008.
97. Gene H. Golub and Charles F. Van Loan, *Matrix Computations*, Johns Hopkins Univ. Press, 1996.
98. R. A. Gonsalves, Phase retrieval and diversity in adaptive optics, *Opt. Eng.* **21**(5): 829–832, 1982.
99. J. W. Goodman, *Statistical Optics*, Wiley, New York, 1985.
100. J. W. Goodman, *Introduction to Fourier Optics*, McGraw-Hill, 1968.
101. S. J. Gortler, R. Grzeszczuk, R. Szeliski, and M. F. Cohen, The lumigraph, *Proc. 23rd Annual Conf. Computer Graphics and Interactive Techniques,* (*SIGGRAPH '96*), ACM, 1996, pp. 43–54.
102. S. R. Gottesman and E. E. Fenimore, New family of binary arrays for coded aperture imaging, *Appl. Opt.* **28**(20):4344–4352, 1989.
103. M. S. Gottlieb, Acousto-optic tunable filters, in A. P. Goutzoulis and D. R. Pape, eds., *Design and Fabrication of Acousto-optic Devices*, Marcel Dekker, 1994, pp. 197–284.
104. H. Grahn and P. Geladi, *Techniques and Applications of Hyperspectral Image Analysis*, Wiley, 2007.
105. A. Greengard, Y. Y. Schechner, and R. Piestun, Depth from diffracted rotation, *Opt. Lett.* **31**(2):181–183, 2006.
106. M. Gu, *Advanced Optical Imaging Theory*, Springer Series in Optical Sciences, Springer-Verlag, 2000.

107. M. G. L. Gustafsson, Nonlinear structured-illumination microscopy: Wide-field fluorescence imaging with theoretically unlimited resolution, *Proc. Natl. Acad. Sci. (USA)* **102**(37):13081–13086, 2005.

108. A. Haar, Zur Theorie der orthogonalen Funktionensysteme, *Mathematische Annalen* **69**:331–371, 1910.

109. A. B. Hamza and D. J. Brady, Reconstruction of reflectance spectra using robust nonnegative matrix factorization, *IEEE Trans. Signal Process.* **54**(9):3637–3642, Sept. 2006.

110. M. W. Haney, Performance scaling in flat imagers, *Appl. Opt.* **45**(13):2901–2910, 2006.

111. P. C. Hansen, *Rank-Deficient and Discrete Ill-Posed Problems: Numerical Aspects of Linear Inversion*, SIAM, Philadelphia, 1997.

112. P. C. Hansen, Deconvolution and regularization with Toeplitz matrices, *Num. Algorithms* **29**(4):323–378, 2002.

113. G. R. Harrison, The production of diffraction gratings: II, Design of Eschelle gratings and spectrographs—(1949), *J. Opt. Soc. Am. A* **39**(7):522, 1949.

114. M. Harwit and N. J. A. Sloane, *Hadamard Transform Optics*, Academic Press, 1979.

115. D. M. Healy and G. K. Rohde, Fast global image registration using random projections, *Proc. 4th Int. IEEE Sym Biomedical Imaging: From Nano to Macro, 2007 (ISBI 2007)*, April 2007, pp. 476–479.

116. S. W. Hell and J. Wichmann, Breaking the diffraction resolution limit by stimulated-emission stimulated-emission-depletion fluorescence microscopy, *Opt. Lett.* **19**(11): 780–782, 1994.

117. G. Hernandez, *Fabry–Perot Interferometers*, Cambridge Univ. Press, 1986.

118. H. Hertz, *Electric Waves, Being Researches on the Propagation of Electric Action with Finite Velocity through Space*, Macmillan, 1893.

119. K. B. Hill, S. A. Basinger, R. A. Stack, and D. J. Brady, Noise and information in interferometric cross correlators, *Appl. Opt.* **36**(17):3948–3958, 1997.

120. H. H. Hopkins, The frequency response of a defocused optical system, *Proc. Roy. Soc. Lond. Ser. A, Math. Phys. Sci.* **231**(1184):91–103, July 1955.

121. H. H. Hopkins, The aberration permissible in optical systems, *Proc. Phys. Soc., Sec. B* **70**(5):449–470, 1957.

122. D. Huang, E. A. Swanson, C. P. Lin, J. S. Schuman, W. G. Stinson, W. Chang, M. R. Hee, T. Flotte, K. Gregory, C. A. Puliafito, and J. G. Fujimoto, Optical coherence tomography, *Science* **254**:1178–1181, 1991.

123. G. Häusler, A method to increase the depth of focus by two step image processing, *Opt. Commun.* **6**(1):38–42, 1972.

124. J. A. Izatt, M. R. Hee, G. M. Owen, E. A. Swanson, and J. G. Fujimoto, Optical coherence microscopy in scattering media, *Opt. Lett.* **19**:590–592, 1994.

125. P. Jacquinot, New developments in interference spectroscopy, *Rep. Progress Phys.* **23**:268–312, 1960.

126. J. F. James and R. S. Sternberg, *The Design of Optical Spectrometers*, Chapman & Hall, 1969.

127. J. L. Jimenez, L. R. C. Fonseca, D. J. Brady, J. P. Leburton, D. E. Wohlert, and K. Y. Cheng, The quantum dot spectrometer, *Appl. Phys. Lett.* **71**(24):3558–3560, 1997.

128. J. D. Joannopoulos, R. D. Meade, and J. N. Winn, *Photonic Crystals: Molding the Flow of Light*, Princeton, Univ. Press, 1995.

129. I. T. Jolliffe, *Principal Component Analysis*, Springer-Verlag, 2002.
130. D. Jungnickel and A. Pott, Perfect and almost perfect sequences, *Discrete Appl. Math.* **95**(1–3):331–359, 1999.
131. A. C. Kak and M. Slaney, *Principles of Computerized Tomographic Imaging*, Society for Industrial and Applied Mathematics, Philadelphia, 2001.
132. A. V. Kanaev, J. R. Ackerman, E. F. Fleet and D. A. Scribner, TOMBO sensor with scene-independent superresolution processing, *Opt. Lett.* **32**(19):2855–2857, 2007.
133. A. V. Kanaev, D. A. Scribner, J. R. Ackerman, and E. F. Fleet, Analysis and application of multiframe superresolution processing for conventional imaging systems and lenslet arrays, *Appl. Opt.* **46**(20):4320–4328, 2007.
134. R. P. Kanwal, *Linear Integral Equations*, Birkhauser, 1997.
135. R. Kautz and W. Singleton, Nonrandom binary superimposed codes, *IEEE Trans. Inform. Theory* **10**(4):363–377, Oct. 1964.
136. C. Kittel and H. Kroemer, *Thermal Physics*, Freeman, 1980.
137. H. Kogelnik, Coupled wave theory for thick hologram gratings, *Bell Syst. Tech. J.* **48**(9):2909–2948, 1969.
138. M. I. Kolobov, *Quantum Imaging*, Springer, New York, 2007.
139. V. P. Koronkevitch and I. G. Palchikova, Kinoforms with increased depth of focus, *Optik* **87**:91–93, 1991.
140. S. Krishna, S. D. Gunapala, S. V. Bandara, C. Hill, and D. Z. Ting, Quantum dot based infrared focal plane arrays, *Proc. IEEE* **95**(9):1838–1852, 2007.
141. P. W. Kruse, Principles of uncooled infrared focal plane arrays, in P. W. Kruse and D. D. Skatrud, eds., *Uncooled Infrared Imaging Arrays and Systems*, Vol. 47 of Semiconductors and Semimetals Series, Academic Press, 1997, Chapter 2.
142. E. Kupce and R. Freeman, Hyperdimensional NMR spectroscopy, *Progress Nuclear Magn. Resonance Spectrosc.* **52**(1):22–30, 2008.
143. H. J. Landau and H. O. Pollak, Prolate spheroidal wave functions, Fourier analysis and uncertainty 2, *Bell Syst. Tech. J.* **40**(1):65, 1961.
144. H. J. Landau and H. O. Pollak, Prolate spheroidal wave functions, Fourier analysis and uncertainty. 3. Dimension of space of essentially time and band-limited signals, *Bell Syst. Tech. J.* **41**(4):1295, 1962.
145. W. P. Latham and M. L. Tilton, Calculation of prolate functions for optical analysis, *Appl. Opt.* **26**(13):2653, 1987.
146. C. L. Lawson and R. J. Hanson, *Solving Least Squares Problems*, Prentice-Hall, 1974.
147. E. L. Lehmann and G. Casella, *Theory of Point Estimation*, Springer, 1998.
148. R. Leitgeb, C. K. Hitzenberger, and A. F. Fercher, Performance of Fourier domain vs. time domain optical coherence tomography, *Opt. Express* **11**:889–894, 2003.
149. E. N. Leith and J. Upatnieks, Reconstructed wavefronts and communication theory, *J. Opt. Soc. Am.* **52**:1123–1130, 1962.
150. P. G. Lemarie, Ondelettes with exponential localization, *Journal De Mathematiques Pures et Appliquees* **67**(3):227–236, 1988.
151. M. Levoy, Light fields and computational imaging, *Computer* **39**(8):46–55, 2006.
152. M. Levoy and P. Hanrahan, Light field rendering, *Proc. 23rd Annual Conf. Computer Graphics and Interactive Techniques, (SIGGRAPH '96)*, ACM, 1996, pp. 31–42.

153. M. Levoy, R. Ng, A. Adams, M. Footer, and M. Horowitz, Light field microscopy, *ACM Trans. Graph.* **25**(3):924–934, 2006.

154. M. Liebling, T. Blu, and M. Unser, Fresnelets: New multiresolution wavelet bases for digital holography, *IEEE Trans. Image Process.* **12**(1):29–43, Jan. 2003.

155. M. Liebling and M. Unser, Autofocus for digital Fresnel holograms by use of a Fresnelet-sparsity criterion, *J. Opt. Soc. Am. A* **21**(12):2424–2430, Dec. 2004.

156. E. Loewen, D. Maystre, E. Popov, and L. Tsonev, Echelles: Scalar, electromagnetic, and real-groove properties, *Appl. Opt.* **34**(10):1707, 1995.

157. A. W. Lohmann, Scaling laws for lens systems, *Appl. Opt.* **28**(23):4996, 1989.

158. L. B. Lucy, An iterative technique for the rectification of observed distributions, *Astron. J.* **79**: 745, June 1974.

159. L. B. Lucy, Resolution limits for deconvolved images, *Astrophys. J.* **104**:1260–1265, 1992.

160. W. Lukosz, Optical systems with resolving powers exceeding the classical limit, II, *J. Opt. Soc. Am. A*, **57**(7).932–941, 1967.

161. B. Lyot, Optical apparatus with wide field using interference of polarized light, *Comptes Rend. Acad. Sci.* (Paris) **197**:1593, 1933.

162. H. A. Macleod, *Thin-film Optical Filters*, Taylor and Francis, 2001.

163. M. Maggioni, G. L. Davis, F. J. Warner, F. B. Geshwind, A. C. Coppi, R. A. DeVerse, and R. R. Coifman, Hyperspectral microscopic analysis of normal, benign and carcinoma microarray tissue sections, *SPIE* **6091**:60910, 2006.

164. S. Mallat, *A Wavelet Tour of Signal Processing*, Academic Press, 1998.

165. L. Mandel and E. Wolf, *Optical Coherence and Quantum Optics*, Cambridge Univ. Press, 1995.

166. E. W. Marchand and E. Wolf, Walther's definitions of generalized radiance, *J. Opt. Soc. Am.* **64**(9):1273–1274, 1974.

167. D. L. Marks, M. Fetterman, R. A. Stack, and D. J. Brady, Spectral tomography from spatial coherence measurements, *SPIE* **3920**:48–55, 2000.

168. D. L. Marks, R. Stack, A. J. Johnson, D. J. Brady, and D. C. Munson, Cone-beam tomography with a digital camera, *Appl. Opt.* **40**(11):1795–1805, 2001.

169. D. L. Marks, R. A. Stack, and D. J. Brady, Digital refraction distortion correction with an astigmatic coherence sensor, *Appl. Opt.* **41**(29):6050–6054, 2002.

170. D. L. Marks, R. A. Stack, D. J. Brady, D. C. Munson, and R. B. Brady, Visible cone-beam tomography with a lensless interferometric camera, *Science* **284**(5423): 2164–2166, 1999.

171. D. L. Marks, R. A. Stack, D. J. Brady, and J. van der Gracht, Three-dimensional tomography using a cubic-phase plate extended depth-of-field system, *Opt. Lett.* **24**(4):253–255, 1999.

172. D. M. Marks, R. A. Stack, and D. J. Brady, Astigmatic coherence sensor for digital imaging, *Opt. Lett.* **25**(23):1726–1728, 2000.

173. D. L. Marks, R. A. Stack, and D. J. Brady, Three-dimensional coherence imaging in the Fresnel domain, *Appl. Opt.* **38**(8):1332–1342, 1999.

174. P. Matousek, I. P. Clark, E. R. C. Draper, M. D. Morris, A. E. Goodship, N. Everall, M. Towrie, W. F. Finney, and A. W. Parker, Subsurface probing in diffusely scattering media using spatially offset Raman spectroscopy, *Appl. Spectrosc.* **59**(4):393–400, 2005.

175. C. L. Matson and D. W. Tyler, Primary and secondary superresolution by data inversion, *Opt. Express*, **14**(2):456–473, 2006.
176. T. Matsuyama, Y. Ohmura, and D. M. Williamson, The lithographic lens: Its history and evolution, *SPIE* **6154**: 615403, 2006.
177. S. B. Mende, E. S. Claflin, R. L. Rairden, and G. R. Swenson, Hadamard spectroscopy with a two-dimensional detecting array, *Appl. Opt.* **32**(34):7095, 1993.
178. M. Mino and Y. Okano, Improvement in OTF of a defocused optical system through use of shaded apertures, *Appl. Opt.* **10**:2219–2226, 1971.
179. T. Mirani, D. Rajan, M. P. Christensen, S. C. Douglas, and S. L. Wood, Computational imaging systems: Joint design and end-to-end optimality, *Appl. Opt.* **47**:B86–B103, 2008.
180. M. G. Moharam and T. K. Gaylord, Rigorous coupled-wave analysis of planar-grating diffraction, *J. Opt. Soc. Am. A* **71**:811, 1981.
181. A. Moini, *Vision Chips*, Springer, 2000.
182. O. Momtahan, C. R. Hsieh, A. Karbaschi, A. Adibi, M. E. Sullivan, and D. J. Brady, Spherical beam volume holograms for spectroscopic applications: Modeling and implementation, *Appl. Opt.* **43**(36):6557–6567, 2004.
183. P. M. Morse and H. Feshbach, *Methods of Theoretical Physics*, Vol. 1, McGraw-Hill, 1953.
184. P. Naulleau and E. Leith, Imaging through optical fibers by spatial coherence encoding methods, *J. Opt. Soc. Am. A* **13**(10):2096–2101, 1996.
185. M. A. Neifeld and J. Ke, Optical architectures for compressive imaging, *Appl. Opt.* **46**(22):5293–5303, 2007.
186. M. A. Neifeld and P. Shankar, Feature-specific imaging, *Appl. Opt.* **42**(17):3379–3389, 2003.
187. K. Nitta, R. Shogenji, S. Miyatake, and J. Tanida, Image reconstruction for thin observation module by bound optics by using the iterative backprojection method, *Appl. Opt.* **45**(13):2893–2900, 2006.
188. S. Ogata, J. Ishida, and T. Sasano, Optical sensor array in an artificial compound eye, *Opt. Eng.* **33**:3649–3655, 1994.
189. J. Ojeda-Castaneda, L. R. Berriel-Valdos, and E. L. Montes, Line-spread function relatively insensitive to defocus, *Opt. Lett.* **8**:458–460, 1983.
190. J. Ojeda-Castaneda, R. Ramos, and A. Noyola-Isgleas, High focal depth by apodization and digital restoration, *Appl. Opt.* **27**:2583–2586, 1988.
191. E. O'Neill, *Introduction to Statistical Optics*, Addison-Wesley, Reading, MA, 1963.
192. M. J. Padgett, A. R. Harvey, A. J. Duncan, and W. Sibbett, Single-pulse, Fourier-transform spectrometer having no moving parts, *Appl. Opt.* **33**(25):6035, 1994.
193. A. Papoulis, Ambiguity function in Fourier optics, *J. Opt. Soc. Am.* **64**:779–788, 1974.
194. A. Papoulis, Generalized sampling expansion, *IEEE Trans. Circuits Syst.* **24**(11):652–654, 1977.
195. S. C. Park, M. K. Park, and M. G. Kang, Super-resolution image reconstruction: A technical overview, *IEEE Signal Process. Mag.* **20**(3):21–36, 2003.
196. S. K. Parks, R. Schowengerdt, and M. A. Kaczynski, Modulation transfer function analysis for sampled image systems, *Appl. Opt.* **23**:2572–2582, 1984.
197. R. G. Paxman, T. J. Schulz, and J. R. Fienup, Joint estimation of object and aberrations by using phase diversity, *J. Opt. Soc. Am. A* **9**(7):1072–1085, 1992.

198. H. M. Pedersen and J. J. Stamnes, Radiometric theory of spatial coherence in free-space propagation, *J. Opt. Soc. Am. A* **17**(8):1413–1420, 2000.

199. M. Pharr and G. Humphreys, *Physically Based Rendering: From Theory to Implementation*, Elsevier/Morgan Kaufmann, 2004.

200. R. Piestun, Y. Y. Schechner, and J. Shamir, Propagation-invariant wave fields with finite energy, *J. Opt. Soc. Am. A* **17**(2):294–303, 2000.

201. R. Piestun, B. Spektor, and J. Shamir, Pattern generation with an extended focal depth, *Appl. Opt.* **37**:5394–5398, 1998.

202. N. P. Pitsianis, D. J. Brady, A. Portnoy, X. Sun, T. Suleski, M. A. Fiddy, M. R. Feldman, and R. D. TeKolste, Compressive imaging sensors, *SPIE* **6232**:62320A, 2006.

203. N. P. Pitsianis, D. J. Brady, and X. Sun, Sensor-layer image compression based on the quantized cosine transform, in *Visual Information Processing XIV*, SPIE, Orlando, FL, 2005, Vol. 5817, p. 250.

204. A. D. Portnoy, N. P. Pitsianis, X. Sun, and D. J. Brady, Multichannel sampling schemes for optical imaging systems, *Appl. Opt.* **47**:B76–B85, 2008.

205. P. Potuluri, *Multiplex Optical Sensors for Reference Structure Tomography and Compressive Spectroscopy*, PhD thesis, Duke Univ., Durham, NC, 2004.

206. P. Potuluri, M. Gehm, M. Sullivan, and D. Brady, Measurement-efficient optical wavemeters, *Opt. Express* **12**(25):6219–6229, 2004.

207. S. Prasad, Digital superresolution and the generalized sampling theorem, *J. Opt. Soc. Am. A* **24**(2):311–325, 2007.

208. K. G. Purchase, D. J. Brady, and K. Wagner, Time-of-flight cross correlation on a detector array for ultrafast packet detection, *Opt. Lett.* **18**(24):2129, 1993.

209. T. S. Ralston, D. L. Marks, P. S. Carney, and S. A. Boppart, Interferometric synthetic aperture microscopy, *Nature Phys.* **3**(2):129–134, 2007.

210. W. H. Richardson, Bayesian-based iterative method of image restoration, *J. Opt. Soc. Am.* **62**(1):55, 1972.

211. A. Rogalski, Infrared detectors: Status and trends, *Progress Quantum Electron.* **27**:59–210, 2003.

212. L. I. Rudin, S. Osher, and E. Fatemi, Nonlinear total variation based noise removal algorithms, *Physica D* **60**(1–4):259–268, 1992.

213. J. S. Sanders and C. E. Halford, Design and analysis of apposition compound eye optical sensors, *Opt. Eng.* **34**:222–235, 1995.

214. Y. Y. Schechner, R. Piestun, and J. Shamir, Wave propagation with rotating intensity distributions, *Phys. Rev. E* **54**(1):R50–R53, July 1996.

215. O. Schmidt, P. Kiesel, and M. Bassler, Performance of chip-size wavelength detectors, *Opt. Express* **15**(15):9701–9706, 2007.

216. G. Schweiger, R. Nett, and T. Weigel, Microresonator array for high-resolution spectroscopy, *Opt. Lett.* **32**(18):2644–2646, 2007.

217. M. Shahram and P. Milanfar, Statistical and information-theoretic analysis of resolution in imaging, *IEEE Trans. Inform. Theory* **52**(8):3411–3437, 2006.

218. M. Shankar, R. Willett, N. Pitsianis, T. Schulz, R. Gibbons, R. T. Kolste, J. Carriere, C. Chen, D. Prather, and D. Brady, Thin infrared imaging systems through multichannel sampling, *Appl. Opt.* **47**(10):B1–B10, 2008.

219. C. E. Shannon, Communications in the presence of noise, *Proc. IRE* **37**:10–21, 1949.

220. J. Sharpe, Optical projection tomography, *Annu. Rev. Biomed. Eng.* **6**:209–228, 2004.

221. J. Sharpe, U. Ahlgren, P. Perry, B. Hill, A. Ross, J. Hecksher-Sorensen, R. Baldock, and D. Davidson, Optical projection tomography as a tool for 3D microscopy and gene expression studies, *Science* **296**(5567):541–545, 2002.

222. S. S. Sherif and T. W. Cathey, Reduced depth of field in incoherent hybrid imaging systems, *Appl. Opt.* **41**(29):6062–6074, 2002.

223. A. Shlivinski, E. Heyman, A. Boag, and C. Letrou, A phase-space beam summation formulation for ultrawide-band radiation, *IEEE Trans. Antennas Propag.* **52**(8):2042–2056, 2004.

224. D. Slepian, Prolate spheroidal wave functions, Fourier analysis and uncertainty—IV: Extensions to many dimensions; generalized prolate spheroidal functions, *Bell Syst. Tech. J.* **43**(6):3009–3057, 1964.

225. D. Slepian and H. O. Pollak, Prolate spheroidal wave functions, Fourier analysis and uncertainty 1, *Bell Syst. Tech. J.* **40**(1):43, 1961.

226. D. L. Snyder, T. J. Schulz, and J. A. O'Sullivan, Deblurring subject to nonnegativity constraints, *IEEE Trans. Signal Process.* **40**:1143–1150, 1992.

227. I. Solc, Birefringent chain filters, *J. Opt. Soc. Am.* **55**(6):621, 1965.

228. D. Takhar, J. N. Laska, M. B. Wakin, M. F. Duarte, D. Baron, S. Sarvotham, K. F. Kelly, and R. G. Baraniuk, A new compressive imaging camera architecture using optical-domain compression, in *Computational Imaging IV*, SPIE, San Jose, CA, 2006, Vol. 6065, p. 606509.

229. J. Tanida, T. Kumagai, K. Yamada, S. Miyatake, K. Ishida, T. Morimoto, N. Kondou, D. Miyazaki, and Y. Ichioka, Thin observation module by bound optics (TOMBO): Concept and experimental verification, *Appl. Opt.* **40**:1806–1813, 2001.

230. P. Thevenaz, T. Blu, and M. Unser, Image interpolation and resampling, in I. N. Bankman, ed., *Handbook of Medical Imaging, Processing and Analysis*, Academic Press, San Diego, CA, 2000, Chapter 25, pp. 393–420.

231. J. L. Tian and W. B. Cook, Recovery of overlapped spectral harmonics in multiplex Fabry–Perot interferometry, *Opt. Eng.* **45**(8):085601, 2006.

232. G. Toraldo di Francia, Degrees of freedom of an image, *J. Opt. Soc. Am.* **59**(7):799, 1969.

233. H. K. Tuy, An inversion formula for cone-beam reconstruction, *SIAM J. Appl. Math.* **43**:546–554, 1983.

234. M. Unser, Sampling—50 years after Shannon, *Proc. IEEE* **88**(4):569–587, April 2000.

235. M. Unser, A. Aldroubi, and M. Eden, Polynomial spline signal approximations: Filter design and asymptotic equivalence with Shannon's sampling theorem, *IEEE Trans. Inform. Theory* **38**(1):95–103, Jan. 1992.

236. M. Unser, A. Aldroubi, and M. Eden, B-Spline signal processing: Part I—Theory, *IEEE Trans. Signal Process.* **41**(2):821–833, Feb. 1993 (this article received the IEEE Signal Processing Society's 1995 Best Paper Award).

237. M. Unser, A. Aldroubi, and M. Eden, B-Spline signal processing: Part II—Efficient design and applications, *IEEE Trans. Signal Process.* **41**(2):834–848, Feb. 1993.

238. E. C. Vail, M. S. Wu, G. S. Li, L. Eng, and C. J. Changhasnain, GaAs micromachined widely tunable Fabry–Perot filters, *Electron. Lett.* **31**(3):228–229, 1995.

239. J. van der Gracht, E. R. Dowski, Jr., M. G. Taylor, and D. M. Deaver, Broadband behavior of an optical-digital focus-invariant system, *Opt. Lett.* **21**(13):919, 1996.

240. A. Vanderlugt, Signal detection by complex spatial filtering, *IEEE Trans. Inform. Theory* **IT-10**(2):139–145, 1964.

241. C. R. Vogel, *Computational Methods for Inverse Problems*, Society for Industrial and Applied Mathematics, Philadelphia, 2002.

242. A. Wagadarikar, R. John, R. Willett, and D. J. Brady, Single disperser design for coded aperture snapshot spectral imaging, *Appl. Opt.* **47**(10):B44–B51, 2008.

243. A. A. Wagadarikar, M. E. Gehm, and D. J. Brady, Performance comparison of aperture codes for multimodal, multiplex spectroscopy, *Appl. Opt.* **46**(22):4932–4942, 2007.

244. A. Walther, Radiometry and coherence, *J. Opt. Soc. Am.* **58**(9):1256, 1968.

245. S-W. Wang, C. Xia, X. Chen, W. Lu, M. Li, H. Wang, W. Zheng, and T. Zhang, Concept of a high-resolution miniature spectrometer using an integrated filter array, *Opt. Lett.* **32**(6):632–634, 2007.

246. E. W. Weisstein, Jacobi-Anger expansion, MathWorld, mathworld.wolfram.com.

247. W. T. Welford, Use of annular apertures to increase focal depth, *J. Opt. Soc. Am.* **50**:749–753, 1960.

248. J. C. Winston, R. and Minano, P. Benitez, and W. T. Welford, *Nonimaging Optics*, Elsevier, 2005.

249. E. Wolf, Coherence and radiometry, *J. Opt. Soc. Am.* **68**(1):6, 1978.

250. E. Wolf, New theory of partial coherence in the space-frequency domain. Part I: Spectra and cross spectra of steady-state sources, *J. Opt. Soc. Am.* **72**:343–350, 1982.

251. E. Wolf, Coherent mode propagation in spatially band-limited wave fields, *J. Opt. Soc. Am. A* **3**:1920–1924, 1986.

252. E. Wolf, *Introduction to the Theory of Coherence and Polarization of Light*, Cambridge Univ. Press, Cambridge, UK, 2007.

253. R. A. Wood, C. J. Han, and P. W. Kruse, Integrated uncooled infrared detector imaging arrays, *Solid-State Sensor and Actuator Workshop, 1992, 5th Technical Digest, IEEE*, June 22–25, 1992, pp. 132–135.

254. S. L. Wood, S. T. Lee, G. Yang, M. P. Christensen, and D. Rajan, Impact of measurement precision and noise on superresolution image reconstruction, *Appl. Opt.* **47**(10): B128–B138, 2008.

255. Z. Xu, Z. Wang, M. Sullivan, D. J. Brady, S. Foulger, and A. Adibi, Multimodal multiplex spectroscopy using photonic crystals, *Opt. Express* **11**(18):2126–2133, 2003.

256. C. C. Yang and M. A. Plonus, Effects of a random phase screen on the performance of an imaging system, *Opt. Lett.* **19**(14):1073, 1994.

257. C. C. Yang and M. A. Plonus, Superresolution effects in weak turbulence, *Appl. Opt.* **32**(36):7528–7531, 1993.

258. P. Yeh, *Optical Waves in Layered Media*, Wiley Interscience, 1988.

259. C. Zachos, D. Fairlie, and T. Curtright, *Quantum Mechanics in Phase Space: An Overview with Selected Papers*, World Scientific series in 20th Century Physics, Vol. 34, World Scientific, 2005.

260. Z. Zalevsky and D. Mendlovic, *Optical Superresolution*, Springer, 2004.

261. Z. Zalevsky, S. Rozental, and M. Meller, Usage of turbulence for superresolved imaging, *Opt. Lett.* **32**(9):1072–1074, 2007.

262. Y. H. Zheng, N. P. Pitsianis, and D. J. Brady, Nonadaptive group testing based fiber sensor deployment for multiperson tracking, *IEEE Sensors J.* **6**(2):490–494, 2006.

INDEX

Bold page number indicates a section reference.